Frontiers of Surface-Enhanced Raman Scattering

Frontiers of Surface-Enhanced Raman Scattering

Single Nanoparticles and Single Cells

Edited by

YUKIHIRO OZAKI
KATRIN KNEIPP
RICARDO AROCA

WILEY

This edition first published 2014

Registered office
John Wiley & Sons Ltd, The Atrium, Southern Gate, Chichester, West Sussex, PO19 8SQ, United Kingdom

For details of our global editorial offices, for customer services and for information about how to apply for permission to reuse the copyright material in this book please see our website at www.wiley.com.

Library of Congress Cataloging-in-Publication Data

Frontiers of surface-enhanced raman scattering : single-nanoparticles and single cells / edited by Yukihiro Ozaki, Katrin Kneipp, Ricardo Aroca.
pages cm
Includes index.
ISBN 978-1-118-35902-0 (cloth)
1. Raman effect, Surface enhanced. 2. Surfaces (Physics) 3. Raman spectroscopy. 4. Spectrum analysis. I. Ozaki, Y. (Yukihiro) editor of compilation. II. Kneipp, Katrin, editor of compilation. III. Aroca, Ricardo, editor of compilation.
QC454.R36F76 2014
543′.57 – dc23

2013037278

A catalogue record for this book is available from the British Library.

ISBN: 9781118359020 (Cloth)

Set in 10/12pt Times by Laserwords Private Limited, Chennai, India
Printed and bound in Singapore by Markono Print Media Pte Ltd

Contents

List of Contributors

Aisha Alsaleh, Materials and Surface Science Group, University of Windsor, Canada

Ricardo Aroca, Materials and Surface Science Group, University of Windsor, Canada

Mohamad Banaee, School of Engineering and Applied Sciences, Harvard University, USA

Lev Chuntonov, Department of Chemical Physics, Weizmann Institute of Science, Israel; Department of Chemistry, University of Pennsylvania, USA

Ying Chen, Department of Chemistry and the Photonics Center, Boston University, USA

Yizhuo Chu, School of Engineering and Applied Sciences, Harvard University, USA

Carlos J. L. Constantino, Faculdade de Ciências e Tecnologia, UNESP Universidade Estadual Paulista, Brazil

Kenneth B. Crozier, School of Engineering and Applied Sciences, Harvard University, USA

Daniela Drescher, Department of Chemistry, Humboldt-Universität zu Berlin, Germany

Yoseph Gebregziabher, Department of Chemistry and the Photonics Center, Boston University, USA

Gilad Haran, Department of Chemical Physics, Weizmann Institute of Science, Rehovot, Israel

Tamitake Itoh, National Institute of Advanced Industrial Science and Technology (AIST), Japan

Satoshi Kawata, Department of Applied Physics, Osaka University, Japan

Yasutaka Kitahama, Department of Chemistry, School of Science and Technology, Kwansei Gakuin University, Japan

Harald Kneipp, Department of Physics, Technical University of Denmark, Denmark

Janina Kneipp, Department of Chemistry, Humboldt-Universität zu Berlin, Germany

Katrin Kneipp, Department of Physics, Technical University of Denmark, Denmark

Paul Lemler, Department of Chemistry and the Photonics Center, Boston University, USA

Jian-Feng Li, State Key Laboratory of Physical Chemistry of Solid Surfaces and College of Chemistry and Chemical Engineering, Xiamen University, China

Kei Murakoshi, Department of Chemistry, Hokkaido University, Japan

Hideki Nabika, Department of Chemistry, Hokkaido University, Japan

Fumika Nagasawa, Department of Chemistry, Hokkaido University, Japan

Yukihiro Ozaki, Department of Chemistry, School of Science and Technology, Kwansei Gakuin University, Japan

W. Ranjith Premasiri, Department of Chemistry and the Photonics Center, Boston University, USA

George C. Schatz, Chemistry Department, Northwestern University, USA

Mai Takase, Department of Chemistry, Hokkaido University, Japan

Zhong-Qun Tian, State Key Laboratory of Physical Chemistry of Solid Surfaces and College of Chemistry and Chemical Engineering, Xiamen University, China

Eric J. Titus, Department of Chemistry and Biochemistry, The University of Texas, USA

Nicholas A. Valley, Chemistry Department, Northwestern University, USA

Diogo Volpati, Faculdade de Ciências e Tecnologia, UNESP Universidade Estadual Paulista, Brazil

Dongxing Wang, School of Engineering and Applied Sciences, Harvard University, USA

Katherine A. Willets, Department of Chemistry and Biochemistry, The University of Texas, USA

Taka-aki Yano, Department of Electronic Chemistry, Tokyo Institute of Technology, Japan

Wenqi Zhu, School of Engineering and Applied Sciences, Harvard University, USA

Lawrence D. Ziegler, Department of Chemistry and the Photonics Center, Boston University, USA

Preface

The discovery of surface-enhanced Raman scattering (SERS) stretches back to the 1970s. It flourished for nearly four decades with a broad range of applications. Today, it is a vibrant quintessential embodiment of nanoscience and nanotechnology. SERS is a plasmon-based spectroscopy at the forefront of the developments in plasmonics, providing the fundamentals for theory and experiment.

SERS continues to grow with remarkable success. There are some major driving forces pushing forward new developments, among them it is worth mentioning three thriving areas. First is the success of single molecule detection by SERS; second the progress in the understanding of SERS fundamentals; and third, the development of nanoscience and nanotechnology. The level of understanding of SERS is currently so advanced that people are beginning to formulate strategies for exploiting SERS as a general platform for chemical and biological analysis with unprecedented routine levels of sensitivity, specificity and reproducibility.

This book is closely related to one of the symposia in Pacifichem 2010 "*Frontiers of Surface-Enhanced Raman Scattering: Single-Nanoparticles and Single Cells*". Thus, the collection is based on the Pacifichem SERS, and most of the contributors have been selected from invited speakers at the symposium. One of the most important purposes of the book is an attempt to convey to the scientific community the enthusiastic discussion on the state-of-the-art SERS during the symposium.

The symposium was, in essence, the first gathering so far convened to examine the "State of the new developments in SERS" with particular emphasis on single nanoparticles and single cells. With that in mind, in this book we have assembled the prime movers in the field worldwide, ensuring that we invited the seminal current contributors to SERS theory, substrate fabrication, applications of SERS to biosystems, chemical analysis, sensing, and of course fundamental innovation through experimentation.

We hope this book will not only be useful but also enjoyable to read. We will be most gratified if the book can inspire readers to try novel and exciting SERS research.

Yukihiro Ozaki,
School of Science and Technology,
Kwansei Gakuin University

Katrin Kneipp,
Department of Physics,
Technical University of Denmark

Ricardo Aroca,
Materials and Surface Science Group,
University of Windsor

Preface

1

Calculation of Surface-Enhanced Raman Spectra Including Orientational and Stokes Effects Using TDDFT/Mie Theory QM/ED Method

George C. Schatz and Nicholas A. Valley
Chemistry Department, Northwestern University, USA

New models combining a quantum mechanical description of a molecular system and an electrodynamics (ED) description of a metal nanoparticle to determine surface-enhanced Raman spectra are described. The models considered involve inclusion of frequency dependence, Stokes shifts, and the effects of surface averaging into previously developed models. Important changes in absolute intensities result from considering these effects while changes in the relative intensities are minor. In a study of metal phthalocyanine, strong variation in the spectrum with orientation of the molecule relative to the nanoparticle surface is noted, and only after orientation averaging is included is there agreement with experiment. We also present calculations with one of the models using a multi-nanoparticle system, and this is found to provide more experimentally realistic enhancements which slightly improve the spectra.

1.1 Introduction: Combined Quantum Mechanics/ Electrodynamics Methods

The surface-enhanced Raman scattering (SERS) response spans multiple orders of magnitude, providing many important challenges to theoretical modeling. Theoretical

Frontiers of Surface-Enhanced Raman Scattering: Single Nanoparticles and Single Cells, First Edition.
Edited by Yukihiro Ozaki, Katrin Kneipp and Ricardo Aroca.

and computational studies have approached this problem in a variety of ways [1–8]. Proper modeling of the full SERS effect must describe both the electromagnetic (EM) enhancement due to the plasmonic particle, and chemical (CHEM) enhancements involving response of the molecule that interacts chemically with the particle. EM enhancement results from oscillation of the conduction band electrons of the metallic structure, referred to as plasmon excitation, in response to an external field. The plasmon excitation enhances the local electric field at both the incident and scattered frequencies, greatly increasing the Raman cross-section of molecules close to the nanoparticle surface. CHEM enhancement is tied to charge transfer between molecule and metal particle, both in the ground and excited states, that results from orbital overlap. Generally, the EM enhancement is large compared to the CHEM enhancement [9], with up to a fourth-power dependence on the electric field (i.e., enhancement $\sim E^4$).

Theoretical modeling of SERS has historically taken one of two paths. One neglects CHEM effects, focusing only on the dominant EM enhancement as determined by the field induced at the position of the molecule by light interacting with the nanoparticle [3]. In the other, the focus can be placed on the CHEM enhancement by doing an electronic structure calculation for the molecule plus a small cluster of metal atoms [4]. To incorporate both mechanisms, it is necessary to combine ED and quantum mechanics (QM) calculations into the optical response, however, this idea has only recently been considered.

Previous work on combining QM and ED has considered the ED calculations for the particle in the absence of the analyte and then the plasmonic excitation is applied as a constant field to the molecule. Corni and Tomasi explicitly coupled electronic structure calculations for the analyte to an ED calculation for the particle in the frequency domain by including effective charges in the molecular Hamiltonian in the quasi-static approximation [10]. Lopata and Neuhauser developed a local, two-level random phase approximation model for density matrix evolution. This evaluated the molecular population transfer rate while the finite-difference time-domain (FDTD) method was used to describe the ED [11]. Masiello and Schatz applied a many body Greens function to evaluate plasmon-enhanced molecular absorption [12]. This enables a better treatment of the effect of the molecule–field interaction, but CHEM effects associated with charge transfer were not included in the applications. A discrete interaction model/QM method to explicitly model nanoparticle interaction has been described by Morton and Jensen [13, 14]. Here an atomistic representation of the nanoparticle (with atoms treated as polarizable dipoles) provides an explicit model for the optical properties of a molecule interacting with the nanoparticle surface. Recently, Chen and Schatz described a hybrid QM/ED method in the time domain using RT-TDDFT (real-time time-dependent density functional theory) in which fields from FDTD are used to provide a plasmon-enhanced driving field for the molecule [15]. More recently, a frequency domain linear response version of this theory has been developed by Mullin and Schatz [16] and applied to calculating SERS spectra for pyridine and Rhodamine 6G. In both studies TDDFT was used to determine frequency-dependent polarizability derivatives for the molecule, with the molecule taken to be in the gas-phase except with a restricted orientation that mimics the molecule on the surface. This approach for the coupling of QM with ED is denoted “model G”.

New multiscale models have recently been developed for the combination of QM and classical ED for describing surface-enhanced Raman and hyper-Raman spectra including CHEM and EM enhancement mechanisms for both [16]. In these models, TDDFT (in the frequency domain) is used for the QM calculation, and Mie theory (restricted to spherical particles) is used for the ED calculations. A new feature, however, is that the surface of a metal cluster treated quantum mechanically is overlaid with the surface of the metal particle to combine the two calculations. In model A, the electric field from plasmon excitation of the metal particle is combined with the CHEM enhancement associated with a static treatment of the molecule-metal structure to determine overall spectra. In model B, the frequency dependence of the Raman spectrum of the isolated molecule is combined with the enhancements determined using model A to refine the enhancement estimate and include resonance Raman effects in the molecule.

Models A, B, and G all neglect the Stokes shifting of the scattered light and use a three point approximation to determine the average over positions of the molecule on the surface. To more faithfully describe Raman scattering, in this paper we present improved models that couple QM calculations and ED calculations to include both EM and CHEM mechanisms, and which remove the previous limitations. In particular, model A has been modified to include Stokes shifting in the EM enhancement, replacing $|E(\omega)|^4$ by $|E(\omega)E(\omega')|^2$ where ω and ω' are the incident and Stokes shifted frequencies, respectively. This leads to model S, and we show how this effect changes both the magnitude and the shape of the Raman spectrum for pyridine. Further modifications to model S to include more extensive surface averaging leads to model V. Comparisons with experiment for metal phthalocyanines and for pyridine demonstrate important improvements in the predictions of these models compared to earlier work. We also show how the EM mechanism can be evaluated for our models for nanoparticle dimers, using Mie theory, leading to much higher enhancement values.

1.2 Computational Details

The calculation of the field for spheres is performed with a locally developed Mie theory code. The silver dielectric parameters used are those from Johnson and Christy [17]. For small particles, electron scattering from the nanoparticle surface becomes an important source of electron dephasing. A factor Av_F/R in the Drude expression for the plasmon width, where v_F is the Fermi velocity, R is the particle radius, and A is an empirical parameter, is included to correct for this. A value of 0.1 for A will be used here.

Quantum mechanical calculations have been performed using the Amsterdam Density Functional (ADF) program package [18]. For systems including pyridine, full geometry optimization and frequency calculations were completed using the Perdew–Wang (PW91) XC-potential and a triple-ζ polarized Slater type (TZP) basis set. All calculations for phthalocyanines were performed with a Becke–Perdew (BP86) XC-potential and a TZP basis set.

Polarizabilities for pyridine were calculated with the AOResponse module, and the asymptotically correct statistical average of orbital potential (SAOP) model XC-potential. The SAOP model potential has been designed to give accurate excited state properties, allowing accurate calculation of response properties. An even-tempered quadruple-ζ plus triple polarization (ET-QZ3P) polar basis set was used for C, H, and N, while a TZP basis set was used for Ag. Scalar relativistic effects for the Ag atoms were accounted for with the zeroth order regular approximation (ZORA). Frequency-dependent polarizabilities were calculated using a global damping parameter of $\Gamma = 0.004$ au (0.1 eV); this is the same as was used in earlier work [19], and is thought to represent a reasonable estimate of the excited state dephasing lifetime.

1.3 Summary of Model Systems

The models discussed here are a few of many that could be imagined where ED and electronic structure calculations are coupled. The molecules are placed in close proximity (less than 1 nm) to the metallic surface, as is common in SERS experiments. The methods used here assume the plasmonic response of the metal nanoparticle considered in the ED calculations is negligibly perturbed by the presence of the molecule(s). This is reasonable in the low concentration regime. Inclusion of a tetrahedral Ag_{20} cluster in the quantum mechanical calculations allows us to add static CHEM enhancement effects to the results using the method of Morton and Jensen [20]. Non-static CHEM contributions can be considered, but caution must be taken due to the underestimation of charge transfer excitation energies calculated with TDDFT [21].

The quantum mechanically treated cluster and larger metal particle are both incorporated by making the particle and cluster surfaces coincident. We will call this an "overlay" structure, as previously described by Mullin *et al.* [16]. This structure for pyridine is shown in Figure 1.1, where we contrast the overlay structure with that used for model G, where the location of the molecule relative to the surface is not determined by a QM calculation. This overlay structure is used in all the models we consider except for model G. Therefore, model G will lack the static CHEM enhancement included in the other models, although orientation and polarization averaging effects that would be associated with a molecule on the nanoparticle surface are included. In model A, the static CHEM enhancement factor (which varies with vibrational mode) is also included in the calculation, and we use the molecule–surface distance from the electronic structure calculation to define the location of the molecule for the EM calculation. We also note that calculating the response of the quantum portion in the static limit will also avoid double counting of plasmon resonance effects when the EM and CHEM enhancements are multiplied.

Models B, S, and V are all extensions of model A. Model B begins with model A and multiplies by the ratio of the Raman intensity for the isolated molecule at the frequency being studied to the static Raman intensity. This model incorporates any frequency dependence

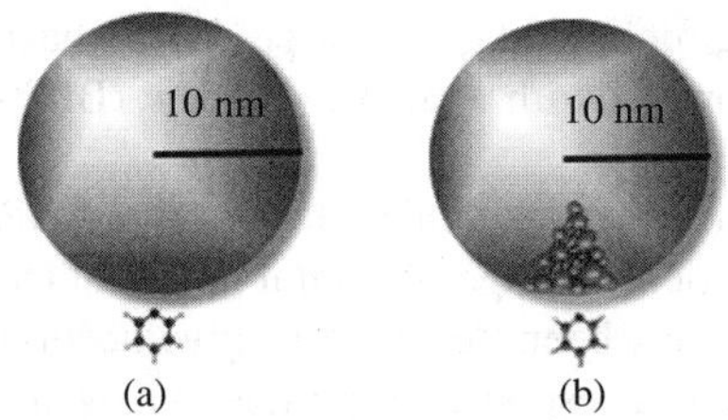

Figure 1.1 *Combined quantum mechanics/electrodynamics model systems (a) gas-phase quantum mechanically treated molecule and electrodynamics treated particle used in Model G and (b) molecule–Ag_{20} cluster treated quantum mechanically where the cluster surface is coincident (overlaid) with that of the particle treated with electrodynamics*

that the molecule polarizability might have, even resonance Raman enhancement, but that of the cluster is not included due to the problem with self-interaction errors. Model S allows the scattered frequency to differ from the incident frequency in the EM component of the enhancement. This model considers the Stokes shift, which takes into account the frequency dependence of the field enhancements caused by the metal particle(s). Model V expands upon model S by including a more complete surface averaging.

In the models involving the Ag_{20} cluster, the molecule orientation relative to the surface is fixed by the ground state structure of the molecule–cluster system. Spectra, which can be compared with experiment, are generated by fixing the position of the molecule relative to the particle surface. Then it is assumed that the azimuthal direction of the molecule relative to the polarization of the EM field is random. Therefore, an average of the intensity over azimuthal directions is used. This type of averaging was performed for Raman spectra for model G by Mullin and Schatz [16]. Expressions for the azimuthal averaging are presented later.

The use of Mie theory is not necessary for the general method used here and, as with any ED method, it comes with strengths and weaknesses. Mie theory provides an exact solution of Maxwell's equations, and therefore can be used down to zero separation between molecule and metal. This provides an improvement over the FDTD method used in other work [16], as fields from FDTD become inaccurate close to the particle surface due to the finite difference approximation. A limitation of Mie theory, however, is it can only be used for single or assemblies of spherical particles. For more general particles, one is forced to use a numerical method.

In summary, the models used and the effects they account for are:

Model G: QM molecule, ED particle
Model A: QM molecule–cluster complex, ED particle
Model B: QM molecule–cluster complex, ED particle, molecular frequency dependence
Model S: QM molecule–cluster complex, ED particle, Stokes scattering
Model V: QM molecule–cluster complex, ED particle, Stokes scattering, surface averaging.

1.4 Azimuthal Averaging

A completely general expression for the Raman intensity is [16]:

$$I = Q^2 |E(\omega')\alpha'_{\mathrm{m}} E(\omega)|^2 \, I_0 / E_0^4 \tag{1.1}$$

Rotational degrees of freedom of the molecule may be considered by rotation of the polarizability tensor of the molecule in the previous expression. As a first approximation, if the molecule is allowed to rotate about the axis along which the molecule is bound to the particle, azimuthal averaging is being included. When the bound axis is the x-axis, the rotation matrices are included as shown:

$$E(\omega') \begin{pmatrix} 1 & 0 & 0 \\ 0 & \cos(\phi) & \sin(\phi) \\ 0 & -\sin(\phi) & \cos(\phi) \end{pmatrix} \alpha'_m \begin{pmatrix} 1 & 0 & 0 \\ 0 & \cos(\phi) & \sin(\phi) \\ 0 & -\sin(\phi) & \cos(\phi) \end{pmatrix} E(\omega) \tag{1.2}$$

The azimuthal angle ϕ is the angle between the lab-fixed y-axis and the rotating molecular y'-axis. Averaging over ϕ leads to the following expression for the azimuthal polarization averaged intensity (here taking the lab-fixed wavevector along z so the polarization is along x and y) [16]:

$$
\begin{aligned}
I(180°) = \frac{d\sigma}{d\Omega}(180°) = \frac{1}{2E_0^2}K_p \Big[& \alpha'^2_{m,xx}\left(\left[E_x^x\right]^4 + \left[E_x^y\right]^4\right) \\
& + \frac{1}{2}\left(\frac{1}{2}\left(\alpha'_{m,yy} + \alpha'_{m,zz}\right)^2 + \alpha'^2_{m,yz} + \frac{1}{4}\left(\alpha'_{m,yy} - \alpha'_{m,zz}\right)^2\right) \\
& \times \left(\left\{\left[E_y^x\right]^2 + \left[E_z^x\right]^2\right\}^2 + \left\{\left[E_y^y\right]^2 + \left[E_z^y\right]^2\right\}^2\right) \\
& + 2\left(\alpha'^2_{m,xy} + \alpha'^2_{m,xz} + \frac{1}{2}\alpha'_{m,xx}\alpha'_{m,yy} + \frac{1}{2}\alpha'_{m,xx}\alpha'_{m,zz}\right) \\
& \times \left\{\left[E_x^x\right]^2\left(\left[E_y^x\right]^2 + \left[E_z^x\right]^2\right) + \left[E_x^y\right]^2\left(\left[E_y^y\right]^2 + \left[E_z^y\right]^2\right)\right\}\Big]
\end{aligned}
\tag{1.3}
$$

Here E_i^j refers to the ith component of the local field, with the initial polarization pointing in the jth direction.

1.5 SERS of Pyridine: Models G, A, B, S, and V

Experimental and theoretical studies of SERS of pyridine generally find around six peaks in the spectrum. The most intense peaks correspond to ring breathing modes at 983 and 1026 cm^{-1} (normal modes will be referred to by their gas-phase DFT energies calculated at the level of theory outlined). Modestly intense ring stretching modes appear at 1208, 1472, and 1583 cm^{-1}, as well as a ring deformation mode at 598 cm^{-1}.

Simulated gas-phase normal Raman spectra of pyridine using static and frequency-dependent polarizability derivatives are shown in Figure 1.2 along with experimental results from Golab *et al.* [22]. Differential cross-sections in both cases have been calculated assuming a wavelength of 357 nm, though only the frequency-dependent results include this wavelength in the response calculations. The specific wavelength chosen corresponds to the plasmon resonance frequency of the small silver particle in the QM/ED models later. Molecular resonance is not expected to occur at this wavelength as the electronic transitions of pyridine are at much higher energies. The calculated spectra are similar in appearance to experimental and other theoretical spectra [22, 23], and the only difference compared to our earlier work [16] is that the reference wavelength used previously was 514.5 nm rather 357 nm. The strongest peaks are observed at 983 and 1026 cm^{-1}. Weaker peaks appear at 598, 651, 1208, 1472, and 1583 cm^{-1}. The absolute intensity of the frequency-dependent spectrum is larger than that of the static spectrum, reflecting pre-resonance enhancement in pyridine. Frequency dependence effects constitute about a factor of 2 at 357 nm.

Previously, Mullin and Schatz have calculated SERS spectra using the QM/ED model G method with FDTD for Raman spectra [16]. Here, the model G approach is applied using Mie theory to describe the enhanced fields around a spherical silver nanoparticle. The spectrum resulting from a QM/ED (G) calculation, assuming a 0.25 nm molecule–metal sphere separation and an incident wavelength of 357 nm, is presented in

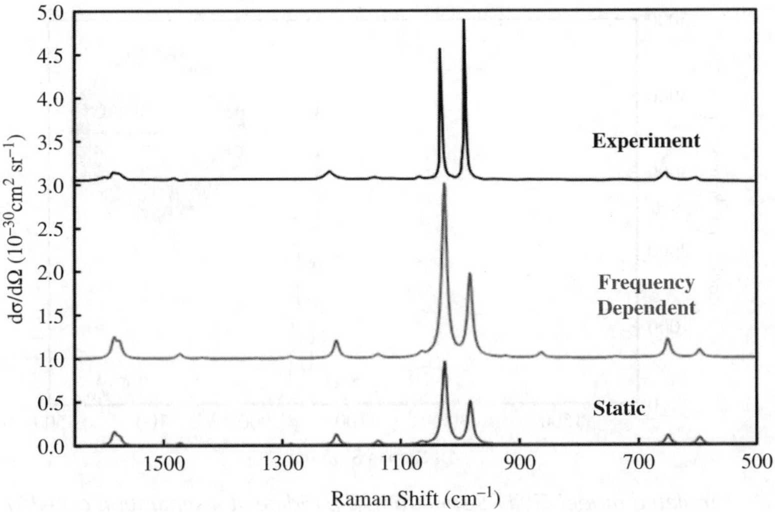

Figure 1.2 *Simulated normal Raman spectrum of pyridine using static and frequency-dependent polarizability derivatives. Differential cross-section is given in units of 10^{-30} cm^2 sr^{-1} and is calculated assuming an incident wavelength of 357 nm. Peaks are broadened by a Lorentzian with a full width at half maximum of 10 cm^{-1}. Scale is for the broadened theoretical spectrum. Reproduced with permission from [23]. Copyright © 2007, American Chemical Society*

Figure 1.3. Only very minor changes in relative intensities are observed in comparison to the gas-phase spectrum. Raman intensities are enhanced by about 4×10^3, which corresponds to purely EM enhancement. The calculated enhancement factor is a factor of 2.5 lower than was found by Mullin and Schatz for a 1 nm separation. This is likely due to the inclusion of electron–surface scattering in the present ED calculations, which slightly dampens the plasmon excitation. Frequency dependence is also not included in the quantum calculations used in the present application, leading to weaker intensities compared to results in Mullin *et al.* [16].

CHEM enhancement effects can be added to the models by including a tetrahedral 20 atom silver cluster (Ag_{20}). This will be done for models A, B, S, and V in an application to pyridine. The geometry and Raman spectra of pyridine bound to this small cluster have been extensively studied in the literature [23, 24]. Calculations here only consider the "S-complex," where the pyridine is bound to the central silver atom on the (111) face. Static CHEM enhancement in this arrangement contributes a factor of 2 to the intensity of the spectrum. Energies of vibrational modes, when bound to the silver cluster, do not change substantially in comparison to the gas phase. In the QM/ED models, the Ag_{20} cluster is embedded in the 20 nm silver sphere, as shown in Figure 1.1b. The actual distance from the nanoparticle surface at which the field is evaluated is slightly larger than in previous work [16] leading to slightly lower enhancements overall.

Model A results are presented in Figure 1.4 along with an experimental SERS spectrum from Golab *et al.* [22]. Compared to the results for the gas-phase molecule and model G, relative intensities of modes at 983 and 1583 cm^{-1} are increased. Better agreement with the

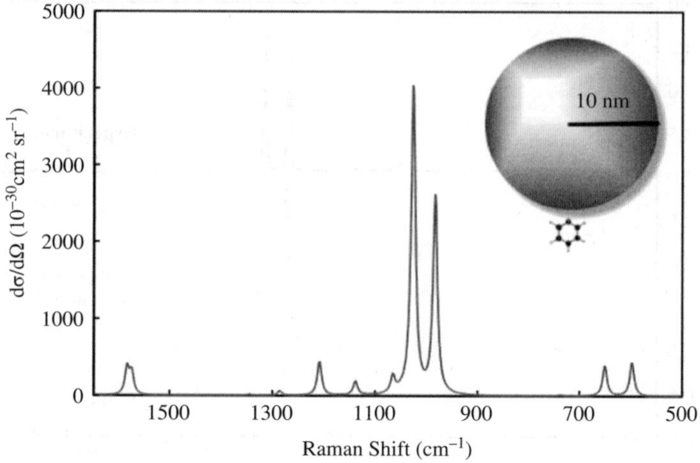

Figure 1.3 *Simulated model G SERS spectrum of pyridine at a separation of 0.25 nm from a 10 nm radius silver sphere at an incident wavelength of 357 nm. Differential cross-section is given in units of* 10^{-30} *cm*^{2} *sr*$^{-1}$*. Peaks are broadened by a Lorentzian with a full width at half maximum of 10 cm*$^{-1}$

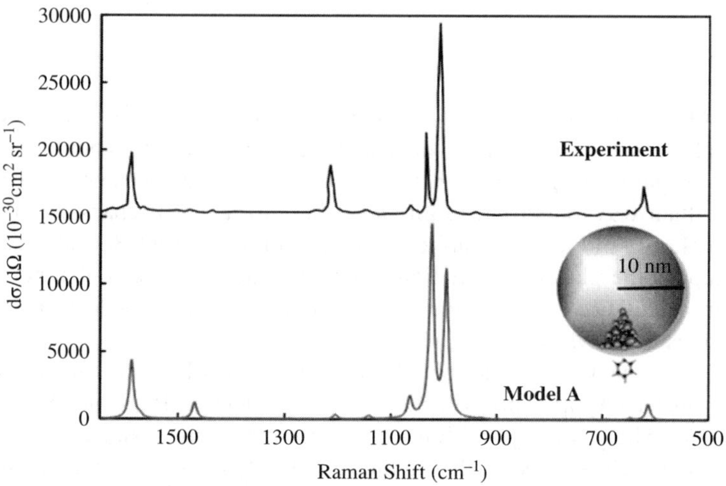

Figure 1.4 *Simulated model A SERS spectrum of pyridine using the static S-complex attached to a 20 nm diameter silver sphere for structure at an incident wavelength of 357 nm. Experimental SERS spectrum digitized from Ref. [23]. Differential cross-section is given in units of* 10^{-30} *cm*^{2} *sr*$^{-1}$*. Peaks are broadened by a Lorentzian with a full width at half maximum of 10 cm*$^{-1}$*. Reproduced with permission from [23]. Copyright © 2007, American Chemical Society*

experimental SERS spectrum is seen with model A than with model G. Absolute intensities are nearly a factor of 4 larger than those obtained in model G, leading to overall enhancement of 1.5×10^4. A factor of $\sim 4 \times 10^3$ is due to EM enhancement, while the remaining factor of 3–4 is due to static CHEM enhancement.

The frequency dependence of the polarizability derivatives at 357 nm is taken into account in model B. Figure 1.5 presents the spectrum resulting from model B. In comparison with model A, the relative intensities of modes at 983 and 1026 cm^{-1} are increased. Again, the spectrum agrees well with the experimental SERS spectrum. Both model A and model B strongly underestimate the peak at 1208 cm^{-1}. The absolute intensity of the model B spectrum is about twice what is seen for model A. This is due to the inclusion of the frequency dependence.

While models G, A, and B result in reasonable results in comparison to experimental SERS spectra, they make unnecessary approximations. They first assume the incident and scattered light have the same frequency in the ED calculations. This leads to an overestimation of the EM enhancement as all the fields are assumed to be at the plasmon resonance frequency where the greatest field enhancement will occur. If the frequency of the scattered light is allowed to be different than the incident frequency, they will no longer experience optimal enhancement, which leads to a weaker signal. This will also lead to modes being enhanced to different degrees, depending on their vibrational energies. Modes with low energies will have smaller Stokes shifts and be nearer the plasmon resonance if the incident frequency is at or near the plasmon resonance. This can change the relative intensities of the peaks in the spectrum and will be dependent on the incident frequency and the frequency dependence of the fields of the specific nanoparticle system.

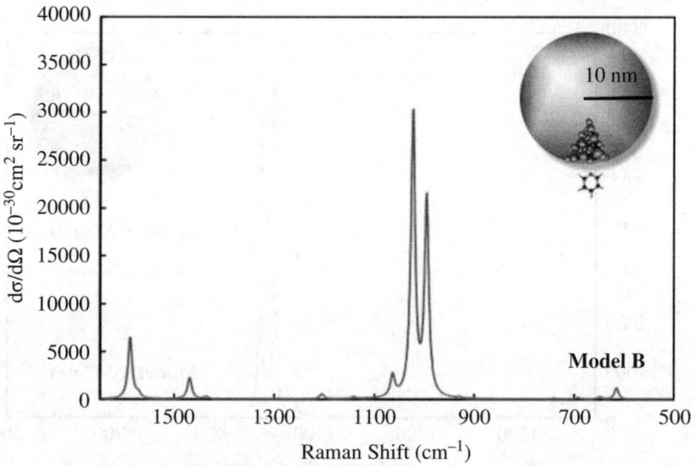

Figure 1.5 *Simulated model B SERS spectrum of pyridine using the static S-complex attached to a 20 nm diameter silver sphere for structure at an incident wavelength of 357 nm. Differential cross-section is given in units of 10^{-30} cm^2 sr^{-1}. Peaks are broadened by a Lorentzian with a full width at half maximum of 10 cm^{-1}*

Another approximation involves the limited amount of field information used from the ED calculations. Models G, A, and B only take into account field enhancements at three points (i.e., along the x-, y-, and z-axes) on the spherical particle. Since the incident k-vector is along z, and the polarizations are along x or y, these points are also special points where the field enhancement is the maximum for a specific Cartesian direction in the molecular frame. It will be shown that this results in an overestimation of the enhancement as well. Models S and V expand upon model A to correct the errors introduced by each of these approximations. Model S uses different field enhancements for the scattered light by performing the ED calculations at the incident frequency and each of the Stokes shifted frequencies. Model V expands on model S by numerically averaging spectra obtained at each point in a grid over the surface of the nanoparticle(s). For each of these models, two incident frequencies will be compared, one (357 nm) corresponding to the plasmon resonance and another (347 nm) that, along with the Stokes shifted 1583 cm^{-1} peak (367 nm), straddles the plasmon resonance peak.

Results using model S with an incident wavelength of 347 and 357 nm are displayed in Figure 1.6. For the system studied, the different incident light energies lead to peaks with roughly the same relative intensities. The absolute intensities are about 50% larger when the incident light is at the plasmon resonance than when the incident and scattered light straddle the plasmon resonance. Better agreement with experimental SERS spectra is obtained with model S than the previous models, as the mode at 598 cm^{-1} increases in intensity relative to the peak at 1583 cm^{-1}. Inclusion of the Stokes shift results in a factor of 3 decrease in the calculated EM enhancement factor, making it $\sim 1 \times 10^3$.

Model V is performed by using the Mie theory calculated field enhancements at points sampled around the entire surface of the spherical particle (i.e., uniformly in azimuthal

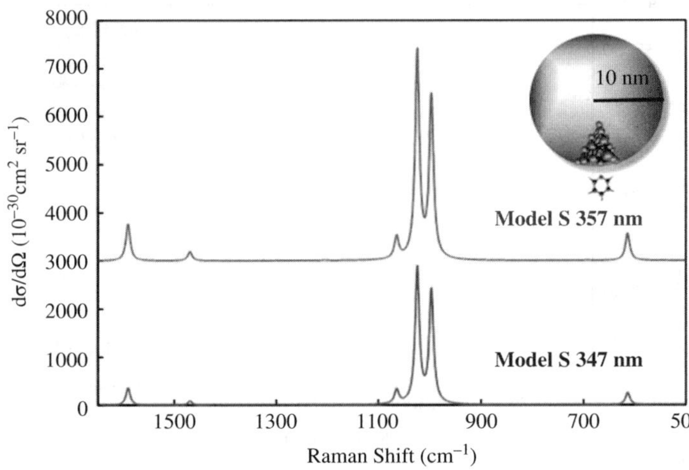

Figure 1.6 *Simulated model S SERS spectra of pyridine using the static S-complex attached to a 20 nm diameter silver sphere for structure at an incident wavelength of 347 and 357 nm. Differential cross-section is given in units of* 10^{-30} *cm*^{2} *sr*$^{-1}$*. Peaks are broadened by a Lorentzian with a full width at half maximum of 10 cm*$^{-1}$

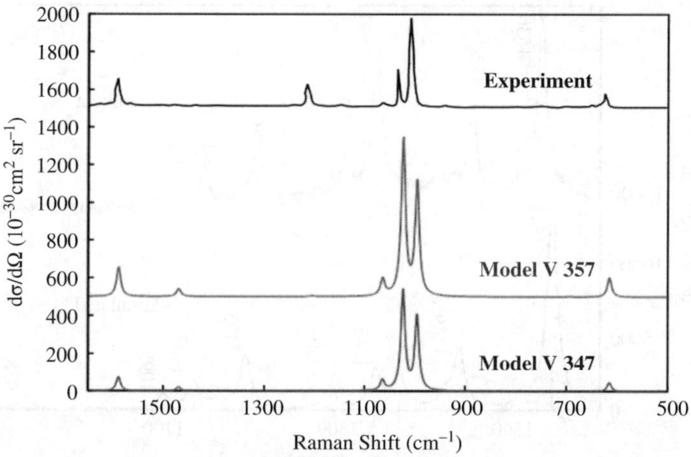

Figure 1.7 *Simulated model V SERS spectra of pyridine using the static S-complex attached to a 20 nm diameter silver sphere for structure at an incident wavelength of 347 and 357 nm (red). Experimental spectrum from Ref. [23] is included. Differential cross-section is given in units of 10^{-30} cm^2 sr^{-1}. Peaks are broadened by a Lorentzian with a full width at half maximum of 10 cm^{-1}. Reproduced with permission from [23]. Copyright © 2007, American Chemical Society*

angle and uniformly in cos(θ) where θ is the polar angle). Specifically we choose 3 nm point spacing in the determination of the Raman spectra. Spectra obtained using Model V are shown in Figure 1.7 along with the experimental spectrum from Golab *et al.* [22]. Only minor changes in relative intensities are noted compared to the model S spectrum. A very minor peak corresponding to the 1208 cm^{-1} mode can be seen. The surface averaging causes a decrease in absolute intensities by around a factor of 5. At 357 nm excitation the total enhancement is 1×10^3, which is reasonable for a single sphere. Excitation at the plasmon resonance still shows 50% stronger intensity than excitation at 347 nm.

1.6 Orientation Effects in SERS of Phthalocyanines

To test the utility of the QM/ED methods discussed, we use them to explain the existence of an extraneous peak in a calculated gas-phase spectrum of a metal phthalocyanine when compared to an experimental surface-enhanced spectrum. The comparison can be seen in Figure 1.8. Metal phthalocyanines have recently [25] been used as targets in tip-enhanced Raman spectroscopy (TERS) experiments and consist of a central metal atom and a surrounding porphyrin-like aromatic structure.

Metal phthalocyanines generally adopt a planar geometry with D_{4h} point group symmetry. Calculated bond lengths and angles were found to be in agreement with previous DFT calculations by Basova *et al.* [26] (for copper phthalocyanine) and Murray *et al.* [27] (for zinc phthalocyanine) as well as crystallographic data [28]. Theoretical gas-phase calculations agree well with the experimental spectrum, except for a calculated peak that appears

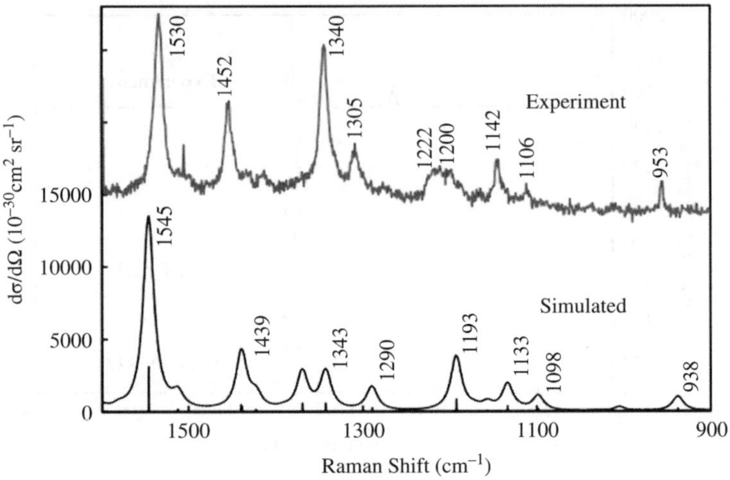

Figure 1.8 *Simulated gas-phase and experimental tip-enhanced resonance Raman spectrum of CuPc. The theoretical spectrum is constructed using ZnPc Raman intensities and CuPc vibrational frequencies. Differential cross-section is given in units of 10^{-30} cm^2 sr^{-1} and is calculated at an incident wavelength of 631.48 nm. Peaks are broadened by a Lorentzian with a full width at half maximum of 15 cm^{-1}. Reproduced with permission from [26]. Copyright © 2009 John Wiley & Sons, Ltd*

near 1370 cm^{-1}. Calculations including a metal cluster (model A) found CHEM effects to be weak and the extra peak persisted.

The QM/ED model G was employed to investigate the dependence of the spectrum on the orientation of the molecule with respect to the surface. Model G allows analysis of the effects of molecular orientation relative to the surface without the need for performing a new quantum mechanical calculation for each orientation. This is not the case for the models that include a metal cluster in the quantum mechanical calculations. Calculations were performed on the closed-shell ZnPc, while experimental results from Ref. [26] are obtained using CuPc. Currently, the available electronic structure codes do not have the ability to calculate resonance response for open shell species, so instead of performing calculations on the open-shell CuPc, calculations are performed on the closed-shell ZnPc. The ZnPc spectrum is very similar to the CuPc spectrum, and the same two peaks at about 1340 and 1370 cm^{-1} appear in the same location. As the angle between the plane of the phthalocyanine and the surface was decreased, only the intensity and not the shape of the spectrum was altered. Experimental evidence pointed to the phthalocyanine lying flat on the surface. When this geometry was studied, the spectrum in Figure 1.9 was the result.

When the molecule is flat on the surface, the peak near 1370 cm^{-1} that is not observed experimentally disappears in the theoretical spectrum. The spectrum as a whole, however, no longer resembles the experimental spectrum, and is much weaker in intensity. Since the plane of the molecule is strictly constrained to be parallel to the surface, only modes with out-of-plane character show up. Modes with only in-plane character, like the 1370 cm^{-1} mode, do not appear.

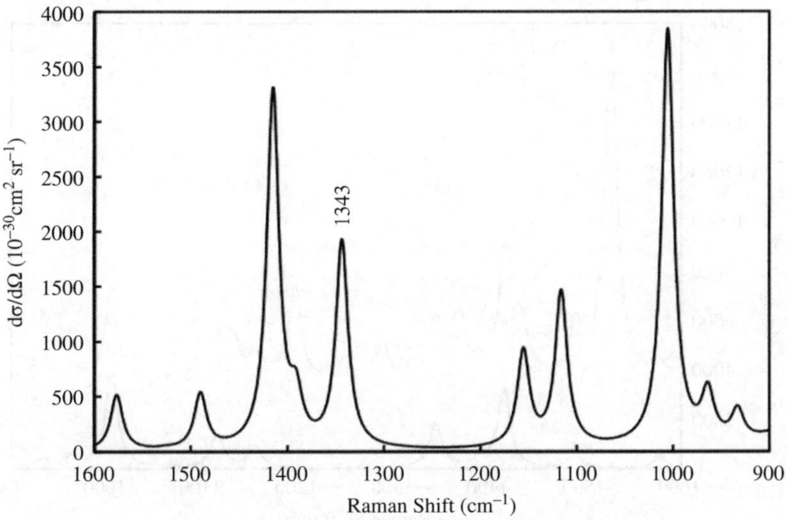

Figure 1.9 *Simulated model G resonance Raman spectrum of ZnPc parallel to the particle surface. Differential cross-section is given in units of 10^{-30} cm^2 sr^{-1} and calculated at an incident wavelength of 631.48 nm. Peaks are broadened by a Lorentzian with a full width at half maximum of 15 cm^{-1}*

Now that there seems to be a possible explanation of why the 1370 cm^{-1} peak is not seen experimentally, an explanation of why the theoretical spectrum of the molecule flat on the surface is not seen in experiments is called for. Considering that molecular vibrations can move the molecule out of true parallel to the surface, and interacting with the surface can bend the molecular framework, calculations are performed using a structure with a one degree angle between the molecule and surface. The calculated spectrum, presented in Figure 1.10, shows that the in-plane modes have returned. Thus even small motions of the molecule out of the plane of the surface results in a Raman spectrum dominated by the in-plane modes that are observed experimentally. Given that it is the in-plane modes that dominate, but the 1370 cm^{-1} mode is not observed, we infer that motional averaging suppresses the 1370 cm^{-1} mode, where there is no intensity for planar structures, but maintains the other, generally stronger, modes.

1.7 Two Particle QM/ED Calculations

Using model V, QM/ED calculations are performed for pyridine on a two-particle system. This system, shown in Figure 1.11 alongside the surface averaged field enhancements, involves a 200 nm diameter particle and a 120 nm particle separated by a distance of 1 nm. Raman QM/ED model V spectra are calculated with an incident wavelength of 870 nm. The k vector of the incident light is perpendicular to the dimer axis and is polarized in the direction of the dimer axis. Effects concerning the dimer orientation, gap size, and grid spacing have not been investigated.

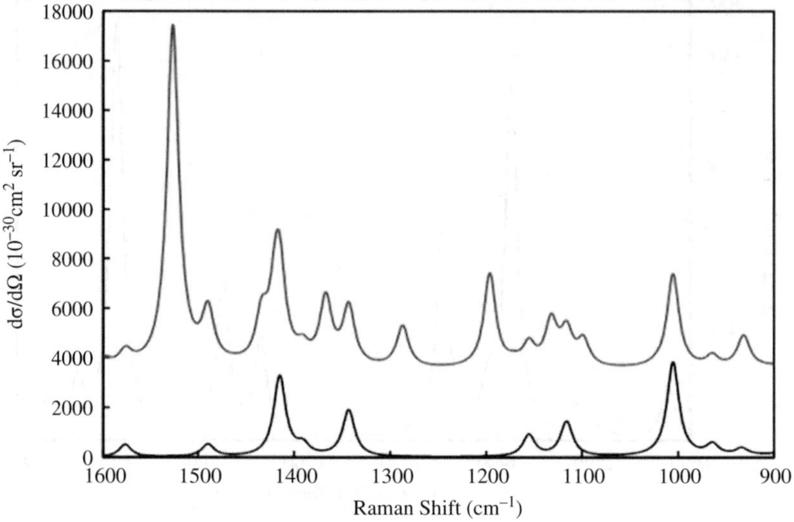

Figure 1.10 *Simulated model G resonance Raman spectrum of ZnPc with a zero (lower trace) and one (upper trace) degree angle between the planes of the molecule and particle surface. Differential cross-section is given in units of 10^{-30} cm^2 sr^{-1} and calculated at an incident wavelength of 631.48 nm. Peaks are broadened by a Lorentzian with a full width at half maximum of 15 cm^{-1}*

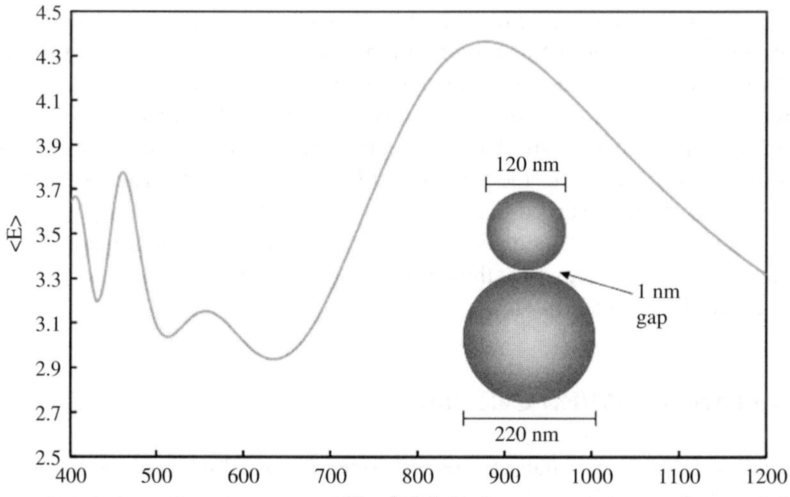

Figure 1.11 *Surface averaged field enhancements calculated for the nanoparticle dimer shown using Mie theory. Incident light is polarized along the dimer axis and the k vector is perpendicular to the dimer axis*

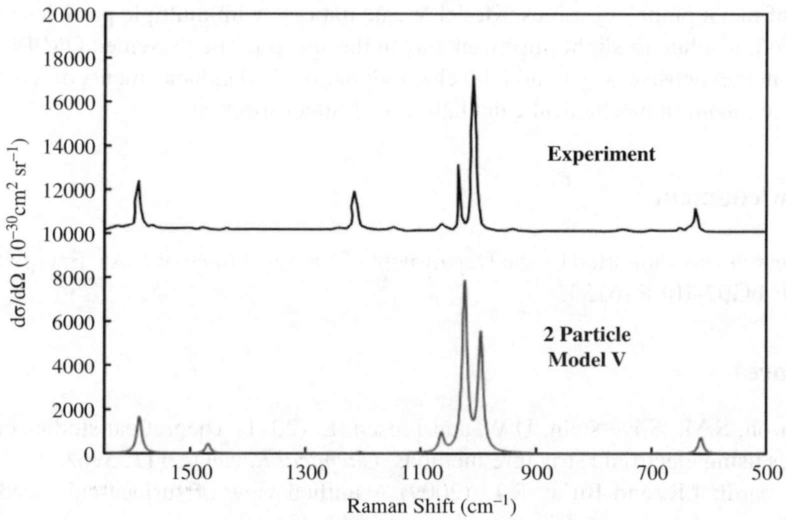

Figure 1.12 *Simulated model V SERS spectrum of pyridine using the static S-complex attached to a two nanoparticle structure at an incident wavelength of 870 nm. Experimental spectrum from Ref. [23] is included. Differential cross-section is given in units of 10^{-30} cm^2 sr^{-1}. Peaks are broadened by a Lorentzian with a full width at half maximum of 10 cm^{-1}. Reproduced with permission from [23]. Copyright © 2007, American Chemical Society*

The calculated model V Raman spectrum is displayed in Figure 1.12. Compared to the one nanoparticle case, the 1208 cm^{-1} mode is slightly more intense, though still not at all as intense as seen experimentally. The absolute intensity of the spectrum increases by a factor of 10 over the one-particle model V result, with enhancements on the order of 10^4 over gas-phase pyridine. This is slightly lower than is observed experimentally for similar systems, but the QM/ED model used here is not accounting for surface roughness.

Though only small changes are seen for pyridine with the different models, inclusion of Stokes scattering, surface averaging, and a second nanoparticle in the models involves a negligible increase in the computational time while making the models more realistic.

1.8 Summary

A series of methods combining QM and ED (Mie theory) calculations to determine SERS spectra have been presented and evaluated for pyridine. Model G involves only the quantum mechanically treated molecule and ED for the nanoparticle. Model A improves the description by including a cluster of silver atoms in the quantum mechanical calculation. Inclusion of frequency dependence in model B has minor effects on the pyridine spectrum, but would be useful for molecules where both static CHEM and molecular resonance effects are important. Model S takes account of the Stokes shift, and model V includes a proper surface average. Including these effects leads to slight decreases in the predicted total enhancement, which is important for quantitative modeling. The ability to easily change molecular orientations relative to the surface of the particle in model G was used to investigate SERS spectra of metal phthalocyanines. Model V calculations, with multiple particles in the ED calculation, resulted in slight improvements in the spectra. The presented QM/ED models provide an inexpensive way to add the electrodynamic field enhancements of nanoparticle systems to quantum mechanical calculations of Raman spectra.

Acknowledgment

This research was supported by the Department of Energy, Office of Basic Energy Science, grant DE-FG02-10ER16153.

References

1. Morton, S.M., Silverstein, D.W. and Jensen, L. (2011) Theoretical studies of plasmonics using electronic structure methods. *Chemical Reviews*, **111**, 3962.
2. Lombardi, J.R. and Birke, R.L. (2009) A unified view of surface-enhanced Raman scattering. *Accounts of Chemical Research*, **42**, 734.
3. Zhao, J., Pinchuk, A.O., McMahon, J.M. *et al.* (2008) Methods for describing the electromagnetic properties of anisotropic silver and gold nanoparticles. *Accounts of Chemical Research*, **41**, 1710.
4. Jensen, L., Aikens, C.M. and Schatz, G.C. (2008) Electronic structure methods for studying surface-enhanced Raman scattering. *Chemical Society Reviews*, **37**, 1061.
5. Tong, L., Zhu, T. and Liu, Z. (2011) Approaching the electromagnetic mechanism of surface-enhanced Raman scattering: from self-assembled arrays to individual gold nanoparticles. *Chemical Society Reviews*, **40**, 1296.
6. Halas, N.J., Lal, S., Chang, W.-S. *et al.* (2011) Plasmons in strongly coupled metallic nanostructures. *Chemical Reviews*, **111**, 3913.
7. Schatz, G.C. (1984) Theoretical-studies of surface enhanced Raman-scattering. *Accounts of Chemical Research*, **17**, 370.
8. Moskovits, M. (1985) Surface-enhanced spectroscopy. *Reviews of Modern Physics*, **57**, 783.
9. Moskovits, M. (2005) Surface-enhanced Raman spectroscopy: a brief retrospective. *Journal of Raman Spectroscopy*, **36**, 485.
10. Corni, S. and Tomasi, J. (2002) Surface enhanced Raman scattering from a single molecule adsorbed on a metal particle aggregate: a theoretical study. *The Journal of Chemical Physics*, **116**, 1156.
11. Lopata, K. and Neuhauser, D. (2009) Multiscale Maxwell-Schrodinger modeling: a split field finite-difference time-domain approach to molecular nanopolaritonics. *The Journal of Chemical Physics*, **130**, 104707.
12. Masiello, D.J. and Schatz, G.C. (2008) Many-body theory of surface-enhanced Raman scattering. *Physical Review A*, **78**, 042505.
13. Morton, S.M. and Jensen, L. (2011) A discrete interaction model/quantum mechanical method to describe. *The Journal of Chemical Physics*, **135**, 134103.
14. Morton, S.M. and Jensen, L. (2010) A discrete interaction model/quantum mechanical method for describing response properties of molecules adsorbed on metal nanoparticles. *The Journal of Chemical Physics*, **133**, 074103.

15. Chen, H., McMahon, J.M., Ratner, M.A. and Schatz, G.C. (2010) Classical Electrodynamics coupled to quantum mechanics for calculation of molecular optical properties: a RT-TDDFT/FDTD approach. *Journal of Physical Chemistry C*, **114**, 14384.
16. Mullin, J., Valley, N., Blaber, M.G. and Schatz, G.C. (2012) Combined quantum mechanics (TDDFT) and classical electrodynamics (Mie Theory) methods for calculating surface enhanced Raman and hyper-Raman spectra. *The Journal of Physical Chemistry. A*, **116**, 9574.
17. Johnson, P.B. and Christy, R.W. (1972) Optical constants of the noble metal. *Physical Review B*, **6**, 4370.
18. Baerends, E.J., Autschbach, J. A. and Berces, A. Amsterdam Density Functional Scientific Computing & Modeling, Amsterdam, 2010.
19. Jensen, L., Zhao, L.L., Autschbach, J. and Schatz, G.C. (2005) Theory and method for calculating resonance Raman scattering from resonance polarizability derivatives. *The Journal of Chemical Physics*, **123**, 174110.
20. Morton, S.M. and Jensen, L. (2009) Understanding the molecule-surface chemical coupling in SERS. *Journal of the American Chemical Society*, **131**, 4090.
21. Moore, J.E., Morton, S.M. and Jensen, L. (2012) Importance of correctly describing charge-transfer excitations for understanding the chemical effect in SERS. *Journal of Physical Chemistry Letters*, **3**, 2470.
22. Golab, J.T., Sprague, J.R., Carron, K.T. *et al.* (1988) A surface enhanced hyper-Raman scattering study of pyridine adsorbed onto silver: experiment and theory. *The Journal of Chemical Physics*, **88**, 7942.
23. Jensen, L., Zhao, L.L. and Schatz, G.C. (2007) Size-dependence of the enhanced Raman scattering of pyridine adsorbed on Ag-N (N=2–8, 20) clusters. *Journal of Physical Chemistry C*, **111**, 4756.
24. Zhao, L.L., Jensen, L. and Schatz, G.C. (2006) Pyridine-Ag_{20} cluster: a model system for studying surface-enhanced Raman scattering. *Journal of the American Chemical Society*, **128**, 2911.
25. Jiang, N., Foley, E.T., Klingsporn, J.N. *et al.* (2012) Observation of multiple vibrational modes in ultrahigh vacuum tip-enhanced Raman spectroscopy combined with molecular-resolution scanning tunneling microscopy. *Nano Letters*, **12**, 5061.
26. Basova, T.V., Kiselev, V.G., Schuster, B.-E. *et al.* (2009) Experimental and theoretical investigation of vibrational spectra of copper phthalocyanine: polarized single-crystal Raman spectra, isotope effect and DFT calculations. *Journal of Raman Spectroscopy*, **40**, 2080.
27. Murray, C., Dozova, N., McCaffrey, J.G. *et al.* (2010) Infra-red and Raman spectroscopy of free-base and zinc phthalocyanines isolated in matrices. *Physical Chemistry Chemical Physics*, **12**, 10406.
28. Brown, C.J. (1968) Crystal structure of β-copper phthalocyanine. *Journal of the Chemical Society A, Inorganic Physical Theoretical*, 2488.

2

Non-resonant SERS Using the Hottest Hot Spots of Plasmonic Nanoaggregates

Katrin Kneipp and Harald Kneipp
Department of Physics, Technical University of Denmark, Denmark

2.1 Introduction

Surface-enhanced Raman scattering (SERS) [1–4] may be one of the most impressive observations to demonstrate the spectroscopic capabilities of new photonic approaches which exploit enhanced local optical fields of plasmonic nanostructures. The effect gained particular interest after the Raman spectrum of a single molecule was measured using SERS [5, 6]. High SERS enhancement level, as exploited in single molecule SERS, is related to so-called "hot spots", that is, areas of dimensions of a nanometer or less, where Raman scattering takes place in strongly enhanced and highly confined local optical fields. So far, aggregates formed by silver or gold nanoparticles provide the best enhancement factors for SERS and are used in most single molecule SERS studies [5, 7–13].

Spectroscopic observations, such as strong anti-Stokes SERS signals due to non-resonant SERS vibrational pumping and non-resonant single molecule SERS using near-infrared (NIR) excitation, indicated that an extremely high SERS enhancement level is only available for very few molecules and showed also that such a high SERS enhancement level is always related to clusters of silver or gold nanoparticles [5, 7, 8]. Such "cluster-based"

Frontiers of Surface-Enhanced Raman Scattering: Single Nanoparticles and Single Cells, First Edition.
Edited by Yukihiro Ozaki, Katrin Kneipp and Ricardo Aroca.

SERS enhancement, which is particularly efficient in the NIR, has also been supported by theory [14].

To date, numerous computations have predicted hot spots at various levels of field enhancement for various nanostructures. There is agreement that the areas of highest local fields exist in narrow gaps between silver or gold nanoparticles, forming aggregates of various sizes and shapes extending from dimers [15–18] and trimers [19] to self-similar structures formed by silver or gold nanospheres [20]. The recently reported super-resolution imaging of SERS on silver nanoaggregates has directly confirmed nanoparticle junctions to be responsible for single molecule SERS [21].

Most single molecule SERS studies use excitation wavelengths which are resonant with electronic transitions in the target molecule and, in this way, exploit a superposition of surface enhancement and resonance Raman scattering (SERRS) [6, 9, 10, 18, 22]. On the other hand, SERS at an extremely high enhancement level also enables Raman measurements from a single molecule without the support of intrinsic molecular resonance Raman enhancement [5, 8, 23]. The most prominent molecules in non-resonant single molecule SERS are probably DNA bases, which exhibit absorption in the ultraviolet [8, 24–27].

Of course, under (pre)resonant conditions, the requirements for the size of surface enhancement can be greatly reduced due to the benefits from a large molecular resonance Raman cross section.

Single molecule SERS under resonance conditions can also be achieved in tip-enhanced Raman scattering where plasmonic enhancement factors are generated for molecules in the gap between a metal tip and a metal surface [28] or for molecules located on silver island films [29].

In general, the level of enhancement required for single molecule SERS also depends on experimental parameters, such as the performance of the Raman system and the background level generated by the enhancing plasmonic structures. Optimizing these parameters could make new promising approaches in SERS-active structures, such as shell-isolated nanoparticle-enhanced Raman spectroscopy (SHINERS) [30], interesting also for single molecule SERS based on a lower enhancement level.

Here we report SERS studies at the single molecule level using non-resonant NIR photons for excitation and aggregates of silver nanoparticles as enhancing plasmonic structures. In Section 2.2, we investigate nanoaggregates formed by silver particles and their hot spots. The aggregates are evaluated regarding their capability to provide extremely high SERS enhancement in their hottest spots based on the observation of SERS vibrational pumping. This test allows identification of "hot" nanoaggregates which provide an extremely high enhancement level in their hottest spots, suitable for measuring single molecule SERS under non-resonant conditions. However, these optical measurements cannot deliver information on the exact position of the hottest hot spots on the aggregate, nor can they provide experimental insight into the dramatic spatial variations in local optical fields within a few nanometers of the plasmonic structure as they are predicted by computations. Exciting capabilities for probing plasmonic structures at a spatial resolution of more than 2 orders of magnitude better than in diffraction limited optical experiments are opened up by using electron beams [31]. We have shown that electron energy loss spectroscopy (EELS) of SERS-active silver nanoaggregates can provide additional insight into the plasmon resonances of these structures and into the spatial distribution of local optical fields at nanometer resolution [32].

Another important question in non-resonant single molecule SERS is to choose an optimum excitation wavelength for achieving the highest local field in the hot spots. We apply SERS using one- and two-photon excitation for probing local fields in hot spots. Tunable laser excitation allows investigation of local fields in hot spots in dependence on photon energy.

In Section 2.3, we use hot nanoaggregates in aqueous solution and non-resonant excitation in the NIR to perform SERS experiments at different concentrations of the target molecule. At very low concentrations of the target molecule, at least a factor of 10 below the concentration of the plasmonic nanoaggregates, we measure SERS spectra using only the hottest hot spots. In this way, we avoid heterogeneity in enhancement factors. We discuss the unique potential of SERS performed in this operating regime for approaching ultimate limits in analytics. A brief summary and conclusions are given in Section 2.4.

2.2 Aggregates of Silver and Gold Nanoparticles and Their Hot Spots

Silver and gold nanoparticles can be prepared chemically by reduction of silver salts [33] or "physically" by laser ablation from bulk metal [34].

Figure 2.1 shows typical nanoaggregates used in this study and formed by silver nanoparticles.

2.2.1 Evaluation of Plasmonic Nanoaggregates by Vibrational Pumping due to a Non-resonant SERS Process

The aggregates shown in Figure 2.1 can provide the highest enhancement level observed in SERS experiments so far. However, the enhancement can vary for different preparation batches and not all nanoaggregates provide a really extremely high enhancement level. Here we evaluate the plasmonic performance of nanoaggregates based on the observation of vibrational pumping due to a strong Raman process. A high SERS enhancement level can be identified by a strong anti-Stokes spectrum which clearly displays deviations from the expected Boltzmann distribution of the anti-Stokes signals of lower and higher energy Raman modes. This is illustrated in Figure 2.2. Silver or gold nanoparticles clearly generate

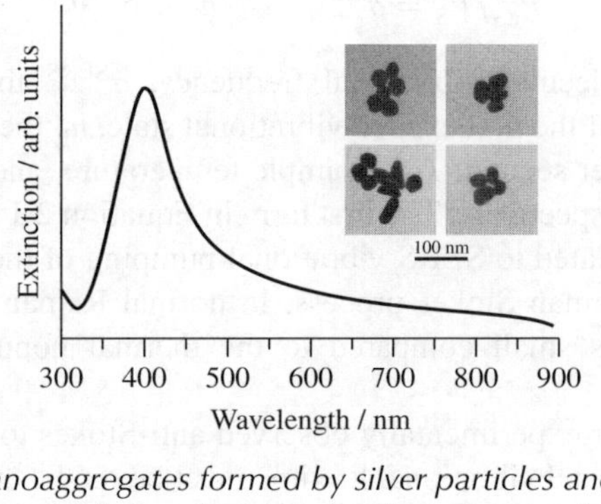

Figure 2.1 *Plasmonic nanoaggregates formed by silver particles and the extinction spectrum measured from their aqueous solution. Reprinted with permission from Ref. [25]*

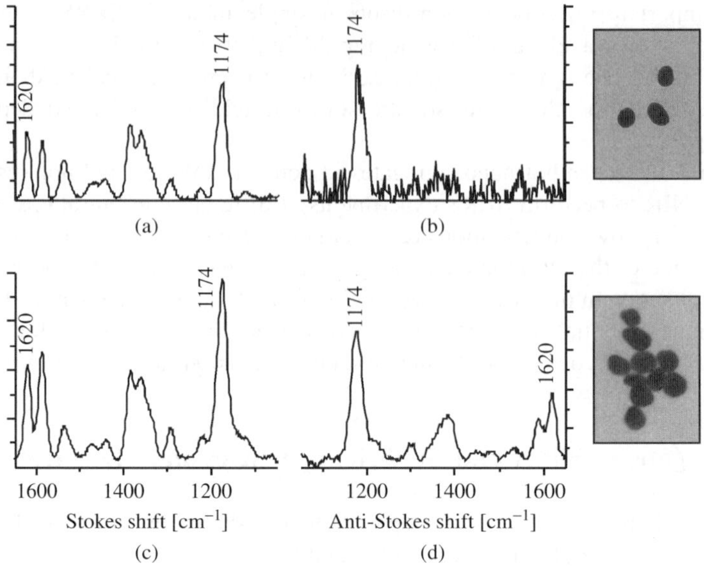

Figure 2.2 *Stokes and anti-Stokes SERS spectra of crystal violet under conditions of "normal" SERS (a,b) and extremely strong SERS (c,d). Spectra were measured using 830 nm non-resonant excitation. Reprinted with permission from Ref. [11], Copyright 2006 American Chemical Society*

a surface-enhanced Raman signal of crystal violet shown in the Stokes spectrum (a). As expected, the anti-Stokes SERS spectrum (b) displays the lower frequency modes only. This situation changes in spectrum (d). Now, with aggregates as enhancing structures, a strong anti-Stokes spectrum appears. In particular, this spectrum shows strong anti-Stokes lines also for the higher energy Raman modes. Information on the effective SERS cross section can be inferred from anti-Stokes to Stokes signal ratios in a stationary experiment. Figure 2.3 illustrates Stokes and anti-Stokes scattering in a molecular level scheme. A simple equation for the anti-Stokes to Stokes signal ratio P_{aS}/P_S can be derived from the rate equation assuming steady state and weak saturation [7]

$$P_{aS}/P_S = \sigma^{SERS} \cdot \tau_1 \cdot n_L + e^{-\frac{h\nu_M}{kT}} \tag{2.1}$$

with ν_M being the molecular vibrational frequency, σ^{SERS} the effective SERS cross section, τ_1 the lifetime of the first excited vibrational state, n_L the number of laser photons per square centimeter per second, T the sample temperature, and h and k the Planck and Boltzmann constants, respectively. The first term in Equation 2.1 describes the anti-Stokes to Stokes signal ratio related to SERS vibrational pumping of the first excited vibrational state due to a strong Raman Stokes process. In normal Raman scattering this term can be neglected since it is small compared to the thermal population described by the second term.

In order to account for experimentally observed anti-Stokes to Stokes signal ratios, the product of cross section and vibrational lifetime must be on the order of 10^{-27} cm^2 s. With vibrational lifetimes on the order of 10 ps [35], effective cross sections which can account

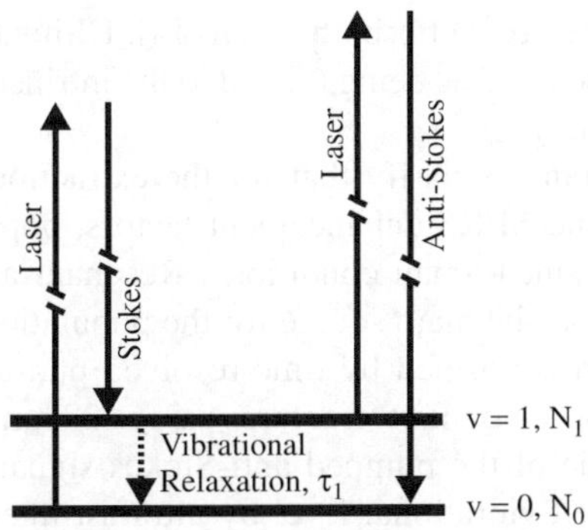

$$\frac{dN_1}{dt} = (N_0 - N_1)\,\sigma^{SERS}\,n_L - \frac{N_1}{\tau_1}$$

Figure 2.3 *Stokes and anti-Stokes SERS. The rate equation describes the population of the first excited vibrational levels due to a strong Raman process*

for Raman pumping of a molecular vibration are estimated to be at least on the order of 10^{-16} cm^2. As an experimental example, effective SERS cross sections on the order of 10^{-16} cm^2 for the UV-absorbing molecule adenine have been inferred from vibrational pumping using 830-nm excitation. Such cross sections imply that a SERS enhancement factor of around 10^{14} is required to overcome the gap between a typical non-resonant Raman cross section on the order of 10^{-30} cm^2 and the effective cross section inferred from observed vibrational pumping. Effective cross sections in SERS experiments on the order of 10^{-16} cm^2 have also been confirmed in single molecule Raman experiments. These experiments show that the SERS signals of a single molecule appear at the same signal level as the non-enhanced Raman signal of the 10^{14} methanol molecules [5]. Enhanced local fields and "chemical" enhancement of the Raman cross section can contribute to non-resonant total enhancement factors of 14 orders of magnitude. Several experimental observations, such as the scaling of enhancement factors between one- and two-photon excited SERS, suggest for the hottest spots on silver- and gold nanoaggregates "electromagnetic enhancement" on the order of 10^{12}, with chemical enhancement effects on the order of 10^2. Achievable enhancement factors on these orders of magnitude are supported also by theoretical studies. Electromagnetic enhancement factors on the order of 10^{12}–10^{13} have been established [36, 37]. Chemical enhancement can result in factors of 10–10^3 [38, 39]. An electromagnetic enhancement on the order of 10^{12} also explains the experimentally observed 6 orders of magnitude higher enhancement level for two-photon excited hyper Raman scattering (HRS) compared with one-photon excited SERS [40, 41].

Anti-Stokes to Stokes signal ratios measured on different structures, from 100 nm aggregates to larger fractal silver structures several micrometers in size, show that enhancement factors on the order of 10^{14} in non-resonant SERS are independent of the size and shape of the cluster but depend critically on the gap size between the nanoparticles forming the aggregate. Despite different SERS enhancement factors for isolated nanoparticles of silver (10^6–10^7) and gold (10^3–10^4) measured at their plasmon resonances at 407 and 514 nm, respectively, the enhancement factors for nanoclusters formed by these particles are on the same order of magnitude (10^{14}) for both metals [8]. Ultimate limitations of plasmonic enhancement have been discussed as being related to the intrinsic nonlocality of the dielectric function of the metal [42].

At this point it is important to clarify that for the extraction of information on effective SERS cross sections and SERS enhancement factors, population pumping has to be performed under non-resonant Raman conditions. Resonant and pre-resonant excitation enable other mechanism(s) as the main source for the population of the first excited vibrational states. This has been confirmed by time-resolved pump–probe anti-Stokes SERS studies on Rhodamine 6G on silver nanostructures using 633 nm excitation [35]. The experiments identified a rise time of the pumped anti-Stokes signal of about 0.8 ps related to the population of the excited vibrational level by ultrafast internal conversion after electronic excitation, whereas population due to a Raman process follows the excitation pulse instantaneously.

Assuming consistency of the inferred effective Raman cross sections with the observed scattering signal, one has to conclude that a very small number of molecules in the SERS sample are involved in the SERS process at such a high enhancement level. We will get back to this issue when we discuss the distribution of the sizes of enhancement factors across the nanoaggregate in Section 2.3.

2.2.2 Probing Plasmonic Nanoaggregates by Electron Energy Loss Spectroscopy

EELS is emerging as a novel characterization tool for plasmonic nanostructures [43–50]. Particularly exciting capabilities for space–time imaging of localized fields have been opened up by photon-induced near-field electron microscopy [51, 52]. We have applied EELS for probing the plasmon resonances and local distribution of plasmonic fields at the nanometer scale for nanoaggregates formed by silver particles [32]. Figure 2.4 shows a nanoaggregate formed by five silver nanoparticles together with typical EEL spectra collected for this structure. These aggregates exhibit bulk plasmon resonance at 3.8 eV (spectrum d) and a broad strong band centered around 3.2 eV. This plasmon resonance also dominates the optical absorption measured from solutions of these silver nanoaggregates around 400 nm (see Figure 2.1). Further surface plasmon resonances of the aggregate have been obtained in the spectra at around 1.8, 1.5, and 1 eV, corresponding to 680, 830, and 1250 nm, respectively. Interestingly, excitation wavelengths used in one- and two-photon excited NIR-SERS experiments at extremely high enhancement level correspond to these plasmon resonances identified in EEL spectra. For example, spectrum c in Figure 2.5 shows the anti-Stokes SERS spectrum of crystal violet measured using 1064 nm excitation. Strong anti-Stokes signals particularly also for high energy Raman modes are a clear indication for the population of vibrational levels by vibrational pumping due to a strong SERS Stokes process [7]. This Stokes pumping process is expected to be efficiently supported by the plasmon resonance around 1 eV. Efficient SERS vibrational pumping using 830 nm excitation (see also Figure 2.2) is supported by the plasmon resonances around 1.5 eV. Plasmon resonance around 1 and 1.5 eV might also support two-photon excited surface-enhanced hyper Raman scattering (SEHRS) which we measured at silver nanoaggregates using excitations at 1064 nm (see spectrum a in Figure 2.5) and in the range of 830 nm. We will discuss HRS as a method for probing local fields in hot spots in more detail later in this chapter.

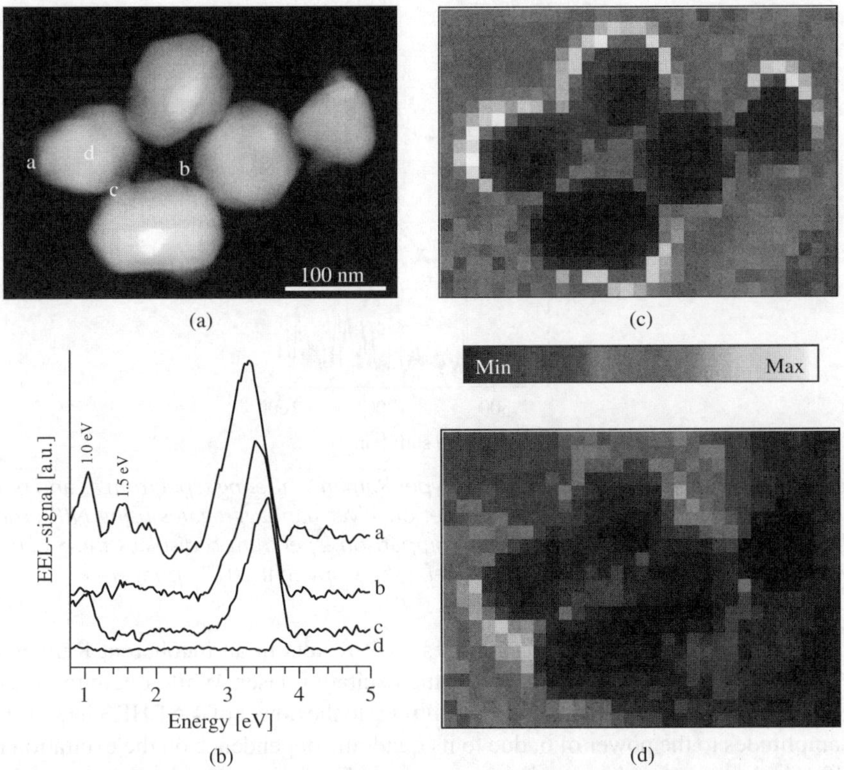

Figure 2.4 *EELS-probing of silver nanoaggregates (a) annular dark-field STEM image, (b) EELS spectra collected from different locations (a)–(d) shown in (a). (c,d) Images of EEL intensities generated in energy windows 1.4–1.6 and 0.8–1.0 eV, respectively. Reprinted with permission from Ref. [32], Copyright 2012 Springer*

Figure 2.4c,d show maps of EELS intensities collected from the nanoenvironment of the aggregate in energy windows in the NIR. The images indicate high EEL signals at the rim of the aggregate and minimum EEL signals in the gaps between the nanoparticles [32]. Minimum EELS signals in gaps between nanoparticles, that is, just in the region where the hot spots of the aggregate are located, have been measured for silver nanoaggregates of various sizes and symmetries [32, 53] and also in the gap of two gold nanoprisms [54]. The inverse relation between EELS intensity and optically excited hot spots has been studied in more detail in Ref. [32]. A comparison between Figure 2.4c,d illustrates changes in the distribution of local optical fields for different plasmon resonances.

2.2.3 Probing Local Fields in Hot Spots by SERS and SEHRS

In particular, hot spots also provide excellent conditions for spectroscopy based on simultaneous action of two photons. Vice versa, two-photon excited SERS at the single molecule level opens up interesting new ways for probing local optical fields in hot spots. Briefly,

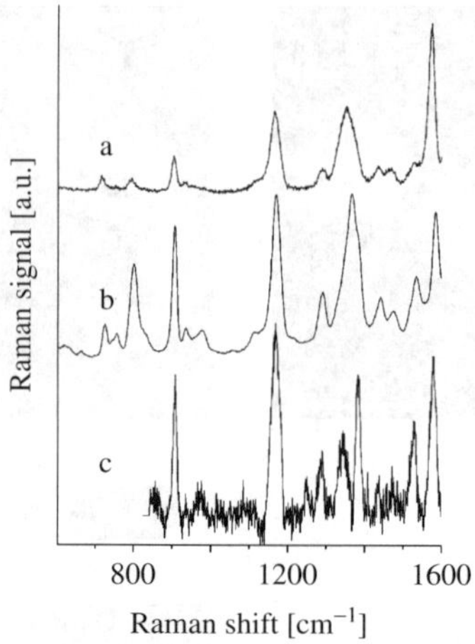

Figure 2.5 *Surface-enhanced two-photon hyper Raman scattering (spectrum a) and pumped anti-Stokes SERS (spectrum c) of crystal violet on silver nanoaggregates using NIR excitation matching plasmon resonance at 1 eV. For comparison, spectrum b displays the Stokes SERS spectrum. Reprinted with permission from Ref. [32], Copyright 2012 Springer*

HRS is a two-photon excited Raman process that results in an incoherent Raman signal shifted relative to the second harmonic of the excitation laser. While one-photon excited SERS depends on the local optical field amplitudes to the power of 4, SEHRS increases with these amplitudes to the power of 6, due to its quadratic dependence on the excitation intensity [40, 41]. In this way, the ratio between two- and one-photon excited surface-enhanced hyper Raman and Raman signals delivers information field enhancement E_{loc}^2/E_0^2 in the hot spots. These measurements are possible at high accuracy. Despite very different signal levels of "normal" Raman scattering and HRS, SERS and SEHRS spectra appear at comparable signal levels and can be measured in the same spectrum by using the first and second diffraction order of the spectral grating [40].

Here we investigate silver nanoaggregates by one-and two-photon excited SERS. Excitation at various wavelengths in the NIR range provided by a tunable picoseconds Ti:Sapphire laser allows one to probe local optical fields in the hot spots in dependence on the photon energy. Figure 2.6 displays the local optical field intensity E_{loc}^2/E_0^2 in the hot spots inferred from the ratio of SEHRS and SERS signals measured for crystal violet on silver nanoaggregates using excitation wavelengths between 780 and 830 nm. The experiments show an increase in the local optical field with increasing wavelength. Increasing intensities for local fields in the hot spots with decreasing photon energy have also been predicted theoretically [36].

2.3 SERS Using Hot Silver Nanoaggregates and Non-resonant NIR Excitation

2.3.1 SERS Signal vs. Concentration of the Target Molecule

In this section we apply plasmonic silver nanoaggregates characterized above as enhancing structures in SERS experiments using non-resonant 830 nm excitation.

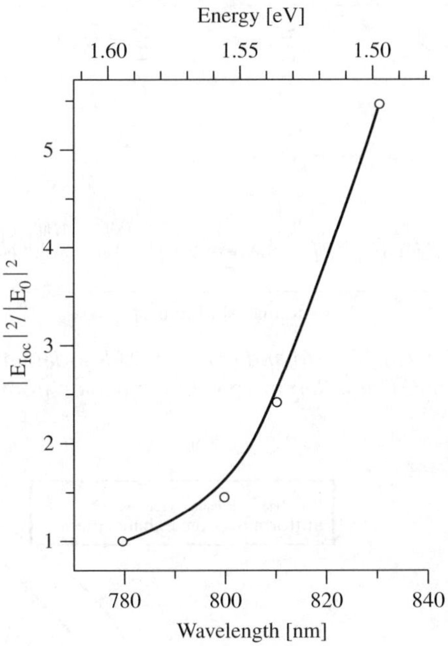

Figure 2.6 *Local optical field intensity $E_{loc}{}^2/E_0{}^2$ in hot spots in dependence on wavelength. Intensities are normalized to the intensity at 780 nm excitation. Reprinted with permission from Ref. [32], Copyright 2012 Springer*

The highest achievable concentration of nanoaggregates in aqueous solution can be inferred from the total amount of silver available from the applied chemical preparation process [33] to be on the order of 10^{-10} M. NaCl was added in 10^{-3} M concentration to the solutions of aggregates to achieve optimum chemical SERS-enhancement. The addition of salt did not give rise to changes in the extinction spectrum of the nanoaggregate solution, that is, no changes in the plasmonic properties occur due to NaCl at low concentrations. SERS samples with different concentrations of Rhodamine 6G were prepared by adding aqueous solutions of the dye at appropriate concentrations in 1 : 15 ratios to the solutions of silver nanoaggreates. Raman scattered light was collected in a 180°-scattering geometry using the same objective for excitation and collection of the scattered light from a 10–20 nl probed volume inside 1 ml sample solution. Dispersion and detection of the scattering signal are achieved using a single stage spectrograph with a deep depletion CCD detector.

Repeated checking of the extinction spectra of the sample solution during and after SERS measurements showed no change, implying no change in the aggregation of the colloidal solutions after addition of the target molecule at low concentrations.

Figure 2.7 shows two SERS spectra measured at 10^{-15} and 10^{-12} M concentrations of the target molecule Rhodamine 6G. Note that these spectra are collected from relatively large probed volumes of about 10 nl, which means that about 6 and 6000 molecules, respectively, contribute to the measured SERS signal.

Figure 2.8 shows a plot of SERS power of the 1510 cm^{-1} Raman line of Rhodamine 6G (see Figure 2.7) versus concentrations of the target molecule between 10^{-15} and 10^{-6} M in

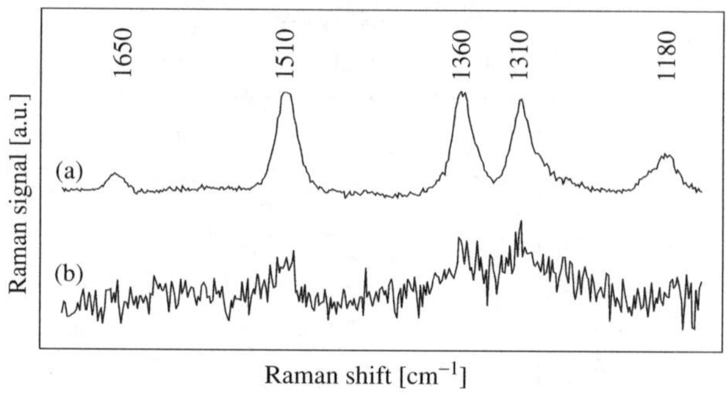

Figure 2.7 *SERS spectra of (a) 10^{-12} M and (b) 10^{-15} M Rhodamine 6G in a solution of silver nanoaggregates, excitation 830 nm. Reprinted with permission from Ref. [25]*

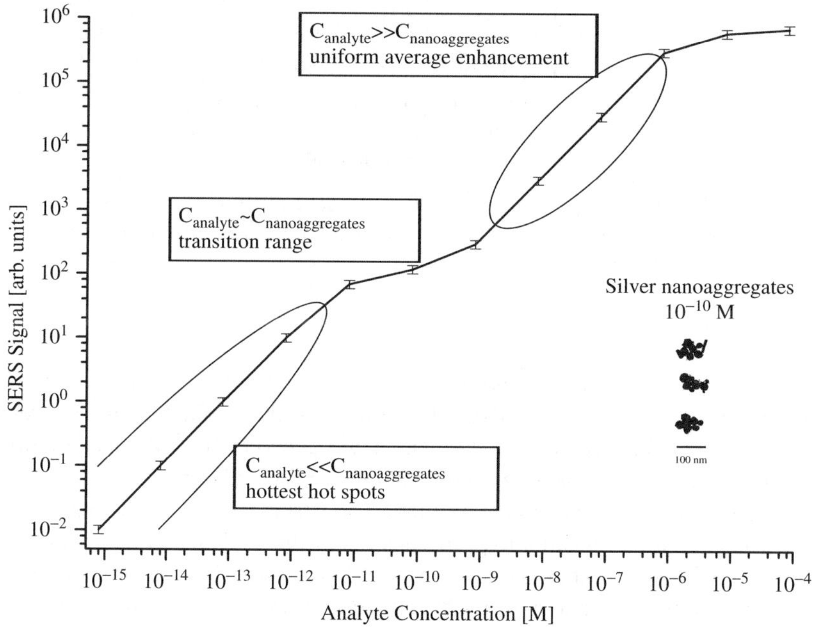

Figure 2.8 *SERS signal power versus concentration of the analyte Rhodamine 6G in a solution of hot silver nanoaggregates, electron microscopy images show typical silver nanoaggregates, for explanation see text. Reprinted with permission from Ref. [25]*

10^{-11} – 10^{-10} M solution of silver nanoaggregates. The electron microscopy images show typical nanoaggregates used in this experiment. Each of these aggregates enables SERS vibrational pumping and can be considered as a "hot aggregate" which provides, in the hottest spot, total enhancement factors on the order of 10^{14} when 830 nm excitation is used.

Brownian motion moves silver nanoaggregates into and out of the probed volume. However, for relatively large probed volumes (>10 nl) used for measuring the concentration series, the number of analyte-loaded nanoaggregates in the probed volume remains statistically constant also at low analyte concentrations. Even for the lowest analyte concentration of 10^{-15} M applied in this measurement, there are six Rhodamine 6G molecules present for generating the SERS signal. This implies that SERS powers measured in time sequence from these samples at volumes ≥10 nl show a Gaussian distribution.

Figure 2.8 shows a linear dependence of SERS signals on the concentration of the analyte between 10^{-15} and 5×10^{-12} M. At these concentrations with $C_{\text{analyte}} \ll C_{\text{nanoagggregate}}$, it is statistically unlikely that two molecules are attached to the same nanoaggregate. Each molecule finds its own enhancing nanoaggregate where it can occupy the hottest hot spots. The chemical preparation of the nanoaggregates makes it likely that the hottest spots on the aggregates of the same batch provide the same enhancement level. The uniform SERS enhancement level is confirmed by a linear relation between the concentration of the analyte and the SERS signal (slope 1 in the double-logarithmic plot) in this concentration range, that is, each analyte molecule added to the solution experiences the same enhancement level. Conducting experiments in this concentration range and reducing probed volumes to a size that only one target molecule or less on average are present in the probed volume, results in a Poisson distribution of the scattering signal powers measured in time sequence. This statistics reflects the probability of the presence of zero, one, two, three molecules in the probed volume during the actual measurement [5, 8, 23, 55]. Both, the linear increase in the SERS signal versus the concentration of the target molecules in a "many-molecule experiment", and the Poisson distribution of scattering signals in single molecule SERS require a uniform enhancement level. Linear plots for SERS signals versus concentration of the target molecule can be obtained below an analyte concentration of about 5×10^{-12} M, as shown in Figure 2.8, that is, for ratios between hot nanoaggregates and target molecules of at least 10–100. Similar ratios have also been applied in statistical studies of single molecule SERS when Poisson statistics of SERS signals measured in time sequence have been obtained [5, 8, 23, 55].

When the concentration of the target molecule approaches the concentration of the nanoaggregates the linear relation between the concentration and the SERS power no longer exists. In this "transition range" statistically more than one molecule can occupy the same nanoaggregate. Experiments show that in the analyte concentration range between about 10^{-11} and 10^{-9} M, that is, around 1 : 1 concentration of target molecules and hot nanoaggregates, SERS signals grow much slower than expected. This indicates that molecules now also start to occupy spots which provide lower enhancement levels at relatively wide distributions. In agreement with losing the linear dependence of SERS power versus analyte concentration in many molecule SERS for $C_{\text{analyte}} \sim C_{\text{nanoagggregate}}$, single molecule SERS experiments performed in this concentration range fail in measuring Poisson statistics in signal distributions.

For higher concentrations of the analyte with $C_{\text{analyte}} \gg C_{\text{nanoagggregate}}$, in Figure 2.8 between about 10^{-9} and 10^{-6} M concentration, SERS power again depends linearly on analyte concentration. Under these conditions, 100–10 000 molecules are attached to the same nanoaggregate and experience an average uniform enhancement. This regime is used in numerous applications of SERS for quantitative studies, such as in chemical analysis (see examples in Ref. [56]) or in quantitative investigations of the reaction kinetics in chemical processes [57].

The linear relation between SERS signal and concentration of the target molecule breaks down at approximately 1 : 1 concentration between nanoaggregates and target molecule.

This shows that only one molecule per nanoaggregate can experience an extremely high enhancement level, that is, there is only one hottest hot spot per nanoaggregate with 100 nm dimension which provides space for only one molecule. Considering the whole surface area of a hot nanoaggregate on the order of 3×10^{-14} m^2 and assuming a hot spot at the dimension of a small molecule, that is, on the order of 10^{-20} m^2, means that only 0.00003% of the surface of the hot nanoaggregates provide electromagnetic SERS enhancement factors on the order of 10^{12}. For more regular AgFON (silver film over nanosphere) substrates, it has been found that 0.0003% of the surface provides an enhancement factor $>10^{10}$ [58] and 0.003% exhibits enhancement factors $>10^9$ [59].

2.3.2 Spectroscopic Potential of Non-resonant SERS Using the Hottest Hot Spots

As we have discussed, optimum enhancement conditions provided in the hottest hot spots can bring effective non-resonant Raman cross sections to the level of fluorescence cross sections of good fluorophores. Compared to fluorescence as a very common spectroscopic tool, SERS provides several methodological advantages. As a process that is not in resonance with electronic transition of the target molecule, SERS avoids problems related to photodecomposition. Superposition of disturbing background signals can be avoided by using strong pumped anti-Stokes signals as spectroscopic signatures. Additionally, hot spots support efficiently two-photon excited optical effects. This allows the collection of surface-enhanced hyper Raman spectra at a signal level comparable with SERS spectra. As a two-photon process, HRS follows different selection rules than "normal" Raman scattering and provides new additional vibrational information about the target molecule. This enables a comprehensive structural characterization using Raman and hyper Raman spectra.

In particular, SERS opens up exciting capabilities for chemical analysis. The effect combines the high molecular structural selectivity and information content of vibrational spectroscopy with single molecule sensitivity.

Moreover, Raman scattering is a very general property of a molecule, and almost every molecule has Raman-active molecular vibrations that can be "seen" by the Raman effect at a cross section of at least 10^{-30} cm^2, whereas far fewer molecules show fluorescence. Due to the primarily "plasmonic" origin of SERS enhancement, it should be possible to achieve an equally strong SERS effect for each molecule which is attached to the hot spots of optimized plasmonic structures, thus making non-resonant SERS a tool for studying a broad range of molecules at the single molecule level.

An interesting question for spectro-analysis of a single molecule involves the total number of photons that can be emitted by one molecule. This is determined by the maximum number of excitation–emission cycles a molecule survives. In fluorescence experiments, this number is limited by the rate of photobleaching of the molecule. As a process that does not involve an electronic transition of the target molecule, SERS photons can be collected from a molecule over an almost unlimited time span.

A number that is of particular interest for the rapid detection and screening of single molecules is the maximum number of photons that can be emitted by a molecule per time interval. Under saturation conditions, this number is inversely proportional to the lifetime of the excited molecular states involved in the optical detection process. Due to the shorter vibrational relaxation times as compared to the electronic relaxation times, a molecule can go through more Raman cycles than fluorescence cycles per time interval. Therefore, the number of Raman photons per time interval that can be emitted by a molecule can be higher than the number of fluorescence photons by a factor of 10^2–10^3.

Limitations of SERS spectroscopy are attributed to the fact that SERS enhancement factors measured at different sites of a SERS-active structure show a wide distribution. The maximum enhancement level is available for only very few molecules. Using SERS as a quantitative analytical tool requires uniform SERS enhancement factors for the target molecules. Due to the distribution of SERS enhancement factors over several orders of magnitude, SERS works as a quantitative analytical tool only in two specific concentration ranges, as is shown in Figure 2.8: at relatively high concentrations, when analyte molecules experience a uniform average enhancement, and at very low concentrations, when target molecules exclusively occupy the hottest hot spots only. In this way one avoids heterogeneity in SERS enhancement and SERS signals scale linearly with the concentration of the analyte. This approach lays the ground for developing SERS as a tool for ultimate chemical analysis where single molecules can be completely characterized based on their surface-enhanced Raman and hyper Raman spectra and small amounts of matter are quantified by counting single molecules based on the SERS signal.

2.4 Summary and Conclusions

We have discussed SERS using non-resonant NIR excitation and hot silver nanoaggregates as enhancing plasmonic structures. The aggregates are evaluated and characterized using one- and two-photon excited SERS and EELS. Our studies show that nanoaggregates at about 100 nm dimension exhibit one hottest hot spot which provides total SERS enhancement of 14 orders of magnitude for one target molecule only. This means that only 0.00003% of the surface of the hot nanoaggregates provides electromagnetic SERS enhancement factors on the order of 10^{12}.The level of enhancement does not depend on the size or shape of the aggregates but is critically dependent on the size of the gap between the nanoparticles forming the aggregate. The optimum wavelengths for achieving the highest enhancement in the hottest spots is not related to the extinction spectrum of the nanoaggregates. Despite the extinction spectrum of the hot nanoaggregates showing a maximum around 400 nm and almost no signal at longer wavelengths, the highest SERS enhancement has been obtained at NIR excitation. However, EEL spectra identify plasmon resonances in the NIR range where SERS at extremely high enhancement has been measured. Experimental studies using tunable excitation in the NIR range show an increase in local field enhancement with decreasing photon energy.

Experiments performed on silver or gold nanoaggregates in solution, with analyte concentrations 1–4 orders of magnitude below the concentration of the enhancing nanoaggregates, avoid heterogeneity in SERS enhancement. Non-resonant SERS based on a uniform and extremely high enhancement level has a unique potential in analytics and sensing. It can surpass fluorescence spectroscopy as a widely used spectroscopic tool and can achieve ultimate limits in chemical analysis.

References

1. Fleischman, M., Hendra, P.J. and McQuillan, A.J. (1974) Raman spectra of pyridine adsorbed at a silver electrode. *Chemical Physics Letters*, **26**, 123.
2. Albrecht, M.G. and Creighton, J.A. (1977) Anomalously intense Raman-spectra of pyridine at a silver electrode. *Journal of the American Chemical Society*, **99**, 5215–5217.
3. Jeanmaire, D.L. and Duyne, R.P.V. (1977) Surface Raman spectroelectrochemistry.1. Heterocyclic, aromatic, and aliphatic-amines adsorbed on anodized silver electrode. *Journal of Electroanalytical Chemistry*, **84**, 1–20.
4. Kneipp, K., Moskovits, M. and Kneipp, H. (2006) *Surface-Enhanced Raman Scattering*, Springer, Heidelberg.
5. Kneipp, K., Yang, W., Kneipp, H. *et al.* (1997) Single molecule detection using surface-enhanced Raman scattering (SERS). *Physical Review Letters*, **78**, 1667–1670.
6. Nie, S. and Emory, S.R. (1997) Probing single molecules and single nanoparticles by surface-enhanced Raman scattering. *Science*, **275**, 1102–1106.
7. Kneipp, K., Yang, W., Kneipp, H. *et al.* (1996) Population pumping of excited vibrational states by spontaneous surface-enhanced Raman scattering. *Physical Review Letters*, **76**, 2444–2447.
8. Kneipp, K., Kneipp, H., Kartha, V.B. *et al.* (1998) Detection and identification of a single DNA base molecule using surface-enhanced Raman scattering (SERS). *Physical Review E*, **57**, R6281–R6284.
9. Xu, H.X., Bjerneld, E.J., Kall, M. *et al.* (1999) Spectroscopy of single hemoglobin molecules by surface enhanced Raman scattering. *Physical Review Letters*, **83**, 4357–4360.
10. Michaels, A.M., Jiang, J. and Brus, L. (2000) Ag nanocrystal junctions as the site for surface-enhanced Raman scattering of single Rhodamine 6G molecules. *The Journal of Physical Chemistry B*, **104**, 11965–11971.
11. Kneipp, K., Kneipp, H. and Kneipp, J. (2006) Surface-enhanced Raman scattering in local optical fields of silver and gold nanoaggregatess – from single-molecule Raman spectroscopy to ultrasensitive probing in live cells. *Accounts of Chemical Research*, **39**, 443–450.
12. Camden, J.P., Dieringer, J.A., Wang, Y.M. *et al.* (2008) Probing the structure of single-molecule surface-enhanced Raman scattering hot spots. *Journal of the American Chemical Society*, **130**, 12616–12617.
13. Kitahama, Y., Tanaka, Y., Itoh, T. *et al.* (2011) Power-law analysis of surface-plasmon-enhanced electromagnetic field dependence of blinking SERS of thiacyanine or thiacarbocyanine adsorbed on single silver nanoaggregates. *Physical Chemistry Chemical Physics*, **13**, 7439–7448.
14. Stockman, M.I., Shalaev, V.M., Moskovits, M. *et al.* (1992) Enhanced Raman scattering by fractal clusters: scale-invariant theory. *Physical Review B*, **46**, 2821–2830.
15. McMahon, J.M., Gray, S.K. and Schatz, G.C. (2011) Fundamental behavior of electric field enhancements in the gaps between closely spaced nanostructures. *Physical Review B*, **83**, 5.
16. Halas, N.J., Lal, S., Chang, W.S. *et al.* (2011) Plasmons in strongly coupled metallic nanostructures. *Chemical Reviews*, **111**, 3913–3961.

17. Lee, J.H., Nam, J.M., Jeon, K.S. *et al.* (2012) Tuning and maximizing the single-molecule surface-enhanced Raman scattering from DNA-tethered nanodumbbells. *ACS Nano*, **6**, 9574–9584.
18. Yoshida, K., Itoh, T., Tamaru, H. *et al.* (2010) Quantitative evaluation of electromagnetic enhancement in SERRS from plasmonic properties and morphologies of individual Ag nanostructures. *Physical Review B*, **81**, 9.
19. Haran, G. (2010) Single-molecule Raman spectroscopy: a probe of surface dynamics and plasmonic fields. *Accounts of Chemical Research*, **43**, 1135–1143.
20. Li, K.R., Stockman, M.I. and Bergman, D.J. (2003) Self-similar chain of metal nanospheres as an efficient nanolens. *Physical Review Letters*, **91**, 227402.
21. Weber, M.L., Litz, J.P., Masiello, D.J. *et al.* (2012) Super-resolution imaging reveals a difference between SERS and luminescence centroids. *ACS Nano*, **6**, 1839–1848.
22. Dieringer, J.A., Lettan, R.B., Scheidt, K.A. *et al.* (2007) A frequency domain existence proof of single-molecule surface-enhanced Raman Spectroscopy. *Journal of the American Chemical Society*, **129**, 16249–16256.
23. Kneipp, K., Kneipp, H., Deinum, G. *et al.* (1998) Single-molecule detection of a cyanine dye in silver colloidal solution using near-infrared suirface-enhanced Raman scattering. *Applied Spectroscopy*, **52**, 175–178.
24. Maruyama, Y., Ishikawa, M. and Futamata, M. (2001) Surface-enhanced Raman scattering of single adenine molecules on silver colloidal particles. *Chemistry Letters*, 834–835.
25. Kneipp, K. and Kneipp, H. (2013) Probing the plasmonic near-field by one- and two-photon excited surface enhanced Raman scattering. *Beilstein J. Nanotechnol.* **13**, 834–842.
26. Blackie, E.J., Le Ru, E.C. and Etchegoin, P.G. (2009) Single-molecule surface-enhanced Raman spectroscopy of nonresonant molecules. *Journal of the American Chemical Society*, **131**, 14466–14472.
27. Treffer, R., Lin, X.M., Bailo, E. *et al.* (2011) Distinction of nucleobases - a tip-enhanced Raman approach. *Beilstein Journal of Nanotechnology*, **2**, 628–637.
28. Pettinger B., Schambach P., Villagomez C.J., *et al.*, Tip-enhanced Raman spectroscopy: near-fields acting on a few molecules, *Annual Review of Physical Chemistry*, **63**, 379–399, 2012.
29. Moula, G., Rodriguez-Oliveros, R., Albella, P. *et al.* (2012) Plasmonics and single-molecule detection in evaporated silver-island films. *Annalen der Physik*, **524**, 697–704.
30. Li, J.F., Huang, Y.F., Ding, Y. *et al.* (2010) Shell-isolated nanoparticle-enhanced Raman spectroscopy. *Nature*, **464**, 392–395.
31. Yurtsever, A. and Zewail, A.H. (2012) Direct visualization of near-fields in nanoplasmonics and nanophotonics. *Nano Letters*, **12**, 3334–3338.
32. Kadkhodazadeh, S., Wagner, J., Joseph, V. *et al.* (2013) Electron energy-loss - and one- and two-photon excited SERS probing of "hot" plasmonic silver nanoaggregates. *Plasmonics*, **8**(2), 763–767.
33. Lee, P.C. and Meisel, D. (1982) Adsorption and surface-enhanced Raman of dyes on silver and gold sols. *The Journal of Physical Chemistry*, **86**, 3391–3395.
34. Kneipp, J., Li, X.T., Sherwood, M. *et al.* (2008) Gold nanolenses generated by laser ablation-efficient enhancing structure for surface enhanced Raman scattering analytics and sensing. *Analytical Chemistry*, **80**, 4247–4251.

35. Kozich, V. and Werncke, W. (2010) The vibrational pumping mechanism in surface-enhanced Raman scattering: a subpicosecond time-resolved study. *Journal of Physical Chemistry C*, **114**, 10484–10488.
36. Stockman, M.I. (2006) Electromagnetic theory of SERS, in *Surface-Enhanced Raman Scattering: Physics and Applications*, Topics in Applied Physics, Vol. **103**, Springer-Verlag, Berlin.
37. Schatz, G.C., Young, M.A. and Van Duyne, R.P. (2006) Electromagnetic mechanism of SERS, in *Surface-Enhanced Raman Scattering: Physics and Applications*, Topics in Applied Physics, Vol. **103**, Springer-Verlag, Berlin.
38. Otto, A., Mrozek, I., Grabhorn, H. *et al.* (1992) Surface-enhanced Raman scattering. *Journal of Physics: Condensed Matter*, **4**, 1143–1212.
39. Lombardi, J.R. and Birke, R.L. (2012) The theory of surface-enhanced Raman scattering. *The Journal of Chemical Physics*, **136**, 11.
40. Kneipp, K., Kneipp, H., Itzkan, I. *et al.* (2002) Nonlinear Raman probe of single molecules attached to colloidal silver and gold clusters, in *Optical Properties of Nanostructured Random Media*, Topics in Applied Physics, Vol. **82** (ed V.M. Shalaev), Springer-Verlag, Berlin, Heidelberg, New York.
41. Kneipp, J., Kneipp, H. and Kneipp, K. (2006) Two-photon vibrational spectroscopy for biosciences based on surface-enhanced hyper-Raman scattering. *Heilkunde - Heilwege*, **103**, 17149–17153.
42. Ciraci, C., Hill, R.T., Mock, J.J. *et al.* (2012) Probing the ultimate limits of plasmonic enhancement. *Science*, **337**, 1072–1074.
43. Nelayah, J., Kociak, M., Stephan, O. *et al.* (2007) Mapping surface plasmons on a single metallic nanoparticle. *Nature Physics*, **3**, 348–353.
44. Koh, A.L., Bao, K., Khan, I. *et al.* (2009) Electron Energy-Loss Spectroscopy (EELS) of surface plasmons in single silver nanoparticles and dimers: influence of beam damage and mapping of dark modes. *ACS Nano*, **3**, 3015–3022.
45. Chu, M.W., Sharma, P., Chang, C.P. *et al.* (2009) Probing surface plasmons in individual Ag nanoparticles in the ultra-violet spectral regime. *Nanotechnology*, **20**, 5.
46. Koh, A.L., Fernandez-Dominguez, A.I., McComb, D.W. *et al.* (2011) High-resolution mapping of electron-beam-excited plasmon modes in lithographically defined gold nanostructures. *Nano Letters*, **11**, 1323–1330.
47. Guiton, B.S., Iberi, V., Li, S.Z. *et al.* (2011) Correlated optical measurements and plasmon mapping of silver nanorods. *Nano Letters*, **11**, 3482–3488.
48. Nicoletti, O., Wubs, M., Mortensen, N.A. *et al.* (2011) Surface plasmon modes of a single silver nanorod: an electron energy loss study. *Optics Express*, **19**, 15371–15379.
49. Mazzucco, S., Stephan, O., Colliex, C. *et al.* (2011) Spatially resolved measurements of plasmonic eigenstates in complex-shaped, asymmetric nanoparticles: gold nanostars. *The European Physical Journal – Applied Physics*, **54**, 9.
50. Alber, I., Sigle, W., Muller, S. *et al.* (2011) Visualization of multipolar longitudinal and transversal surface plasmon modes in nanowire dimers. *ACS Nano*, **5**, 9845–9853.
51. Barwick, B., Flannigan, D.J. and Zewail, A.H. (2009) Photon-induced near-field electron microscopy. *Nature*, **462**, 902–906.
52. Yurtsever, A., van der Veen, R.M. and Zewail, A.H. (2012) Subparticle ultrafast spectrum imaging in 4D electron microscopy. *Science*, **335**, 59–64.

53. Mirsaleh-Kohan, N., Iberi, V., Simmons, P.D. *et al.* (2012) Single-molecule surface-enhanced Raman scattering: can STEM/EELS image electromagnetic hot spots? *Journal of Physical Chemistry Letters*, **3**, 2303–2309.
54. Duan, H.G., Fernandez-Dominguez, A.I., Bosman, M. *et al.* (2012) Nanoplasmonics: classical down to the nanometer scale. *Nano Letters*, **12**, 1683–1689.
55. Zhou, Z.G., Wang, G.Y. and Xu, Z.Z. (2006) Single-molecule detection in a liquid by surface-enhanced resonance Raman scattering. *Applied Physics Letters*, **88**, 034104.
56. McNay, G., Eustace, D., Smith, W.E. *et al.* (2011) Surface-Enhanced Raman Scattering (SERS) and surface-enhanced resonance Raman scattering (SERRS): a review of applications. *Applied Spectroscopy*, **65**, 825–837.
57. Joseph, V., Engelbrekt, C., Zhang, J.D. *et al.* (2012) Characterizing the kinetics of nanoparticle-catalyzed reactions by surface-enhanced Raman scattering. *Angewandte Chemie International Edition*, **51**, 7592–7596.
58. Fang, Y., Seong, N.H. and Dlott, D.D. (2008) Measurement of the distribution of site enhancements in surface-enhanced Raman scattering. *Science*, **321**, 388–392.
59. Chien, F.C., Huang, W.Y., Shiu, J.Y. *et al.* (2009) Revealing the spatial distribution of the site enhancement for the surface enhanced Raman scattering on the regular nanoparticle arrays. *Optics Express*, **17**, 13974–13981.

3

Effect of Nanoparticle Symmetry on Plasmonic Fields: Implications for Single-Molecule Raman Scattering

Lev Chuntonov[1,2] *and Gilad Haran*[1]

[1]*Department of Chemical Physics, Weizmann Institute of Science, Israel*
[2]*Department of Chemistry, University of Pennsylvania, USA*

3.1 Introduction

Surface-enhanced Raman scattering [1, 2] (SERS), a spectroscopy technique with a sensitivity level down to one molecule [3–5], has recently opened up exciting opportunities for studying single-molecule behavior at the interface with noble metals. Such studies employ single molecules as miniature probes of the nanoscale environment, and have also led to devices for ultrasensitive diagnostics and nano-optical applications [6–11]. The strong electromagnetic field localization required for the observation of single-molecule surface-enhanced Raman scattering (SMSERS) is readily obtained when the studied molecule is adsorbed inside the gap between closely adjacent nanoparticles possessing a localized plasmon resonance of frequency close to that of the excitation source and the Raman band [4, 12–16]. The standard model of SERS distinguishes between the two processes occurring at the excitation and scattered energy [17–19]. The first one involves high local electromagnetic field enhancement at the molecular location due to the localized plasmon excitation on the surface of the nanoparticles. The second process involves scattering of the light that the molecule emits by the localized plasmons excited on the nanoparticles. While the first step of the interaction is important for understanding the SERS enhancement process, it is the second step that determines the characteristics of the emitted light, for example, its directivity and polarization [18].

Frontiers of Surface-Enhanced Raman Scattering: Single Nanoparticles and Single Cells, First Edition.
Edited by Yukihiro Ozaki, Katrin Kneipp and Ricardo Aroca.

In the long-wavelength limit the response of a single spherical plasmonic nanoparticle is approximated as an induced dipole located in the center of a sphere whose polarizability depends on the metal dielectric function [20, 21]. The optical properties of clusters of interacting nanoparticles in the near- and far- radiation zones strongly depend on the way the nanoparticles are arranged [13, 14, 22–29]. These unique properties have been recently utilized to build new materials with improved optical characteristics [30–33]. Clearly, the interactions of adsorbed molecules with the electric field of light are also influenced by the nanoparticle arrangement. Our interest in better understanding the mechanisms behind such interactions has motivated a series of spectroscopic and theoretical studies summarized in this chapter [18, 34–36]. The underlying mechanisms are introduced here through two phenomena. First, we show how the shape of the plasmon spectrum of a cluster can be correlated with the shape of the Raman spectrum of an adsorbed molecule. Second, we examine the relation between the polarization of the Raman-scattered light and the plasmon mode structure of the cluster. Before discussing these two aspects, however, we present a systematic summary of the plasmon mode picture of the smallest nanoparticle clusters that are relevant to the discussion: dimers and trimers of spherical silver nanoparticles.

3.2 Methodology

In our experiments the studied molecules are adsorbed on the surface of nanoparticles that are aggregated into dimers and trimers and deposited on a surface [18, 37]. SMSERS spectra are measured upon excitation with a CW laser. The plasmon modes of the nanoparticle clusters are studied with single-particle plasmon scattering optical spectroscopy [35, 36, 38]. The structure of each cluster is imaged immediately following optical measurements by either scanning electron microscopy (SEM) or transmission electron microscopy (TEM).

The plasmon spectra of clusters are analyzed based on their structural symmetry using the theoretical framework provided by group theory. Symmetry-adapted linear combinations (SALCs) of the individual nanoparticle plasmon modes, derived with the aid of the irreducible representations of the relevant point group, are known to serve as a good basis for understanding cluster plasmon modes [29]. When the symmetry of the cluster is lowered the irreducible representations of the symmetry group are transformed, as summarized in the relevant correlation tables. However, correlation tables alone provide only information on the symmetry changes and are not sufficient to track the plasmon mode transformation in detail, as each symmetry point group can generally include several SALCs of the same symmetry. The explicit shapes and energies of the plasmon modes for arbitrary cluster geometries are obtained from a full theoretical description based on hybridization theory, which provides the eigenstates of the system as solutions of the Euler–Lagrange equation for interacting surface charge distributions induced on the individual nanoparticles [26, 27, 29]. This theory makes it possible, therefore, to track the changes in the plasmon modes that follow a change in the cluster symmetry.

For simplicity we discuss in this chapter only the dipolar ($l = 1$) modes of the multipole expansion, such that the solutions can be represented by point dipoles, one positioned in the center of each nanoparticle. The dipolar model is valid in the long wavelength limit ($ka \ll 1$, where k is the scattering wave vector and a is the radius of a nanoparticle), which holds approximately for silver nanoparticles of about 50 nm excited with visible light. Better agreement with the experimental data is obtained when plasmon modes of up to the order of $l = 10$ are considered. However, including the higher order terms significantly complicates the intuitive understanding of the interaction picture. When the nanoparticles are in close proximity to each other and strongly interact, terms of multipole orders higher than dipolar may admix into the dipolar modes and modify their character. For example, such admixing of the quadrupolar modes into the dark dipolar mode in a nanoparticle dimer makes this mode weakly visible in the plasmon spectrum [27, 29, 39]. Nevertheless, the energies of such modes are significantly higher than the energies of the low-energy (so-called bonding) modes of dimers and trimers lying in the spectral range of typical SERS signals and, therefore, to a first approximation the former can be omitted in the analysis. Although the actual eigenvalues obtained by the plasmon hybridization calculation within the dipole limit are higher in energy than their experimental and numerical counterparts, the qualitative trends in eigenvalue evolution are preserved by this simplified approach.

Full electromagnetic simulations with the Multiple Multipole Program (MMP) [40, 41] and MNPBEM (metallic nanoparticles boundary element method) [42] were used to simulate the actual fields near the nanoparticles surface, from which the plasmon and Raman spectra and surface charge distributions were derived. In numerical simulations, dielectric functions from Ref. [43] were used. In the plasmon hybridization calculation and harmonic oscillator model the optical response of silver was modeled by the Drude dielectric function with $\omega_B = 9.5$ eV, $\varepsilon_S = 5$, and $\varepsilon_E = 1.8$, where ω_B is the bulk plasmon frequency, ε_S is the background polarizability of the metal ions, and ε_E is the permittivity of the medium surrounding the nanoparticles.

3.3 Plasmon Mode Structure of Nanoparticle Clusters

3.3.1 Dimers

A nanoparticle dimer is the simplest cluster with an interparticle gap suitable for the enhancement of the SERS signal to the level needed for single molecule experiments. The plasmon mode structure of dimers has been extensively studied both experimentally and theoretically [10, 12, 13, 18, 34]. Therefore, only a brief reminder is given here. Two parameters strongly affect the plasmon mode resonance energies, intensities, and spectral widths, and, consequently, the SERS enhancement factors. These are the size of the nanoparticles and the interparticle distance. A dimer of two identical nanoparticles is described by the symmetry point group $D_{\infty h}$. Collective plasmon modes of such a dimer are constructed with SALCs of individual nanoparticle modes. Each pair of plasmon modes coming from the two individual nanoparticles forms a new pair of cluster modes, termed bonding and anti-bonding, similar to molecular orbital formation from atomic orbitals [27]. When the nanoparticles are far apart, the interparticle interaction is weak and the new collective modes are degenerate and have the same energy as the individual nanoparticle plasmon modes. As the interparticle distance decreases, the energies of the bonding and antibonding pair split due to the interparticle interaction.

In the context of SMSERS discussed here only two bonding plasmon modes of the dimer are of relevance, namely the transverse Π_g mode and the longitudinal Σ_u^+ mode, shown

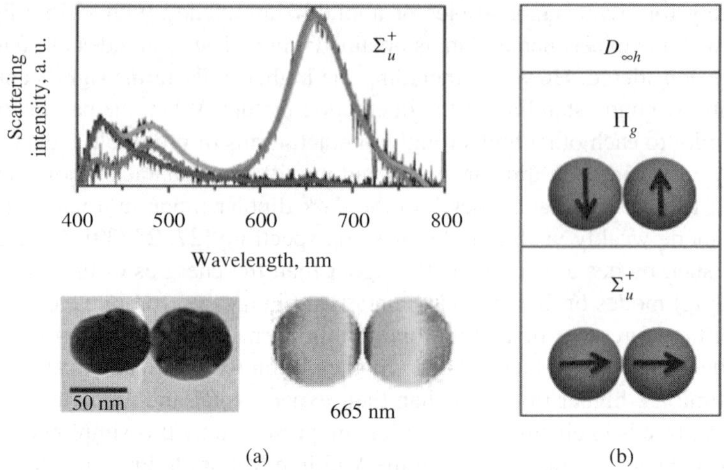

Figure 3.1 *Plasmonic dimer. (a) Plasmon scattering spectrum. Thin black lines – raw experimental data; other lines – low pass filter applied. The clusters are excited by either non-polarized light (green) or light linearly polarized parallel (red) and perpendicular (blue) to dimer axis. The polarized spectra are normalized to the corresponding peaks of the non-polarized spectra. Lower panel – TEM image of the dimer and numerically simulated surface charge distribution (the scale bar is 50 nm). (b) Schematic presentation of bonding plasmon modes. The symbols of the corresponding irreducible representations are shown. Adapted with permission from: Ref. [36]. Copyright (2011) American Chemical Society. (*See insert for color representation of the figure.*)*

schematically in Figure 3.1. The former mode has the higher energy and within the dipole limit is "dark", that is, its net dipole moment is zero, and it does not interact with light. The latter mode is the lowest energy bright mode observed in the scattering spectrum, as shown in Figure 3.1. The larger the nanoparticles and the smaller the interparticle distance, the stronger the interparticle interaction, resulting in stronger localization of the surface charge near the gap and stronger splitting between the energies of anti-bonding and bonding modes. In particular, the Σ_u^+ plasmon resonance red shifts, is spectrally broadened, and its spectral intensity increases. SERS signal enhancement also increases monotonically as interparticle distances decrease down to about 1 nm. For distances below 1 nm, in addition to the electromagnetic interaction, the quantum tunneling of electrons between the nanoparticles should be considered, which leads to a blue shift of the plasmon spectrum and a decrease of the SERS signal enhancement [10, 44–46].

3.3.2 Trimers

3.3.2.1 D_{3h} Symmetry

Consider a cluster of three silver nanoparticles located in the corners of an equilateral triangle. This cluster has the highest possible symmetry for a trimer – D_{3h}. An example of such a cluster is shown in Figure 3.2a, together with the plasmon scattering spectra, corresponding to two orthogonal orientations of the linearly polarized excitation light: longitudinal and transverse. In the former orientation the polarization is parallel to one of the bases of

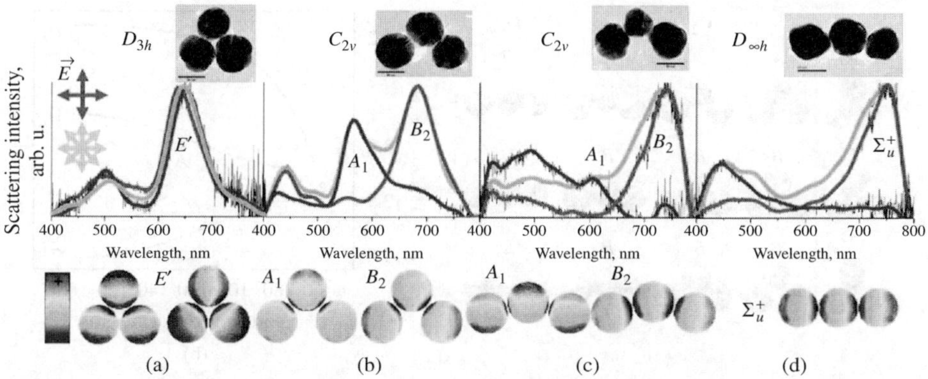

Figure 3.2 *(a–d) Scattering spectra of silver nanoparticle trimers showing mode splitting due to gradual opening of the vertex angle. Thin black lines, raw experimental data; color lines, low pass filter applied. The clusters are excited by either non-polarized light (green) or light linearly polarized along the longitudinal (red) and transverse (blue) directions (which are indicated by arrows of matching colors). The polarized spectra are normalized to the corresponding peaks of the non-polarized spectra. Plasmon mode symmetries of the bonding modes are marked. TEM images of the clusters are shown in the insets (the bars are 50 nm). Calculated surface charge distributions of the selectively excited bonding modes are shown below the spectra. Adapted with permission from: Ref. [35]. Copyright (2011) American Chemical Society. (See insert for color representation of the figure.)*

the triangle, while in the latter it is perpendicular to this base. It is apparent from Figure 3.2a that plasmon resonances excited with the two orthogonal polarizations have identical energies and spectral shapes. Indeed, the symmetry group D_{3h} includes doubly degenerate in-plane SALCs corresponding to the irreducible representations of symmetry E′, as dictated by the presence of symmetry operations constituting the rotational subgroup C_3 [47, 48]. For the current discussion we focus on the low-energy bonding modes of the cluster only, as they are the only ones that contribute in the spectral range explored by SERS experiments. The surface charge distributions calculated numerically for the selectively excited degenerate transverse and longitudinal bonding resonances of the D_{3h} cluster are shown in the lower panel of Figure 3.2a. These charge distributions represent physical plasmon states of the cluster with the corresponding symmetry. The physical states mostly involve E′ dipolar SALCs (see Figure 3.3), which contribute more than 60% of the charge distributions. Some additional contributions, responsible for the deviation between the calculated physical states of Figure 3.2 and the dipolar SALCs, may originate from admixtures of higher order multipoles [26, 29]. Mode assignment was, therefore, made based on the larger dipolar contribution, while coupling to higher order multipoles remains beyond the scope of the present discussion.

The dipolar SALCs which coincide with the modes obtained from plasmon hybridization theory are illustrated in the left column of the correlation table in Figure 3.3. The degenerate pair includes one mode having its net dipole moment aligned along the longitudinal direction of the cluster, and another along the transverse direction. In agreement with their non-zero net dipole moments the E′ modes are optically active (bright) as seen from the corresponding peaks in the scattering spectrum. In addition to the degenerate bonding E′

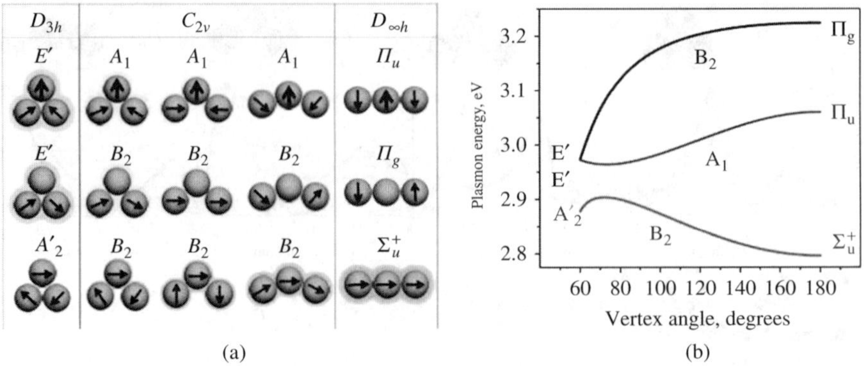

Figure 3.3 Correlation table of bonding dipolar plasmon eigenmodes of trimers for varying vertex angles, obtained using the plasmon hybridization theory. (a) Schematic illustration of the modes and (b) plasmon energy diagram. The symbols of the corresponding irreducible representations are shown. The thickness of halos surrounding the particles represents the relative "brightness" of plasmon modes. Adapted with permission from: Ref. [35]. Copyright (2011) American Chemical Society

modes, it is important to consider an additional mode of the D_{3h} trimer that actually has the lowest energy – the ring-like A'_2 mode. This mode is dark owing to its zero net dipole moment. It will be shown below that when the symmetry of the D_{3h} trimer is broken, this mode plays an important role in the interaction with light.

The symmetry of the D_{3h} cluster can be broken in several ways. The first series of the symmetry-broken nanoparticle clusters discussed here corresponds to gradually opening one of the vertex angles of the triangle such that the resulting cluster has C_{2v} symmetry. Further opening of the vertex angle leads to a linear chain of three nanoparticles and the corresponding $D_{\infty h}$ symmetry. The second series involves gradually "resizing" the vertex nanoparticle such that the resulting series of trimers of C_{2v} symmetry gradually converges to a dimer, which also belongs to the $D_{\infty h}$ point group. Finally, a third series corresponds to a gradual simultaneous resizing of two nanoparticles at the triangle base. In this case the series of trimers of C_{2v} symmetry converges to a single nanoparticle described by the K_h point group. Below we show in detail the evolution of the plasmon modes of the cluster within each series.

3.3.2.2 *Symmetry Breaking by Vertex Angle Opening*

The degree of symmetry breaking by opening the vertex angle of a trimer is quantified by measuring the vertex angle ϕ of the triangle: $\phi = \pi/3$ stands for the equilateral triangle of D_{3h} symmetry, $\pi/3 < \phi < \pi$ for the Λ-shaped cluster of C_{2v} symmetry, and $\phi = \pi$ for a linear chain of three nanoparticles having $D_{\infty h}$ symmetry. Measured plasmonic spectra that are representative for the corresponding trimers are shown in Figure 3.2b–d. Using the group theoretical framework we constructed a group correlation table for the plasmon modes in the sequence of shapes as shown in Figure 3.3. We used plasmon hybridization theory to evaluate the ϕ-dependence of the plasmon mode shapes of the symmetry-broken clusters (Figure 3.3a) and their energies (Figure 3.3b). As the C_3 rotation operation becomes invalid in symmetry-broken clusters, in agreement with the correlation table we find that increasing ϕ beyond $\pi/3$ lifts the degeneracy of the E′ plasmon resonances and results in two new plasmon bands, selectively excited by light linearly polarized along the cluster's longitudinal and transverse directions. The transverse and longitudinal in-plane E′ modes transform to modes A_1 and B_2 of C_{2v}, respectively, which are symmetric and anti-symmetric with respect to the σ_v reflection plane. Also here the mode assignment is based on the analysis of the surface charge distribution induced in the cluster. The net dipoles of the A_1 and B_2 modes are oriented along the cluster's transverse and longitudinal directions.

The splitting between the A_1 and B_2 modes increases monotonically as the vertex angle increases, while the intensity of the red-shifted longitudinal resonance increases and that of the blue-shifted transverse resonance decreases. The evolution of the A_1 mode naturally follows the correlation table, which rationalizes the decrease in the corresponding spectral peak intensity by a decrease in the net dipole moment of this plasmon mode as it evolves into the dark Π_u mode at the $D_{\infty h}$ limit. Both longitudinal E′ and A'_2 modes of D_{3h} transform to longitudinal B_2 modes of C_{2v}. The evolution of these modes, however, is *not* readily understood from the measured plasmon spectrum. Rather than the observed increase in the scattering intensity and the red spectral shift of the longitudinal plasmon band, the correlation table suggests that an E′ mode of the D_{3h} symmetry group transforms into the dark Π_g mode of the $D_{\infty h}$, which is higher in energy than the original mode, while it is the dark A'_2 mode of D_{3h} which transforms into the bright longitudinal Σ_u^+ mode of the linear chain. While the dark A'_2 mode cannot be excited by the linearly-polarized light, the lower-energy B_2 mode originating from it has a non-zero dipole moment in the longitudinal direction. As ϕ increases the dipole moment of the lowest energy B_2 mode gradually increases, whereas the dipole moment of the higher-energy B_2 mode decreases. This gradual interchange of the total dipole moments of the B_2 modes of the Λ-shaped cluster is reflected in a corresponding change in the scattering intensity of the two plasmon resonances: the intensity of the former resonance increases, while the intensity of the latter decreases. The energy diagram in Figure 3.3b shows, in fact, that the energy of the B_2 mode originating from E′ increases with ϕ, while the energy of the B_2 mode originating from A'_2 increases for small values of ϕ, up to 80°, then decreases for higher values.

It is evident from Figure 3.3 that the indicated change in the direction of the plasmon energy shift of the bonding B_2 modes and their coupling to the excitation light occurs within a small range of angles, $60° < \phi < 80°$. As a consequence of such a "fast" evolution of the plasmon modes, the B_2 mode originating from E′ cannot be observed in the spectrum beyond these values of ϕ and the only bonding longitudinal resonance observed corresponds to the B_2 mode originating from A'_2. This is the reason that only a single red-shifting mode is seen in the experimental spectra. (Precise control over the cluster geometry is still an experimental challenge, so that the range $60° < \phi < 80°$ is missed.) Our conclusions are in full agreement with a series of detailed electromagnetic numerical simulations that were performed for all possible cluster geometries.

As ϕ increases further, the cluster geometry finally attains the symmetry of the group $D_{\infty h}$ which corresponds to a linear chain of nanoparticles with $\phi = \pi$. Here, the higher energy B_2 mode of C_{2v}, which at this point is already dark, transforms to the mode Π_g of $D_{\infty h}$. On the other hand, the lower energy B_2 mode transforms into a bright mode of symmetry Σ_u^+, corresponding to a parallel alignment of the individual dipoles, which is the only bright bonding mode for this symmetry group.

3.3.2.3 Symmetry Breaking by Resizing a Vertex Nanoparticle

Another way to break the D_{3h} symmetry of the nanoparticle trimer is by reducing the size of one of the nanoparticles. Representative clusters are shown in Figure 3.4. Similar to the case discussed above, the new symmetry point group is C_{2v} where the degeneracy of the E′ modes is lifted and new peaks corresponding to A_1 and B_2 modes appear in the plasmon spectrum. Simulated charge distributions at the resonant wavelengths in the middle column of Figure 3.4 are used to assign the corresponding plasmon modes. The transformation of the new modes resembles at first glance the previous case: the A_1 mode gradually shifts to the blue and loses its intensity, while the B_2 mode does not change its spectral position. The change in the cluster geometry is quantified in this case by taking the ratio R_1/R_2, where R_1 and R_2 are the radii of the vertex and base nanoparticles, respectively. While R_1 is gradually reduced, R_2 is kept constant at 32 nm, such that $R_1/R_2 = 1$ corresponds to a D_{3h} cluster, $0 < R_1/R_2 < 1$ – to a C_{2v} cluster, and $R_1/R_2 = 0$ – to a dimer having $D_{\infty h}$

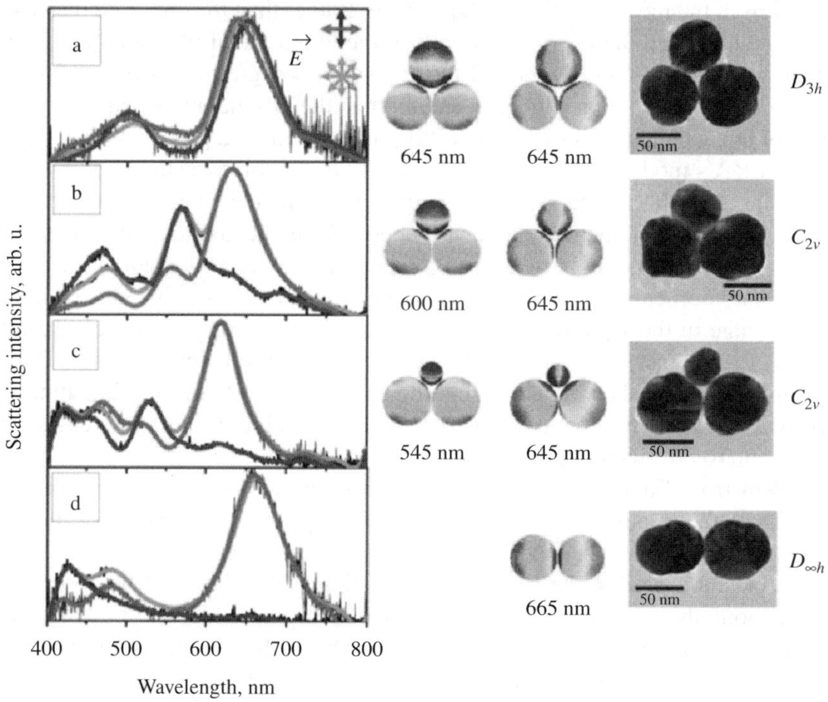

Figure 3.4 *Experimental scattering spectra, numerical simulations of surface charge distribution, and TEM images (scale bar is 50 nm) of the plasmon spectra of symmetry-broken trimers for different sizes of the vertex nanoparticle. Green lines denote excitation by non-polarized light, red lines excitation by linearly polarized light along the trimer's base, and blue lines excitation by light polarized perpendicular to the trimer's base. Polarized spectra are normalized to the corresponding peaks of the non-polarized spectra. The sizes of vertex nanoparticle (R_1) used in the simulations are 32, 24, 16, and 0 nm corresponding to R_1/R_2 values of 1, 0.75, 0.5, and 0 respectively. The corresponding symmetry point groups are also indicated. Adapted with permission from: Ref. [36]. Copyright (2011) American Chemical Society. (*See insert for color representation of the figure.*)*

symmetry. Let us first discuss the evolution of the A_1 mode briefly. With the decrease in the size of the vertex nanoparticle the net dipole moment of the A_1 mode decreases and at the dimer limit it transforms to the null mode, where there is no charge induced on the surface of the nanoparticles. The dimer cluster has three degrees of freedom less than the trimer. Therefore, this artificial null mode appears here due to the application of a basis set suitable for a trimer, which has more degrees of freedom.

Next, we discuss the evolution of the B_2 modes originating from the E′ and A'_2 of the D_{3h} cluster. The correlation table (Figure 3.5a) and plasmon energy diagram (Figure 3.5c) demonstrate the corresponding transformation of the plasmon modes, which is found to depend on interparticle separation. Energy diagrams for three different interparticle separations are shown: left panel – 33 nm, middle panel – 14 nm, and right panel – 1 nm. When the nanoparticles are placed far apart, the interaction between them is weak. In such a case the energy curves of the same symmetry can cross as in the left panel, where the two B_2 curves cross at $R_1/R_2 = 0.6$. Here, the B_2 mode originating from E′ transforms to the Σ_u^+ mode of $D_{\infty h}$, while the B_2 mode originating from A'_2 transforms to the Π_g mode. Reducing the interparticle separation increases the strength of the interaction between the B_2 modes, and the corresponding curves now present an avoided crossing. The smaller the interparticle

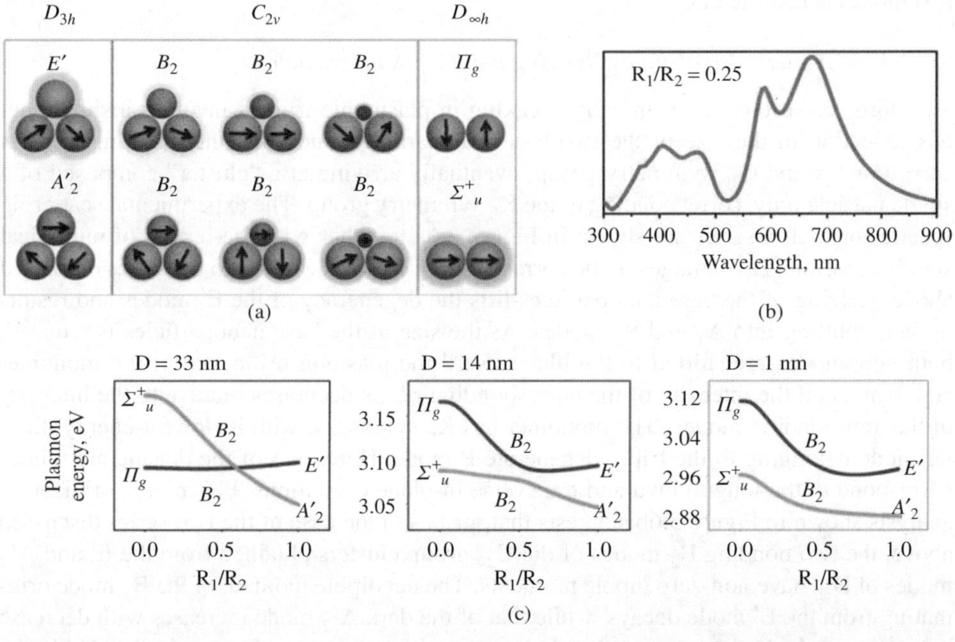

Figure 3.5 *Correlation table of bonding dipolar plasmon eigenmodes of trimers for varying size of the vertex nanoparticle, obtained using the plasmon hybridization theory. (a) Schematic illustration of the modes: the symbols of the corresponding irreducible representations are shown, the thickness of halos surrounding the particles represents the relative "brightness" of plasmon modes, while their color corresponds approximately to the resonance wavelength. (b) An example of the plasmon spectrum for* $R_1/R_2 = 0.25$*, excitation with longitudinally polarized light, and interparticle distance of 1 nm. (c) Plasmon energy diagram for interparticle separations of 33 (left), 14 (middle), and 1 nm (right). Adapted with permission from: Ref. [36]. Copyright (2011) American Chemical Society*

distance, the stronger the interaction between the nanoparticles and the larger the splitting between the curves. The avoided crossing implies that the evolution of the plasmon modes is different than in the case of a weak interaction, such that the B_2 mode originating from E′ transforms to the Π_g mode of $D_{\infty h}$, while the B_2 mode originating from A'_2 transforms to the Σ_u^+ mode. While the Σ_u^+ mode is the bright bonding mode of the dimer, the Π_g mode is a dark transverse mode, which is not observed at all in the scattering spectrum.

The case of strong interaction is directly relevant to our experimental results. Here, the total dipole moment of the B_2 mode originating from E′ gradually decreases and that of the B_2 mode originating from A'_2 gradually increases, as also seen in Figures 3.4 and 3.5. The magnitude of the total dipole moments of the two B_2 modes of C_{2v} becomes equal approximately at $R_1/R_2 = 0.35$ and the energy separation between the modes for this ratio is sufficient for them to be, in principle, clearly distinguished in the scattering spectrum. We performed detailed numerical simulations that confirmed the spectral shapes predicted by the symmetry and plasmon hybridization analyses. The corresponding simulated spectrum is presented in Figure 3.5b, showing that two longitudinal bonding plasmon resonances can be resolved in the scattering spectrum, however, only in the narrow range of $0.2 < R_1/R_2 < 0.4$. For R_1/R_2 values outside this range, either the total dipole moment magnitude of one of the modes is too low to be observed, or the energy separation between the two modes is insufficient.

3.3.2.4 *Symmetry Breaking by Resizing the Base Nanoparticles*

The third scenario of the symmetry breaking in plasmonic trimers involves a simultaneous reduction in the size of the two base nanoparticles, such that the resulting clusters correspond to the C_{2v} symmetry group, eventually arriving at a "cluster" composed of a single particle only, corresponding to the K_h symmetry group. The experimental scattering spectra for such a series are shown in Figure 3.6a together with the results of numerical simulations, and TEM images of the corresponding clusters. Similar to the series discussed above, resizing of the base nanoparticles lifts the degeneracy of the E′ modes and results in their splitting into A_1 and B_2 modes. As the size of the base nanoparticles is reduced, both new modes are shifted to the blue toward the plasmon resonance of the monomer at 450 nm, and the intensity of the corresponding peaks decreases relative to the intensity of the anti-bonding modes. The monomer has K_h symmetry, with its lowest-energy dipolar mode belonging to the triply degenerate P mode. Here, two of the degenerate modes correspond to the longitudinal and transverse in-plane excitations. Plasmon hybridization analysis shown in Figure 3.6b suggests that, just as in the case of the two series discussed above, the two bonding B_2 modes of the C_{2v} group clusters obtained from the E′ and A'_2 modes of D_{3h} have non-zero dipole moments. The net dipole moment of the B_2 mode originating from the E′ mode decays, while that of the dark A'_2 mode increases with decrease in the size of the base nanoparticles. The modes obtain comparable magnitudes within the narrow range of $0.4 < R_2/R_1 < 0.6$. The broadening of the peaks and insufficient spectral separation, however, can hinder their explicit spectral separation.

To summarize this section, we have found that the experimental findings in general confirm changes in mode energies and characters predicted by symmetry analysis. However, in each case studied it was found that changes that are critical for correct understanding of mode evolution occur within a narrow range of geometric parameters and could be easily overlooked. The next section is dedicated to a discussion of how the symmetry dependence of the plasmon modes is manifested in SMSERS spectra.

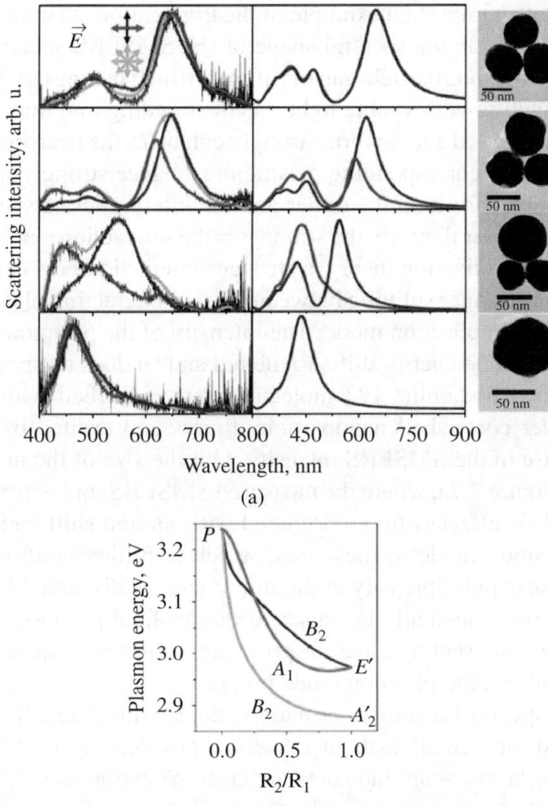

Figure 3.6 (a) Experimental scattering spectra, corresponding TEM images (scale bar is 50 nm), and numerical simulations of the plasmon spectra of symmetry-broken trimers for varying sizes of the base nanoparticles. Green lines represent excitation by the non-polarized light, red lines excitation by linearly polarized light along the trimer's base, and blue lines excitation by light polarized perpendicularly to the trimer's base. The sizes of the base nanoparticles R_2 used in the simulations are 32, 24, 8, and 0 nm corresponding to R_2/R_1 ratios of 1, 0.75, 0.25, and 0 respectively. Polarized spectra are normalized to the corresponding peaks of the non-polarized spectra. The corresponding symmetry point groups are indicated. (b) Plasmon energy diagram of the in-plane bonding dipolar eigenmodes calculated by plasmon hybridization theory. The interparticle separation is 1 nm. The symmetry symbols of the plasmon modes are marked in agreement with the symmetry group correlation tables: $R_2 = 1$ corresponds to D_{3h}, $0 < R_2/R_1 < 1$ corresponds to C_{2v}, and $R_2/R_1 = 0$ corresponds to K_h. Adapted with permission from: Ref. [36]. Copyright (2011) American Chemical Society. (See insert for color representation of the figure.)

3.4 Effect of Plasmon Modes on SMSERS

3.4.1 Effect of the Spectral Lineshape

In the previous section we have shown in detail the symmetry correlations between the different plasmon modes of a single nanoparticle, dimers and trimers of various geometries.

In this section we present an important example of the implications that the actual plasmon mode spectral lineshape has on the spectral shape of the SMSERS spectrum. Consider a molecule, located within the interparticle gap of a dimer, which belongs to the $D_{\infty h}$ symmetry group. Under illumination with visible light it is the bonding longitudinal Σ_u^+ plasmon mode of the dimer that is excited and, in turn, strongly enhances the near-field sensed by the molecule. The energy of the corresponding plasmon resonance strongly depends on both the size of the nanoparticles forming the dimer and the interparticle distance. The bigger the nanoparticles and the closer they are, the stronger is the interaction between them. In the framework of plasmon hybridization theory, a stronger interaction between the individual nanoparticles results in a stronger splitting between the two interacting plasmon modes and a stronger red shift of the Σ_u^+ plasmon mode. The intensity of the plasmon resonance peak increases as the plasmon mode energy shifts to the red and so does its spectral width.

To observe this effect, Rhodamine 123 molecules were adsorbed within the interparticle gaps of gold–silver core–shell nanoparticle dimers and excited by a CW laser at 532 nm [34]. An increase in the SMSERS intensity with the size of the nanoparticles was observed, as shown in Figure 3.7a, where the measured SMSERS intensity is plotted versus the nanoparticle size. This effect is fully correlated with the red shift and increase in the intensity of the Σ_u^+ plasmon mode of the dimer, which is further confirmed by numerical simulations of the near-field intensity at the dimer gap, as discussed below. Although the spectral profiles of the near-field intensity and the far-field plasmon scattering spectra discussed in the previous section have, in principle, different shapes, the two can be alternatively used to evaluate the plasmon mode properties.

Simulated near-field spectra for dimers of nanoparticles with gradually increasing size (Figure 3.7b) show a red shift, an intensity increase, and broadening of the Σ_u^+ plasmon resonance. The dashed line in the figure indicates the laser excitation wavelength used in the experiments and the dashed area marks the Raman bands region. For small nanoparticles, the Σ_u^+ mode is at a shorter wavelength than the excitation laser, and the overlap between

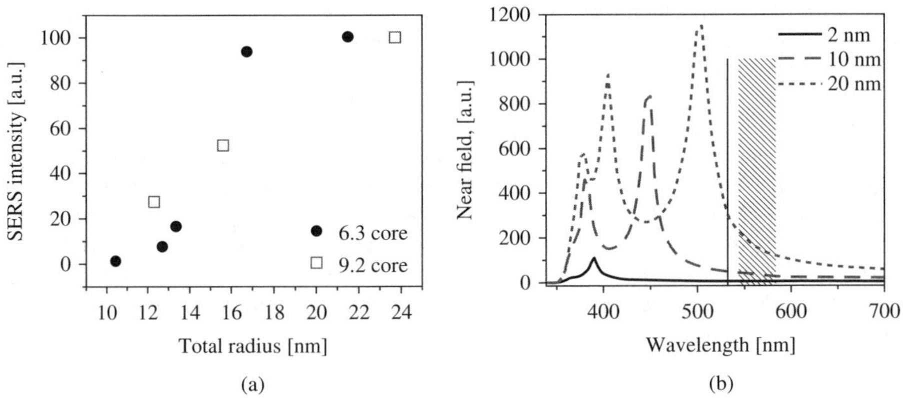

Figure 3.7 (a) Measured SERS intensity as a function of the overall radius of core–shell gold–silver nanoparticles for two different gold core radii. (b) Calculated normalized near-field within an interparticle gap of 1 nm for Ag dimers with radii of 2, 10, and 20 nm. Dashed line indicates 532 nm excitation wavelength. The dashed area represents the range of the Raman spectrum of Rhodamine 123. Adapted with permission from: Ref. [34]. Copyright (2009) American Chemical Society

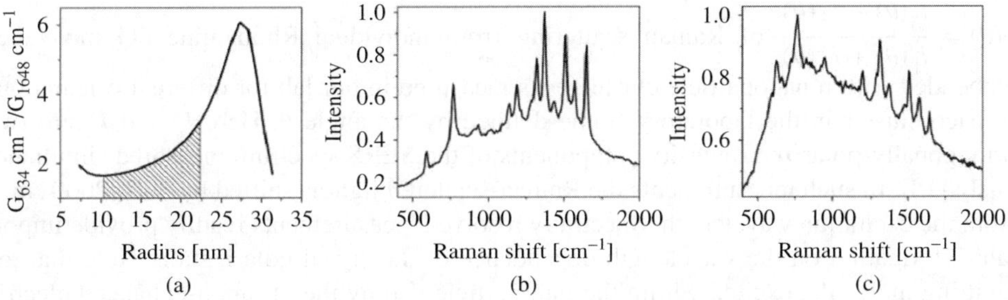

Figure 3.8 (a) Calculated SERS enhancement at the interparticle gap of a dimer as a function of nanoparticle radius for Raman bands at 550 and 582 nm (Raman shifts of 634 and 1648 cm^{-1}, respectively). The interparticle distance is 1 nm. (b,c) SERS spectra of Rhodamine 123 from dimers of 14 nm (b) and 22 nm radius (c). Adapted with permission from: Ref. [34]. Copyright (2009) American Chemical Society

the plasmon spectrum, the excitation laser, and Raman bands is very small. As the particle size increases, the Σ_u^+ mode is shifted to longer wavelengths. As a result, the overlap of the plasmon spectrum with the excitation and Raman band wavelengths increases and results in an increased intensity of the SERS signal. In the nanoparticle size range studied experimentally ($10 < R < 22$ nm) the Raman signal increases monotonically with increase in nanoparticle size.

The spectral region of Raman bands is found to sit on the sloping red side of the plasmon spectrum. The spectral width of the plasmon mode determines the slope of the plasmon spectral line in this region. The presence of such a slope implies different SERS enhancement for different vibrational bands. The ratio between the SERS enhancement for the lowest energy (634 cm^{-1}) and highest energy (1648 cm^{-1}) Raman bands in the spectrum of Rhodamine 123 as a function of particle size (estimated from the numerical simulations) is shown in Figure 3.8a. As the size increases, the lowest energy mode experiences a stronger enhancement as compared to the highest energy mode. This effect is indeed observed experimentally, as shown in Figures 3.8b,c, where SMSERS spectra from dimers of nanoparticles of 14 nm are compared to those of 22 nm. Thus, while $I(634\ \mathrm{cm}^{-1})/I(1648\ \mathrm{cm}^{-1}) \simeq 0.77$ for 14 nm nanoparticles, it is approximately 1.4 for nanoparticles of 22 nm. The underlying mechanism of this effect is understood by examining Figure 3.7: the low energy Raman band, which is closer in energy to the laser line, is also closer to the Σ_u^+ plasmon mode resonance and experiences, therefore, stronger enhancement. For nanoparticles of the small size, however, the spectral line of the plasmon is flat and enhancement of the high energy Raman band is similar to that of the low energy one. When the size of the nanoparticles increases and the plasmon resonance shifts to the red the spectral line of the resonance is steep and the difference in the enhancement of low and high energy bands becomes significant. A similar effect in nanoparticle dimers was observed also for fluorescent molecules by Feldmann and coworkers [49].

3.4.2 Effect of Multiple Normal Modes

In this section, we use the polarization state of the Raman-scattered light to understand the interaction of a molecule with the plasmon modes of a nanoparticle cluster, and we focus the discussion on clusters made of three nanoparticles. The polarization function

$\rho(\theta) = \dfrac{I_{\parallel}(\theta) - I_{\perp}(\theta)}{I_{\parallel}(\theta) + I_{\perp}(\theta)}$ of Raman scattering from individual Rhodamine 6G molecules embedded within nanoparticle clusters was measured in our lab for different orientations of each cluster in the laboratory frame defined by the angle θ. Here $I_{\parallel}$ and $I_{\perp}$ are two orthogonally-polarized intensity components of the SERS spectrum measured simultaneously [18]. In such measurements the Raman-scattered light is shifted by 300–2000 cm^{-1} from the excitation wavelength. Spectrally resolved measurements readily provide important information on the wavelength dependence of the signal polarization. Note that the emitting molecule, excited within the interparticle gap by the plasmon-enhanced electric field normal to the surface, is the only source of light observed at the Raman-scattered wavelength [17]. Therefore, only the second step of the SERS process (i.e., interaction of the Raman emission with the plasmonic cluster) affects the polarization of the far-field signal [18].

We can analyze this process in the following way. A molecule adsorbed in the gap between two nanoparticles excites plasmon modes of the cluster, typically off-resonance. Therefore, the far-field measured in SMSERS experiments can be modeled as a superposition of electromagnetic fields emitted by each plasmon mode as excited by the molecule. At the zenith observation point (out-of-plane just above the emitter), which adequately describes the detection geometry in a typical single-molecule experiment, the electric field can be written simply as

$$\begin{pmatrix} E_x(\omega) \\ E_y(\omega) \end{pmatrix} \propto \begin{pmatrix} \sum_i a_i D_{\mathrm{T},i}(\omega) \\ \sum_j a_j D_{\mathrm{L},j}(\omega) \end{pmatrix} \tag{3.1}$$

where $E_x(\omega)$ and $E_y(\omega)$ are the in-plane electric field components, $D_{\mathrm{T},i}(\omega)$ and $D_{\mathrm{L},j}(\omega)$ are the complex amplitudes of the cluster plasmon modes i and j in the transverse and longitudinal direction, respectively, and a_i and a_j are coefficients arising from the projection of the molecular dipole on the plasmon mode direction. The plasmon modes $D(\omega)$ can be represented as harmonic oscillators, where

$$D(\omega) = A(\omega)e^{i\delta(\omega)},$$

$$A(\omega) = \frac{A_0\omega_0\gamma}{\sqrt{(\omega_0^2 - \omega^2)^2 + (\omega\gamma)^2}},$$

$$\delta(\omega) = \tan^{-1}\left(\frac{\gamma\omega}{\omega_0^2 - \omega^2}\right).$$

ω_0 is the resonance frequency of the plasmon mode, γ is the resonance spectral width and amplitude, and A_0 is associated with the net dipole of the plasmon mode. The summations in Equation 3.1 are over all possible transverse or longitudinal modes of the cluster.

In general, the light emitted by the plasmonic cluster has elliptical polarization, as described by Equation 3.1. When such optical activity is observed in chiral molecules, it is associated with their Raman optical activity (ROA) tensor. ROA is defined as $\Delta = \dfrac{I_\mathrm{R} - I_\mathrm{L}}{I_\mathrm{R} + I_\mathrm{L}}$, where $I_{\mathrm{R,L}}$ is the intensity of right- (left-)handed circularly polarized light. Below, we show that, upon excitation with linearly polarized light, symmetry-broken trimers show giant ROA, with values as high as the maximal $\Delta = \pm 1$, which corresponds to pure circular polarization at the Raman-scattered wavelength. Notably, the ROA observed in the present case originates from the interaction between the molecular electric dipole and multiple electric dipole resonances of the nanoparticle cluster, as opposed to the case of optical activity observed in bare plasmonic nanostructures that originates from the interaction between electric and magnetic modes. Also, no magnetic dipole of the molecule is involved , which is unique to the present effect when compared to Raman-optically-active chiral molecules.

Consider first a simple example of a C_{2v} trimer of equal size nanoparticles, where the symmetry is broken by imposing a vertex angle of $\phi = \pi/2$ and a molecule is adsorbed within the gap between adjacent nanoparticles. The light emitted by the molecule excites two bright bonding plasmon modes, $D_T(\omega)$ of A_1 symmetry and $D_L(\omega)$ of B_2 symmetry, as shown in the numerically simulated emission spectrum of such a cluster in Figure 3.9a. Here, a dipolar emitter positioned in the gap between the nanoparticles is assumed as the excitation source. Figures 3.9b,c show the simulated relative magnitudes of the electric far-field components and their relative phase, respectively. Note that the out-of-plane component E_z of the electric field is negligible for all wavelengths. As expected from Equation 3.1, the ensuing elliptical polarization of the emitted light depends strongly on the excitation wavelength. The ROA spectrum is shown in Figure 3.9d for both dextrorotatory (d) and levorotatory (l) configurations of the particles. (Here we use the terms dextro- and levorotatory for structures that convert linear polarization into right-hand and left-hand circular polarization, respectively.) At 640 nm Δ attains maximal values of +1 and −1 for the d and l clusters, respectively, such that the Raman-scattered light has purely left- or right-handed circular polarization. In these simulations we again assumed that the observation point is situated at the zenith.

The ROA can be expressed in terms of the two plasmon modes as follows:

$$\Delta = 2\frac{b_T\, b_L}{b_T^2 + b_L^2}\sin(\delta_{TL}) \tag{3.2}$$

where $b_i(\omega) = a_i A_i(\omega)$, and $\delta_{TL}(\omega) = \delta_T(\omega) - \delta_L(\omega)$. $\Delta = \pm 1$ when the two oscillators are of the same amplitude ($b_T = b_L$) and in quadrature phase, corresponding to circular polarization, and Δ vanishes when the oscillators are either in- or out-of-phase or one of the amplitudes is significantly larger than the other. The parameters of the two harmonic oscillators can be found from fitting the spectral resonances of the cluster's emission spectrum in Figure 3.9a to Lorentzian lineshapes, as shown in the figure by green lines. Using the parameters obtained from the fit ($\omega_T = 2.04$ eV (608 nm), $\gamma_T = 0.27$ eV (80 nm), $\omega_L = 1.76$ eV (705 nm), $\gamma_L = 0.18$ eV (70 nm), and $b_T/b_L = 0.56$) we evaluated the electric field and the corresponding ROA, which are in a very good agreement with the numerical results, as compared in Figure 3.9.

In order to compare the harmonic oscillator model to experimental SMSERS results reported in Ref. [18], we write the polarization function $\rho(\theta)$ in terms of the model parameters as

$$\rho(\theta) = \frac{(1 - r^2)\cos(2\theta) + 2r\sin(2\delta_{TL})}{r^2 + 1} \tag{3.3}$$

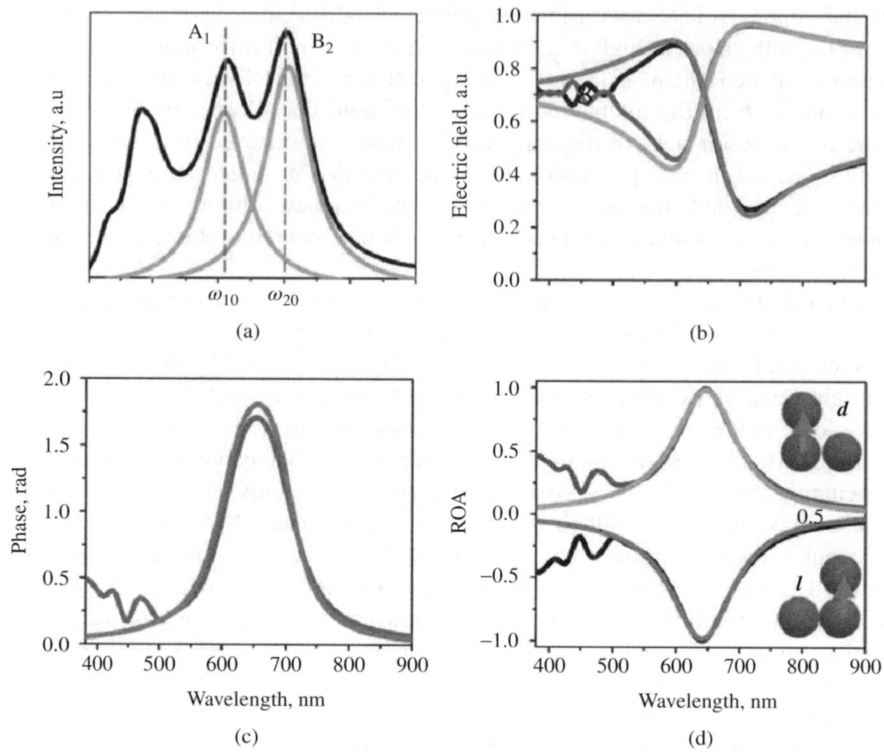

Figure 3.9 *SMSERs from a symmetry-broken silver trimer ($\phi = 90°$, $R_1 = R_2 = R_3 = 32$ nm, interparticle distance of 1 nm). (a) Emission spectrum with the symmetries of the excited plasmon modes marked. Green lines show Lorentzian fits of the spectral peaks. (b) Relative magnitudes of the electric field components of Raman-scattered light obtained from an electromagnetic simulation of a trimer of* d *configuration (E_z is negligible) and from the harmonic oscillators model based on the parameters of the Lorentzian fit (see(a)). Blue and black lines – simulated results for the E_x and E_y components, respectively, green and red lines – harmonic oscillator model results. (c) Relative phase of the electric field components from (b). Blue line – simulation and red line – model. (d) ROA of the two trimers with* d *and* l *configurations. Black and blue lines – electromagnetic simulation, red and green lines – harmonic oscillator model. (*See insert for color representation of the figure.*)*

where $r = b_T/b_L$. Below, we analyze one example of a trimer by evaluating the ROA from measured values of the polarization function and comparing them to the simulated values and those obtained from the harmonic oscillator model. The cluster geometry used in the simulations was evaluated from the SEM image shown in Figure 3.10a. The parameters of the model (Equations 3.1 and 3.2) were obtained from a fit to the simulated emission spectrum shown in Figure 3.10b.

The measured polarization function $\rho(\theta)$ values for an adsorbed Rhodamine 6G molecule at two wavelengths, 555 and 583 nm, are shown in Figure 3.10c as black and green dots, respectively and solid lines show the fit. While the pattern measured at 555 nm shows a clear indication that the polarization is very close to linear, the polarization at

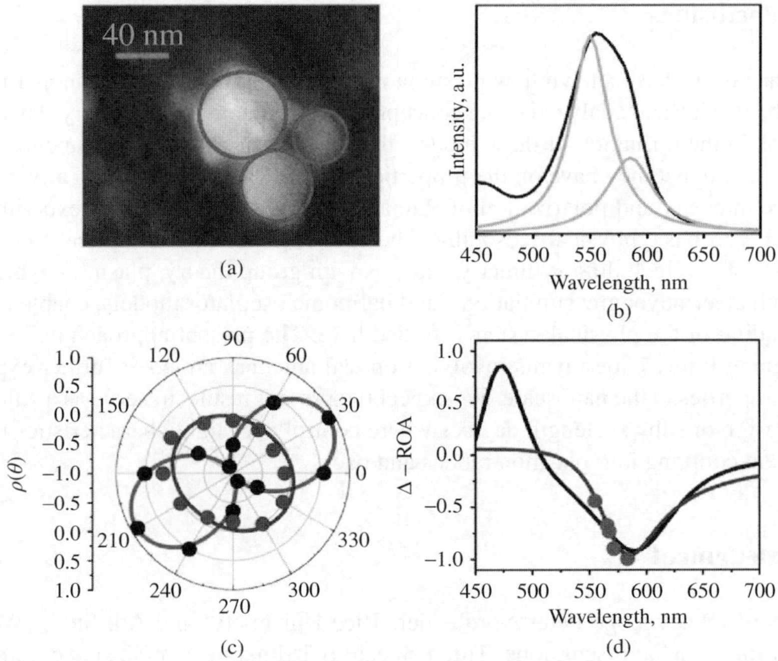

Figure 3.10 *Experimental observation of the effect of a symmetry broken trimer on SERS polarization. (a) SEM image of a trimer. Estimated nanoparticle sizes are $R_1 = 20$ nm, $R_2 = 27$, and $R_3 = 32$ nm, and the distances between nanoparticles are 1 nm (for particles 1–2 and 2–3), and 8 nm (for particles 1–3). (b) Simulated emission spectrum with the nanoparticle dimensions and gap sizes estimated from the image (see text). The molecule is assumed to be located between particles 1 and 2. (c) Experimental polarization functions of the Raman-scattered light measured at 555 nm (black dots) and 583 nm (green dots). Red and blue lines show a fit to Equation 3.3. (d) ROA calculated from the experimental data (dots), from the electromagnetic simulation (black line), and from the harmonic oscillator model (blue line).* (See insert for color representation of the figure.)

583 nm is essentially circular. The ROA values calculated from the experimental data are plotted for several wavelengths in Figure 3.10d as red dots. In particular, the values obtained for the 555 and 583 nm bands are $\Delta \approx -0.4$ and -1, respectively. Figure 3.10d shows the simulated ROA spectrum in black line, which is in a very good agreement with the experimental points. The harmonic oscillator model was then tested with values obtained from a fit: $\omega_T = 2.25$ eV (550 nm), $\gamma_T = 0.12$ eV (~30 nm), $\omega_L = 2.1$ eV (590 nm), $\gamma_L = 0.23$ eV (~65 nm), and $b_T/b_L = 2$. The wavelength-dependent ROA calculated from this model is shown in Figure 3.10d as a blue line, and is in very good agreement with both experimental and numerical results. While for excitation at 550 nm one plasmon resonance clearly dominates the spectrum and results in an almost linear polarization of the scattered light, for excitation at 583 nm the contributions of both resonances are of the same magnitude and their spectral distance from the excitation results in amplitudes and phases of the electric field components corresponding to circularly polarized scattered light.

3.5 Conclusions

In this chapter we have shown how plasmon modes of small plasmonic nanoparticle clusters can be studied and analyzed using concepts from the theory of symmetry. This analysis helps explain the formation of these modes, their evolution with changes in cluster geometry, and the impact they have on the properties of SMSERS light. Specifically, variations in relative intensity and polarization of Raman bands in single-molecule experiments are explained with basic physical reasoning. The broad spectrum of analytical tools used in the present study, including symmetry analysis with group theory, plasmon hybridization theory, full electrodynamic simulations, and harmonic oscillator models, enabled in-depth understanding of the physical systems studied here. The present approach paves the way for designing future experiments involving optical antennas aimed at further exploration of light properties at the nanoscale. We expect the present results to serve as a valuable aid for the design of subwavelength devices where control over light characteristics is crucial for efficient coupling into plasmon modes and back.

Acknowledgment

The authors acknowledge Peter Nordlander (Rice University) and Adi Stern (Weizmann Institute) for valuable discussions, Timur Shegai (Chalmers University) and Tali Dadosh (University of Pennsylvania) for help with the data presentation, and Takumi Sannomiya (Tokyo Institute of Technology) for help with MMP simulations. This research was made possible by a grant to G.H. from the Israel Science Foundation (450/10).

References

1. Moskovits, M. (1985) Surface-enhanced spectroscopy. *Reviews of Modern Physics*, **57**(3), 783–826.
2. Moskovits, M. (2005) Surface-enhanced Raman spectroscopy: a brief retrospective. *Journal of Raman Spectroscopy*, **36**(6–7), 485–496.
3. Le Ru, E.C., Meyer, M. and Etchegoin, P.G. (2006) Proof of single-molecule sensitivity in surface enhanced Raman scattering (SERS) by means of a two-analyte technique. *The Journal of Physical Chemistry B*, **110**(4), 1944–1948.
4. Michaels, A.M., Nirmal, M. and Brus, L.E. (1999) Surface enhanced Raman spectroscopy of individual rhodamine 6G molecules on large Ag nanocrystals. *Journal of the American Chemical Society*, **121**, 9932–9939.
5. Dieringer, J.A., Lettan, I.I.R.B., Scheidt, K.A. and Van Duyne, R.P. (2007) A frequency domain existence proof of single-molecule surface-enhanced Raman spectroscopy. *Journal of the American Chemical Society*, **129**(51), 16249–16256.
6. Haran, G. (2010) Single-molecule Raman spectroscopy: a probe of surface dynamics and plasmonic fields. *Accounts of Chemical Research*, **43**(8), 1135–1143.
7. Camden, J.P., Dieringer, J.A., Zhou, J. and Van Duyne, R.P. (2008) Controlled plasmonic nanostructures for surface-enhanced spectroscopy and sensing. *Accounts of Chemical Research*, **41**(12), 1653–1661.

8. Kneipp, K., Wang, Y., Kneipp, H. *et al.* (1997) Single molecule detection using Surface-Enhanced Raman Scattering (SERS). *Physical Review Letters*, **78**(9), 1667.
9. Nie, S. and Emory, S.R. (1997) Probing single molecules and single nanoparticles by surface-enhanced Raman scattering. *Science*, **275**(5303), 1102–1106.
10. Zuloaga, J., Prodan, E. and Nordlander, P. (2009) Quantum description of the plasmon resonances of a nanoparticle dimer. *Nano Letters*, **9**(2), 887–891.
11. Taminiau, T.H., Stefani, F.D., Segerink, F.B. and van Hulst, N.F. (2008) Optical antennas direct single-molecule emission. *Nature Photonics*, **2**(4), 234–237.
12. Hao, E. and Schatz, G.C. (2004) Electromagnetic fields around silver nanoparticles and dimers. *The Journal of Chemical Physics*, **120**(1), 357–366.
13. Romero, I., Aizpurua, J., Bryant, G.W. and García De Abajo, F.J. (2006) Plasmons in nearly touching metallic nanoparticles: singular response in the limit of touching dimers. *Optics Express*, **14**(21), 9988–9999.
14. Li, Z., Shegai, T., Haran, G. and Xu, H. (2009) Multiple-particle nanoantennas for enormous enhancement and polarization control of light emission. *ACS Nano*, **3**(3), 637–642.
15. Jiang, J., Bosnick, K., Maillard, M. and Brus, L. (2003) Single molecule Raman spectroscopy at the junctions of large Ag nanocrystals. *The Journal of Physical Chemistry B*, **107**(37), 9964–9972.
16. Michaels, A.M., Jiang and Brus, L. (2000) Ag nanocrystal junctions as the site for surface-enhanced Raman scattering of single rhodamine 6G molecules. *The Journal of Physical Chemistry B*, **104**(50), 11965–11971.
17. Kerker, M., Wang, D.-S. and Chew, H. (1980) Surface enhanced Raman scattering (SERS) by molecules adsorbed at spherical particles: errata. *Applied Optics*, **19**(24), 4159–4174.
18. Shegai, T., Li, Z.P., Dadosh, T. *et al.* (2008) Managing light polarization via plasmon–molecule interactions within an asymmetric metal nanoparticle trimer. *Heilkunde - Heilwege*, **105**(43), 16448–16453.
19. Alonso-González, P., Albella, P., Schnell, M. *et al.* (2012) Resolving the electromagnetic mechanism of surface-enhanced light scattering at single hot spots. *Nature Communications*, **3**, 684.
20. Bohren, C.F. and Huffman, D.R. (1998) *Absorption and Scattering of Light by Small Particles*, John Wiley & Sons, Inc., New York.
21. Kreibig, U. (1995) *Optical Properties of Metal Clusters*, Springer Series in Materials Science, Springer-Verlag, Berlin, Heidelberg.
22. Quinten, M. and Kreibig, U. (1986) Optical properties of aggregates of small metal particles. *Surface Science*, **172**, 557–577.
23. Fan, J.A., Wu, C., Bao, K. *et al.* (2010) Self-assembled plasmonic nanoparticle clusters. *Science*, **328**(5982), 1135–1138.
24. Halas, N.J., Lal, S., Chang, W.-S. and Link, S. (2011) Plasmons in strongly coupled metallic nanostructures. *Chemical Reviews*, **111**(6), 3913–3961.
25. Hentschel, M., Saliba, M., Giessen, H. *et al.* (2010) Transition from isolated to collective modes in plasmonic oligomers. *Nano Letters*, **10**, 2721–2726.
26. Prodan, E., Radloff, C., Halas, N.J. and Nordlander, P. (2003) A hybridization model for the plasmon response of complex nanostructures. *Science*, **302**, 419–422.

27. Nordlander, P., Oubre, C. and Prodan, E. (2004) Plasmon hybridization in nanoparticle dimers. *Nano Letters*, **4**(5), 899–903.
28. Wang, H., Brandl, D.W., Nordlander, P. and Halas, N.J. (2007) Plasmonic nanostructures: artificial molecules. *Accounts of Chemical Research*, **40**, 53–62.
29. Brandl, D.W., Mirin, N.A. and Nordlander, P. (2006) Plasmon modes of nanosphere trimers and quadrumers. *The Journal of Physical Chemistry B*, **110**(25), 12302–12310.
30. Yu, N., Aieta, F., Genevet, P. *et al.* (2012) A broadband, background-free quarter-wave plate based on plasmonic metasurfaces. *Nano Letters*, **12**(12), 6328–6333.
31. Noginov, M.A., Zhu, G., Belgrave, A.M. *et al.* (2009) Demonstration of a spaser-based nanolaser. *Nature*, **460**(7259), 1110–1112.
32. Luk'yanchuk, B., Zheludev, N.I., Maier, S.A. *et al.* (2010) The Fano resonance in plasmonic nanostructures and metamaterials. *Nature Materials*, **9**(9), 707–715.
33. Hentschel, M., Schäferling, M., Weiss, T. *et al.* (2012) Three-dimensional chiral plasmonic oligomers. *Nano Letters*, **12**(5), 2542–2547.
34. Dadosh, T., Sperling, J., Bryant, G.W. *et al.* (2009) Plasmonic control of the shape of the Raman spectrum of a single molecule in a silver nanoparticle dimer. *ACS Nano*, **3**(7), 1988–1994.
35. Chuntonov, L. and Haran, G. (2011) Trimeric plasmonic molecules: the role of symmetry. *Nano Letters*, **11**, 2440–2445.
36. Chuntonov, L. and Haran, G. (2011) Effect of symmetry breaking on the mode structure of trimeric plasmonic molecules. *The Journal of Physical Chemistry C*, **115**(40), 19488–19495.
37. Weiss, A. and Haran, G. (2001) Time-dependent single-molecule Raman scattering as a probe of surface dynamics. *The Journal of Physical Chemistry B*, **105**(49), 12348–12354.
38. Mock, J.J., Barbic, M., Smith, D.R. *et al.* (2002) Shape effects in plasmon resonance of individual colloidal silver nanoparticles. *The Journal of Chemical Physics*, **116**, 6755–6759.
39. Brown, L.V., Sobhani, H., Lassiter, J.B. *et al.* (2010) Heterodimers: plasmonic properties of mismatched nanoparticle pairs. *ACS Nano*, **4**(2), 819–832.
40. Hafner, C. (2007) Boundary methods for optical nano structures. *Physica Status Solidi B: Basic Solid State Physics*, **244**(10), 3435–3447.
41. Sannomiya, T. and Hafner, C. (2010) Multiple multipole program modelling for nano plasmonic sensors. *Journal of Computational and Theoretical Nanoscience*, **7**(8), 1587–1595.
42. Hohenester, U. and Trügler, A. (2012) MNPBEM – a Matlab toolbox for the simulation of plasmonic nanoparticles. *Computer Physics Communications*, **183**(2), 370–381.
43. Lynch, D.W. and W.R. Hunter, Comments on the optical constants of metals, in *Handbook of Optical Constants of Solids* (ed. E.D. Palik) 1998, pp. 275–367, Academic Press: San Diego, CA.
44. Esteban, R., Borisov, A.G., Nordlande, P. *et al.* (2012) Bridging quantum and classical plasmonics with a quantum-corrected model. *Nature Communications*, **3**, 825.
45. Savage, K.J., Hawkeye, M.M., Esteban, R. *et al.* (2012) Revealing the quantum regime in tunnelling plasmonics. *Nature*, **491**, 574.

46. Marinica, D.C., Kazansky, A.K., Nordlander, P. *et al.* (2012) Quantum plasmonics: nonlinear effects in the field enhancement of a plasmonic nanoparticle dimer. *Nano Letters*, **12**(3), 1333–1339.
47. Cotton, F.A. (1990) *Chemical Applications of Group Theory*, Wiley-Interscience, New York.
48. Bunker, P. and Jensen, P. (1998) *Molecular Symmetry and Spectroscopy*, NRC Press, Ottawa.
49. Ringler, M., Schwemer, A., Wunderlich, M. *et al.* (2008) Shaping emission spectra of fluorescent molecules with single plasmonic nanoresonators. *Physical Review Letters*, **100**(20), 203002.

4

Experimental Demonstration of Electromagnetic Mechanism of SERS and Quantitative Analysis of SERS Fluctuation Based on the Mechanism

Tamitake Itoh

National Institute of Advanced Industrial Science and Technology (AIST), Japan

4.1 Experimental Demonstration of the EM Mechanism of SERS

4.1.1 Introduction

The local mode density of electromagnetic (EM) fields in the vicinity of metal nanoparticles (NPs) is increased by plasmon resonance, resulting in enhancement of optical transition rates of molecules close to NP surfaces [1–10]. These phenomena are collectively referred to as EM effects [11]. A broad plasmon resonance line width (~200 meV) enables EM effects to enhance both the excitation and the emission transition rates of a molecule located in the fields [1, 3, 7, 8, 10]. The molecular optical responses enhanced by plasmon resonance are known as surface-enhanced Raman scattering (SERS), surface-enhanced hyper Raman scattering (SEHRS), surface-enhanced fluorescence (SEF), surface-enhanced absorption, and so on [12]. In particular, the enhancement factor of SERS of molecules located at the junction of metal NP dimers becomes up to 10^{14}, allowing us to measure Raman spectra at the single molecule (SM) level [13–16]. Raman spectra have distinct vibrational bands known as "molecular fingerprints", which enable us to attribute molecular species in detail. Thus, SERS has motivated researchers to extend it to ultrasensitive analysis of biomolecules [17–20]. This extension was mostly triggered by the introduction

Frontiers of Surface-Enhanced Raman Scattering: Single Nanoparticles and Single Cells, First Edition.
Edited by Yukihiro Ozaki, Katrin Kneipp and Ricardo Aroca.

of SMSERS detections, the developments in nanotechnology that enabled the preparation of sophisticated plasmonic substrates [20–22], and the advances in theoretical calculations [15, 23–26].

Despite the impact of ultra-sensitivity in SERS, the mechanism of SERS, which should be a guideline for researchers involved in its application, has been unclear. We have attempted to clarify the mechanism by identifying correlations between SERS and its mechanism. Two SERS origins have been proposed by many researchers. They are called the EM mechanism and the chemical mechanism [27, 28]. The EM mechanism is characterized by twofold EM enhancement in the Raman process; the first enhancement is due to coupling of plasmon with incident light and the second to coupling of plasmon with Raman scattering light [1, 8, 12, 27, 29]. Chemical enhancement is characterized by shifting of Raman scattering in nonresonance to that in resonance through the formation of charge transfer complexes between adsorbed molecules and a metal surface [28, 30–33]. Both mechanisms have been investigated experimentally in detail and found to be correct. Thus, in an effort to find out which mechanism is dominant, quantitative evaluation of enhancement based on one mechanism exclusively is important. The EM mechanism has universality, not being dependent on the molecular species. Thus, in this section we focus on our recent experimental investigations of the EM mechanism.

4.1.2 Observations of the EM Mechanism in SERS Spectral Variations

First, we give an outline of the EM mechanism of SERS. A Raman process is composed of an excitation and an emission process. In these processes, a molecule and a field exchange light energy. The rates of the exchange are increased by an increase in the mode density of the field. Plasmon resonance of Au and Ag increases the mode density greatly because of their large conduction electron density and long oscillation time, namely, plasmon polariton. Thus, plasmon makes the Raman excitation and emission process efficient by coupling with them. In other words, Raman excitation is enhanced due to coupling between incident light and plasmon (so-called, first EM enhancement), and Raman emission is enhanced due to coupling between Raman light and plasmon (so-called, second EM enhancement) [1, 3, 8, 15, 23, 24, 34–40]. The relationship among plasmon resonance, SERS intensity, and Ag and Au nanostructures has been independently examined to analyze EM enhancement in SERS [34–42]. On the other hand, the first and the second EM enhancement were not independently treated in the previous work, which therefore failed to provide conclusive evidence for twofold EM enhancement in SERS. The twofold EM enhancement factor of SERS $|M|^4$ (EM enhancement factor M^4 up to 10^{14}) has made detection of SMs realistic at interparticle junctions in Ag and Au dimers [15, 43, 44]. Thus, twofold EM enhancement is important for verifying SERS that enables us to detect SMs.

The twofold EM enhancement provides us with a simple expression for total SERS enhancement because it is the product of EM enhancement of the incident and scattered light coupled with plasmon [1, 3, 8, 15, 23, 24, 34–41, 45]. The total enhancement factor $|M_{\text{total}}|^2$ of SERS is given by

$$|M_{\text{total}}|^2 = \left|\frac{E^{\text{Loc}}\left(\lambda_{\text{L}}\right)}{E^{\text{I}}(\lambda_{\text{L}})}\right|^2 \times \left|\frac{E^{\text{Loc}}\left(\lambda_{\text{L}} \pm \lambda_{\text{R}}\right)}{E^{\text{I}}(\lambda_{\text{L}} \pm \lambda_{\text{R}})}\right|^2 = |M_1(\lambda_{\text{L}})|^2 \times |M_2(\lambda_{\text{L}} \pm \lambda_{\text{R}})|^2 \quad (4.1)$$

where E^{I} and E^{Loc} are the amplitudes of the incident and local electric fields, respectively; λ_L is the excitation wavelength; $+\lambda_R$ and $-\lambda_R$ are the wavelengths of the anti-Stokes and Stokes-shifted Raman scattering, respectively; and $|M_1|^2$ and $|M_2|^2$ are the first and second EM enhancement factors, respectively. The spectrum of $|M_2|^2$ is expected to be similar to that of plasmon resonance, because the second enhancement was produced by scattering of Raman light through plasmon resonance [37–40]. Thus, the dependence of SERS spectra on plasmon resonance maxima can be considered to be a key for identifying the second enhancement. However, EM fields on larger Ag NP aggregates are complex due to overlapping between dipoles and multipoles [46]. Thus, characterization of the dependence of SERS spectra on plasmon resonance maxima has been difficult. This difficulty was resolved by selecting Ag NP aggregates with mainly dipolar plasmon resonance that satisfies the following two criteria: polarization dependence of a plasmon resonance maximum follows a cosine-squared law, and SERS maxima and plasmon resonance maxima have the same polarization dependence. As a consequence of applying these two criteria, we identified coupling of dipolar plasmon resonance with SERS [34–36]. The experimental set-up for selecting such Ag NP aggregates is reported in Ref. [47]. We detected Ag NP aggregate-by-aggregate variations in the SERS spectra of Rhodamine 6G (R6G). In the context of second EM enhancement in SERS, the SERS spectral variations can be explained in terms of the dependence of the plasmon resonance wavelength on the Ag NP aggregate, because the coupling efficiency between SERS light and plasmon resonance depends on the plasmon resonance wavelengths. Thus, careful comparison between SERS spectra and plasmon resonance spectra would provide convincing evidence for twofold EM enhancement.

Figure 4.1a–d show plasmon resonance spectra and SERS spectra of R6G from five Ag NP aggregates selected using the above two criteria. We can notice that the SERS bands close to the plasmon resonance maxima are selectively enhanced. This selective enhancement indicates a correlation between the spectra of SERS enhancement factors and plasmon resonance maxima, verifying the selective enhancement, which provides us with evidence of second EM enhancement. Figure 4.1e summarizes the plasmon resonance maxima dependence of the anti-Stokes to Stokes intensity ratios for SERS maxima at 553 nm ($+634\,\mathrm{cm}^{-1}$) and 580 nm ($-634\,\mathrm{cm}^{-1}$). According to Equation 4.1 this ratio is given by $|M_1(\lambda_L)|^2|M_2(\lambda_L-\lambda_R)|^2/|M_1(\lambda_L)|^2|M_1(\lambda_L+\lambda_R)|^2$. Thus, the first EM enhancement factor M_1 cancels. This reduction in the ratio, $|M_2(\lambda_L-\lambda_R)|^2/|M_2(\lambda_L+\lambda_R)|^2$, can be correlated with a decrease in the coupling efficiency of anti-Stokes light and plasmon resonance due to separation of the plasmon resonance maxima from the anti-Stokes bands. This correlation supports the conjecture that the second EM enhancement is associated with the origin of variations in anti-Stokes and Stokes spectra. Figure 4.1f–h show the excitation-wavelength dependence of SERS spectra and the plasmon-resonance maxima dependence of SERS spectra. SERS spectra of the aggregate excited at 568 and 647 nm are shown in Figure 4.1g,h, respectively. The intensities of the SERS bands around 630 nm in Figure 4.1g, and 660 nm in Figure 4.1h are larger than those around 588 and 710 nm. It is apparent that the SERS bands close to the plasmon resonance maxima are selectively enhanced and that this effect is dependent on the excitation wavelength. These observations demonstrate that second EM enhancement is involved in the enhancement of SERS intensity.

In this subsection, we investigated evidence of twofold EM enhancement as NP aggregate-to-aggregate variations in anti-Stokes and Stokes SERS spectra by comparing SERS

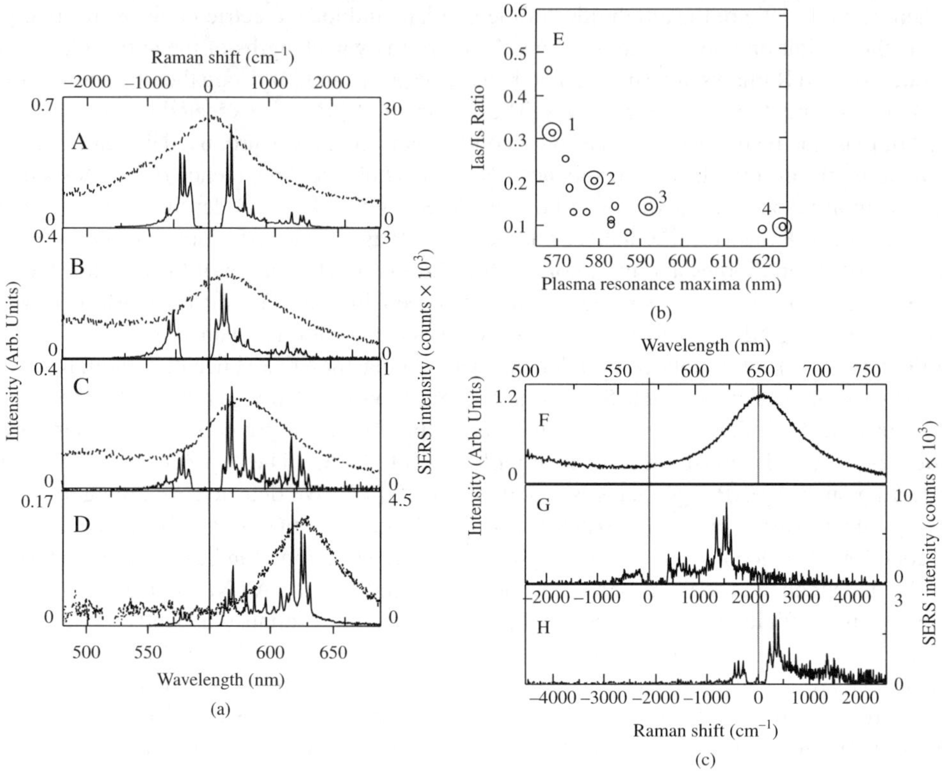

Figure 4.1 (a–d) Anti-Stokes and Stokes SERS spectra (solid lines) and plasmon resonance spectra (dotted lines) from four representative Ag NP aggregates. The vertical dashed line indicates the excitation laser wavelength. Plasmon resonance maxima in (a–d) were observed at 568, 577, 588, and 623 nm. (e) Plasmon resonance maxima dependence of anti-Stokes to Stokes intensity ratios. Doubly-circled points indicated by (a–d) correspond to the data in (a–d). (f) Plasmon resonance spectra, (g) SERS spectra excited at 568 nm, and (h) SERS spectra excited at 647 nm. All spectra in (f–h) were obtained from the same Ag NP aggregate. Reproduced with permission from Ref. [39]. © 2007, The American Physical Society

spectra and plasmon resonance spectra from single Ag NP aggregates. By investigating the selective enhancement of SERS bands whose maxima were close to the plasmon resonance maxima, we found that variations in the SERS spectra are induced by the second enhancement in the SERS process.

4.1.3 Observations of the EM Mechanism in the Refractive Index Dependence of SERS Spectra

By correlation between SERS spectra and plasmon resonance spectra we established that variations in SERS spectra originate from selective enhancement of SERS bands by plasmon resonance. In other words, the origin of the variations in the second EM enhancement is coupling of Raman light with plasmon. However, this consideration does not completely exclude intrinsic uncertainty due to inhomogeneity in the measurements of multiple NP aggregates. To exclude such uncertainty in the demonstration of the EM mechanism, we use the optical property that the plasmon resonance maximum depends on the refractive index of the surrounding media [48, 49]. From the above relation between SERS and plasmon resonance spectra we expect that SERS spectra will show refractive index dependence, because twofold EM enhancement factors depend on the refractive index. The experimental confirmation of this logical connection does not need spectral measurements on multiple Ag NP aggregates, because the plasmon resonance of single Ag NP aggregates can be changed by simply replacing the surrounding medium. To investigate the dependence of SERS spectra on the refractive index of the surrounding medium, we measured both plasmon resonance and SERS spectra from single Ag NP aggregates without and with an index matching 20 μm thick oil layer ($n = 1.5$, Immersion oil TYPE-F, Olympus, Tokyo). Ag NP aggregates in such samples were located at the interface between air ($n = 1.0$) and glass ($n = 1.5$). The effective refractive index n_{eff} of the interface is ~1.3 [48, 49]. Thus, Ag NP aggregates (~100 nm) were totally covered with oil for which the refractive index is the same as that of glass, thus, n_{eff} of the interface is 1.5.

Figure 4.2a–f show plasmon resonance and SERS spectra of three Ag NP aggregates without and with an oil layer. Here we can observe red shifts in the plasmon resonance maxima. The red shifts are induced by an increase in the refractive index [48, 49]. As shown in Figure 4.2a–f, large changes are common to SERS spectra and plasmon resonance maxima. Such spectral shifts are commonly observed for all single Ag NP aggregates. This observation indicates that the spectral changes in SERS are caused by red shifts in the plasmon resonance maxima. The Ag NP aggregate whose plasmon resonance maximum is located around 700 nm in Figure 4.2e does not show any SERS bands when covered with an oil layer. The disappearance of the SERS bands may be due to the lack of spectral overlap between the SERS and plasmon resonance generating EM enhancement.

The changes in the SERS spectra induced by the surrounding media are analyzed based on the twofold EM enhancement, as indicated in Equation 4.1. SERS spectra are calculated as the products of spectra of $|M_{\text{total}}|^2$ and Raman derived from ensemble Ag NP aggregates. We calculated the spectra of plasmon resonance and $|M_{\text{total}}|^2$ for an Ag NP dimer using $n_{\text{eff}} = 1.3$ and 1.5 by the FDTD (finite-difference time-domain) method. Figure 4.3a,b show that an increase in n_{eff} causes a red shift in both the spectral maximum of the plasmon resonance and $|M_{\text{total}}|^2$. Figure 4.3c,d show the calculated SERS spectra using $n_{\text{eff}} = 1.3$ and 1.5, respectively. The calculated changes in SERS spectra are consistent with the experimental ones in Figure 4.2a–d. This consistency verifies that the SERS spectral changes in Figure 4.2a–d are induced by red shifts in the spectra of EM enhancement factors. Figure 4.3e shows the plot of the intensity ratios of anti-Stokes (619 nm) to Stokes SERS (707 nm) bands versus the plasmon resonance maxima λ_{p}. The inset of Figure 4.3e shows the calculated ratios versus λ_{p} based on twofold EM enhancement. The calculated ratios using $n_{\text{eff}} = 1.3$ and 1.5 are well consistent with the experimental ones.

In this subsection, the refractive index dependence of SERS spectra was examined by comparing SERS spectra with plasmon resonance spectra from single Ag NP aggregates, excluding uncertainty due to inhomogeneity in the measurements of multiple NP aggregates. The results can be well explained by the selective enhancement of SERS bands whose maxima are close to the plasmon resonance maxima. The selective enhancement is a direct evidence of the twofold enhancement in the EM mechanism of SERS.

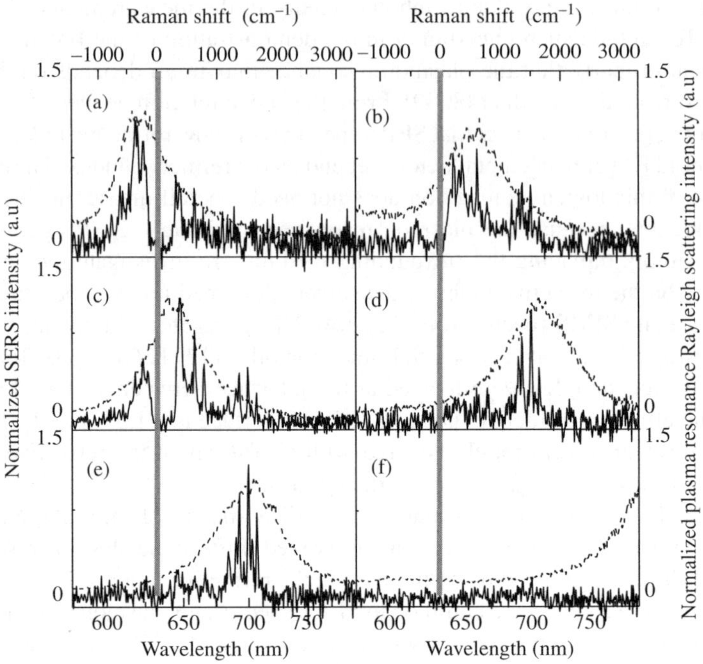

Figure 4.2 *Plasmon resonance and SERS spectra of three representative single Ag NP aggregates without (a,c,e) and with (b,d,f) covering oil. Reproduced with permission from Ref. [40]. Copyright © 2009, American Institute of Physics*

4.1.4 Quantitative Evaluation of theEM Mechanism of SERS

Evaluation of SERS spectral variations based on the EM mechanism in the previous subsections leads us to believe that twofold EM enhancement plays an important role in SERS. This evaluation provided us with comprehensive guidelines to design SERS substrates. However, twofold EM enhancement factors in the previous evaluation were fitting parameters to reproduce experimental spectra [8]. Twofold EM enhancement factors are functions of plasmon resonance, and plasmon resonance is a function of the morphology of the Ag NP aggregates. Thus, without measurements of the morphology of Ag NP aggregates, rigorous calculation of twofold EM enhancement factors is difficult. In this subsection, plasmon resonance and SERS spectra from single isolated Ag NP aggregates are combined with scanning electron microscopy (SEM) images to accomplish one-to-one correspondence between the optical properties of plasmon resonance, that of SERS, and the morphology of Ag NP aggregates. The experimental observations are compared with FDTD calculations of EM fields based on the individual morphology of the Ag NP aggregates. In this subsection, we use a Raman spectrum derived from ensemble Ag NP aggregates as $\sigma_{RS}(\lambda_L,\lambda_L \pm \lambda_R) + (1/|M_d|^2)\sigma_{FL}(\lambda_L,\lambda_L \pm \lambda_R)$, where $\sigma_{RS}(\lambda_L,\lambda_L \pm \lambda_R)$, $\sigma_{FL}(\lambda_L,\lambda_L \pm \lambda_R)$, and $1/|M_d|^2$ are the Raman cross section of a molecule in a free space, the fluorescence cross section of a molecule in a free space, and the quenching factor due to energy transfer,

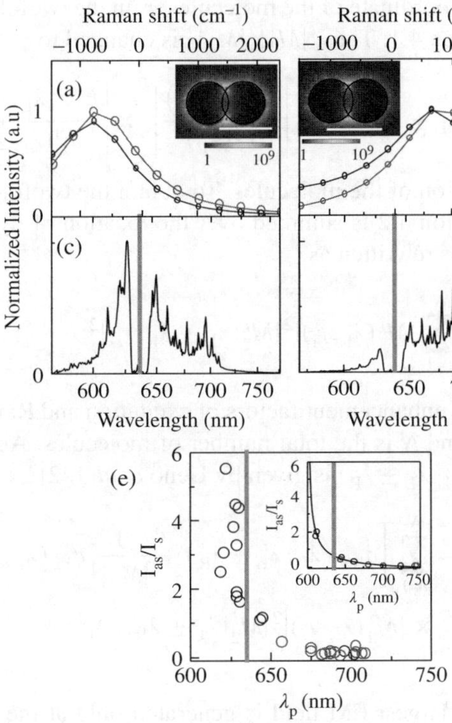

Figure 4.3 *Calculated spectra of plasmon resonance (red lines) and SERS enhancement factors (blue lines) at (a)* n_{eff} *= 1.3 and (b)* n_{eff} *= 1.5. Insets: spatial distribution of SERS enhancement factors at plasmon resonance maxima. Scale bars are 50 nm. Calculations of SERS spectra at (c)* n_{eff} *= 1.3 and (d)* n_{eff} *= 1.5. (e) Experiments of* λ_p *dependence of intensity ratios of anti-Stokes to Stokes SERS bands at 619 and 707 nm. Insets: calculations. Blue and red circles show the ratios of anti-Stokes to Stokes SERS bands at 619 and 707 nm without and with covering oil, respectively. Reproduced with permission from Ref. [40]. Copyright © 2009, American Institute of Physics.* (See insert for color representation of the figure.)

discussed in the next section [3, 50]. Here, for simplicity of discussion we do not separate Raman σ_{RS} from $(1/|M_d|^2)\sigma_{FL}$ in the evaluation.

We selected SERS-active Ag NP aggregates that show dipolar plasmon resonance whose maxima are located around 600–680 nm. Such Ag NP aggregates generate detectable SERS spectra and were always dimers [26]. The key to the selection is identification of dipolar plasmon resonance coupled with SERS [34–40]. To confirm the criteria for the selection we evaluated the plasmon resonance and morphology of 12 Ag NP aggregates that show SERS and SEF activity. All of them were dimers and have plasmon resonance maxima from 600 to 680 nm, fulfilling the criteria. Several reports demonstrate theoretically and experimentally that the positions are limited to junctions between two Ag NPs [8, 15, 23, 24]. Indeed, Imura *et al.* observed experimentally that SERS and SEF came from the junctions formed by two Ag NPs [43].

Here we add the position coordinate of the molecules, *r*, in the twofold EM enhancement factor $|M_1|^2|M_2|^2$ to Equation 4.1. Thus, $|M_1|^2|M_2|^2$ is changed to

$$|M_1(\lambda_L, r)|^2|M_2(\lambda_L \pm \lambda_R, r)|^2 = \left|\frac{E^{loc}(\lambda_L, r)}{E^I(\lambda_L)}\right|^2 \times \left|\frac{E^{loc}(\lambda_L \pm \lambda_R, r)}{E^I(\lambda_L \pm \lambda_R)}\right|^2 \tag{4.2}$$

where *r* is an arbitrary position of the molecules. To obtain the twofold EM enhancement factor, $|M_1|^2|M_2|^2$ in Equation 4.2 is summed over the location of dye molecules on the Ag NP aggregate, and is thus rewritten as

$$\sum_{i=1}^{N} |M_1^i(\lambda_L, r_i)|^2|M_2^i(\lambda_L \pm \lambda_R, r_i)|^2 \tag{4.3}$$

$|M_1^i|^2$ and $|M_2^i|^2$ are the EM enhancement factors of excitation and Raman at the center of gravity r_i of a molecule *i*, and *N* is the total number of molecules. Accordingly, the total cross section spectra $\sigma_{total}(\lambda_L,\lambda_L \pm \lambda_R)$ is given by Ueno *et al.* [22]

$$\sigma_{total}(\lambda_L, \lambda_L \pm \lambda_R) = \sum_{i=1}^{N}\left[\sigma_{RS}\left(\lambda_L, \lambda_L \pm \lambda_R\right) + \frac{1}{|M_d|^2}\sigma_{FL}(\lambda_L, \lambda_L \pm \lambda_R)\right] \times |M_1^i(\lambda_L, r_i)|^2|M_2^i(\lambda_L \pm \lambda_R, r_i)|^2 \tag{4.4}$$

For an Ag NP dimer, the largest EM field is generated only at the junction of Ag NP aggregates [3, 15, 26], and this largest EM field at the junction dominates $|M_1|^2|M_2|^2$ [3, 15, 26]. Thus, $\sum_{i=1}^{N}|M_1^i|^2|M_2^i|^2$ is approximately written as $|M_1(\lambda_L,r_o)|^2|M_2(\lambda_L \pm \lambda_R,r_o)|^2$ for a single molecule in the junction, even if some additional molecules exist at positions other than the junction. Here, r_o denotes the position at the junction.

We have identified that most Ag NP aggregates showing both dipolar plasmon resonance and SERS activity are dimers [26]. Both experimental and theoretical work have predicted that the structural anisotropy of Ag dimers induces anisotropy in both the plasmon resonance and SERS [34–41, 51, 52]. To verify this prediction, we measured the anisotropy of both the plasmon resonance and SERS spectra, and then compared them with FDTD calculations of the plasmon resonance spectra and $|M_1|^2$ using a mesh size 1.5 nm. Figure 4.4a–c show the SEM image, the plasmon resonance spectra, and the SERS spectra obtained from an Ag dimer. Figure 4.4f shows that the FDTD calculations of polarization dependence of a plasmon resonance spectrum of the Ag dimer in Figure 4.4e reproduce well the experimental polarization dependence in Figure 4.4b. Figure 4.4f,g show that the structural anisotropy of the Ag dimer induces splitting of the plasmon resonance into the longitudinal mode at 620 nm and the transverse mode at <400 nm, and coupling of the longitudinal plasmon resonance with light polarized along the long axis of the Ag dimer generated the largest $|M_1|^2$ at the dimer junctions. The agreement between experimental and calculated polarization dependence suggests that the calculated spatial distribution of $|M_1|^2$ in Figure 4.4g reflects local EM fields around an Ag NP dimer. Thus, the SERS in Figure 4.4c is generated by an EM field coupled with the longitudinal plasmon resonance at the junctions. Indeed, Figure 4.4d,h show that the experimental anisotropy of the plasmon resonance and $|M_1|^2$ agreed well with the calculated anisotropy. Note that *r* was selected at an arbitrary position in the junction in Figure 4.4i. The agreement is consistent with the prediction by the EM mechanism that anisotropy of Ag NP aggregates causes anisotropy of SERS.

We must consider the spectral shape of $|M_2|^2$ to evaluate the total spectral shape of $|M_1|^2|M_2|^2$ induced by plasmon resonance. Figure 4.5a–c show an SEM image of an Ag NP dimer, an experimental plasmon resonance spectrum, and an experimental SERS spectrum, respectively. Figure 4.5d–f show the boundary condition for an Ag NP dimer in the FDTD calculation, a calculated plasmon resonance spectrum, and a spectrum of $|M_1(\lambda_L, r_o)|^2|M_2(\lambda_L \pm \lambda_R, r_o)|^2$, respectively. The cross in Figure 4.5h indicates the position r_o selected for the calculation of the spectrum in Figure 4.5f. Note that we selected r_o at an arbitrary position in the junction in Figure 4.5h simply because we do not consider rigorously the absolute value of $|M_1(\lambda_L, r_o)|^2|M_2(\lambda_L \pm \lambda_R, r_o)|^2$ in the calculations in Figure 4.5e–g. Figure 4.5b,e show that the experimental plasmon resonance spectrum agreed well with the calculation. The agreement makes us confident that the use of the SEM morphology is valid for the boundary condition in the FDTD calculation. Figure 4.5e,f show that the spectral shape of the plasmon resonance was quite similar to that of $|M_1(\lambda_L, r_o)|^2|M_2(\lambda_L \pm \lambda_R, r_o)|^2$. The similarity is reasonable considering that the EM enhancement is induced by the plasmon resonance [3, 8, 26, 39, 40]. Figure 4.5c shows that the SERS spectrum around 610 nm was selectively enhanced by the plasmon resonance in Figure 4.5b. We found that the selectively enhanced SERS spectra are common to all our experimental results.

To quantitatively evaluate twofold EM enhancement factors using the FDTD calculation with Yee cells, in which all components of an electric and a magnetic field are set at the center of a side and of a plane in a cubic cell [53], we examined the accuracy in our calculations by comparing FDTD calculations with self-consistent calculation under the common calculation conditions [24]. Note that we did not employ self-consistent calculation because our calculation is limited to asymmetrical Ag nanostructures while self-consistent calculation offers an exact solution to symmetrical Ag nanostructures [24]. We further address an issue about the use of Yee cells in FDTD calculation. The intensity of the EM field is maximized at a surface and decays exponentially outside with increasing distance from the surface. However, for a surface having a curvature comparable to the mesh size, the calculated exponential decay and the peak intensity of the EM field will be smoothed and smaller than the real intensity. This issue leads to underestimation of the EM field intensity at the surface, and is not negligible for a surface with small curvature [24]. Fortunately, this underestimation can be reduced by the use of smaller meshes. We selected the morphology of an Ag dimer equivalent to that in Figure 2A of Ref. [24], and evaluated a factor of underestimation for the dimer with mesh sizes of 0.98, 0.5, and 0.2 nm; these values are smaller than the gap of the dimer. Figure 4.6a shows the morphology of the selected Ag dimer. The x-axis is set to overlap the long axis of the dimer. An EM field reaches the theoretical maximum at the local face-to-face surfaces along the x-axis, as shown in the square in Figure 4.6a. Note that we used electric flux density instead of an EM field to avoid discontinuity of the field across the surface. Figure 4.6b shows the magnified view of the square in Figure 4.6a. Figure 4.6c shows a spectrum of $|M_1|^2$ calculated using mesh size 0.98 nm. The spectral maximum locates at ~418 nm, which agrees with Figure 2A in Ref. [24]. This agreement provides evidence that the plasmon resonance for the FDTD calculation of $|M_1|^2$ is correct.

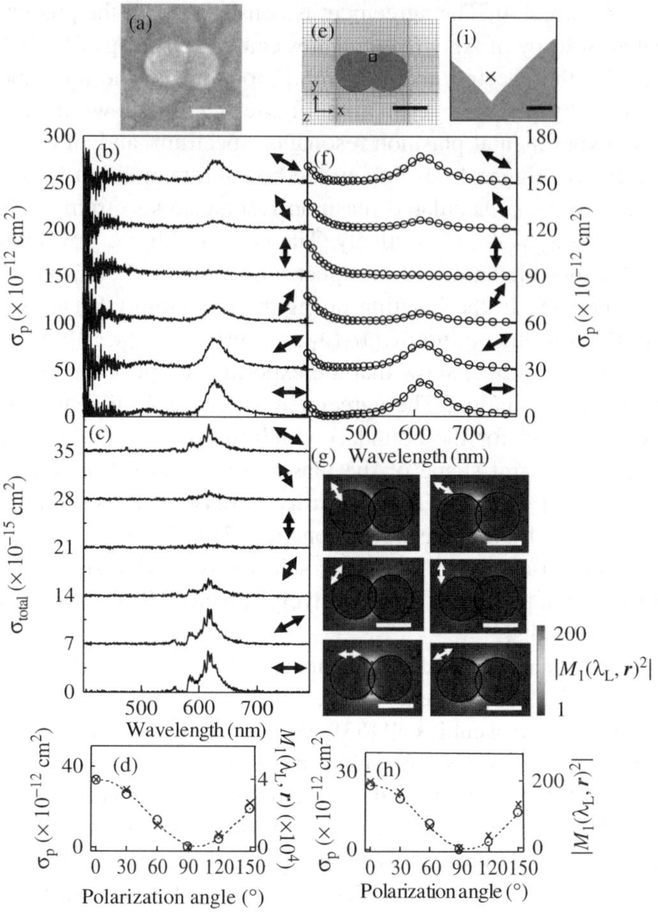

Figure 4.4 (a) SEM image, (b) polarization dependence of experimental plasmon resonance spectra, and (c) polarization dependence of experimental SERS spectra of a Ag NP dimer. (d) Polarization angle dependence of cross sections of experimental plasmon resonance (open circles with a dashed curve) and experimental EM enhancement factors (crosses with a dashed curve). (e) Modeled structure of an Ag NP dimer for the FDTD calculation, (f) polarization dependence of a calculated plasmon resonance spectrum, and (g) polarization dependence of the calculated spatial distribution of first EM enhancement factors. (h) Polarization angle dependence of cross sections of calculated plasmon resonance (open circles with a dashed curve) and calculated EM factors (crosses with a dashed curve) as a function of polarization angles. (i) Magnified view in the vicinity of the crevice of the Ag NP dimer in (e). Scale bars in (a, e, g, and i) are 50, 50, 50, and 1.5 nm, respectively. Mesh size 1.5 nm, refractive index 1.3 of the surrounding medium, and circularly-polarized incident light (532 nm) were selected in the FDTD calculation. Reproduced with permission from Ref. [26]. Copyright © 2010, The American Physical Society

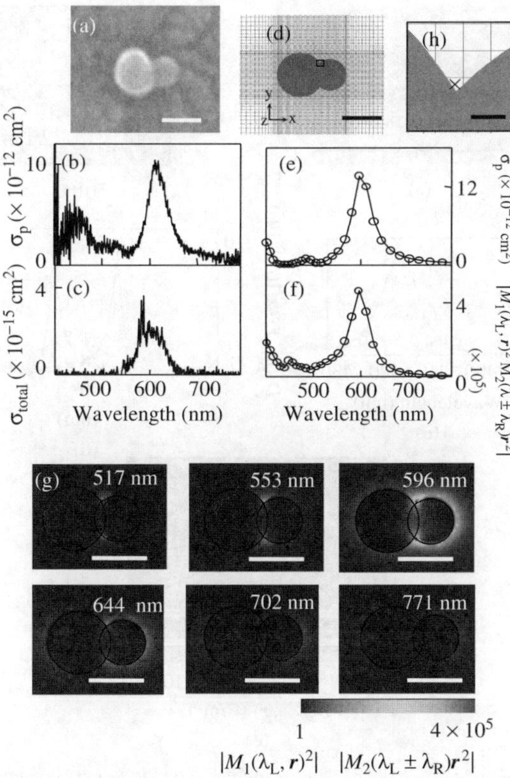

Figure 4.5 *(a) SEM image, (b) experimental plasmon resonance spectrum, and (c) experimental SERS spectrum, (d) modeled structure in the FDTD calculation, (e) calculated plasmon resonance spectrum, and (f) calculated spectrum of the EM enhancement factor of an Ag NP dimer. (g) Calculated spectrum of spatial distribution of twofold EM enhancement factors. (h) Magnified view in the vicinity of the crevice of the Ag NP dimer in (d). The SEM image, the plasmon resonance spectrum, and the SERS spectrum in (a–c) were observed from the identical Ag NP. Spectrum of twofold EM enhancement factors in (f) was calculated at the position of the cross in (h). Scale bars in (a, d, g, and h) are 50, 50, 50, and 1.5 nm, respectively. Mesh size 1.5 nm, refractive index 1.3 of the surrounding medium, and circularly-polarized incident light (532 nm) were selected in the FDTD calculation. Reproduced with permission from Ref. [26]. Copyright © 2010, The American Physical Society*

Here, we explain the criterion for the accuracy in the calculation of $|M_1|^2$. The distance d outside from the surface, where the electric flux density is reduced by a factor of $|1/e|^2$, is defined as $d = \lambda/(2\pi\sqrt{-\varepsilon(\lambda)})$. Here ε is the dielectric constant of Ag (or a vacuum) [54, 55]. Along the long axis of a dimer, the ratio of the decay length of bulk Ag d_{Ag} to that of vacuum d_v remains constant ($d_v/d_{Ag} \sim 2.36$) [54, 55]. Thus, we can evaluate the accuracy in the FDTD calculation from the deviation of the ratio from ~2.36. Figure 4.6d shows the decay curves of the electric flux density at 418 nm with several mesh sizes. The ratios evaluated are 0.18, 0.4, and 2.0 with mesh size 0.98, 0.5, and 0.2 nm, respectively. The value of 2.0 is satisfactory considering that FDTD calculation using 0.2 nm mesh requires a computational time longer than 20 days. The calculated electric flux density at the center

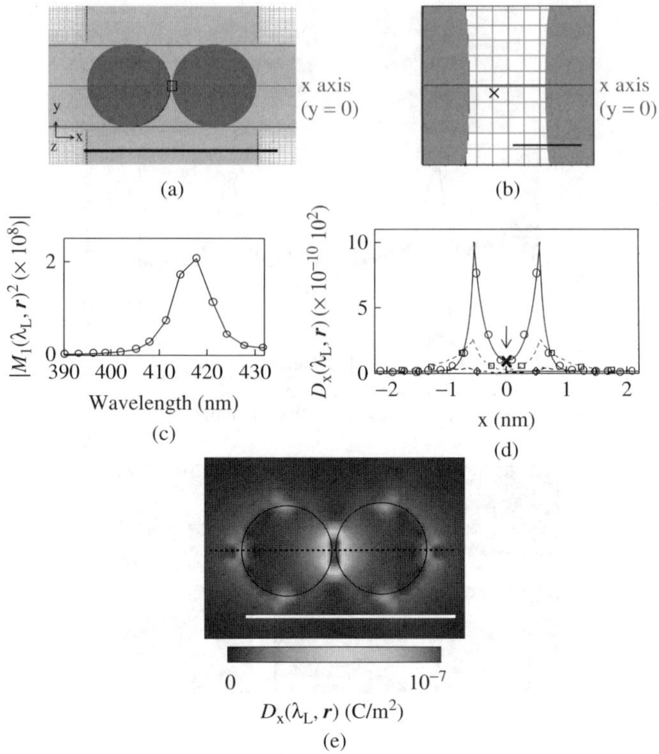

Figure 4.6 *(a) Modeled structure of an Ag NP dimer in the FDTD calculation. (b) Magnified view of the vicinity of the gap in the modeled structure. (c) Calculation spectrum of the twofold EM enhancement factor at the cross in (b). (d) Calculated cross-section of spatial distribution of the electric flux density (diamonds, squares, and circles; mesh size is 1.5, 0.5, and 0.2 nm respectively). (e) Spatial distribution of the electric flux density* D_x*. The cross indicated by the arrow in (d) indicates the calculated enhancement factor in Figure 2A of Ref. [24]. Note that the cross section in (d) indicates the electric flux density along the dashed line in (e). Scale bars in (a, b, and e) are 50, 1.0, and 50 nm, respectively. Reproduced with permission from Ref. [26]. Copyright © 2010, The American Physical Society. (*See insert for color representation of the figure.*)*

of the gap is well consistent with the self-consistent calculation of Figure 2B in Ref. [24] as shown by the cross in Figure 4.6d. Figure 4.6e shows the spatial distribution of the electric flux density.

Finally, we calculate SERS cross-section spectra by the use of mesh size 0.2 nm to quantitatively evaluate twofold enhancement factors. In Figure 4.7, we show a set of SERS cross-section spectra acquired from the same dimer with three different incident wavelengths (Figure 4.7c–e). The SERS cross sections were calculated by multiplying $|M_1(\lambda_L, r_o)|^2 |M_2(\lambda_L \pm \lambda_R, r_o)|^2$ by $\sigma_{RS} + (1/|M_d|^2)\sigma_{FL}$ using three laser lines (Figure 4.7h–j), where r_o (the cross in Figure 4.7k) is the location of the center of a R6G molecule. The center locates at ~0.8 nm from the Ag surface. This distance is sufficiently large so as to block charge transfer interaction between R6G molecules and Ag surfaces [7, 56–58]. The shape

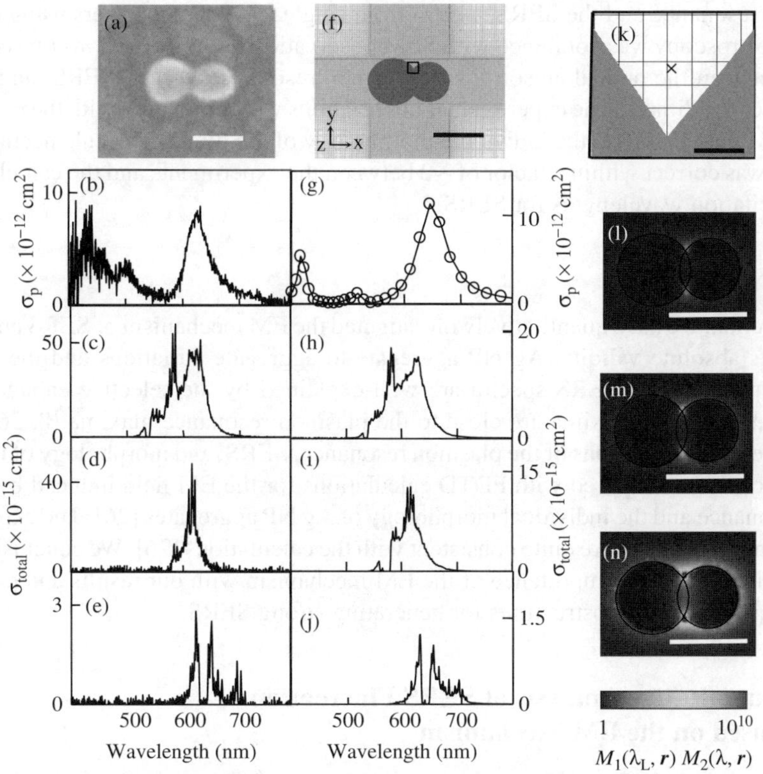

Figure 4.7 *(a) SEM image of an Ag NP dimer, (b) experimental plasmon resonance spectrum, and experimental SERS spectra excited at (c) 532 nm, (d) 561 nm, and (e) 633 nm. (f) Modeled structure of an Ag NP dimer in the FDTD calculation, (g) calculated plasmon resonance spectrum and calculated SERS spectra excited at (h) 532 nm, (i) 561 nm, and (j) 633 nm. (k) Magnified view in the vicinity of the crevice of the model structure in (f). Spatial distribution of calculated twofold EM enhancement factor excited at (l) 532 nm, (m) 561 nm, and (n) 633 nm. Experimental plasmon resonance maximum in (b) was ~620 nm, its FWHM was ~60 nm, and its cross section was ~1 × 10^{-11} cm^2. Circularly-polarized incident light (532, 561, and 633 nm), the mesh size 0.2 nm, and the refractive index 1.3 of the surrounding medium were selected in the FDTD calculation. Scale bars in (a, f, l, m) are 50 nm. Scale bar in (k) is 0.5 nm. Reproduced with permission from Ref. [26]. Copyright © 2010, The American Physical Society. (See insert for color representation of the figure.)*

of the SERS spectra is correctly reproduced in the calculations and, moreover, the experimental cross sections are consistent with the calculations within a factor of ~2, which is acceptable considering the huge enhancement factor of 10^9 involved in the SERS process. This quantitative consistency of the experimental SERS spectra with the calculations for three incident wavelengths provided us with convincing evidence for the EM mechanism, in which SERS is described as the product of Raman scattering and plasmon resonance of the nanostructure.

In this subsection, we have studied the quantitative nature of the EM enhancement effect on SERS, and demonstrated its absolute validity. Polarization-controlled observation of the plasmon resonance and the SERS spectra from single isolated Ag dimers using dark-field microspectroscopy was combined with SEM observation to achieve one-to-one correspondence between the optical anisotropy of plasmon resonance, that of SERS, and the morphology of the dimers. The experimental observations were compared with the calculations of the EM field based on the individual morphology of the dimer. The enhancement factor of $\sim 10^9$ was correct within a factor of ~ 2 between the experiments and the calculations for three excitation wavelengths for SERS.

4.1.5 Summary

In this section, we have quantitatively investigated the EM mechanism of SERS and demonstrated its absolute validity. Ag NP aggregate-to-aggregate variations and the refractive index dependence of SERS spectra are well-explained by the selective enhancement of SERS bands whose maxima are close to the plasmon resonance maxima [8, 26, 39, 40]. Experimental observations of the plasmon resonance, SERS, and morphology of the Ag NP aggregates were compared with FDTD calculations for the EM field induced by the plasmon resonance and the individual morphology of Ag NP aggregates [26]. The experimental enhancement factors were quite consistent with the calculations [26]. We conclusively confirm the indispensable importance of the EM mechanism with our results and extend it to the design of metal nanostructures for generating strong SERS.

4.2 Quantitative Analysis of SERS Fluctuation Based on the EM Mechanism

4.2.1 Introduction

SERS and SEF, which usually coexists with SERS as a background, have been treated by several mechanisms including the EM mechanism [1, 3, 8]. The EM mechanism is explained as follows: the local mode density of EM fields in the vicinity of metal NPs is increased by plasmon resonance, resulting in enhancement of the optical transition rates of molecules close to NP surfaces [1, 3, 8]. The fluctuation in SERS and SEF has been studied extensively [7, 59–67]. The analysis of SERS and SEF fluctuation based on the EM mechanism is advantageous to the quantitative understanding of the fluctuation mechanism.

Here, representative fluctuations for SERS and SEF are briefly summarized: molecules activated to metastable non-absorbing and non-emissive states [61, 62], molecules thermally diffusing in and out of hot spots [63], molecules thermally diffusing on the NP surface coupled with photo-induced electron transfer, the structural relaxation of surface active sites [64], a thermally-stimulated molecular reorientation and chemical process [65], photoionization via charge-transfer states [66], and rearrangement of the morphology of the metallic substrate [67]. However, analysis of SERS and SEF fluctuation has been limited to qualitative approaches owing to the complexity of their mechanisms, which prevents us from examining the direct correlation between the fluctuation and its origins. Furthermore unclearness in the SERS and SEF mechanism has hindered quantitative interpretation of fluctuation.

In this section we analyze fluctuation in SERS and SEF in terms of intensity and spectral fluctuation using a system composed of Ag NP dimer and R6G molecules (Ag dimer/R6G).

We controlled the fluctuation from single Ag NP dimers by intense NIR laser pulse irradiation. This control enabled us to systematically observe SERS and SEF from stable to unstable, excluding Ag NP dimer-by-dimer variations in fluctuation. The highlights are twofold: the intensity fluctuation in SERS and SEF is correlated with the distance between the molecules and the metal surfaces within a few angstroms; and the spectral fluctuation in SERS and SEF is correlated with the plasmon resonance blue shifts. Analyzing the correlations revealed that the fluctuation in SERS and SEF is quantitatively explained within the framework of the EM mechanism [50].

4.2.2 Intensity and Spectral Fluctuation in SERS and SEF

Intensity and spectral fluctuation in SERS and SEF have large Ag dimer-by-dimer variations. The variations have prevented us from systematically analyzing the mechanism of the fluctuation. The variations in intensity fluctuation can approximately be categorized into three phases (stable, unstable, and intermittent) as shown in Figure 4.8. Figure 4.8a1–a3 show temporally stable, unstable, and intermittent total intensity of SERS and SEF. Figure 4.8b1–b3 show temporally stable, unstable, and intermittent SERS and SEF spectra. Not only SERS and SEF but also plasmon resonance spectra show spectral changes during SERS and SEF measurements [35, 37]. Reproducible control of the variations using one Ag dimer will exclude the dimer-by-dimer variations and allow us to systematically analyze the fluctuation. Fortunately, we found that irradiation with intense NIR laser pulses reproducibly controls the three phases. Thanks to this finding we can observe changes in SERS and SEF from a stable phase to an unstable phase using single Ag dimers.

To examine the fluctuation in SERS and SEF within the framework of the EM mechanism, we measured the relationship among plasmon resonance spectra, SERS and SEF spectra, and the temporal trajectories of SERS and SEF intensity from single Ag dimers adsorbed by R6G molecules, before and after irradiating with 1064-nm laser pulses (1.5 ns FWHM, 27×10^{11} W cm^{-2}, SPOT Series Laser, Elforlight, Northants). We first observed plasmon resonance spectra, then SERS and SEF spectra, and finally the temporal trajectories of SERS and SEF intensity after irradiating with each laser pulse. Figure 4.9a1–a4 show that the plasmon resonance maxima were blue shifted by ~50 nm after irradiating with six pulses. Figure 4.9b1–b4 show that temporally stable SERS and SEF spectral shapes were changed after irradiating laser pulses; finally, temporal instability was observed in the spectral shape around 560 nm, as indicated by two panels in Figure 4.9b4. Figure 4.9c1–c4 show that the temporal trajectories of total SERS and SEF intensity were changed from stable in c1 and c2 to unstable in c3, and finally intermittent "on" and "off" in c4. These results are common to most of the Ag dimers we selected.

4.2.3 Framework for Analysis of Fluctuation in SERS and SEF

We have reproduced SERS and SEF spectra by the EM mechanism [8]. This achievement motivated us to quantitatively analyze the intensity and spectral fluctuation in SERS and SEF in Figure 4.9c1–c4 by the EM mechanism. Within the framework of the EM mechanism, the molecules in unstable adsorption on Ag dimers are expected to cause fluctuation in the SERS and SEF intensity, because the unstable adsorption changes the "effective" distance d_{av} between a molecule and an Ag dimer surface. Here, "effective" means that d_{av}

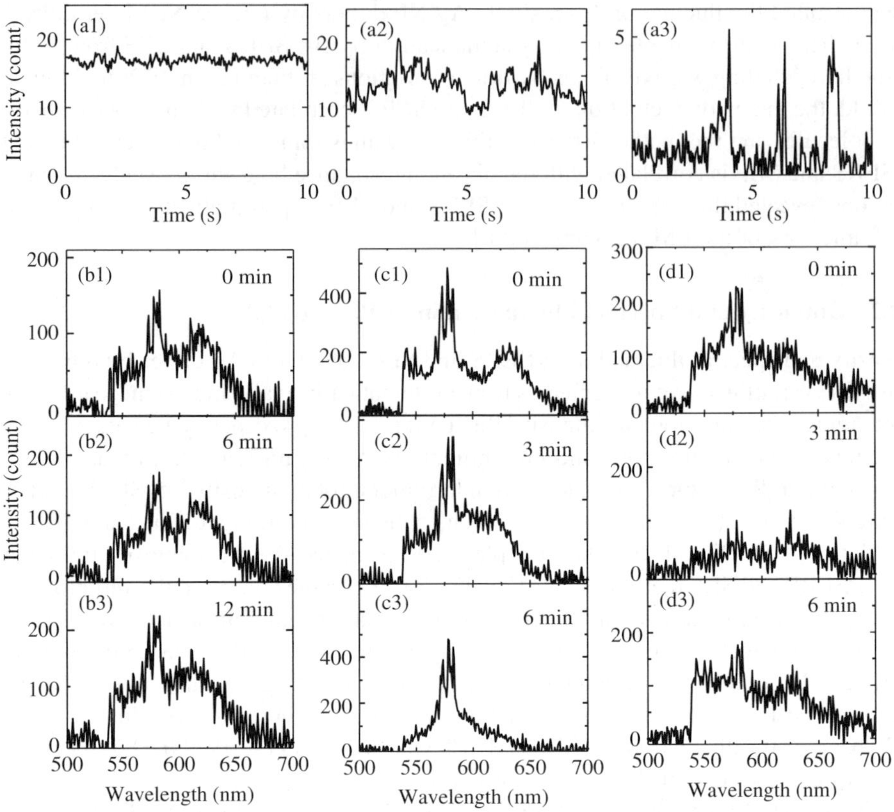

Figure 4.8 *(a1) Temporally stable, (a2) fluctuating, and (a3) intermittent total intensity of SERS and SEF from three representative Ag NP dimers. The intensity is integrated from 540 to 700 nm. SERRS and SEF from three representative Ag NP dimers: (b1–b3) temporally stable, (c1–c3) fluctuating, and (d1–d3) intermittent. Time lapse for the measurement is indicated in each figure. Exposure time for each spectrum is 5 s. Reproduced with permission from Ref. [50]. Copyright © 2012, American Institute of Physics*

includes several factors, not only distance but also molecular orientation on a metal surface. An increase in d_{av} reduces both enhancement factors $|M_{total}|^2$ and $|M_d|^2$, where $|M_d|^2$ is an enhancement factor of the decay rate for a molecule in the excited state due to resonance energy transfer from a molecule to an Ag surface [3, 7, 8, 51]. For a molecule close to the bottom of a crevasse in an Ag dimer d_{av} is ~1.0 nm.

Here, we outline the EM mechanism of SERS and SEF. The total cross section of SERS and SEF $\sigma_{total}(\lambda_L, \lambda_L \pm \lambda_R, d_{av}) = \sigma_{RS}(\lambda_L, \lambda_L \pm \lambda_R) + (1/|M_d|^2)\sigma_{FL}(\lambda_L, \lambda_L \pm \lambda_R)$, as indicated by Equation 4.4. $|M_{total}|^2$ is a product of the first and second EM enhancement factors, each of which is induced by coupling of plasmon with incident light and Raman plus fluorescence light [3, 7, 8]. Thus, σ_{SERS} is given by

$$\sigma_{SERS}(\lambda_L, \lambda_L \pm \lambda_R, d_{av}) = |M_{total}(\lambda_L, \lambda_L \pm \lambda_R, d_{av})|^2 \sigma_{RS}(\lambda_L, \lambda_L \pm \lambda_R) \tag{4.5}$$

In the case of σ_{SEF}, the decay rate of the excited molecular electronic state is also enhanced by a factor of $|M_d|^2$ through energy transfer from a molecule to a metal surface

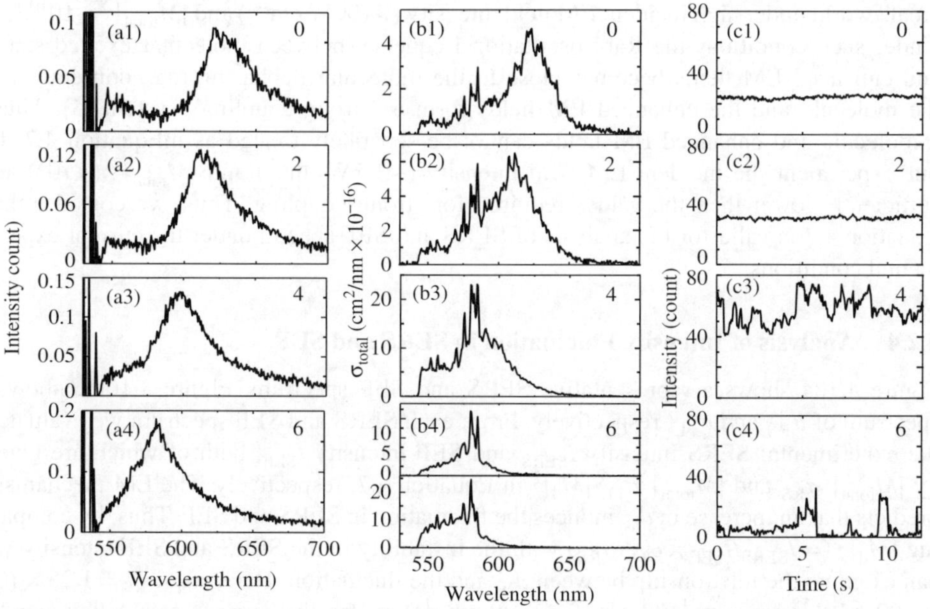

Figure 4.9 *Temporal changes in (a1–a4) plasmon resonance spectra and (b1–b4) SERS and SEF spectra. (c1–c4) Total SERS and SEF intensity integrated from 540 to 700 nm, all of which were measured from a single Ag NP dimer after irradiating with1064-nm laser pulses. The number of the irradiated laser pulses is indicated in each figure. Note that two panels in (b4) shows spectral fluctuation observed just after irradiating six pulses without further laser pulse irradiation. Reproduced with permission from Ref. [50]. Copyright © 2012, American Institute of Physics*

[3, 7, 8]. Thus, $1/|M_d|^2$ is multiplied by the fluorescence cross-section of a molecule in a free space. Thus,

$$\sigma_{SEF}(\lambda_L, \lambda_L \pm \lambda_R, d_{av}) = \frac{|M_{total}(\lambda_L, \lambda_L \pm \lambda_R, d_{av})|^2}{|M_d(d_{av})|^2}\sigma_{FL}(\lambda_L, \lambda_L \pm \lambda_R) \quad (4.6)$$

Note that the use of σ_{FL} is a first approximation for evaluation of SEF, because a spectral shape of σ_{FL} may be modulated by an increased contribution of ultrafast electronic dynamics in the vicinity of a metallic object when enhanced decay rates exceed molecular vibrational decay rates [3, 60, 68]. Equation 4 assumes that a quantum yield of R6G is unity. Combining Equations 4.5 and 4.6 yields

$$\sigma_{total}(\lambda_L, \lambda_L \pm \lambda_R, d_{av}) = |M_{total}(\lambda_L, \lambda_L \pm \lambda_R, d_{av})|^2 \times \left(\sigma_{RS}\left(\lambda_L, \lambda_L \pm \lambda_R\right) + \frac{\sigma_{FL}(\lambda_L, \lambda_L \pm \lambda_R)}{|M_d(d_{av})|^2}\right) \quad (4.7)$$

Equation 4.7 indicates that an increase in d_{av} is associated with a decrease in the ratio between σ_{RS} and $\sigma_{FL}/|M_d|^2$, because $|M_d|^2$ is a decreasing function against d_{av}. Note that the theoretical work predicts that this model followed by Equation 4.7 deviates from the "real" world under high incident EM field intensity (>10^5 W cm^{-2}) and $|M_{total}|^2 > 10^{10}$ [3]. Under such conditions the Rabi oscillation frequency between molecular excited states and enhanced EM fields becomes close to the molecular dephasing rate; consequently, the molecule and the enhanced EM fields form a "strong coupling" system [3]. Thus, a molecule and enhanced EM fields cannot be separately treated as inEquation 4.7. In our experiment the incident EM field intensity (~3.5 W cm^{-2}) and $|M_{total}|^2$ (<10^8) are sufficiently lower than the values required for strong coupling. Thus, we consider that Equation 4.7 is valid for the analysis of SERS and SEF spectra under the current experimental conditions.

4.2.4 Analysis of Intensity Fluctuation in SERS and SEF

Figure 4.10a shows a representative SERS and SEF spectrum. Figure 4.10b,c show a spectrum of σ_{RS} and σ_{FL}, respectively. From each SERS and SEF spectrum we evaluated the experimental SERS intensity I_{SERS} and SEF intensity I_{SEF}, both of which are given by $|M_{total}|^2\sigma_{RS}$ and $|M_{total}|^2\sigma_{FL}/|M_d|^2$ in Equation 4.7, respectively. The EM mechanism predicts that an increase in d_{av} induces the fluctuation in SERS and SEF. Thus, by comparing $|M_d|^2$ $(= I_{SERS}/I_{SEF} \times \sigma_{FL}/\sigma_{RS})$ with the instability in the SERS and SEF intensity we can examine the relationship between d_{av} and the fluctuation. Here, $\sigma_{FL}/\sigma_{RS} = 1.25 \times 10^5$ $((2.99 \times 10^{-18}\ \text{cm}^2\ \text{nm}^{-1})/(2.4 \times 10^{-23}\ \text{cm}^2\ \text{nm}^{-1}))$ under the current calculation condition.[1] To evaluate the fluctuation we used the normalized standard deviation score S

$$S = \frac{1}{\langle I \rangle}\sqrt{\frac{\sum_{n=1}^{N}(I_n - \langle I \rangle)^2}{N}} \tag{4.8}$$

for temporal trajectories of SERS and SEF intensity to cancel Ag dimer-by-dimer intensity variations. Here, I_n and $<I>$ are the total intensity of the SERS and SEF from each detection point n with a bin time 0.32 s and the total intensity of SERS and SEF averaged over all the detection time (~2.7 min), respectively; and N is the number of detection points (~500). The larger S the stronger the fluctuation in the total intensity of the SERS and SEF.

We analyzed the relationship between $|M_d|^2$ and S for eight dimers. Figure 4.10d shows the probability that $|M_d|^2$ values become lower at higher S values. By comparison of temporal trajectories of total SERS and SEF intensity with the S values, the above probability was approximately classified into three regions: a region showing stable intensity, as shown in Figure 4.9c1,c2 ($S <$ ~0.05), another showing instability, as shown in Figure 4.9c3 (~0.05 < S < ~0.6), and the third showing intermittence, as shown in Figure 4.9c4 ($S >$ ~0.6). Figure 4.10d shows that $|M_d|^2$ decreased with increase in the S values for $S <$ ~0.6. From Equation 4.7 this relation indicates that intensity fluctuation in SERS and SEF is induced by increase in d_{av}. Thus, we estimated the difference in d_{av} between the stable and fluctuating regions. In Figure 4.10d, $|M_d|^2 \sim 12 \times 10^4$ in the stable region ($S <$ ~0.05) was about twice as large as $|M_d|^2 \sim 6 \times 10^4$ in the unstable

[1] The spectrum of σ_{RS} in Figure 4.10b is made by comparison of Table 1 of Ref. [69] with an SERS spectrum from an ensemble of Ag NP aggregates, because both $|M_{EM}|^2$ and $1/|M_d|^2$ in the SERS spectrum are constant over all the wavelengths involved. The spectrum of σ_{FL} in Figure 4.10c is set to be from the absorption cross-section of R6G (3.0×10^{-16} cm^2 at 532 nm)and quantum yield ~1.

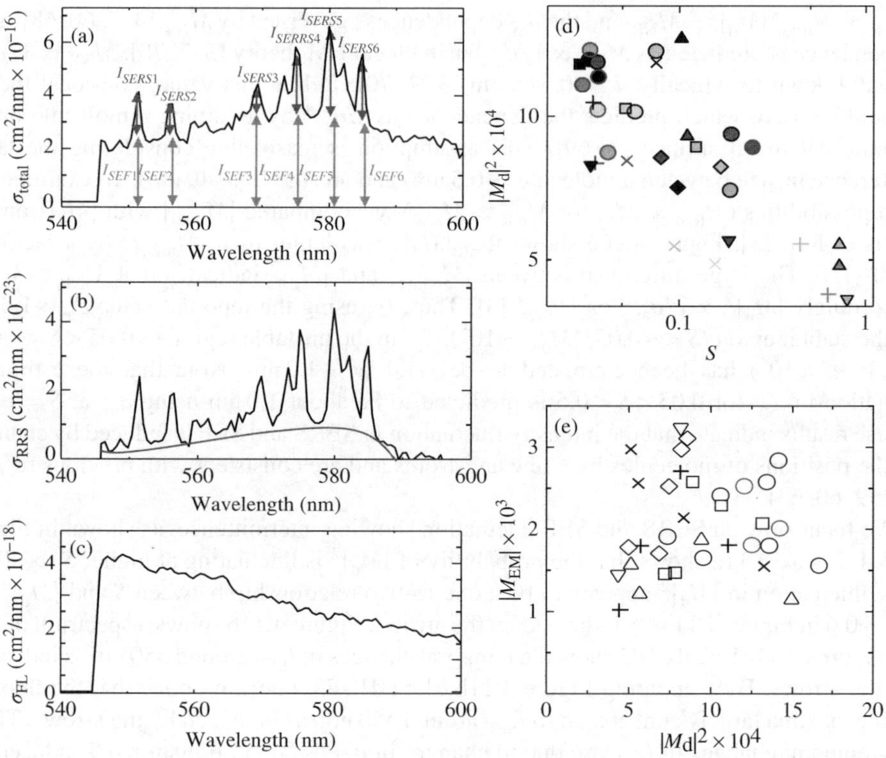

Figure 4.10 *(a) Evaluating* I_{SERS}/I_{SEF} *for panels (d). Total SERS intensity* I_{SERS} *and SEF intensity* I_{SEF} *were averaged over* I_{SERS1} *to* I_{SERS6} *(upper arrows) and that of* I_{SEF1} *to* I_{SEF6} *(lower arrows), respectively. (b)* σ_{RS} *of spectrum R6G in an aqueous solution. Raman intensity is defined as an upper part of the red curve that is considered as back ground fluorescence. The unit of the y-axis is converted into a cross section (cm*^{2} *nm*$^{-1}$*). (c)* σ_{FL} *spectrum of R6G in an aqueous solution detected with the same experimental set-up. The unit of the y-axis is converted into a cross section (cm*^{2} *nm*$^{-1}$*). (d) Normalized standard deviation scores S dependence of* $|M_d|^2$ *in* $0 < S < 0.6$ *for eight dimers denoted by eight symbols (○, ○, □, Δ, ∇, ◇, +, and ×). Filled circles are taken from the Ag NP dimer used for this figure. Number of irradiated laser pulses is indicated by colors. Gray color indicates before laser pulse irradiation, and faint blue, green, orange, and red indicates after first, second, third, and fourth, laser pulse irradiation. (e)* $|M_{EM}|$ $(= (I_{SERS}/\sigma_{RS})^{1/2})$ *dependence of* $|M_d|^2$ $(= I_{SERS}/I_{SEF} \times \sigma_{RS}/\sigma_{FL})$ *for the same eight dimers as those in (d). Reproduced with permission from Ref. [50]. Copyright © 2012, American Institute of Physics. (See insert for color representation of the figure.)*

region ($\sim 0.05 < S < \sim 0.6$). In a simple sphere approximation, $|M_{\mathrm{total}}|^2$ varies according to $(1/(1 + d_{\mathrm{av}}{}^4/R))^6$, where an Ag NP radius R is much larger than d_{av} and therefore $|M_{\mathrm{total}}|^2$ is approximately constant when d_{av} is small compared with the size of the molecule [24]. $|M_{\mathrm{d}}|^2$ is separated into two contributions, from radiative and non-radiative emission: $|M_{\mathrm{d}}|^2 = M_{\mathrm{Rad}} + M_{\mathrm{NR}}$ [3, 7, 60, 68]. The dependence of $|M_{\mathrm{d}}|^2$ on d_{av} depends on which of these dominates. If $M_{\mathrm{Rad}} \gg M_{\mathrm{NR}}$, $|M_{\mathrm{d}}|^2$ does not depend much on d_{av} and $|M_{\mathrm{d}}|^2 \sim |M_{\mathrm{total}}|$ by optical reciprocity, which applies to the radiative part of the emission [3, 7, 60, 68]. If

$M_{NR} \gg M_{Rad}$, $|M_d|^2 \sim M_{NR}$ and the d_{av} dependence is governed by M_{NR} [3, 7, 60, 68]. This dependence is predicted as $M_{NR} \propto 1/d_{av}^3$ using local EM theory [3, 7, 70]. $M_{NR} \propto 1/d_{av}^3$ is valid down to typically d_{av} ~0.5–1 nm [3, 7, 70]. Below this value, non-local theory should be used, which predicts the dependence as $1/d_{av}^4$ by assuming a molecule on an infinite flat metal surface [3, 69]. This assumption is reasonable considering the large difference in size between a molecule (~0.5 nm) and an Ag NP (~40 nm). To examine the two possibilities ($M_{Rad} \gg M_{NR}$ or $M_{NR} \gg M_{Rad}$), we compared $|M_{total}|$ with $|M_d|^2$ on the basis of Ref. [3]. Figure 4.10e shows that $|M_d|^2$ was larger than $|M_{total}|^2$ by a factor of ~10–120. The large difference between $|M_{total}|^2$ and $|M_d|^2$ indicates that $M_{NR} \gg M_{Rad}$, accordingly $|M_d|^2 \propto 1/d_{av}^3$ or $1/d_{av}^4$ [3]. Thus, by using the reported value d_{av} ~1.0 nm in the stable region ($S <$ ~0.05, $|M_d|^2 \sim 10^5$), d_{av} in the unstable region (~0.05 $< S <$ ~0.6, $|M_d|^2 \sim 5 \times 10^4$) has been estimated to be ~1.4 or ~1.2 nm. Note that the estimated variation in d_{av} for $0.05 < S < 0.6$ is predicted to be about 1.0 nm using d_{av} at $S = 0$ [7]. These results indicate that the intensity fluctuation in SERS and SEF is induced by changes in the positions of molecules by a few angstroms and are consistent with previous reports [7, 59, 60, 63].

We focus now on SERS and SEF fluctuation showing intermittence, as shown in Figure 4.9c4. Figure 4.11a shows that the probability of $|M_d|^2$ is fluctuating at higher S (>~0.6). The fluctuation in $|M_d|^2$ appears to be a one-to-two relationship between S and $|M_d|^2$ for $S >$ ~0.6 in Figure 4.11a, as indicated by the arrows. Figure 4.11b shows a spectrum of σ_{FL} and Figure 4.11c1,c2,d1–d3 show the temporal changes in I_{SEF} around 550 nm as indicated by the arrows. By comparing Figure 4.11b,c1,c2,d1–d3, one can notice that the fluorescence maxima largely contributed to I_{SEF} around 550 nm, as indicated by the arrows. Thus, the temporal changes in I_{SEF} are due to changes in $\sigma_{FL}/|M_d|^2$ in Equation 4.7 induced by temporal changes in d_{av}. The one-to-two relationship is occasionally observed for $S <$ ~0.6, but is minor compared with that for $S >$ ~0.6.

4.2.5 Analysis of Spectral Fluctuation in SERS and SEF

We analyzed the spectral fluctuation in SERS and SEF spectra, an example of which is shown in Figure 4.9b1–b4. The relationship between plasmon resonance spectra and SERS and SEF spectra has been shown in Ref. [8] and quantitatively reproduced by the EM mechanism. The quantitative reproduction indicates that the spectral fluctuation in SERS and SEF spectra is also quantitatively reproduced by the EM mechanism. However, several other mechanisms for SERS spectral fluctuation, including changes in charge transfer interaction between molecules and Ag surfaces andmolecular photo-decomposition have been reported [61–66]. These reports enhance the necessity for quantitative evaluation of spectral fluctuations in SERS and SEF exclusively using the EM mechanism.

The spectral changes cannot be explained by an increase in d_{av}, because the increase changes the values of $|M_d|^2$ and thus spectral shapes are independent of d_{av}. Equation 4.7 indicates that the spectral changes in $|M_{EM}|^2$ cause the spectral changes in SERS and SEF. To confirm the indication we experimentally derived changes in $|M_{total}|^2$ spectra from SERS and SEF spectra. From Equation 4.7,

$$|M_{total}(\lambda_L, \lambda_L \pm \lambda_R, d_{av})|^2 = \frac{\sigma_{total}(\lambda_L, \lambda_L \pm \lambda_R, d_{av})}{\sigma_{RS}(\lambda_L, \lambda_L \pm \lambda_R) + \frac{\sigma_{FL}(\lambda_L, \lambda_L \pm \lambda_R)}{|M_d(d_{av})|^2}} \tag{4.9}$$

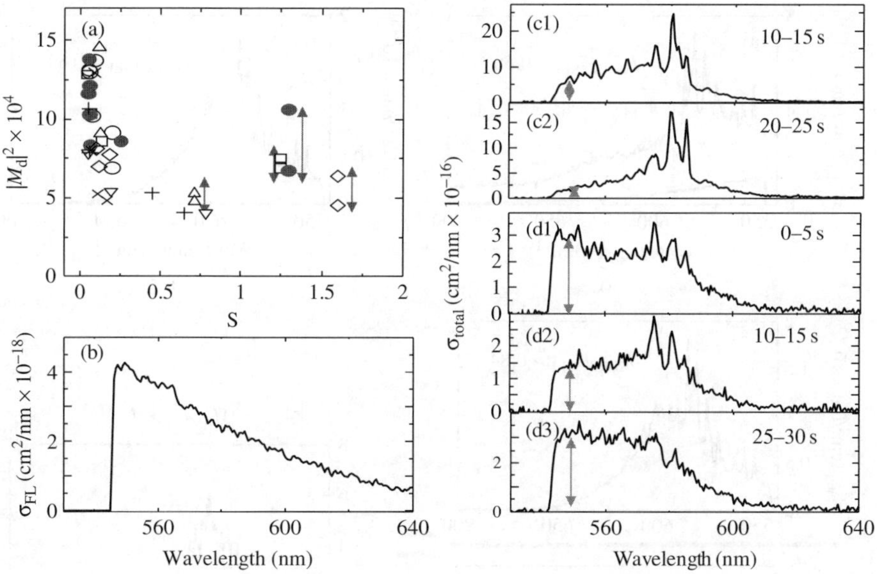

Figure 4.11 *(a) S dependence of $|M_d|^2$ in $0 < S < 2.0$ for the same eight Ag NP dimers as in Figure 4.10. Fluctuation in $|M_d|^2$ is highlighted by binary points (arrows). (b) σ_{FL} spectrum of R6G in an aqueous solution detected with the same experimental set-up. (c1,c2, d1–d3) Intensity fluctuation in SERS and SEF spectra from two Ag NP dimers just after irradiating six and four pulses without irradiating further pulse, respectively. Detection periods are indicated in each figure. The fluorescence intensity that dominantly fluctuated is highlighted by arrows. Reproduced with permission from Ref. [50]. Copyright © 2012, American Institute of Physics.* (See insert for color representation of the figure.)

Here, $\sigma_{RS} + \sigma_{FL}/|M_d|^2$ in Equation 4.9 is the SERS and SEF spectrum not enhanced by $|M_{total}|$: $\sigma^o{}_{total}$. However, direct experimental measurement of $\sigma^o{}_{total}$ is difficult to do. As an alternative we used the SERS and SEF spectrum $\sigma_{en}(\lambda_L, \lambda)$ experimentally measured from an ensemble of Ag NP aggregates in place of $\sigma^o{}_{total}$, because both $|M_{total}|^2$ and $1/|M_d|^2$ in σ_{en} are constant over all the wavelengths involved [8, 39, 40]. Thus, the spectral shape of σ_{en} is the same as that of $\sigma^o{}_{total}$. The value of σ_{en} was converted into that of $\sigma^o{}_{total}$ for a single R6G molecule using the reported cross sections σ_{RS}, σ_{FL} (see footnote 1), and $|M_d|^2$ of $\sigma^o{}_{total}$ was calculated to be 7.3×10^4 using $|M_d|^2 = I_{SERS}/I_{SEF} \times \sigma_{RS}/\sigma_{FL}$ (from Equations 4.5 and 4.6), where $I_{SERS}/I_{SEF} = 0.58$, as noted in Figure 4.12a. However, one concern remains for the evaluation of $|M_{total}|^2$ using Equation 4.9. The concern is changes in I_{SEF} by irradiating laser pulses, as shown in Figure 4.9b1–b4. The effect of the changes on the calculation using Equation 4.9 should be removed before evaluating $|M_{total}|^2$. Thus, we removed the fluctuation of I_{SEF} from $\sigma^o{}_{total}(\lambda_L, \lambda)$: $\sigma^o{}_{total} - k\sigma_{FL}/|M_d|^2$. Here, k is selected so that I_{SERS}/I_{SEF} for $\sigma^o{}_{total} - k\sigma_{FL}/|M_d|^2$ is the same as $I_{SERS}/I_{SEF} = 0.84$ from σ_{total} as noted in Figure 4.12b. In other words, we set $|M_d|^2$ in $\sigma^o{}_{total} - k\sigma_{FL}/|M_d|^2$ to be the same as $|M_d|^2$ in experimentally measured σ_{total} for the alternative of $\sigma_{RS} + \sigma_{FL}/|M_d|^2$. Figure 4.12c shows the graphical evaluation of $|M_{total}|^2$ by substituting σ_{total} and $\sigma^o{}_{total} - k\sigma_{FL}/|M_d|^2$ into Equation 4.9. In fact, such evaluation of $|M_{total}|^2$ considering changes in I_{SEF} is overdelicate, because we found that the difference in the values of $|M_{total}|^2$ is only within several

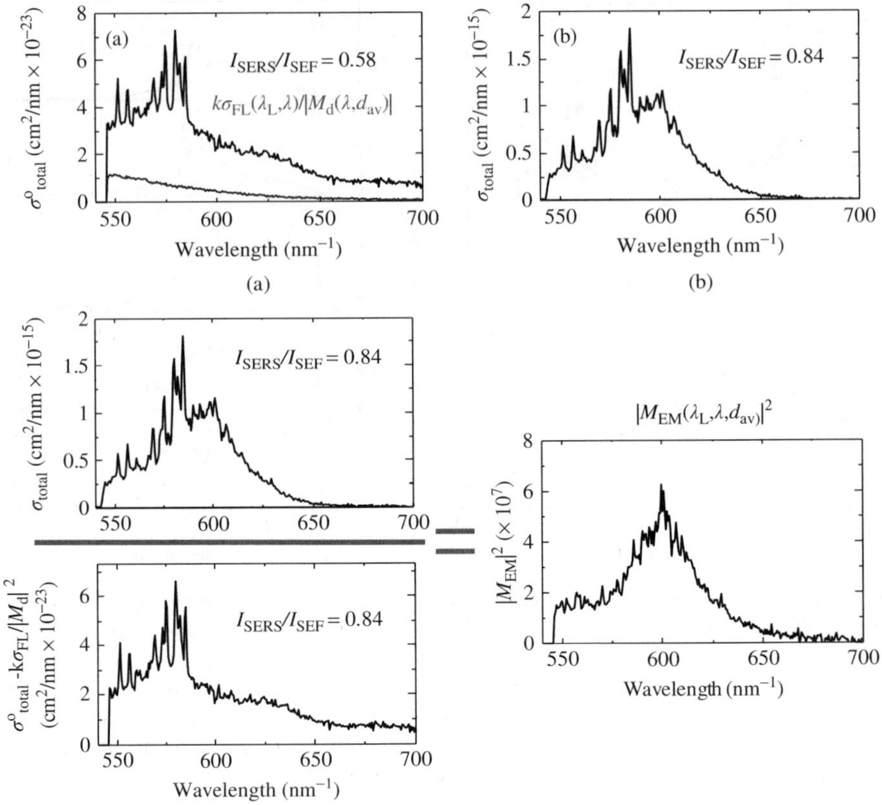

(c)

Figure 4.12 *(a) σ^o_{total} of an ensemble of Ag NPs/R6G aggregates (upper trace). Here, $I_{SERS}/I_{SEF} = 0.58$ ($|M_d|^2 = 7.25 \times 10^4$) from σ^o_{total}. Lower trace represents $k\sigma_{FL}/|M_d|^2$, where $k = 1-0.58/0.84$. (b) σ_{total} of a Ag NP dimer/R6G. Here, $I_{SERS}/I_{SEF} = 0.84$. (c) Graphical presentation of deriving $|M_{EM}|^2$ from Equation 4.9. Note that I_{SERS}/I_{SEF} in $\sigma^o_{total} - k\sigma_{FL}/|M_d|^2$ is adjusted to 0.84, which is the same as I_{SERS}/I_{SEF} in σ_{total}, using $k = 1-0.58/0.84$. Reproduced with permission from Ref. [50]. Copyright © 2012, American Institute of Physics*

percent by comparing $\sigma_{total}/\sigma^o_{total}$ with $\sigma_{total}/(\sigma^o_{total} - k\sigma_{FL}/|M_d|^2)$. The small difference indicates that the effect of $k\sigma_{FL}/|M_d|^2$ on $|M_{total}|^2$ is negligibly small in the current experiments. However, the procedure should be kept for calculating $|M_{total}|^2$ in the case of molecules showing large SEF against SERS.

Figure 4.13a1–a6 show the evaluated $|M_{total}|^2$ spectra (red lines) and plasmon resonance spectra (black lines). Both spectral shapes are quite similar to each other. The similarity holds the possibility that the spectral fluctuation in SERS and SEF in Figure 4.9b1–b4 is induced by the changes in plasmon resonance spectra. To enhance this possibility, we examined the relationship between maximum wavelength λ_{EM} of $|M_{total}|^2$ and maximum wavelength λ_p of the plasmon resonance for eight Ag dimers. Figure 4.13b shows that all the Ag dimers clearly gave simultaneous blue shifts in both λ_{EM} and λ_p. The observations reveal that the spectral fluctuation in SERS and SEF is induced by plasmon resonance, which is

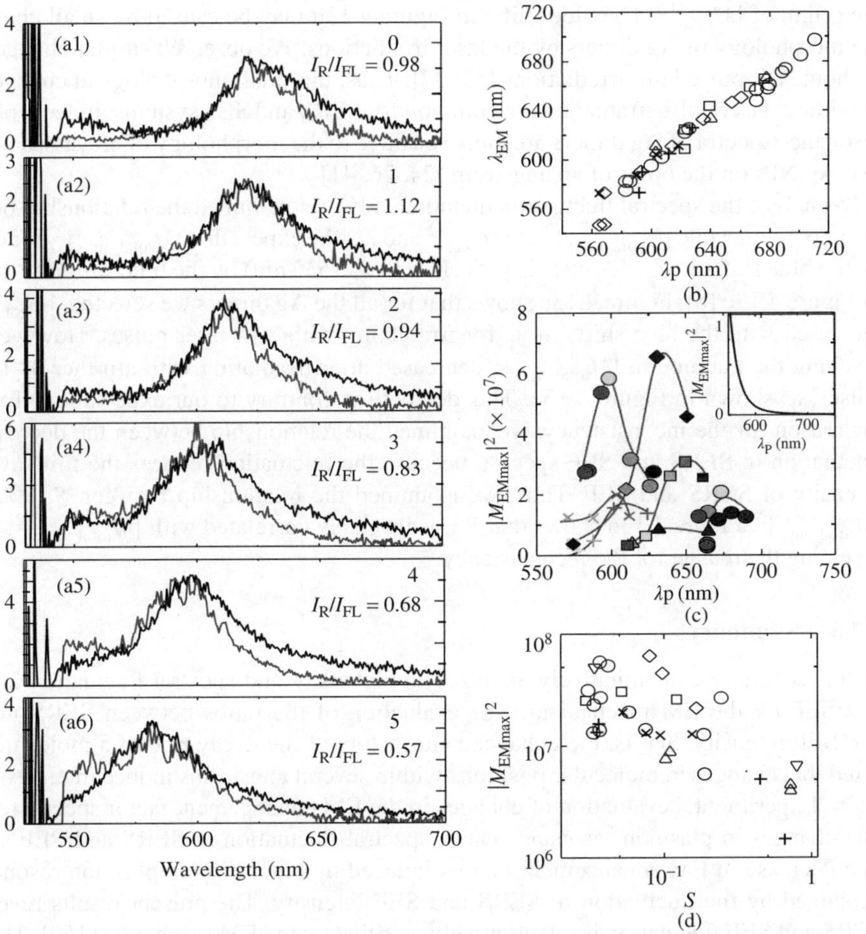

Figure 4.13 *(a1–a6) Spectral changes in plasmon resonance (black lines) and $|M_{EM}|^2$ (red lines) from an Ag dimer measured after irradiating with1064-nm laser pulses. The number of the NIR pulses and I_{SERS}/I_{SEF} are indicated in each figure. I_{SERS}/I_{SEF} is obtained in the same manner as that for Figure 4.11a. Spectra of σ_{total} in Figure 4.9b1–b4 are used to obtain I_{SERS}/I_{SEF}. (b) Maximum wavelength λ_p of plasmon resonance dependence of maximum wavelength λ_M of $|M_{EM}|^2$ for eight Ag NP dimers denoted by eight symbols (○, ○, □, Δ, ∇, ◇, +, and ×). (c) λ_p dependence of maximum value $|M_{EMmax}|^2$ of $|M_{EM}|^2$ for the eight Ag NP dimers. Inset is calculated $|M_{EMmax}|^2$ against λ_p assuming a Lorentzian line shape for plasmon resonance. The lines are fitted curves. Number of irradiated laser pulses is indicated by colors. Gray color indicates before NIR laser pulse irradiation, and faint blue, green, orange, red, brown, and moss green indicate after first, second, third, fourth, fifth, and sixth NIR laser pulse irradiation. (d) S dependence of $|M_{EMmax}|^2$ for the eight dimers. Filled circles in (b–d) are taken from (a1–a6). Reproduced with permission from Ref. [50]. Copyright © 2012, American Institute of Physics.* (See insert for color representation of the figure.)

the origin of $|M_{total}|^2$. The blue shifts in Figure 4.13b may be caused by small changes in the morphology of Ag dimers by the laser irradiations. We observed similar changes even without NIR pulse laser irradiations [35, 37]. Thus, the small morphological changes with or without laser pulse irradiation are common for SERS and SEF systems. Indeed, plasmon resonance spectra of Ag dimers are quite sensitive to the morphology of crevasses between two Ag NPs on the order of an angstrom [24, 26, 41].

To analyze the spectral fluctuation quantitatively, we examined the relationship between the maximum value $|M_{total\_max}|^2$ of $|M_{total}|^2$ and λ_p. We expect that $|M_{total\_max}|^2$ is increased by the blue shifts in λ_p, because λ_p gets close to λ_L (532 nm) by the blue shifts, as the inset of Figure 4.13c [8]. Figure 4.13c shows that for all the Ag dimers we selected, $|M_{total\_max}|^2$ increased with the blue shifts in λ_p for irradiating with 2–4 laser pulses. However, after reaching the maximum, $|M_{total\_max}|^2$ decreased after irradiating with afurther 3–10 laser pulses, as shown in Figure 4.13c. This decrease is contrary to our expectation. To reveal the reason for the inconsistency, we examined the relationship between the decrease and fluctuation in SERS and SEF spectra, because the fluctuation reduces the time-averaged intensity of SERS and SEF. Thus, we examined the relationship between S (<1.0) and $|M_{total\_max}|^2$. Figure 4.13d shows that S was inversely correlated with $|M_{total\_max}|^2$, clearly revealing the reason for the inconsistency.

4.2.6 Summary

In this section we quantitatively analyzed the intensity and spectral fluctuation in SERS and SEF by the EM mechanism. The evaluation of the ratios between SERS intensity and SEF intensity, that is, the enhancement factors of the decay rate of a molecule, indicated that changes in molecular positions within several angstroms induce intensity fluctuation. Experimental evaluation of changes in the EM enhancement factor spectra revealed that changes in plasmon resonance cause spectral fluctuation in SERS and SEF spectra. The decrease in EM enhancement factors induced by blue shifts in plasmon resonance is explained by the fluctuation in SERS and SEF intensity. The present results reveal that SERS and SEF fluctuation is quantitatively clarified by the EM mechanism [50]. The evaluation does not mean that the fluctuation in SERS and SEF can be totally explained by the EM mechanism excluding other mechanisms. Indeed, blinking in the system composed of Ag NPs and pyridine molecules showing a strong chemical effect may not be explained exclusively by the EM mechanism. The reason for selecting the current system dominated by EM enhancement is to evaluate quantitatively the blinking phenomena by only one mechanism. The present achievement could help us to evaluate the degree of involvement of other mechanisms in the blinking.

4.3 Conclusion

In this chapter we quantitatively analyzed EM enhancement and fluctuation in SERS. Ag NP aggregate-by-NP aggregate variations and refractive index dependence of SERS spectra are well-explained by the selective enhancement of SERS bands whose maxima are close to the plasmon resonance maxima. The selective enhancement is evidence of the second EM enhancement. Experimental observations of plasmon resonance, SERS, and the morphology of the Ag NP dimers were compared with FDTD calculations for the EM field induced by plasmon resonance and the individual morphology of Ag NP dimers. The experimental enhancement factors ~10^9 were consistent with calculations within an error factor of ~2. We fortify the importance of the EM mechanism with our results. In SERS fluctuation the evaluation of enhancement factors of the decay rate of a molecule indicated that changes in molecular positions within several angstroms induce intensity fluctuation. Evaluation of changes in EM enhancement factor spectra revealed that changes in plasmon resonance cause spectral fluctuation in SERS and SEF spectra. The present work reveals that SERS and SEF fluctuation is quantitatively clarified by the EM mechanism.

These evaluations do not mean that the SERS can be totally explained by the EM mechanism excluding other mechanisms. The reason for selecting the current system dominated by EM enhancement is to evaluate SERS phenomena quantitatively by only one mechanism. The present achievement could help us to evaluate the degree of involvement of other mechanisms in SERS.

Acknowledgments

The author thanks Prof. Y. Ozaki, Dr K. Yoshida, Dr Y. S. Yamamoto, Prof. H. Tamaru, Dr V. Biju, Prof. M. Ishikawa for their discussion and comments. The current study was partly supported by JSPS KAKENHI Grant-in-Aid for Scientific Research (C) Number 20510111 from the Ministry of Education, Culture, Sports, Science and Technology of Japan. This research is partly supported by the Japan Society for the Promotion of Science (JSPS) through its "Funding Program for World-Leading Innovative R&D on Science and Technology (FIRST Program)."

References

1. Pettinger, B. (1986) Light scattering by adsorbates at Ag particles: quantum-mechanical approach for energy transfer induced interfacial optical processes involving surface plasmons, multipoles, and electron-hole pairs. *The Journal of Chemical Physics*, **85**, 7442.
2. Dulkeith, E., Morteani, A.C., Niedereichholz, T. *et al.* (2002) Fluorescence quenching of dye molecules near gold nanoparticles: radiative and nonradiative effects. *Physical Review Letters*, **89**, 203002.
3. Johansson, P., Xu, H. and Käll, M. (2005) Surface-enhanced Raman scattering and fluorescence near metal surfaces. *Physical Review B*, **72**, 035427.
4. Anger, P., Bharadwaj, P. and Novotny, L. (2006) Enhancement and quenching of single-molecule fluorescence. *Physical Review Letters*, **96**, 113002.
5. Tam, F., Goodrich, G.P., Johnson, B.R. and Halas, N.J. (2006) Plasmonic enhancement of molecular fluorescence. *Nano Letters*, **7**, 496–501.
6. Yannopapas, V., Paspalakis, E. and Vitanov, N.V. (2009) Plasmon-induced enhancement of quantum interference near metallic nanostructures. *Physical Review Letters*, **103**, 063602.
7. Galloway, C.M., Etchegoin, P.G. and Le Ru, E.C. (2009) Ultrafast nonradiative decay rates on metallic surfaces by comparing surface-enhanced Raman and fluorescence signals of single molecules. *Physical Review Letters*, **103**, 063003.

8. Yoshida, K., Itoh, T., Biju, V. *et al.* (2009) Experimental evaluation of the twofold electromagnetic enhancement theory of surface-enhanced resonance Raman scattering. *Physical Review B*, **79**, 085419.
9. Vandenbem, C., Brayer, D., Froufe-Pérez, L.S. and Carminati, R. (2010) Controlling the quantum yield of a dipole emitter with coupled plasmonic modes. *Physical Review B*, **81**, 085444.
10. Sun, G. and Khurgin, J.B. (2012) Origin of giant difference between fluorescence, resonance, and nonresonance Raman scattering. *Physical Review A*, **85**, 063410.
11. Purcell, E.M. (1946) Spontaneous emission probabilities at radio frequencies. *Physics Reviews*, **69**, 681.
12. Moskovits, M. (1985) Surface-enhanced spectroscopy. *Reviews of Modern Physics*, **57**, 783.
13. Kneipp, K., Wang, Y., Kneipp, H. *et al.* (1997) Single molecule detection using surface-enhanced Raman scattering (sers). *Physical Review Letters*, **78**, 1667.
14. Nie, S. and Emory, S. (1997) Probing single molecules and single nanoparticles by surface-enhanced Raman scattering. *Science*, **275**, 1102.
15. Xu, H., Bjerneld, E., Käll, M. and Borjesson, L. (1999) Spectroscopy of single hemoglobin molecules by surface enhanced Raman scattering. *Physical Review Letters*, **83**, 4357.
16. Micheals, A., Nirmal, M. and Brus, L. (1999) Surface enhanced Raman spectroscopy of individual rhodamine 6G molecules on large Ag nanocrystals. *Journal of the American Chemical Society*, **121**, 9932.
17. (a) Dieringer, J.A., Lettan, R.B. II,, Scheidt, K.A. and Van Duyne, R.P. (2007) A frequency domain existence proof of single-molecule surface enhanced Raman spectroscopy. *Journal of the American Chemical Society*, **129**, 16249. (b) Cao, Y.C., Jin, R. and Mirkin, C.A. (2002) Nanoparticles with Raman spectroscopic fingerprints for DNA and RNA detection. *Science*, **297**, 1536.
18. Qian, X., Peng, X.H., Ansari, D.O. *et al.* (2007) In vivo tumor targeting and spectroscopic detection with surface-enhanced Raman nanoparticle tags. *Nature Biotechnology*, **26**, 83.
19. Anker, J.N., Hall, W.P., Lyandres, O. *et al.* (2008) Biosensing with plasmonic nanosensors. *Nature Materials*, **7**, 442.
20. Li, J.F., Huang, Y.F., Ding, Y. *et al.* (2010) Shell-isolated nanoparticle-enhanced Raman spectroscopy. *Nature*, **464**, 392.
21. Hicks, E.M., Zou, S.L., Schatz, G.C. *et al.* (2005) Controlling plasmon line shapes through diffractive coupling in linear arrays of cylindrical nanoparticles fabricated by electron beam lithography. *Nano Letters*, **5**, 1065.
22. Ueno, K., Juodkazis, S., Mizeikis, V. *et al.* (2008) Nanoparticle plasmon-assisted two-photon polymerization induced by incoherent excitation source. *Advanced Materials*, **20**, 26.
23. Inoue, M. and Ohtaka, K. (1983) Surface enhanced Raman scattering by metal spheres. I. Cluster effect. *Journal of the Physical Society of Japan*, **52**, 3853.
24. Xu, H., Aizpurua, J., Käll, M. and Apell, P. (2000) Electromagnetic contributions to single-molecule sensitivity in surface-enhanced Raman scattering. *Physical Review E*, **62**, 4318.

25. Kuwata, H., Tamaru, H., Esumi, K. and Miyano, K. (2003) Resonant light scattering from metal nanoparticles: practical analysis beyond rayleigh approximation. *Applied Physics Letters*, **83**, 4625.
26. Yoshida, K., Itoh, T., Tamaru, H. *et al.* (2010) Quantitative evaluation of electromagnetic enhancement in surface-enhanced resonance Raman scattering from plasmonic properties and morphologies of individual Ag nanostructures. *Physical Review B*, **81**, 115406.
27. Jeanmaire, D.L. and Van Duyne, R.P. (1977) Surface Raman electrochemistry. Part 1. Heterocyclic, aromatic and aliphatic amines adsorbed on the anodised silver electrode. *Journal of Electroanalytical Chemistry*, **84**, 1.
28. Grant Albrecht, M. and Alan Creighton, J. (1977) Anomalously intense Raman spectra of pyridine at a silver electrode. *Journal of the American Chemical Society*, **99**, 5215.
29. Wang, D. and Kerker, M. (1981) Enhanced Raman scattering by molecules adsorbed at the surface of colloidal spheroids. *Physical Review B*, **24**, 1777.
30. Otto, A., Mrozek, I., Grabhorn, H. and Akemann, W. (1992) Surface-enhanced Raman scattering. *Journal of Physics: Condensed Matter*, **4**, 1143.
31. Campion, A. and Kambhampati, P. (1998) Surface-enhanced Raman scattering. *Chemical Society Reviews*, **27**, 241.
32. Birke, R.L., Znamenskiy, V. and Lombardi, J.R. (2010) A charge-transfer surface enhanced Raman scattering model from time-dependent density functional theory calculations on a Ag_{10}-pyridine complex. *The Journal of Chemical Physics*, **132**, 214707.
33. Wu, D., Li, J., Ren, B. and Tian, Z. (2008) Electrochemical surface-enhanced Raman spectroscopy of nanostructures. *Chemical Society Reviews*, **37**, 1025.
34. Itoh, T., Hashimoto, K. and Ozaki, Y. (2003) Polarization dependences of surface plasmon bands and surface-enhanced Raman bands of single Ag nanoparticles. *Applied Physics Letters*, **83**, 2274.
35. Itoh, T., Hashimoto, K., Ikehata, A. and Ozaki, Y. (2003) Direct demonstration for changes in surface plasmon resonance induced by surface-enhanced Raman scattering quenching of dye molecules adsorbed on single Ag nanoparticles. *Applied Physics Letters*, **83**, 5557.
36. Itoh, T., Hashimoto, K., Ikehata, A. and Ozaki, Y. (2004) Changes in excitation profiles of surface-enhanced resonance Raman scattering induced by changes in surface plasmon resonance of single Ag nano-aggregates. *Chemical Physics Letters*, **389**, 225.
37. Itoh, T., Kikkawa, Y., Biju, V. *et al.* (2006) Scattering-active single Ag nanoaggregates. *The Journal of Physical Chemistry B*, **110**, 21536.
38. Itoh, T., Biju, V., Ishikawa, M. *et al.* (2006) Surface-enhanced resonance Raman scattering and background light emission coupled with plasmon of single Ag nanoaggregates. *The Journal of Chemical Physics*, **124**, 134708.
39. Itoh, T., Yoshida, K., Biju, V. *et al.* (2007) Second enhancement in surface-enhanced resonance Raman scattering revealed by an analysis of anti-Stokes and Stokes Raman spectra. *Physical Review B*, **76**, 085405.
40. Yoshida, K., Itoh, T., Biju, V. *et al.* (2009) Spectral shapes of surface-enhanced resonance Raman scattering sensitive to the refractive index of media around single Ag nanoaggregates. *Applied Physics Letters*, **95**, 263104.

41. Hao, E. and Schatz, G.C. (2004) Electromagnetic fields around silver nanoparticles and dimers. *The Journal of Chemical Physics*, **120**, 357.
42. Khan, I., Cunningham, D., Graham, D. *et al.* (2005) Identification and characterization of active and inactive species for surface-enhanced resonance Raman scattering. *The Journal of Physical Chemistry. B*, **109**, 3454.
43. Imura, K., Okamoto, H., Hossain, M. and Kitajima, M. (2006) Visualization of localized intense optical fields in single gold-nanoparticle assemblies and ultrasensitive Raman active sites. *Nano Letters*, **6**, 2173.
44. Le Ru, E. and Etchegoin, P. (2006) Rigorous justification of the |E|4 enhancement factor in surface enhanced Raman spectroscopy. *Chemical Physics Letters*, **423**, 63.
45. Garciá-Vidal, F.J. and Pendry, J.B. (1996) Collective theory for surface enhanced Raman scattering. *Physical Review Letters*, **77**, 1163.
46. Kreibig, U. and Vollmer, M. (1995) *Optical Properties of Metal Clusters*, Springer Series in Materials Science, Vol. **25**, Springer-Verlag, Berlin.
47. Itoh, T., Kikkawa, Y., Yoshida, K. *et al.* (2006) Correlated measurements of plasmon resonance Rayleigh scattering and surface-enhanced resonance Raman scattering using a dark-field microspectroscopic system. *Journal of Photochemistry and Photobiology, A: Chemistry*, **183**, 322.
48. Itoh, T., Asahi, T. and Masuhara, H. (2002) Direct demonstration of environment-sensitive surface plasmon resonance band in single gold nanoparticles. *Japanese Journal of Applied Physics*, **41**, L76.
49. Itoh, T., Uwada, T., Asahi, T. *et al.* (2007) Analysis of localized surface plasmon resonance by elastic light-scattering spectroscopy of individual Au nanoparticles for surface-enhanced Raman scattering. *Canadian Journal of Analytical Sciences and Spectroscopy*, **52**, 130.
50. Itoh, T., Iga, M., Tamaru, H. *et al.* (2012) Analysis of localized surface plasmon resonance by elastic light scattering spectroscopy of individual Au 266 nanoparticles for surface-enhanced Raman scattering. *The Journal of Chemical Physics*, **136**, 024703.
51. Ambjörnsson, T., Mukhopadhyay, G., Apell, S. and Käll, M. (2006) Resonant coupling between localized plasmons and anisotropic molecular coatings in ellipsoidal metal nanoparticles. *Physical Review B*, **73**, 085412.
52. Tally, C., Jackson, J., Oubre, C. *et al.* (2005) Surface-enhanced Raman scattering from individual Au nanoparticles and nanoparticle substrates. *Nano Letters*, **5**, 1569.
53. Taflove, A. and Hagness, S. (2005) *Computational Electrodynamics*, 3rd edn, Artech House, Norwood, OH.
54. Ong, T., Celli, V. and Marvin, A. (1994) General relation between surface impedance and surface curvature. *Journal of the Optical Society of America A*, **11**, 759.
55. See, for example Reather, H. (1988) *Surface Plasmons on Smooth and Rough Surfaces and on Gratings*, Chapter 2, Springer-Verlag.
56. Sen, T., Sadhu, S. and Patra, A. (2007) Surface energy transfer from rhodamine 6G to gold nanoparticles: a spectroscopic ruler. *Applied Physics Letters*, **91**, 043104.
57. Sen, T. and Patra, A. (2008) Resonance energy transfer from rhodamine 6G to gold nanoparticles by steady-state and time-resolved spectroscopy. *Journal of Physical Chemistry C*, **112**, 3216.

58. Saini, S., Srinivas, G. and Bagchi, B. (2009) Modeling the nanophase structural dynamics of phenylated sulfonated poly ether ether ketone ketone (Ph-SPEEKK) membranes as a function of hydration. *The Journal of Physical Chemistry B*, **113**, 1817.
59. Kitahama, Y., Tanaka, Y., Itoh, T. and Ozaki, Y. (2010) Power-law statistics in blinking SERS of thiacyanine adsorbed on a single silver nanoaggregate. *Physical Chemistry Chemical Physics*, **12**, 7457.
60. Etchegoin, P.G. and Le Ru, E.C. (2010) Resolving single molecules in surface-enhanced Raman scattering within the inhomogeneous broadening of Raman peaks. *Analytical Chemistry*, **82**, 2888.
61. Dickson, R.M., Cubitt, A.B., Tsien, R.Y. and Moerner, W.E. (1997) Toward nanometer-scale optical photolithography: utilizing the near-field of bowtie optical nanoantennas. *Nature*, **388**, 355.
62. Yu, J., Hu, D. and Barbara, P.F. (2000) Unmasking electronic energy transfer of conjugated polymers by suppression of O-2 quenching. *Science*, **289**, 1327.
63. Bosnick, K.A., Jiang, J. and Brus, L.E. (2002) Fluctuations and local symmetry in single-molecule rhodamine 6G Raman scattering on silver nanocrystal aggregates. *The Journal of Physical Chemistry B*, **106**, 8096.
64. Emory, S.R., Jensen, R.A., Wenda, T. *et al.* (2006) Re-examining the origins of spectral blinking in single-molecule and single-nanoparticle SERS. *Faraday Discussions*, **132**, 249.
65. Wang, Z. and Rothberg, L.J. (2005) Origins of blinking in single-molecule Raman spectroscopy. *The Journal of Physical Chemistry B*, **109**, 3387.
66. Weiss, A. and Haran, G. (2001) Time-dependent single-molecule Raman scattering as a probe of surface dynamics. *The Journal of Physical Chemistry B*, **105**, 12348.
67. Moskovits, M., Tay, L.L., Yang, J. and Haslett, T. (2002) SERS and the single molecule. *Topics in Applied Physics*, **82**, 215.
68. Le Ru, E.C., Etchegoin, P.G., Grand, J. *et al.* (2007) Mechanisms of spectral profile modification in surface-enhanced fluorescence. *Journal of Physical Chemistry C*, **111**, 16076.
69. Shim, S., Stuart, C.M. and Mathies, R.A. (2008) Resonance Raman cross-sections and vibronic analysis of rhodamine 6G from broadband stimulated Raman spectroscopy. *ChemPhysChem*, **9**, 697.
70. Larkin, I.A., Stockman, M.I., Achermann, M. and Klimov, V.I. (2004) Dipolar emitters at nanoscale proximity of metal surfaces: Giant enhancement of relaxation in microscopic theory. *Physical Review B*, **69**, 121403.

5

Single-Molecule Surface-Enhanced Raman Scattering as a Probe for Adsorption Dynamics on Metal Surfaces

Mai Takase, Fumika Nagasawa, Hideki Nabika, and Kei Murakoshi
Department of Chemistry, Hokkaido University, Japan

5.1 Introduction

The use of of surface-enhanced Raman scattering (SERS) for single-molecule detection may offer novel possibilities for the development of various plasmonic devices [1, 2] by helping researchers understand how a single molecule behaves when subjected to temperature, pressure, electrochemical potential, and illuminated light [3, 4]. The coherence of inter- and intra-molecule responses of a single molecule under applied perturbation is also important information to have [5]. Although the use of various spectroscopic techniques often requires a well-established selection rule, our present understanding of the single-molecule SERS phenomenon is quite limited, in spite of a number of intensive studies [6, 7]. An explanation for the significant Raman enhancement that is based only on the localization of the electromagnetic (EM) field induced by surface plasmons fails to account for the fact that there are SERS-active and SERS-inactive molecules. Thus, there is still controversy concerning the concept of single-molecule SERS [8, 9].

Experimental verification of the detection of a single molecule using SERS has been carried out using a wide variety of metal nanostructures. In addition to the pioneering works using unique tools in novel ways [10–12], *in situ* measurements of SERS using a system with a controlled surface coverage of bi-analytes have been carried out as a superior method

Frontiers of Surface-Enhanced Raman Scattering: Single Nanoparticles and Single Cells, First Edition.
Edited by Yukihiro Ozaki, Katrin Kneipp and Ricardo Aroca.

to confirm single-molecule detection [13–15]. In these systems, SERS spectra that can be assigned to just one of the two analytes are frequently observed, even when the bi-analyte is co-adsorbed as a monomolecular layer on the sample surface. These observations suggest that the SERS-active sites are comparable in size to single molecules. The observation frequency and histograms provide the information not only on the composition of analytes, but also on the adsorption states of molecules on the surface. The use of bi-analytes provides a simple method of observing single molecules using SERS.

Explaining the Raman enhancement as a result of EM field localization alone does not take into consideration the very small size of the SERS-active sites It has, however, been widely accepted that charge transfer (CT) resonance between a molecule and a metal surface, the so-called chemical effect, plays an important role in the molecular specificity of SERS [16–20]. It is possible that this CT mechanism can contribute to the Raman enhancement. The influence of the energy of the excitation light and the electrochemical potential of the Fermi level of the metals on the Raman enhancement is proof of the contribution of the CT effect. Experimental observations of specific Raman modes in SERS spectra also agree with the model of vibronic coupling with state-to-state resonance [21]. Thus, mode selectivity, the enhancement factor, and substrate metal dependence have been discussed based on the CT effect.

In spite of extensive theoretical and experimental SERS studies, there are still serious discrepancies between the observed results of single-molecule SERS spectra and those obtained by theory based on the expected selection rules of SERS [19, 21–24]. In addition to our limited understanding of the phenomenon, a poor understanding of molecular structures resulting from a variety of experimental conditions, such as the intensity of highly localized EM fields, the adsorption strength onto the surface, and induced chemical reactions, also makes it difficult to assign the vibrational modes observed by SERS. To further understand SERS, we need information about the target molecules simultaneously with that obtained from SERS measurements.

In this chapter, we first introduce the simultaneous measurement of SERS and the conductivity of a single-molecule junction consisting of an Au–Ag heterodimer. We can correlate the electronic and geometrical structures of a single-molecule junction with flowing current *in situ* by combining the information obtained from the SERS spectra and the conductivity of a single molecule. Based on precise assignment of the observed Raman modes, the adsorption dynamics of the bridging molecule at the gap is clarified [25]. The characteristic behavior of the SERS spectra of an adsorbed molecule at the gap of the Au–Ag heterodimer is then summarized. Time-dependent changes in the SERS spectra reflect the adsorption dynamics of the molecules by one of the metals on a single-molecule level [26]. Careful comparison of these spectra with those of Au or Ag homodimer systems proves the formation of a single-molecule bridge at the gap of the Au–Ag heterodimer. Based on these observations, the possibility of applying single-molecule SERS to the study of the adsorption dynamics on a single-molecule level is discussed.

5.2 Simultaneous Measurements of Conductance and SERS of a Single-Molecule Junction

Simultaneous measurements of conductance and SERS of a single-molecule junction were carried out using a mechanically controllable break junction (MCBJ) system combined

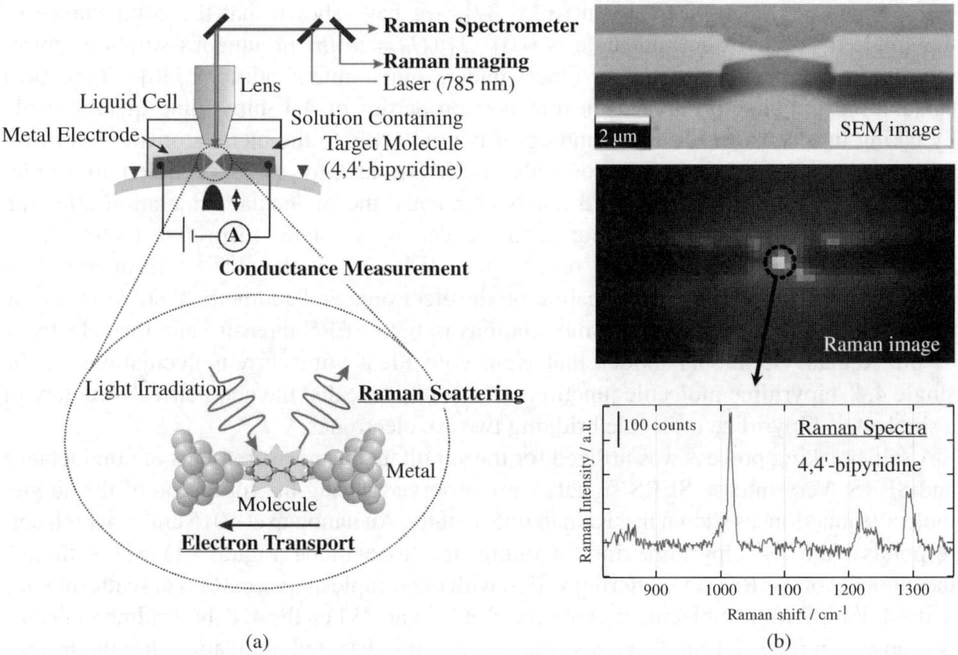

Figure 5.1 *(a) Schematic view of the MCBJ-SERS system. (b) SEM image of an Au nanobridge in the MCBJ sample and Raman imaging of the Au nanogap observed at 1015 cm^{-1}. (*See insert for color representation of the figure.*)*

with a near-infrared confocal Raman spectroscope (Figure 5.1). It has been shown that simultaneous measurements on conductance and SERS of a single-molecule junction provide valuable information on the structure of the junctions [27–29]. Tian *et al.* combined SERS and a MCBJ method to observe an enhancement of SERS intensity with decreasing gap size [27, 30]. Ward *et al.* reported simultaneous measurements of conductance and SERS of a molecular junction where they observed b_2 modes that are not observed in the bulk crystal [28]. Using "fishing mode" tip-enhanced Raman spectroscopy, Ren *et al.* observed the splitting of a Raman peak for a single-molecule junction, as well as a shift in peak energy induced by the bias voltage. The peak energy shift was attributed to the change in the binding energy between the molecule and the metal electrodes [29]. While simultaneous SERS and conductance measurements are a powerful technique, correlations between the geometrical structure and the SERS spectra of the *in situ* single-molecule junction during flowing current are not yet fully understood.

We adopted a 4,4′-bipyridine molecule as the unit of the single-molecule junction, because its vibrational characteristics and electron transport properties have been thoroughly investigated. We fully assigned the Raman modes of the molecule based on a comparison of SERS observations and Raman spectra of the bulk single crystal by referring to previously documented results [31]. The electron transport properties of this molecule have also been under intense investigation. Although previously documented conductance values range from 1.6×10^{-4} to 0.01 G_0 [29, 32–34], these wide ranges may reflect differences in the experimental conditions (e.g., type of solvent, concentrations, bias voltage, and sample treatment) [35–37]. We have shown that the conductance of the single 4,4′-bipyridine molecule is 0.01 G_0 ($G_0 = 2e^2/h$) in aqueous solution, corresponding to the value reported by others under comparable conditions [30]. Theoretical calculations explained the electron transport properties of 4,4′-bipyridine quantitatively by taking into consideration the binding of the molecule to the metal atoms to hybridize the highest occupied molecular orbitals (HOMOs) and lowest unoccupied molecular orbitals (LUMOs) with s, p, and d bands of metals, the molecular orientation affecting the resonance electron transmission, and the density of states respective to the Fermi level of the metal [30, 38]. The results provided confirmation of the formation of a single-molecule bridge and information on the electronic and geometrical structure of the junction. We observed significant modulations in both SERS intensity and the selectivity of the Raman vibrational modes that were coincident with current fluctuations in the single 4,4′-bipyridine molecule junction. The results revealed the geometrical structure of a single 4,4′-bipyridine molecule bridging two Au electrodes.

A self-breaking process was utilized for the simultaneous measurements of conductance and SERS. Very intense SERS signals were observed during the formation of the single-molecule junction, as shown in a Raman image of the Au nanogap at 1015 cm^{-1}, which corresponds to the 4,4′-bipyridine ring-breathing vibrational mode (Figure 5.1) [31]. Although the intensity of the Raman scattering varied with the samples, intense Raman scattering due to the 4,4′ bipyridine molecule was observed at the gap. When the 4,4′-bipyridine molecule was absent in the solution, Raman scattering was not detected, indicating that the Raman signal was solely due to the molecule in the gap.

The intensity of the scattering was very sensitive to the polarization direction of the 785 nm laser light. Polarization parallel to the Au nanogap resulted in the most intense scattering signal, indicating that the 4,4′-bipyridine signal was due largely to the localized surface plasmon excitation of the Au nanoelectrode [28]. Finite-difference time-domain (FDTD) calculations were carried out with PLANC-FDTD software to estimate the effect of the localization of the EM field as shown in Figure 5.2. For the structural model used in the calculation, the edges of Au electrodes were approximated as a rectangular parallelpipe assembly. The interelectrode gap distance was set to 1 nm. Figure 5.2b shows the local electric field (E) map around the gap region, which is normalized by the incident field (E_0). The wavelength used in the calculation was 785 nm, which is the excitation wavelength of the present experiment. When the incident electric field polarization was parallel to the electrode axis, a huge enhancement of up to 140 nm was observed at the gap, which corresponded to a Raman enhancement factor of $(140)^4 = 3.8 \times 10^8$. These results indicate that the localized EM field does contribute to the SERS detection of a single molecule.

A single 4,4′-bipyridine molecule junction shows three types of SERS spectra shown in Figure 5.3a. Possible structures of the bridging molecule based on the conductivity measurements are shown in Figure 5.3b. The observed SERS bands showing distinct wavenumbers were assigned according to the results of a density functional theory calculation at the B3LYP level with a LanL2DZ basis set for Au atoms and a 6-31G** set for other atoms, as well as by experimental observation using a polarized Raman measurement on a single crystal of 4,4′-bipyridine [31]. In the calculation, the vibration symmetry type is described by using D_2 symmetry of the 4,4′-bipyridine molecule.

The bands at 1025, 1062, 1217, and 1296 cm^{-1} (bottom of Figure 5.3a) observed just before or after the breaking of the junction can be assigned to totally symmetric a modes. The number of a modes and relative intensities obtained by calculations agree well with the

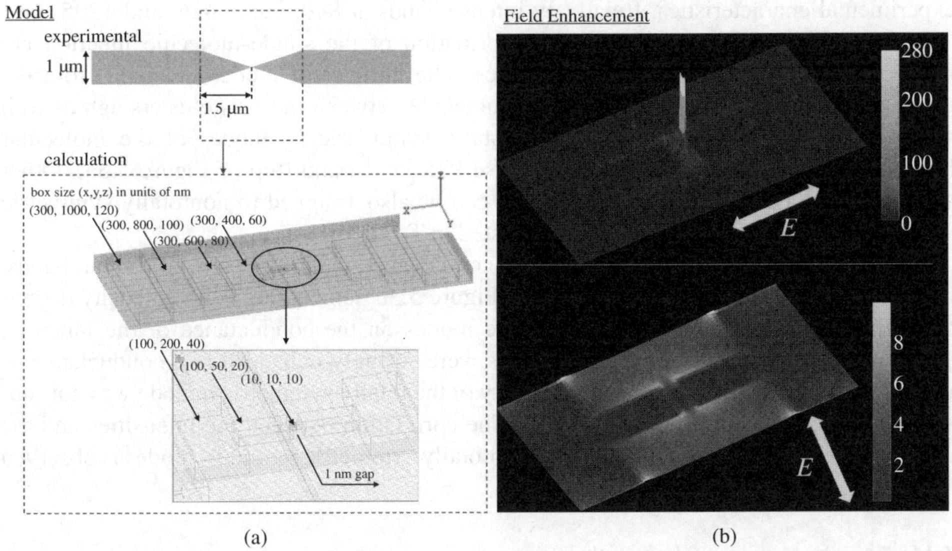

Figure 5.2 *Structural model of metal nanojunction (a) and calculated relative intensity of electromagnetic field under the light illumination with distinct polarization directions (b).* (See insert for color representation of the figure.)

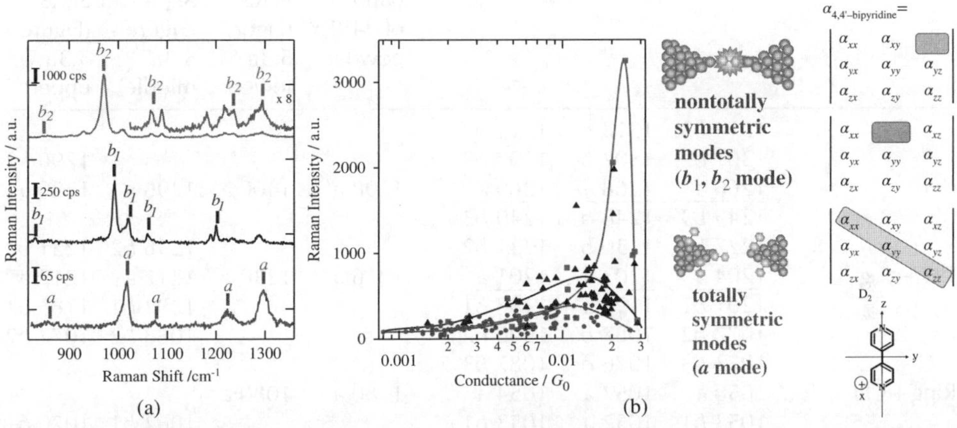

Figure 5.3 *(a) Three types of SERS spectra of a single 4,4′ bipyridine molecule junction. Assignment of the totally symmetric* a *mode , and the nontotally symmetric* b_1 *mode and* b_2 *mode are shown as dotted lines. (b) SERS intensity dependence of* a *(circles),* b_1 *(triangles), and* b_2 *(squares) modes on the conductance of the junction. Illustrations on the right-hand side depict possible structures of the junction and Raman tensors.* (See insert for color representation of the figure.)

experimental characteristics. Relatively intense bands at 840, 998, 1026, and 1205 cm^{-1} (middle of Figure 5.3a) observed upon formation of the single-molecule junction are assigned to nontotally symmetric b_1 modes. The number of modes obtained from the calculation for a bridging 4,4′-bipyridine molecule between the Au_2 clusters agrees well with the experimental observation. Nontotally symmetric b_2 modes of the molecular bridging could be assigned to a very intense 975 cm^{-1} band (top of Figure 5.3a). Other weak bands at 1072, 1221, and 1299 cm^{-1} can be also assigned to nontotally symmetric b_2 modes. Details of the assignments are listed in Table 5.1.

In order to correlate the single-molecule conductance with the observed Raman bands, more than 700 samples were investigated. Figure 5.3b depicts the SERS intensity dependence for totally and nontotally symmetric modes on the conductance of the junction. The nontotally symmetric b_1 and b_2 modes were strongly enhanced at a conductance of approximately 0.01 G_0, whereas the intensity of the totally symmetric a mode was not sensitive to the conductance of the junction. The correlation between the intensities and the conductance shows that the SERS of the nontotally symmetric b_1 and b_2 modes is observed

Table 5.1 *Peak wavenumbers of the Raman bands and their assignments determined by the calculation and the experiment*

Assigned	Calculation			Experimental			
	BPY	Au-BPY	Au-BPY-Au	Raman band of 44BPY powder	MCBJ SERS (Figure 5.3a lower)	MCBJ SERS (Figure 5.3a middle)	MCBJ SERS (Figure 5.3a upper)
	1314 *b3*	1312 *b*	1311 *b3*				
	1301 *b2*	1302 *b*	1302 *b2*				1299 *b2*
CH	1266 *a*	1268 *a*	1269 *a*	1298 *a*	1300 *a*	1296 *a*	1280 *a*
	1245 *b3*	1248 *b*	1249 *b3*				
	1227 *b2*	1230 *b*	1231 *b2*			1236 *b2*	1221 *b2*
CH + ring	1204 *a*	1203 *a*	1201 *a*	1216 *a*	1230 *a*	1217 *a*	1212 *a*
	1201 *b1*	1197 *a*	1197 *b1*			1202 *b1*	1186 *b1*
	1077 *b2*	1087 *b*	1090 *b2*			1088 *b2*	1072 *b2*
	1073 *b3*	1076 *b*	1087 *b3*				
Ring + CH	1058 *a*	1057 *a*	1054 *a*	1080 *a*	1080 *a*		
	1053 *b1*	1052 *a*	1053 *b1*			1062 *b1*	1070 *b1*
	1017 *b1*	1018 *a*	1018 *b1*			1025 *b1*	1015 *b1*
Ring	974 a	996 a	997 a	999 *a*	1020 *a*	(1015 *a*)	
	968 *b1*	971 *a*	991 *b1*			1002 *b1*	
	958 *a*	962 *a*	960 *a*				
	958 *b1*	957 *a*	959 *b1*				
	945 *b2*	949 b	951 *b2*				975 *b2*
	942 *b3*	947 *b*	948 *b3*				934 *b3*
CH deformation	856 *a*	855 *a*	852 *a*	854 *a*	860 *a*		906 *a*
	856 *b1*	851 *a*	851 *b1*			840 *b1*	862 *b1*
	831 *b2*	836 *b*	838 *b2*				848 *b2*

upon formation of a single-molecule junction. The b_2 mode becomes more dominant than the b_1 mode in a higher conductance regime.

Selective enhancement in nontotally symmetric modes of molecules has been explained by a CT resonance between the adsorbed molecule and the metal [16–20]. It has been recognized that the contribution of a highly localized intense EM field alone is not enough to explain the selective enhancement of a specific vibrational mode, especially nontotally symmetric modes described by relatively small non-diagonal terms in a polarizability tensor [21]. If a resonance process via optical excitation of intermediate electronic states between the molecule and the metal is involved in the Raman scattering process, additional polarization due to the induced dipole of an excitation changes the observable terms in a Raman tensor, which are different in non-resonant cases. The observed nontotally symmetric mode provides information on the orientation of the molecule in relation to an anisotropic EM field at the Au gap and on the induced dipole caused by the CT resonance. In the present system, resonance of the excitation of electrons at the Fermi level to LUMO with the B_3 symmetry can be considered under the 785 nm light illumination, due to an energy shift of the molecular orbitals caused by hybridization of the molecular and metal electronic states [30]. The resonance leads to the formation of an induced dipole moment perpendicular to the pyridine ring (x-direction shown in Figure 5.3). Thus the change in the selectivity from the b_1 mode to the b_2 mode in the present system should correspond to the change in the polarization of the EM field from perpendicular to the bipyridine ring (y-direction) to parallel to the long axis of the bipyridine molecule (z-direction), because the nondiagonal terms, α_{xy} and α_{xz}, in the polarizability tensor of the bipyridine molecule are responsible for b_1 and b_2 modes, respectively (see Figure 5.3). Thus, this change in observed mode reflects the orientation change of the Au molecule in the gap.

An intense b_2 mode is observed when a molecule vertically bridges the gap. Whereas the b_1 mode is enhanced when a molecule becomes tilted. Mode switching of b_1 from/to b_2 reflects changes in the orientation of molecules parallel or perpendicular to the direction of anisotropic EM fields (Table 5.2), respectively. The present system suggests that a molecule that vertically bridges the gap shows higher conductivity.

As for the intensity of the SERS bands, we need additional information on the resonance process. Selective observation of an intense 975 cm^{-1} band attributed to the b_2 mode may involve a change in the Raman polarizability, which originates from the electronic localization of the molecule in the CT process as well as from changes in the local environment

Table 5.2 *Possible excitation and enhanced vibrations of a single 4,4′-bipyridine molecule at the gap of electrodes*

CT transition	CT state	Dipole moment	Orientation (direction)	Enhanced vibration
Metal E_F to LUMO	B_3(LUMO)	μ_x(X)	x	$a(\alpha_{xx}\alpha_{yy}\alpha_{zz})$
			y	$b_1(\alpha_{xy})$
			z	$b_2(\alpha_{xz})$
HOMO to metal E_F	B_1(HOMO)	μ_z(Z)	x	$b_2(\alpha_{xz})$
			y	$b_3(\alpha_{yz})$
			z	$a(\alpha_{xx}\alpha_{yy}\alpha_{zz})$

of the molecules, the metal nanostructure. In addition, modulation of the selection rules of the optical absorption process at a highly localized EM field must also be addressed [39]. It has been demonstrated theoretically that forbidden transitions can be allowed by nanostructured EM fields [40]. A change in the selection rules of the optical excitation in a highly localized EM would contribute to a change in the SERS intensity of a specific mode. If the resonance of the excitation becomes significant, strong coupling of photons and excitons, which drastically alters optical absorption properties of the system, should be considered [41]. We should carefully consider these possibilities for further precise analysis of the selectivity of the mode and intensity of single-molecule SERS spectra.

5.3 SERS Observation Using Heterometallic Nanodimers at the Single-Molecule Level

In order to explore potential applications of the single-molecule SERS phenomenon, we may extend the system of metal nanostructure. Particular nanostructures, such as gap or dimer structures, dot arrays [14, 42], colloidal aggregations [10, 11, 13], and dot–surface pairs [23, 43], can be selected in order to produce an intense EM field. However, heterometallic nanostructures are another important application of the single-molecule SERS technique. Too investigate the molecular process that takes place at the surface of an atom and at a single-molecule level, information on the electronic structure of the absorbed sites of substrate metals provided by single-molecule SERS can be used. The use of heterometallic structures may lead to an understanding of the dynamic process that occurs during adsorption. Recently, the use of dot–surface pairs enabled the construction of heterodimeric structures by considering the optical properties of the respective metals. For example, Park *et al.* fabricated and compared two independent systems: Au dot–Au surface and Ag dot–Au surface pairs [43]. Their results demonstrated strong metal dependence in the spectral features of the SERS signal, which was explained by taking into account the difference in the Fermi energies of Au and Ag dots. This finding resulted from the fact that the SERS signal from single molecules contains information on their environment, for example, the nearby material, local field anisotropy, local temperature, local ionic strength, and so on [44–46]. However, this type of information from heterodimeric systems has not been discussed in detail and is quite limited, especially for dot array or colloidal aggregation systems, despite the fact that most SERS-based applications rely on them.

In this section, we show an example of SERS spectral features from a dot array system with well-defined heterodimeric structures. For comparison, homodimeric structures were also fabricated. In this method, the dimer array is obtained by successive deposition of a metal while changing the deposition angle. The second layer is deposited in the same fashion using a different metal. It is thus possible to fabricate heterodimers composed of different metals. The SERS signal from the Au–Ag heterodimer system was compared with those from the Au–Au and Ag–Ag homodimer systems.

Figure 5.4 presents typical AFM images of the Au–Ag hetero-dimer structure. The size of the perpendicular bisectors of the equilateral Au triangles was approximately 60 nm, and the diameter of the Ag dot was approximately 20 nm (Figure 5.4a). The long axes of the individual dimers were well aligned (Figure 5.4b). The optical properties of the metal dimers were optimized for the SERS measurements using the 785 nm (1.58 eV) light

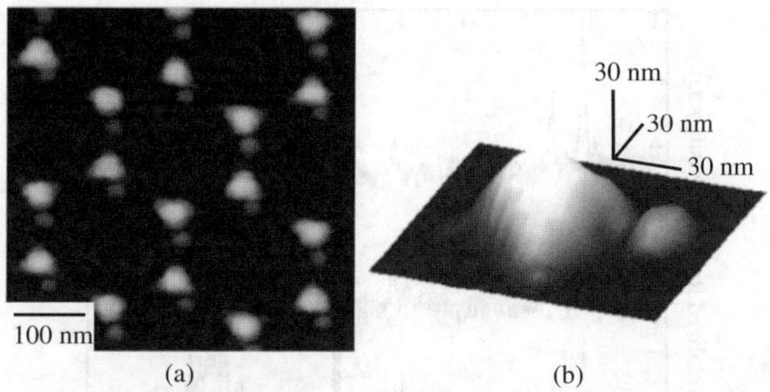

Figure 5.4 *(a) Two-dimensional and (b) three-dimensional AFM images of Au–Ag heterodimer*

excitation. The polarized excitation along the long axes of the dimers led to the appearance of a peak at approximately 785 nm. When the polarization of the illuminated light was aligned with the short axes of the metal dimers, the peak at approximately 785 nm disappeared. This apparent dependence of the polarization direction on structural anisotropy of the dimer suggests successful excitation of the localized surface plasmon localized at the gap of the metal dimer.

Relatively intense SERS was observed from the substrates with metal dimers immersed in aqueous solution containing either 2,2′-bipyridine or 4,4′-bipyridine. Figure 5.5a shows the SERS spectra measured using the Au- and Ag-homodimers, and the Au–Ag heterodimers, immersed in an aqueous solution containing 100 μM 2,2′-bipyridine. The observed SERS bands are attributable to the vibrational modes of the ring breathing at approximately 1016 cm^{-1}, the in-plane ring deformation coupled with the C-H bending at 1060 cm^{-1}, and the inter-ring stretching at 1300 cm^{-1} [31, 47, 48]. The expanded plot of the SERS spectra at approximately 980–1040 cm^{-1} shows that the 1012–1016 cm^{-1} band of the ring-breathing mode observed for the Au-homodimer (Figure 5.5a(i)) shifts to 1008–1012 cm^{-1} in the system of the Ag-homodimer (Figure 5.5a(ii)). This difference in the wavenumber of the ring-breathing mode may reflect the change in the electron density of the ring due to the difference in the adsorption strength of 2,2′-bipyridine molecules on the surfaces of the Au- and the Ag-homodimers. At the Au–Ag hetero dimer (Figure 5.5a(iii)), the vibrational band of the ring-breathing mode at approximately 1012 cm^{-1} seems to be superimposed on those observed at the Au- and Ag-homodimers.

Quite similar behavior in terms of the change in the wavenumber of the ring-breathing mode at the metal dimers was also observed in an aqueous solution of 4,4′-bipyridine. Figure 5.5b shows SERS spectra of 4,4′-bipyridine with the metal dimers. The vibration modes are assigned to the ring breathing at approximately 1020 cm^{-1}, the in-plane ring deformation coupled with the C-H bending at 1080 cm^{-1}, the in-plane C-H bending at 1240 cm^{-1} and the inter-ring stretching at 1300 cm^{-1} [48, 49]. The ring-breathing mode observed at 1018 cm^{-1} for the Au-homodimer (Figure 5.5b(i)) was at a higher wavenumber than that at 1015 cm^{-1} for the Ag-homodimer (Figure 5.5b(ii)). The band at the Au–Ag heterodimer (Figure 5.5b(iii)) was slightly broadened at approximately 1018 cm^{-1} comparable to results for the Au-homodimer. The SERS spectra observed in the present system may reflect the information on the adsorption site at a single-molecule level. We have estimated

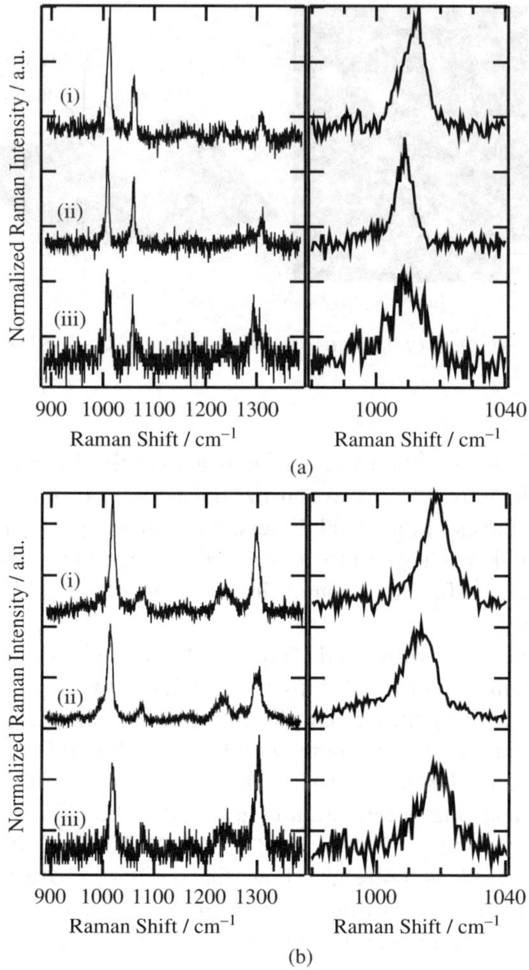

Figure 5.5 *SERS spectra of (a) 2,2′-bipyridine and (b) 4,4′-bipyridine. The substrates were (i) Au-homodimer, (ii) Ag-homodimer, and (iii) Au–Ag heterodimer. The spectra were acquired in 100 μM bipyridine solution*

the size of SERS hotspot using the bi-analyte technique under the condition in which the surface coverage of the adsorbed molecules is less than one monolayer thick on the metal surfaces [14]. In the present concentration, the size of active SERS site is comparable with that of a single molecule, which is significantly smaller than a domain containing a few tens of molecules.

Sequential observation of the SERS spectrum using the Au–Ag heterodimer provides information on the dynamics of the absorption. The wavenumber of the ring-breathing mode at the Au–Ag heterodimer showed characteristic time-dependent change. In Figure 5.6a, SERS spectra sequentially recorded every 1 s are shown using the Au–Ag heterodimer in 100 μM 2,2′-bipyridine solution. Although the relative intensities of the respective bands are somewhat stable, the spectral features of the ring-breathing mode indicate time-dependent changes. Broad bands at $t = 7$ and 8 s become sharp at $t = 9$ s with

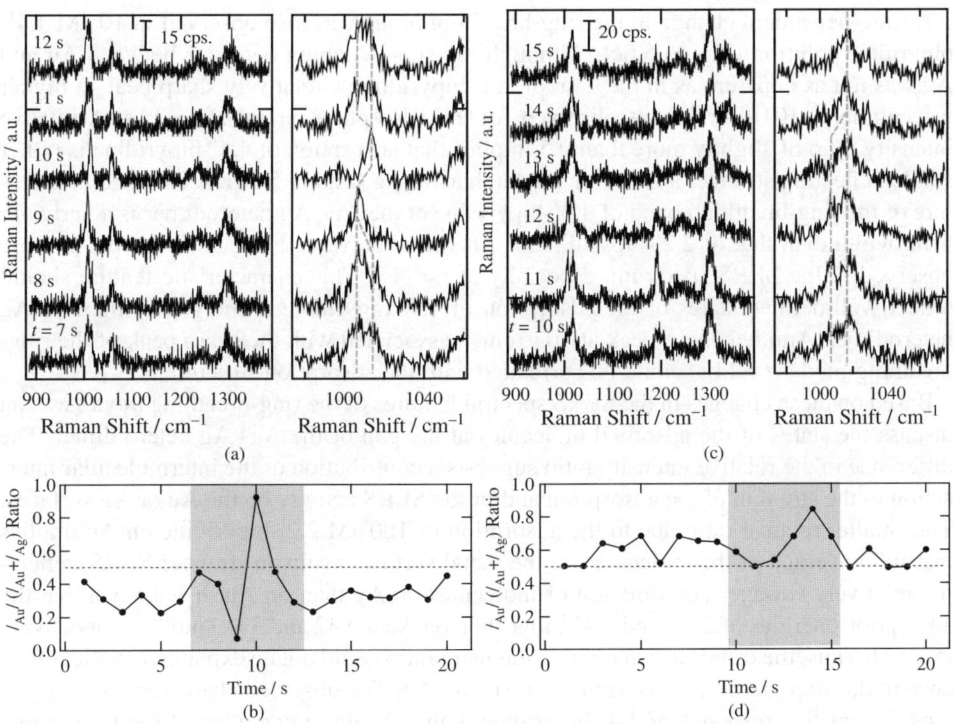

Figure 5.6 *(a) Successive SERS spectra of 2,2′-bipyridine on Au–Ag heterodimer substrate and (b) time course of SERS intensity of the ring-breathing mode. (c) Successive SERS spectra of 4,4′-bipyridine on Au–Ag heterodimer substrate and (d) time course of SERS intensity of the ring-breathing mode. The gray regions in (b) and (d) indicate the time for SERS measurements shown in (a) and (c)*

a peak at 1010 cm^{-1}. The peak shifts to 1016 cm^{-1} while remaining sharp at $t = 10$ s. The peak broadens again after $t = 11$ s. These observed time-dependent changes in the band features reflect the adsorption site at the gap of the Au–Ag heterodimer at the single-molecule level. The intensity of the peak was deconvoluted to estimate the possibility of adsorption on the Au and Ag surface based on the difference in the wavenumbers of the ring-breathing mode of Au and Ag. Figure 5.6b shows the time course of the relative ratio of SERS intensities from 2,2′-bipyridine molecules on Au as compared to those on the Ag surface. A value of 0.5 corresponds to the possibility of equal adsorption on the Au and Ag surfaces in this plot. Changes in the spectral feature at $t = 9$ and 10 s are successful observations of the adsorption site of 2,2′-bipyridine switching from the Ag to the Au surface at the gap of the Au–Ag heterodimer. The significant change in the values of the ratio from 0 to 1 suggests adsorbed-site switching from Au to Ag surfaces. Values between 0.2 and 0.4 before and after the site switching indicate that the possibility of adsorbing on Ag is slightly higher than that on Au. These results also suggest that the timescale of the site switching should be on the order of 1 s. However, this adsorption time was limited when the single molecule was at the same metal dot during 1 s of Raman exposure time. In other words, if the molecular adsorption time is shorter than 1 s when molecular switching is on the same metal structure, we will observe a spectral change in Figure 5.5.

A time-dependent change in the ring-breathing mode was also observed in 10 μM 4,4′-bipyridine solution (Figure 5.6c). Although the site-switching behavior between Au and Ag was not as apparent as in the case of 2,2′-bipyridine, a relatively sharp peak at higher wavenumber (1018 cm^{-1}) was observed for the adsorption on Au at $t = 14$ s. A relative intensity ratio of slightly more than 0.5 implies that adsorption of 4,4′-bipyridine is rather stable, reflecting stronger adsorption on Au than on Ag (Figure 5.6d). A characteristic feature of the ring-breathing band of 4,4′-bipyridine at the Au–Ag heterodimer is a narrower band width than that of 2,2′-bipyridine. Even when the value is 0.5, two sharp peaks are observed in the SERS spectrum, especially at $t = 11$ s. This characteristic feature should reflect well-defined states of the adsorption of 4,4′-bipyridine at the gap of the Au–Ag heterodimer. Another sharp peak at 1020 cm^{-1} associated with the sharp peak of the ring-breathing mode ($t = 11$ s) could be correlated with the adsorption structure.

Based on these changes in the SERS spectral features of the ring-breathing mode, we can discuss the states of the adsorbed molecules at the gap of the Au–Ag hetero dimer. The difference in the relative intensity ratio suggests a contribution of the intermolecular interaction to the strength of the adsorption and/or the SERS activity on the Au or Ag surfaces. The smaller relative ratio due to the adsorption of 100 μM 2,2′-bipyridine on Ag implies that full coverage of the molecules on the metal surface results in stronger SERS, reflecting relatively stronger coordination of molecules on Ag than on Au. It is known that the adsorption energies of 2,2′- and 4,4′-bipyridine on Au are 42 and 35 kJ mol^{-1}, respectively [50, 51]. Thus, the difference in relative intensity ratio could not be explained by the difference in the strength of the adsorption on Au and Ag. The observed characteristics suggest a more sensitive response of 4,4′-bipyridine than 2,2′-bipyridine. One of the factors that may cause the insensitive response of the relative intensity ratio of 4,4′-bipyridine on the solution concentration is the formation of a bridging structure at the gap between Au and Ag by the 4,4′-bipyridine molecule. Both ends of a 4,4′-bipyridine molecule adsorbed on Au and Ag may contribute to the observation of characteristic SERS band features of the ring-breathing mode on Au and Ag. The bridging of the molecule at the gap should lead to restriction of the motion on the surface. The doubled sharp peaks shown in Figure 5.6c at $t = 11$ s could be considered a characteristic feature of the bridging molecule. Another sharp peak simultaneously observed at 1018 cm^{-1} is attributable to the ring-breathing mode of 4,4′-bipyridine on the Au surface and is always observed. On the other hand, the peak at 1015 cm^{-1} for the ring-breathing mode at the Ag surface is not always observed. Detailed analysis of the peak intensities provides information on the number of molecules on the Ag and Au surfaces.

It is noteworthy that a new signal at 1200 cm^{-1} was observed. The possibility that this signal indicates a bridging structure has been suggested [52–55]. Generally, vibrational bands at 1200 cm^{-1} are not observed even with SERS measurements using Au- and Ag- homodimers, as shown in Figure 5.5. The relatively sharp and intense feature of the 1200 cm^{-1} band in the spectrum at $t = 11$ s (Figure 5.6) reflects a very characteristic structure of the adsorbed molecule. In the system of the Au–Ag heterodimer, a bridging structure between Ag and Au at the gap of the dimer is the most probable structure. There is no other possible structure showing double peaks at 1018 and 1020 cm^{-1} at the peak-switching behavior. It is expected, therefore, that this is a very characteristic feature of the SERS spectrum of molecules bridging the Au–Ag heterodimer.

5.4 Conclusion

Adsorption dynamics is discussed based on single-molecule SERS measurement. A detailed analysis of Raman modes to determine the orientation of molecules and the characteristics of the adsorption sites was possible after the observation of a single molecule was confirmed. Examples in this chapter prove that simultaneous measurement of single-molecule conductivity and control of the coverage of target molecules on metal surfaces are useful for the confirmation. The present study shows that a single molecule at the metal gap produces Raman spectra containing detailed information on the electronic and geometrical structures of the molecule. In these systems, nonfluorescent molecules such as bipyridine can be used in single-molecule detection, if the electronic states of the molecule and metal respond to optical resonance with the localized plasmons.

It should be noted that a *true* selection rule of SERS at the single-molecule level has not been clarified yet, even after the studies shown here. Although the numbers and the wavenumbers of observed Raman modes are explained based on previously documented theories, we cannot discuss the origin of the difference in the relative intensity of the modes. Selective enhancement of the b_2 mode with more than one order higher intensity than the a mode at a single-molecule junction requires further detailed studies for full understanding of the origins of the changes in the Raman intensity. We should consider additional resonance as with the CT effect. Localized plasmons induce a very intense EM field with a steep gradient on a nanoscale. It has been pointed out that steep field gradients comparable to the size of molecules result in a breakdown in the standard selection rules for *electronic excitations* [39, 56]. Such a breakdown has been theoretically demonstrated for single-molecule excitations in nanogaps [57, 58] and recently simulated for excitons in a metal-tipped nanorod [40]. Although changes in electronic excitation selection rules have not yet been proved experimentally, even for plasmon-active metal nanostructures [11], their contribution to the SERS process should be considered. If one can obtain revised selection rules of SERS, true single-molecule SERS may open up novel applications based on a variety of unconventional molecular photofunctions.

Acknowledgments

This work was partially supported by a Grant-in-Aid for scientific research 18750001, from the Ministry of Education, Science and Culture, Japan. In particular, support from the Priority Area "Strong Photon-Molecule Coupling Fields (No. 470)" is acknowledged.

References

1. Mahmoud, M.A., Chamanzar, M., Adibi, A. and El-Sayed, M.A. (2012) Effect of the dielectric constant of the surrounding medium and the substrate on the surface plasmon resonance spectrum and sensitivity factors of highly symmetric systems: silver nanocubes. *Journal of the American Chemical Society*, **134**, 6434–6442.
2. Halas, N.J., Lal, S., Chang, W.S. *et al.* (2011) Plasmons in strongly coupled metallic nanostructures. *Chemical Reviews*, **111**, 3913–3961.

3. Nabika, H., Takase, M., Nagasawa, F. and Murakoshi, K. (2010) Toward plasmon-induced photoexcitation of molecules. *Journal of Physical Chemistry Letters*, **1**, 2470–2487.
4. Takimoto, B., Nabika, H. and Murakoshi, K. (2010) Force applied to a single molecule at a single nanogate molecule filter. *Nanoscale*, **2**, 2591–2595.
5. Chizhik, A.M., Jager, R., Chizhik, A.I. *et al.* (2011) Optical imaging of excited-state tautomerization in single molecules. *Physical Chemistry Chemical Physics*, **13**, 1722–1733.
6. Brike, R.L. and Lombardi, J.R. (1988) in *Spectroelectrochemistry* (ed R.J. Gale), Plenum Press, New York, pp. 263–348.
7. Ru, L.E.C. and Etchegoin, P.G. (2008) *Principles of Surface-Enhanced Raman Spectroscopy and Related Plasmonic Effects*, Elsevier Science, Amsterdam.
8. Otto, A. (2002) What is observed in single molecule SERS, and why? *Journal of Raman Spectroscopy*, **33**, 593–598.
9. Etchegoin, P.G. and Ru, E.C.L. (2008) A perspective on single molecule SERS: current status and future challenges. *Physical Chemistry Chemical Physics*, **10**, 6079–6089.
10. Nie, S. and Emory, S.R. (1997) Probing single molecules and single nanoparticles by surface-enhanced Raman scattering. *Science*, **275**, 1102–1106.
11. Kneipp, K., Wang, Y., Kneipp, H. *et al.* (1997) Single molecule detection using surface-enhanced Raman scattering. *Physical Review Letters*, **78**, 1667–1670.
12. Kneipp, K., Kneipp, H., Itzkan, I. *et al.* (1999) Ultrasensitive chemical analysis by Raman spectroscopy. *Chemical Reviews*, **99**, 2957–2976.
13. Ru, L.E.C., Meyer, M. and Etchegoin, P.G. (2006) Proof of single-molecule sensitivity in surface enhanced Raman scattering (SERS) by means of a Two-analyte technique. *Journal of Physical Chemistry, B*, **110**, 1944–1948.
14. Sawai, Y., Takimoto, B., Nabika, H. *et al.* (2007) Observation of a small number of molecules at a metal nanogap arrayed on a solid surface using surface-enhanced Raman scattering. *Journal of the American Chemical Society*, **129**, 1658–1662.
15. Dieringer, J.A., Wustholz, K.L., Masiello, D.J. *et al.* (2009) Surface-enhanced Raman excitation spectroscopy of a single Rhodamine 6G molecule. *Journal of the American Chemical Society*, **131**, 849–854.
16. Moskovits, M. (1985) Surface-enhanced spectroscopy. *Reviews of Modern Physics*, **57**, 783–826.
17. Lombardi, J.R. and Birke, R.L. (2008) A unified approach to surface-enhanced Raman spectroscopy. *Journal of Physical Chemistry C*, **112**, 5605–5617.
18. Lombardi, J.R. and Birke, R.L. (2009) A unified view of surface-enhanced Raman scattering. *Accounts of Chemical Research*, **42**, 734–742.
19. Lombardi, J.R., Birke, R.L. and Haran, G. (2011) Single molecule SERS spectral blinking and vibronic coupling. *Journal of Physical Chemistry C*, **115**, 4540–4545.
20. Lombardi, J.R. and Birke, R.L. (2012) The theory of surface-enhanced Raman scattering. *Journal of Chemical Physics*, **136**.
21. Osawa, M., Matsuda, N., Yoshii, K. and Uchida, I. (1994) Charge-transfer resonance Raman process in surface-enhanced Raman-scattering from p-aminothiophenol adsorbed on silver – Herzberg-Teller contribution. *Journal of Physical Chemistry*, **98**, 12702–12707.

22. Moskovits, M. (1982) Surface selection rules. *Journal of Chemical Physics*, **77**, 4408–4416.
23. Ikeda, K., Suzuki, S. and Uosaki, K. (2011) Crystal face dependent chemical effects in surface-enhanced Raman scattering at atomically defined gold facets. *Nano Letters*, **11**, 1716–1722.
24. Huang, Y.F., Zhu, H.P., Liu, G.K. *et al.* (2010) When the signal is not from the original molecule to be detected: chemical transformation of para-aminothiophenol on Ag during the SERS measurement. *Journal of the American Chemical Society*, **132**, 9244–9246.
25. Konishi, T., Kiguchi, M., Takase, M. *et al.* (2013) Single molecule dynamics at a mechanically controllable break junction in solution at room temperature. *Journal of the American Chemical Society*, **135**,1009–1014.
26. Takase, M., Sawai, Y., Nabika, H. and Murakoshi, K. (2011) Detection of adsorption sites at the gap of a hetero-metal nano-dimer at the single molecule level. *Journal of Photochemistry and Photobiology A*, **221**,169–174.
27. Tian, J.H., Liu, B., Li, X. *et al.* (2006) Study of molecular junctions with a combined surface-enhanced Raman and mechanically controllable break junction method. *Journal of the American Chemical Society*, **128**, 14748–14749.
28. Ward, D.R., Halas, N.J., Ciszek, J.W. *et al.* (2008) Simultaneous measurements of electronic conduction and Raman response in molecular junctions. *Nano Letters*, **8**, 919–924.
29. Liu, Z., Ding, S.Y., Chen, Z.B. *et al.* (2011) Revealing the molecular structure of single-molecule junctions in different conductance states by fishing-mode tip-enhanced Raman spectroscopy. *Nature Communications*, **2**, 305-1–305-6.
30. Li, Z.Y. and Kosov, D.S. (2006) Dithiocarbamate anchoring in molecular wire junctions: a first principles study. *Journal of Physical Chemistry B*, **110**,(9893–9898.
31. Nagasawa, F., Takase, M., Nabika, H. and Murakoshi, K. (2011) Polarization characteristics of surface-enhanced Raman scattering from a small number of molecules at the gap of a metal nano-dimer. *Chemical Communications*, **47**, 4514–4516.
32. Xu, B. and Tao, N.J. (2003) Measurement of single-molecule resistance by repeated formation of molecular junctions. *Science*, **301**, 1221–1223.
33. Quek, S.Y., Kamenetska, M., Steigerwald, M.L. *et al.* (2009) Mechanically controlled binary conductance switching of a single-molecule junction. *Nature Nanotechnology*, **4**, 230–234.
34. Horiguchi, K., Kurokawa, S. and Sakai, A. (2009) Conductance and I-V characteristics of Au/BPY/Au single molecule junctions. *Journal of Chemical Physics*, **131**.
35. Cuevas, J.C. and Scheer, E. (2010) *Molecular Electronics: An Introduction to Theory and Experiment*, World Scientific, Singapore, Hackensack, NJ.
36. Kiguchi, M., Takahashi, T., Kanehara, M. *et al.* (2009) Effect of end group position on the formation of a single porphyrin molecular junction. *Journal of Physical Chemistry C*, **113**, 9014–9017.
37. Kiguchi, M., Takahashi, T., Takahashi, Y. *et al.* (2011) Electron transport through single molecules comprising aromatic stacks enclosed in self-assembled cages. *Angewandte Chemie International Edition*, **50**, 5707–5710.

38. Kiguchi, M. and Murakoshi, K. (2009) Highly conductive single molecular junctions by direct binding of pi-conjugated molecule to metal electrodes. *Thin Solid Films*, **518**, 466–469.
39. Burstein, E., Lundqvist, S. and Mills, D.L. (1982) in *Surface Enhanced Raman Scattering* (eds R.K. Chang and T.E. Furtak), Plenum Press, New York, pp. 67–89.
40. Jain, P.K., Ghosh, D., Baer, R. *et al.* (2012) Near-field manipulation of spectroscopic selection rules on the nanoscale. *Proceedings of the National Academy of Sciences of the United States of America*, **109**, 8016–8019.
41. Schwartz, T., Hutchison, J.A., Genet, C. and Ebbesen, T.W. (2011) Reversible switching of ultrastrong light-molecule coupling. *Physical Review Letters*, **106**, 196405-1–196405-4.
42. Ueno, K., Juodkazis, S., Shibuya, T. *et al.* (2008) Nanoparticle Plasmon-assisted two-photon polymerization induced by incoherent excitation source. *Journal of the American Chemical Society*, **130**, 6928–6929.
43. Park, W.H. and Kim, Z.H. (2010) Charge transfer enhancement in the SERS of a single molecule. *Nano Letters*, **10**, 4040–4048.
44. Lim, B., Kobayashi, H., Yu, T. *et al.* (2010) Synthesis of Pd-Au bimetallic nanocrystals via controlled overgrowth. *Journal of the American Chemical Society*, **132**, 2506–2507.
45. Kalsin, A.M., Pinchuk, A.O., Smoukov, S.K. *et al.* (2006) Electrostatic aggregation and formation of core-shell suprastructures in binary mixtures of charged metal nanoparticles. *Nano Letters*, **6**, 1896–1903.
46. Takase, M., Nabika, H., Hoshina, S. *et al.* (2013) Local thermal elevation probing of metal nanostructure during laser illumination utilizing surface-enhanced Raman scattering from a single-walled carbon nanotube. *Physical Chemistry Chemical Physics*, **15**, 4270–4274.
47. Joo, S.W. (2006) Adsorption of bipyridine compounds on gold nanoparticle surfaces investigated by UV–vis absorbance spectroscopy and surface enhanced Raman scattering. *Spectroscopy Letters*, **39**, 85–96.
48. Joo, S.W. (2004) Surface-enhanced Raman scattering of 4,4′-bipyridine on gold nanoparticle surfaces. *Vibrational Spectroscopy*, **34**, 269–272.
49. Suzuki, M., Niidome, Y. and Yamada, S. (2006) Adsorption characteristics of 4,4′-bipyridine molecules on gold nanosphere films studied by surface-enhanced Raman scattering. *Thin Solid Films*, **496**, 740–747.
50. Uosaki, K. and Hill, H.A.O. (1981) Adsorption behavior of 4,4′-bipyridyl at the gold/water interface and its role in the electron transfer reaction between cytochrome c and gold electrode. *Journal of Electroanalytical Chemistry and Interfacial Electrochemistry*, **122**, 321–326.
51. Yang, D., Bizzotto, D., Lipkowski, J. *et al.* (1994) Electrochemical and second harmonic generation studies of 2,2′-bipyridine adsorption at the Au(111) electrode surface. *Journal of Physical Chemistry*, **98**, 7083–7089.
52. Eddowes, M.J. and Hill, H.A.O. (1977) Novel method for investigation of electrochemistry of metalloproteins - cytochrome-C. *Journal of the Chemical Society, Chemical Communications*, 771–772.

53. Albery, W.J., Eddowes, M.J., Hill, H.A.O. and Hillman, A.R. (1981) Mechanism of the reduction and oxidation reaction of cytochrome c at a modified gold electrode. *Journal of the American Chemical Society*, **103**, 3904–3910.
54. Rodriguez, J.A. (1990) The adsorption of nitric oxide, pyridine, and sulfur dioxide on silver: a quantum-chemical study. *Surface Science*, **226**, 101–118.
55. Mayer, D., Dretschkow, T., Ataka, K. and Wandlowski, T. (2002) Structural transitions in 4,4′-bipyridine adlayers on Au(111) – an electrochemical and in-situ STM-study. *Journal of Electroanalytical Chemistry*, **524**, 20–35.
56. Turro, N.J. (1991) *Modern Molecular Photochemistry*, University Science Books, Herndon, VA.
57. Iida, T. and Ishihara, H. (2009) Unconventional control of excited states of a dimer molecule by a localized light field between metal nanostructures. *Physica Status Solidi A: Applications and Materials Science*, **206**, 980–984.
58. Iida, T., Aiba, Y. and Ishihara, H. (2011) Anomalous optical selection rule of an organic molecule controlled by extremely localized light field. *Applied Physics Letters*, **98**, 053108-1–053108-3.

6

Analysis of Blinking SERS by a Power Law with an Exponential Function

Yasutaka Kitahama and Yukihiro Ozaki
Department of Chemistry, School of Science and Technology, Kwansei Gakuin University, Japan

6.1 Introduction

Surface-enhanced Raman scattering (SERS) is sensitive enough to measure the Raman spectrum of a single molecule adsorbed on a noble metal nanoaggregate [1–46]. On such nanoaggregates, the incident light is resonant with plasmon because of the dipolar oscillation of the conduction band electrons (localized surface plasmon resonance, LSPR). As a result, the local electromagnetic (EM) field can be very much enhanced at a junction in the nanoaggregate. The surface-plasmon-enhanced EM field may remarkably enhance the effective Raman scattering cross section ($10^8 - 10^{14}$-fold) of molecules adsorbed on a noble metal nanoaggregate [1–6]. Thus, a very small amount of analyte can be identified by SERS by the corresponding vibrational fingerprints. Furthermore, SERS provides detailed information regarding molecular structure and orientation on the metal surface. In the case of a single molecule on an Ag nanoaggregate, blinking of SERS could be observed. The peaks of the blinking SERS signal disappear, fluctuate, and slightly shift from the corresponding Raman peaks. They cause deterioration of the reproducibility and signal-to-noise ratio of the spectra [6–33], making it difficult to identify the analyte and investigate its molecular structure. The blinking SERS is probably induced by a single molecule that enters and leaves an enhanced EM field at a junction that measures just a few nanometers (a "hot site") [9–12]. Thus, the blinking is considered as evidence for single molecule detection, like digitalized SERS intensity and the bi-analyte technique, in which unique vibrational

Frontiers of Surface-Enhanced Raman Scattering: Single Nanoparticles and Single Cells, First Edition.
Edited by Yukihiro Ozaki, Katrin Kneipp and Ricardo Aroca.

signatures are observed from mixed analytes [7, 41–46]. However, there is a possibility that the blinking SERS is due to fluctuation of the enhanced EM field at the junction of an Ag nanoaggregate [15, 35]. Moreover, SERS of amorphous carbon, which is formed by photo- and thermo-degradation of the adsorbed molecules, shows similar behavior [47–52]. Thus, the mechanism of blinking SERS is still a contentious issue.

Although the blinking SERS and single molecule SERS have been studied considerably [6–52], there have been only a few quantitative investigations on the dependence of blinking SERS [14–20, 32, 38]. One of them is an investigation using an autocorrelation

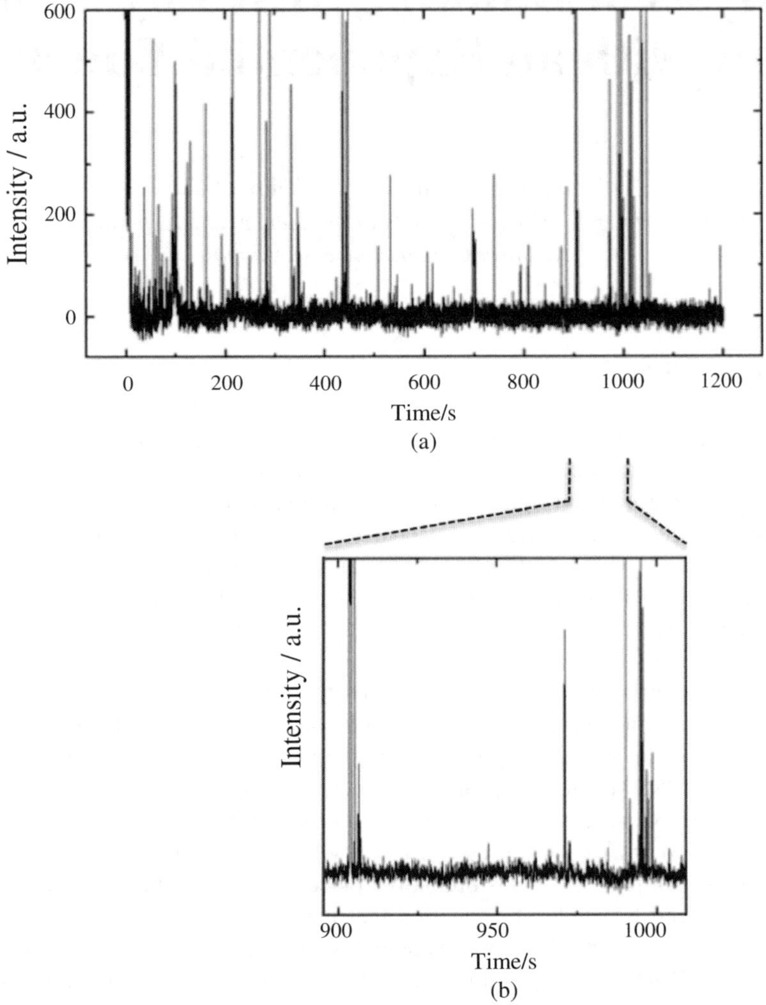

Figure 6.1 (a) Time-profile of SERS intensities of thiacyanine adsorbed on single Ag nanoaggregate excited at 458 nm from 0 to 1200 s. (b) The expanded time-profile of (a) from around 900 to 1000 s. © 2013, Y. Kitahama et al.

function, which indicates some periodicity [13–15]. It has also been applied to fluorescence correlation spectroscopy, which can measure the diffusion coefficient and concentration of molecules that go in and out of a focal area [53, 54]. However, the autocorrelation function was not reproduced by any simple function, suggesting a complex process in the blinking SERS [15].

The blinking SERS from a single Ag nanoaggregate with an adsorbed dye molecule can be analyzed by a power law in a manner similar to that employed in studies of blinking fluorescence from a single semiconductor quantum dot (QD) [55–59]. Power law statistics have been widely applied to the analysis of long-range ordered non-exponential behaviors, especially "self-similarity" or where a feature of objects does not change if length scales are expanded [60]. An instance of "self-similarity" or "fractal" is represented in Figure 6.1, which shows time-profiles of blinking SERS intensity from the same single Ag nanoaggregate. The time-profile from 400 to 1100 s (in Figure 6.1a) is similar to its expansion from 900 to 1000 s (Figure 6.1b). The power law in blinking statistics of the fluorescence from a QD can be explained by the distribution of the first passage time required for a random walker to return to its starting point, for a random walk of a photo-excited electron in a QD [55]. Figure 6.2 illustrates one- and two-dimensional random walk models. The number and width of the arrows represent the first passage time to return to the starting point, which corresponds to the duration time in the emissive or non-emissive state, and the distribution, respectively. Thus, the probability distribution decreases as the first passage time, namely, the duration time, lengthens. A three-dimensional random walker does not always return to its starting point, unlike one- and two-dimensional random walkers (Polya's theorem) [55]. They may be similar to the blinking statistics of SERS due to a random walk of an adsorbed molecule on a noble metal nanoaggregate. To the best of our knowledge, analysis of blinking SERS using a power law has been very limited [16–20].

In this chapter, the dependences of the blinking SERS on LSPR wavelength [19], excitation laser intensity [18, 19], and temperature are introduced in terms of the power law

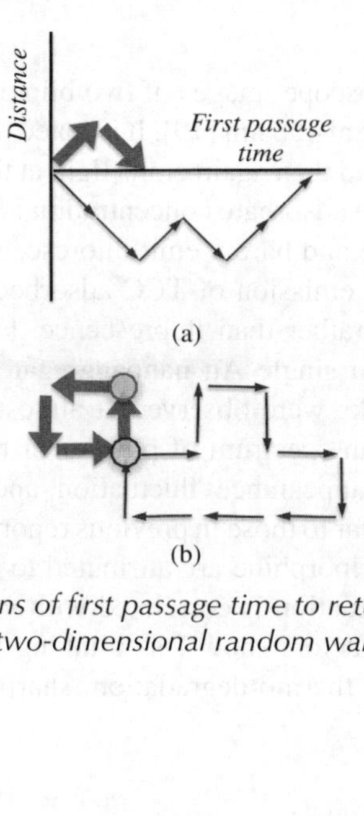

Figure 6.2 Schematic illustrations of first passage time to return to the starting point and the distribution for (a) one- and (b) two-dimensional random walkers

exponents and truncation in the power law. These are compared with the power law analysis of the blinking fluorescence from a single QD and other analyses of blinking SERS, such as the autocorrelation function or averaged duration times of the bright and dark SERS events. In contrast to blinking fluorescence from a single QD, blinking SERS is influenced not only by the diffusion of the molecule adsorbed on the metal, but also by the light-induced emissive state, or the enhanced EM field on the metal. These are quite significant both in terms of exploration of the essence and the suppression of blinking for possible application to single molecule SERS.

6.2 Materials and Methods

The details of the experimental procedure and set-up have been described elsewhere [17–20]. Briefly, colloidal Ag nanoparticles were prepared using the Lee–Meisel method [61]. We used 3,3′-diethylthiacyanine (TC) iodine salt, 3,3′-disulfopropylthiacyanine (anionic TC) triethylamine, or 3,3′-diethyl-5,5′-dichloro-9-methylthiacarbocyanine (TCC) iodine salt as analyte dye molecules. A stock aqueous solution of the dye (several μM), an NaCl aqueous solution, and an Ag colloidal solution were mixed. The sample solution was spin-coated onto a glass plate. An aliquot of a 1 M NaCl solution was dropped onto the glass plate to immobilize the sample Ag nanoaggregates on the surface. The glass plate was sandwiched with another glass plate to prevent the solution from evaporating. The Ag nanoaggregates with adsorbed dye molecules on an inverted microscope (Olympus, IX-70) were excited using an unfocused beam of the 458, 514, or 568 nm line of an Ar or Kr ion laser. Videos of the blinking SERS were taken for 5–20 min by the inverted microscope coupled with a cooled digital CCD camera (Hamamatsu, ORCA-AG) whose time-resolution was set at ∼60 ms.

6.3 Power Law Analysis

Figure 6.3a shows video microscope images of two blinking SERS-active Ag nanoaggregates with adsorbed TC excited at 458 nm [19]. It is noted that a bright spot in the upper left of the video images vanishes and then again emits light at the same position (the upper left) during the excitation. When the adsorbate concentration is low, the molecules stay in close proximity to a metallic surface and barely emit fluorescence due to energy transfer to the metal [62]. Thus, the blinking emission of TCC adsorbed on a single Ag nanoaggregate may be mainly due to SERS rather than fluorescence. Figure 6.3b exhibits temporally-fluctuating SERS spectra from single Ag nanoaggregates with adsorbed TC excited at 458 nm. These prominent peaks were observed at almost the same wavenumbers as the peaks in a conventional Raman spectrum of powder of the dye. These behaviors of the SERS peaks, namely, their disappearance, fluctuation, and the slight shift from the corresponding Raman peak are similar to those in previous reports [6–33]. The fluctuating SERS peaks of metal-free tetraphenylporphine are attributed to structural change [22]. For several analytes, the spectral fluctuation is correlated with vibronic coupling, and the peaks governed by Frank–Condon factors show less blinking [23]. In the case of amorphous carbon formed by photo- and thermo-degradation, sharp peaks are observed at random

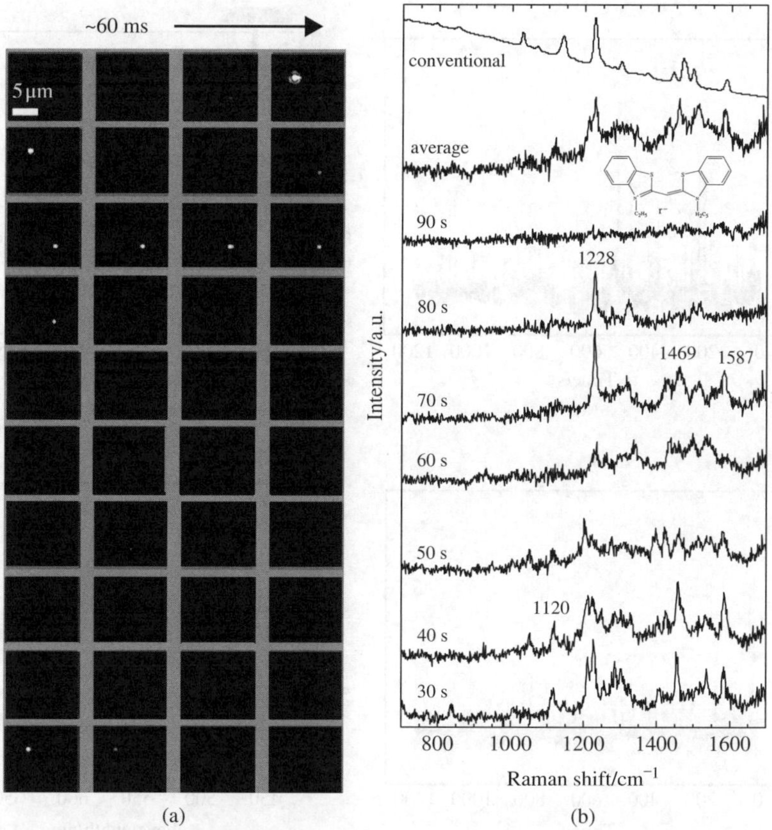

Figure 6.3 *(a) Video microscope images and (b) temporally-fluctuating spectra of SERS from a single Ag nanoaggregate with adsorbed TC excited at 458 nm. In (a), the scale bar is 5 μm and the exposure time per image was ~60 ms. The integration times per spectrum in (b) was 10 s. The top panel in (b) shows a conventional Raman spectrum of TC excited at 785 nm. The inset in (b) shows the molecular structure of 3,3′-diethylthiacyanine (TC) iodine salt. Reproduced from [19] by permission of The Royal Society of Chemistry*

positions and average out as two broad maxima around 1300 and 1550 cm^{-1} [47–52]. In this case, however, it was thought that the heat of the molecule adsorbed on the junction because of the enhanced EM field is quickly transferred to the Ag nanoaggregate [62, 63]. Indeed, the anti-Stokes to Stokes intensity ratios for the SERS peaks were estimated not by a Boltzmann distribution at high temperature, but by the selective enhancement of the SERS peaks that are close to the LSPR band [64]. Thus, these indicated that the blinking may be attributed not to SERS of amorphous carbon, but rather to SERS of the dye molecules. The blinking of the SERS intensity can be represented by a time-profile of the emission intensity from a single Ag nanoaggregate with adsorbed molecule, because it has been reported that SERS signals are strongly related to the background emission [15, 65].

Figure 6.4a,b present time-profiles of blinking SERS intensities of TC adsorbed on single Ag nanoaggregates with LSPR bands at ~460 and ~590 nm, respectively, excited at

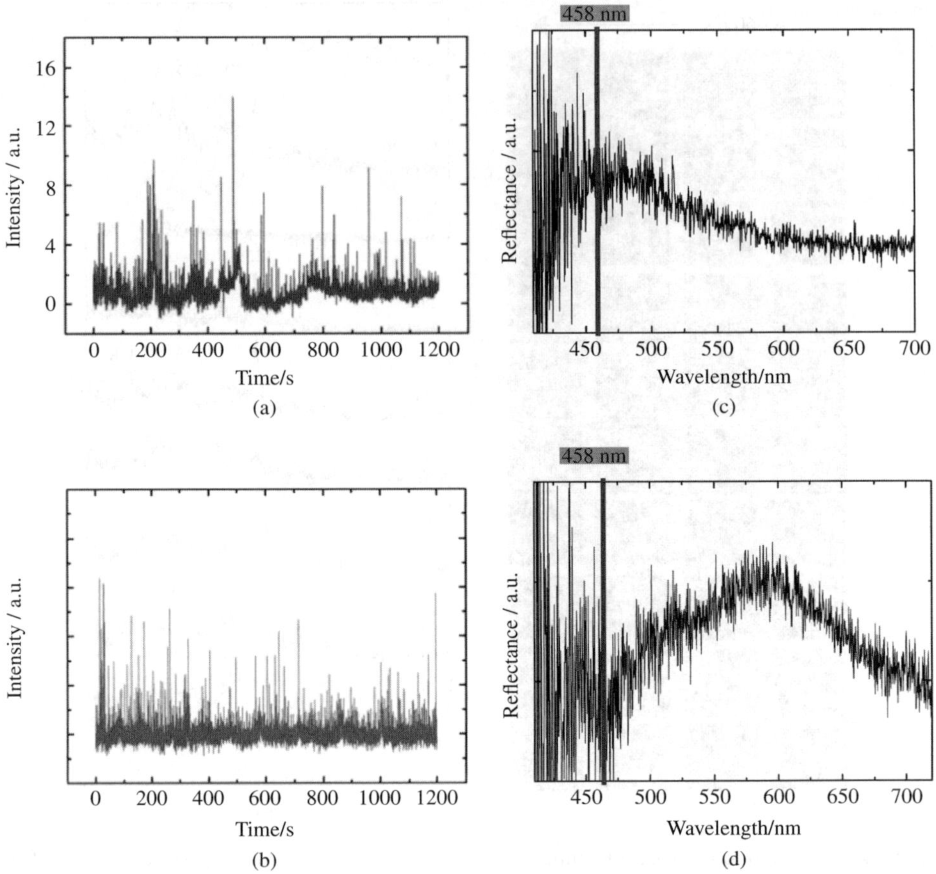

Figure 6.4 *Time-profiles of 458 nm-excited SERS intensities of TC adsorbed on single Ag nanoaggregates with LSPR at (a) 460 nm and (b) 590 nm [19]. (c,d) The LSPR spectra from the single Ag nanoaggregates which show the time-profiles in (a,b), respectively. Reproduced from [19] by permission of The Royal Society of Chemistry. (*See insert for color representation of the figure.*)*

458 nm [19]. The single Ag nanoaggregate with a LSPR band at ∼460 nm (Figure 6.4c) displays a bumpy baseline in the time-profile, as shown in Figure 6.4a. For example, the SERS was emitted for a long time, ∼200, 500, and 750 s. In the case of a nanoaggregate with the LSPR band at ∼590 nm (Figure 6.4d), the baseline is almost flat, as shown in Figure 6.4b. The characteristics of these time-profiles were quantified by a power law. From the baseline, an averaged intensity, I_{base}, and a standard deviation, σ, were evaluated. We defined bright and dark SERS events as the events showing larger and smaller intensities than a threshold of $I_{base} + 3\sigma$, respectively. The probability distribution for a time duration t_1 of the bright or dark SERS events is given by

$$P(t_1) = \sum_{t=t_1} n(t)/t_1 \tag{6.1}$$

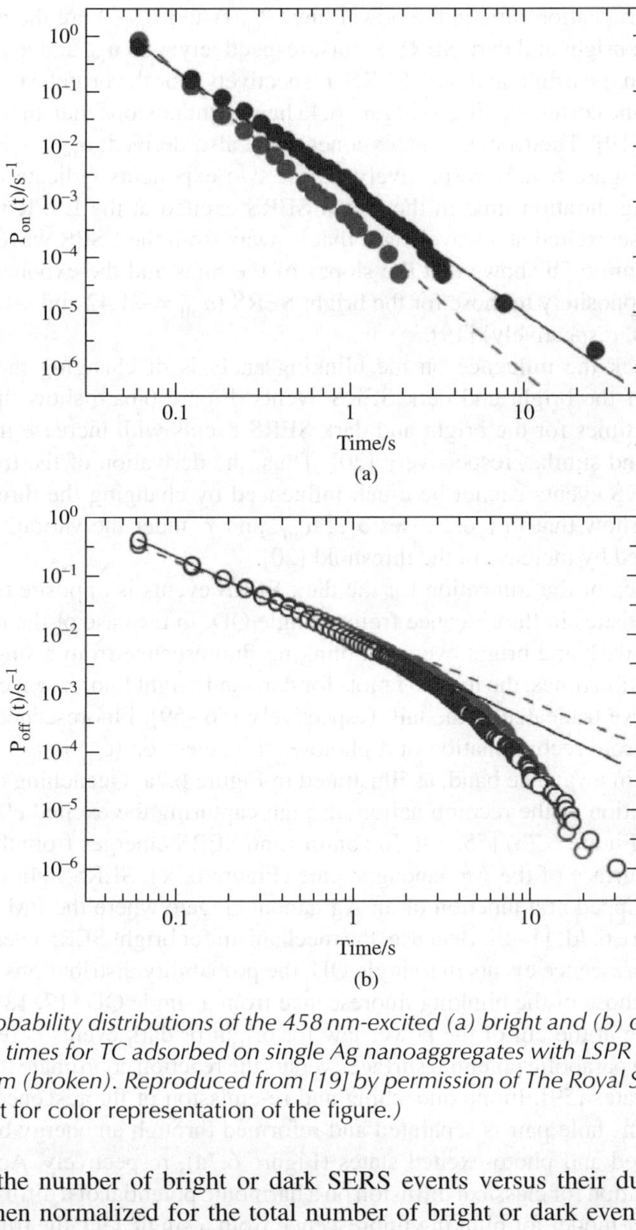

Figure 6.5 *Probability distributions of the 458 nm-excited (a) bright and (b) dark SERS events versus duration times for TC adsorbed on single Ag nanoaggregates with LSPR at 460 nm (solid line) and 590 nm (broken). Reproduced from [19] by permission of The Royal Society of Chemistry. (*See insert for color representation of the figure.*)*

where $n(t)$ is the number of bright or dark SERS events versus their duration times, a value that is then normalized for the total number of bright or dark events, respectively. Figure 6.5a,b show the probability distributions of the bright and dark SERS events, respectively, versus their duration times. In the bright SERS, the log–log plot shows the line given by a power law as

$$P_{\text{on}}(t) = At^{\alpha_{\text{on}}} \tag{6.2}$$

In the dark SERS, the log–log plots were truncated at their tails and were reproduced by a power law with an exponential function as

$$P_{\text{off}}(t) = At^{\alpha_{\text{off}}} \exp(-t/\tau) \tag{6.3}$$

where τ is the truncation time in the power law; $P_{on}(t)$ and $P_{off}(t)$ are the probability distributions of the bright and dark SERS events, respectively; and α_{on} and α_{off} are the power law exponents in the bright and dark SERS, respectively. For the bright SERS, Figure 6.5a shows that the line corresponding to Figure 6.4a has a gentler slope than that corresponding to Figure 6.4b [19]. Their power law exponents are also derived: $\alpha_{on} = -2.24$ and -2.72 correspond to Figure 6.4a,b, respectively. These two exponents indicate that the probabilities of a long duration time in the bright SERS excited at the LSPR wavelength are higher than those excited at a wavelength that is away from the LSPR wavelength. On the other hand, Figure 6.5b shows that the slopes of the lines and the exponents in the dark SERS behave oppositely to those for the bright SERS ($\alpha_{off} = -1.42$ and -1.12 correspond to Figure 6.4a,b, respectively) [19].

Here, we check the influence on the blinking analysis of changing the threshold for the definition of the bright and dark SERS events. Figure 6.6a,b show that the number of the duration times for the bright and dark SERS events with increase in the threshold are decreased and similar, respectively [20]. Thus, the derivation of the truncation times in the dark SERS events cannot be much influenced by changing the threshold. Indeed, Figure 6.6c–e show that in most cases α_{on}, α_{off}, and τ under the various conditions are slightly increased by increase in the threshold [20].

The appearance of the truncation for the dark SERS events is opposite to the result for the blinking statistics in fluorescence from a single QD. In the case of the probability distribution of the dark and bright events in blinking fluorescence from a single QD plotted against the duration times, the log–log plots for dark and bright fluorescence events display a line and a curve truncated at the tail, respectively [56–59]. Fluorescence from a single QD originates from recombination of a photo-excited electron (e^-) in a conduction band and a hole (h^+) in a valence band, as illustrated in Figure 6.7a. Quenching of fluorescence is due to prevention of the recombination through capturing the excited electron at a surface trap state (Figure 6.7b) [55, 59]. In contrast, no SERS emerges from the molecule on the rest of the surface of the Ag nanoaggregate (Figure 6.7c). SERS light is emitted from the molecule trapped at a junction of an Ag nanoaggregate where the EM field is greatly enhanced (Figure 6.7d) [1–6]. Because, the mechanism for bright SERS events is similar to that of dark fluorescence events in a single QD, the probability distributions of the blinking SERS contrast those of the blinking fluorescence from a single QD [17, 18].

The truncation at the tail of the power law for bright or dark events is reproduced by a random walk on parabolic potential surfaces versus the reaction coordinate for emissive and non-emissive states [59]. In the quenching and re-emission of fluorescence from a single QD, the electron–hole pair is separated and reformed through an energy barrier between the photo-ionized and photo-excited states (Figure 6.7a), respectively. According to the differential equation for classical diffusion on a harmonic potential of a diffusion-controlled electron-transfer model for blinking fluorescence from a single QD, the truncation time is derived by

$$\tau = \frac{\Gamma}{E_a/2k_B T} \tag{6.4}$$

where E_a is an energy barrier between emissive and non-emissive states and Γ is the diffusion time to overcome the energy barrier [59]. When there is no energy barrier or a very long diffusion time (slow diffusion), the truncation time becomes much longer, in fact, truncation does not occur. For dark fluorescence events from a single QD, no truncation

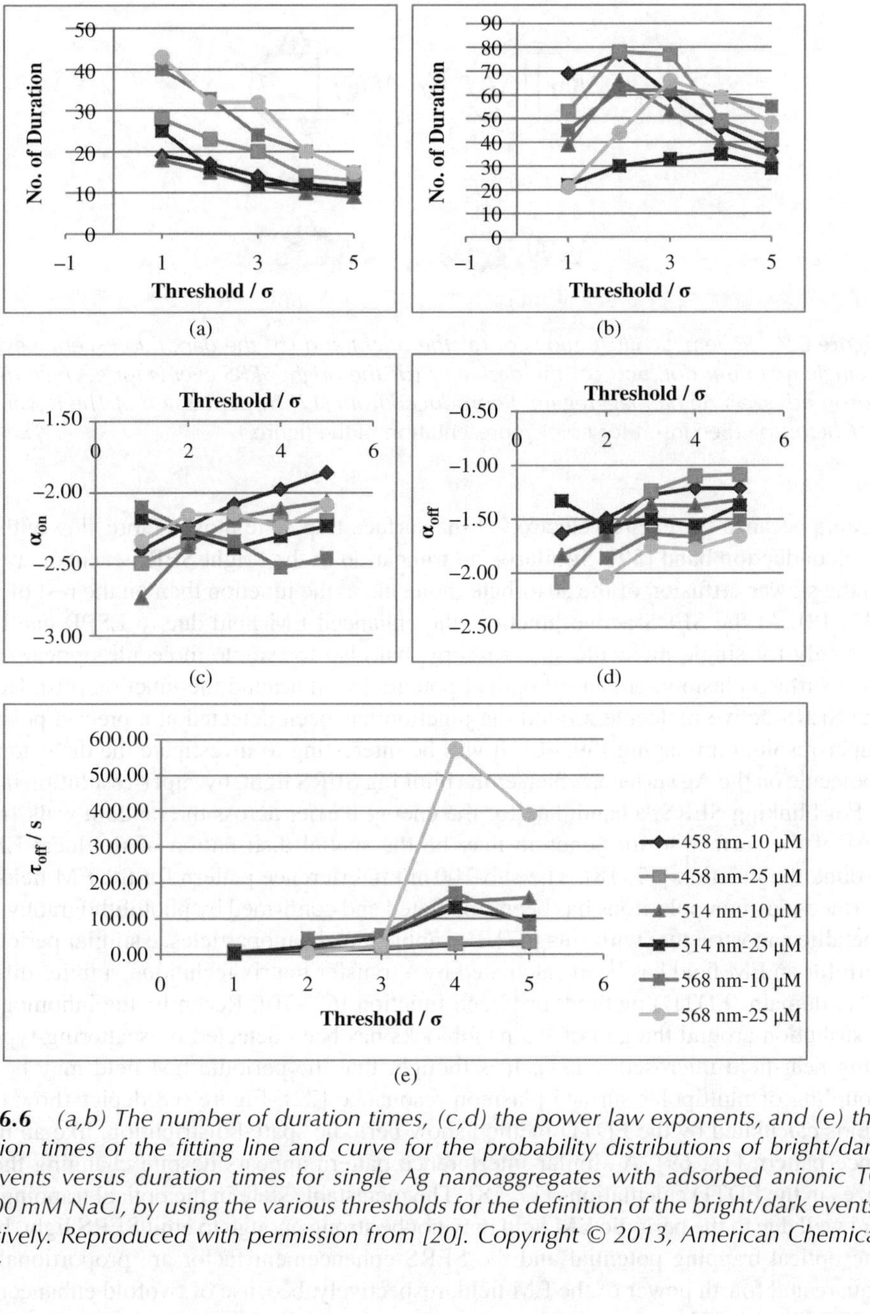

Figure 6.6 *(a,b) The number of duration times, (c,d) the power law exponents, and (e) the truncation times of the fitting line and curve for the probability distributions of bright/dark SERS events versus duration times for single Ag nanoaggregates with adsorbed anionic TC with 100 mM NaCl, by using the various thresholds for the definition of the bright/dark events, respectively. Reproduced with permission from [20]. Copyright © 2013, American Chemical Society*

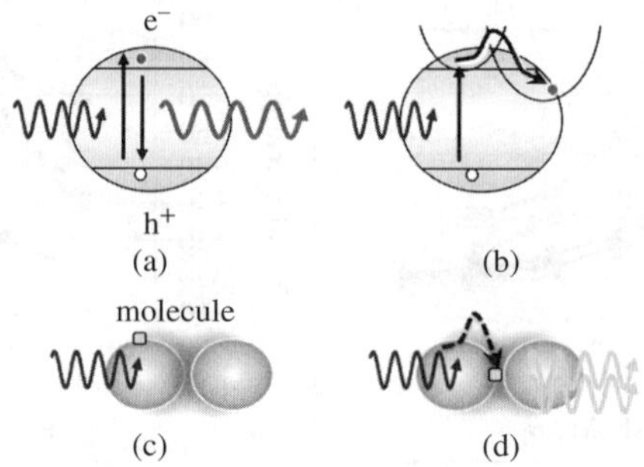

Figure 6.7 *Schematic illustrations of (a) the bright and (b) the dark fluorescence events for a single quantum dot, and (c) the dark and (d) the bright SERS events for a single molecule adsorbed on an Ag nanoaggregate. Reproduced from [17] by permission of The Royal Society of Chemistry. (*See insert for color representation of the figure.*)*

occurs because the excited electron at the surface trap state moves more slowly than that in a conduction band [59]. Similarly, no truncation in the bright SERS events may be due to the slower diffusion of the adsorbate molecule at the junction than on the rest of surface [17, 18]. At the SERS-active junction, the enhanced EM field due to LSPR can be used not only for single molecule spectroscopy, but also for single molecule optical trapping by a surface-plasmon-enhanced optical potential well around the junction [66]. Recently, the SERS-active molecule around the junction has been detected at a precise position by super-resolution imaging [39, 40]. It will be interesting to investigate the diffusion of the molecule on the Ag surface, which emits blinking SERS light, by super-resolution imaging.

For blinking SERS, a candidate for the energy barrier across the random walk from the rest of the surface to the junction may be the spatial distribution of the local EM field around the junction [17, 18]. The sub-100 nm interference pattern for the EM field due to surface-plasmon polaritons has been calculated and confirmed by photolithography using a metallic nanostructure with slits [67]. For noble metal nanoparticles, a similar periodic pattern for an EM field has been calculated by a transfer matrix technique, a finite-difference time-domain (FDTD) method, or Green function [68–70]. Recently, the inhomogeneous distribution around the gap of Au nanoblocks has been detected by scattering-type scanning near-field microscopy [71]. It is thought that the periodic EM field may be due to coupling of multipolar surface plasmon resonance [72]. Figure 6.8 depicts how the EM fields calculated by the FDTD method show periodic spatial distribution, like an interference pattern [18, 69]. A similar interference pattern appears despite changing the mesh sizes in the FDTD calculations [17, 18]. The metastable state in the optical trapping potential well due to the periodic EM field may not be strong enough to emit SERS light, because the optical trapping potential and the SERS enhancement factor are proportional to the square and fourth power of the EM field, respectively, because of twofold enhancement of SERS [63, 64, 73].

We make a comparison between this result and another power law analysis of blinking SERS peaks of Fe-protoporphyrin IX (10^{-11} M) on Ag excited at 514 nm [16]. The α_{off} was reported to be −1.56, −1.51, and −1.46, like blinking fluorescence from a single QD [55–58], although the analysis for TC derived $\alpha_{off} = -1.42$ and −1.12 from

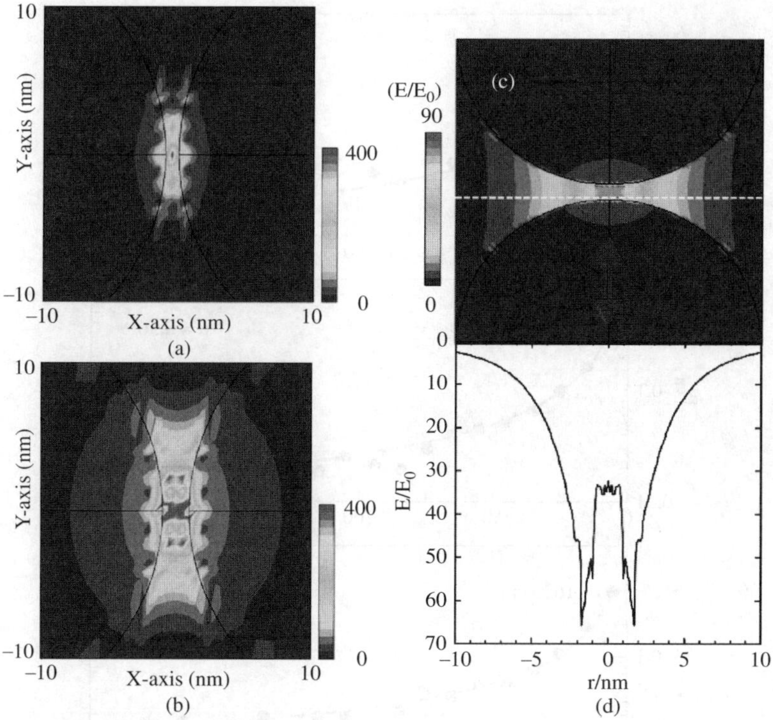

Figure 6.8 *Calculated spatial distribution of the EM field around the gap of Ag nanoparticles. Linearly arrayed particles, 16 nm in radii with gap size (a) 1 nm and (b) 2 nm were used. The excitation wavelength was 470 nm. Reprinted with permission from [69]. Copyright 2008 by Springer. (c) The Ag nanodimer with a diameter of 20 nm and a gap of 1 nm in 0.1 nm mesh was used. The 514-nm polarization along the long axis was used. (d) Calculated intensity of the EM field along a white broken line in (c). Reproduced from [18] by permission of The Royal Society of Chemistry.* (See insert for color representation of the figure.)

Figure 6.5b [19]. Figure 6.9 shows semi-logarithm plots of the probability distribution of dark SERS events from Fe-protoporphyrin IX [16]. They are reproduced by a power law without an exponential function. Thus, the truncation in the power law was not observed. The absence of the truncation may originate from the strong laser power density (≤1 mW [16], corresponding to a power density of ≤130 kW cm^{-2}) which changes the metastable state into an SERS-emissive state. In consequence, the energy barrier between the emissive and non-emissive states, which induces the truncation, possibly vanishes.

6.4 Plasmon Resonance Wavelength Dependence

6.4.1 Power Law Exponents for the Bright and Dark Events

We investigated the dependence of the power law exponents on LSPR wavelength [19]. Figure 6.10 plots the exponents of the power law that reproduce probability distributions

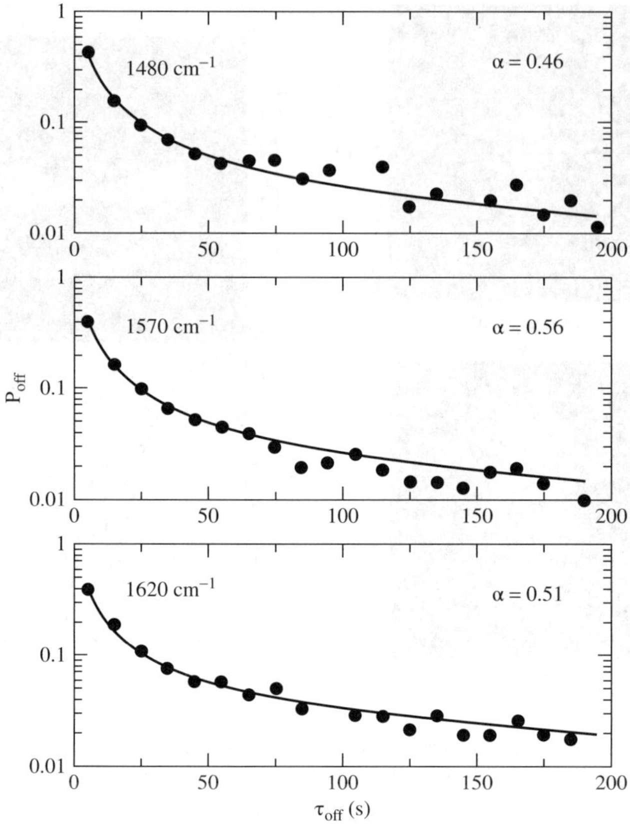

Figure 6.9 *Probability distribution of dark SERS events at the peaks of Fe-protoporphyrin IX (10^{-11} M) adsorbed on an Ag nanoaggregate excited at 514 nm versus the duration time (τ_{off}). The fitting curves are given by $1/\tau_{off}{}^{1+\alpha}$. Reprinted with permission from [16]. Copyright 2005 by the American Physical Society*

of the bright and dark SERS events against their duration times at the various LSPR wavelengths [19]. There are only a few results for the Ag nanoaggregates whose LSPR bands appear at shorter wavelengths than the excitation wavelength, because the LSPR bands at shorter wavelengths enhance only weak anti-Stokes Raman scattering [64]. As the LSPR wavelength approaches the excitation wavelength, the power law exponents in the bright and the dark SERS seem to increase and decrease, respectively. In other words, the probabilities of a long duration time in the bright SERS and a short duration time in the dark SERS increase as a result of the resonance of the plasmon with the excitation light.

Figure 6.11 shows the correlation between the power law exponents in the bright and the dark SERS [19]. The exponents in the bright SERS seem to increase with a decrease in those in the dark SERS. Note that the former must be independent of the latter, because the probability distributions of the bright and the dark SERS events are derived not from the total of both events, but from the total of bright or dark events, respectively. As the LSPR wavelength approaches the excitation wavelength, the exponents were plotted as shown

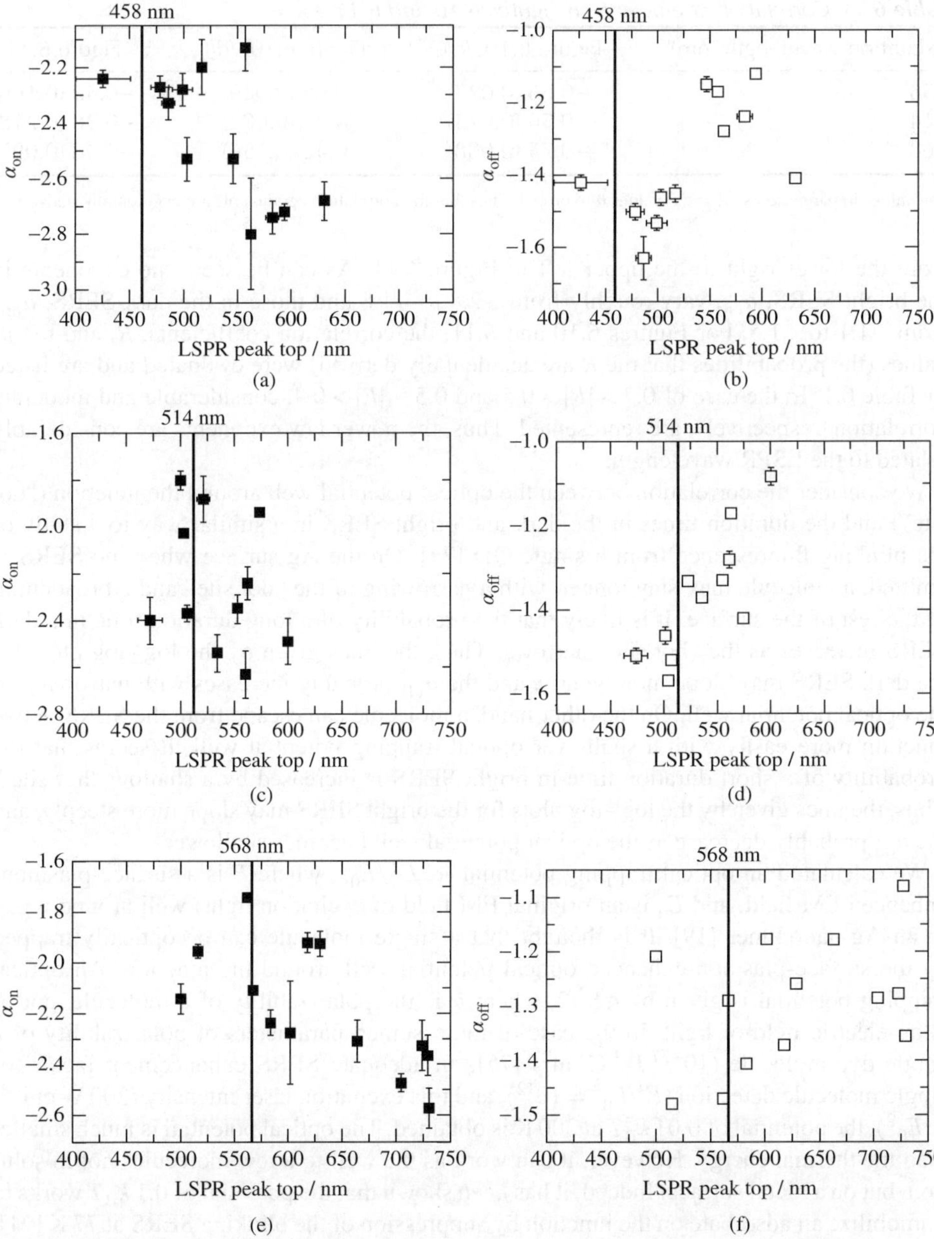

Figure 6.10 *Power-law exponents of the fitting lines for the probability distributions of bright/dark SERS events versus duration times at various LSPR wavelengths for single Ag nanoaggregates with adsorbed TC excited at (a,b) 458 nm and (c,d) 514 nm, and adsorbed TCC excited at (e,f) 568 nm. Reproduced from [19] by permission of The Royal Society of Chemistry*

Table 6.1 *Correlation coefficients in Figures 6.10 and 6.11*

Excitation wavelength (nm)	Figure 6.10a/c/e	Figure 6.10b/d/f	Figure 6.11
458	−0.66 (0.027)	0.65 (0.029)	−0.42 (0.169)
514	−0.56 (0.073)	0.82 (0.002)	−0.38 (0.218)
568	−0.75 (0.008)	0.46 (0.156)	−0.48 (0.097)

The values in parentheses are the *p*-values (the probabilities that the correlation coefficients are accidentally derived).

from the lower right to the upper left in Figure 6.11. As can be seen, the exponents in the bright SERS, α_{on}, vary roughly from −2.8 to −1.8 and those in the dark SERS, α_{off}, from −1.1 to −1.5. For Figures 6.10 and 6.11, the correlation coefficients, *R*, and the *p*-values (the probabilities that the *R* are accidentally derived) were evaluated and are listed in Table 6.1. In the case of $0.7 > |R| > 0.5$ and $0.5 > |R| > 0.4$, considerable and moderate correlation, respectively, are represented. Thus, the power law exponents are considerably related to the LSPR wavelength.

We consider the correlation between the optical potential well around the junction ("hot site") and the duration times in the dark and bright SERS in a similar way to a study of the blinking fluorescence from a single QD [74]. On the Ag surface where no SERS is emitted, a molecule may stay longer, with a narrowing of the "hot site" and a broadening of the rest of the surface. It is likely that the probability of a long duration time in a dark SERS increases as the "hot site" narrows. Thus, the lines given by the log–log plots for the dark SERS may slope more gently, and the α_{off} probably increases with narrowing of the optical potential well. On the other hand, a molecule can escape from the SERS-active junction more easily with a shallower optical trapping potential well. It seems that the probability of a short duration time in bright SERS is increased by a shallow "hot site." Thus, the lines given by the log–log plots for the bright SERS may slope more steeply, and the α_{on} probably decrease as the optical potential well becomes shallower.

We calculated an optical trapping potential ($\propto E^2/E_0{}^2$, where E is a surface-plasmon-enhanced EM field, and E_0 is an original EM field of excitation light) well around a gap in an Ag nanodimer [19]. It is thought that a single molecule can be optically trapped by the surface-plasmon-enhanced optical potential well around the junction. An optical trapping potential is given by $\alpha E^2/2$, where α is the polarizability of a molecule, and E is an electric field of light. In the case of the assumed parameters of polarizability of a single dye molecule (10^{-37} J^{-1} C^2 m^2) [75], an adequate SERS enhancement factor for single molecule detection ($E^4/E_0{}^4 = 10^{12}$), and this excitation laser intensity (200 W cm^{-2}, $\propto E_0{}^2$), the potential of 0.01 k_BT at 300 K is obtained. The optical potential is much smaller than the thermal energy. However, it can work as the well to a few molecules not in solution, but on a metal surface. Indeed, it has been shown that the potential of 0.1 k_BT works to immobilize an adsorbate on the junction by suppression of the blinking SERS at 77 K [21].

Figure 6.12 shows the results of the FDTD calculation for (a) a spatial distribution of the enhanced EM field, (b) an intensity of the optical trapping potential well, (c) the LSPR spectrum, and (d) the depth and (e) the width of the optical trapping potential wells at various excitation wavelengths [19]. As shown in Figure 6.12b, the calculated optical trapping potential well becomes shallower and narrows as the excitation wavelengths (458 or 488 nm) leave the LSPR wavelength (455 nm, see Figure 6.12c) for the Ag nanodimer in Figure 6.12a. Note that Figure 6.12d,e are roughly consistent with Figure 6.10a/c/e and

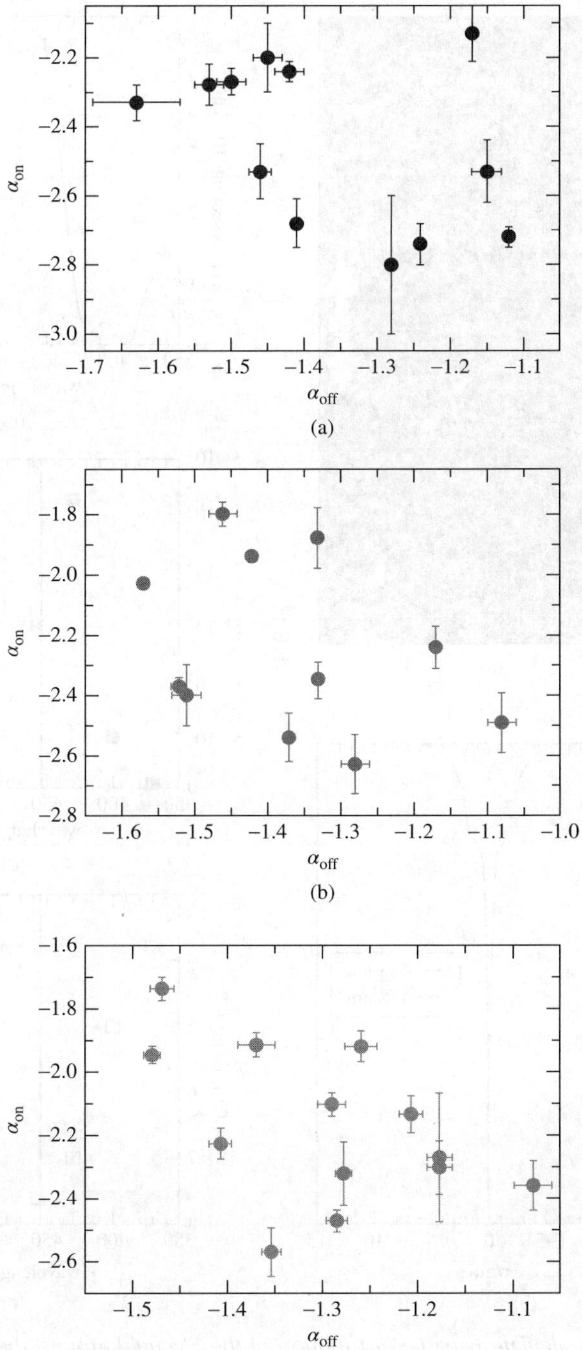

Figure 6.11 *The power-law exponents of the fitting lines for the probability distributions of bright SERS events versus those of dark SERS events from single Ag nanoaggregates with adsorbed TC excited at (a) 458 nm and (b) 514 nm, and adsorbed TCC excited at (c) 568 nm. Reproduced from [19] by permission of The Royal Society of Chemistry*

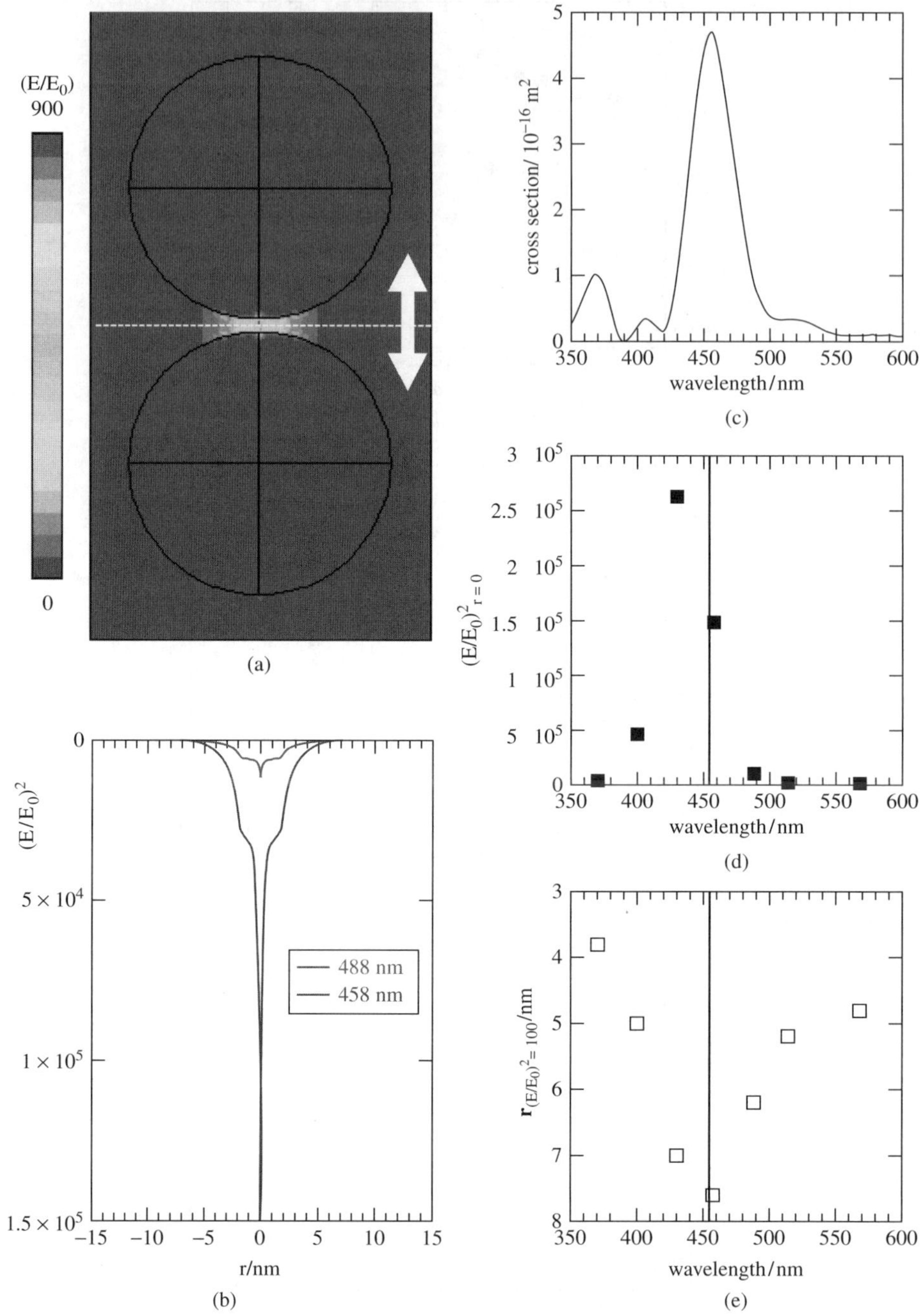

Figure 6.12 *(a) Calculated spatial distribution of the EM field around the gap (1 nm) in Ag nanoparticles with a diameter of 20 nm excited by vertical polarizations at 430 nm. (b) Calculated intensity of the optical trapping potential well along the horizontal broken line in (a). (c) Calculated far-field scattering efficiency spectrum for the Ag nanoparticles referred to in (a). (d) Calculated depths and (e) calculated half-widths where there is a 100-fold enhancement of optical trapping potential wells, like that in (b), at various excitation wavelengths. Reproduced from [19] by permission of The Royal Society of Chemistry. (*See insert for color representation of the figure.*)*

b/d/f, respectively; that is, the depth and width of the calculated optical trapping potential well at various excitation wavelengths correspond to the α_{on} and α_{off} plotted against the LSPR wavelength, respectively. The depth of the calculated optical potential well and the α_{on} decrease simultaneously as the LSPR wavelength leaves the excitation wavelength. In the case of Ag nanoaggregates whose LSPR band is located at a wavelength that is away from the excitation wavelength, the decrease in width of the calculated optical trapping potential well coincides with an increase in the α_{off}.

6.4.2 Truncation Time for the Dark Events

Figure 6.13 shows the LSPR wavelength dependence of the truncation times in the power law that reproduces the probability distributions of the dark SERS events against their duration times. It is noted that the truncation times are independent of LSPR wavelength for each of the SERS-active Ag nanoaggregates. There is the possibility that the photo-thermal conversion due to LSPR causes the short truncation times via fast diffusion of adsorbed molecules (see Equation 6.4) by increase in the local temperature. However, the possibility is not confirmed by this result.

6.5 Energy Density Dependence

6.5.1 Power Law Exponents for the Bright and Dark Events

Figure 6.14 shows histograms of the exponent for the power law for the bright and dark SERS events under different excitation laser intensities [19]. The LSPR bands of the SERS-active single Ag nanoaggregates appear at various wavelengths. As the laser

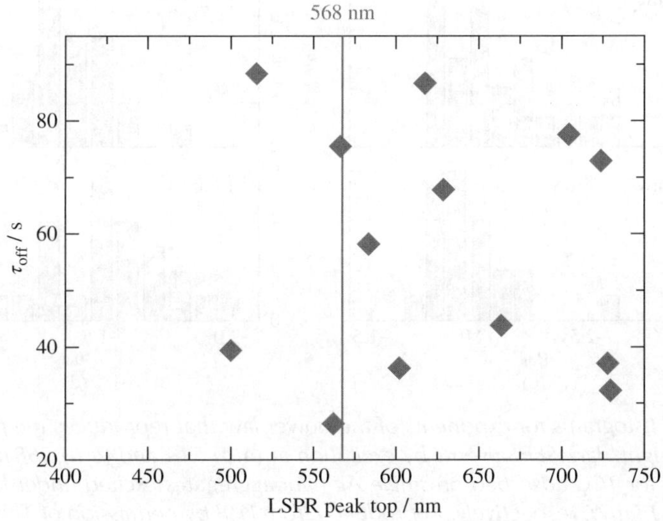

Figure 6.13 *Truncation times of the fitting curves for the probability distributions of dark SERS events versus duration times at various LSPR wavelengths for single Ag nanoaggregates with adsorbed TCC excited at 568 nm. © 2013, Y. Kitahama et al.*

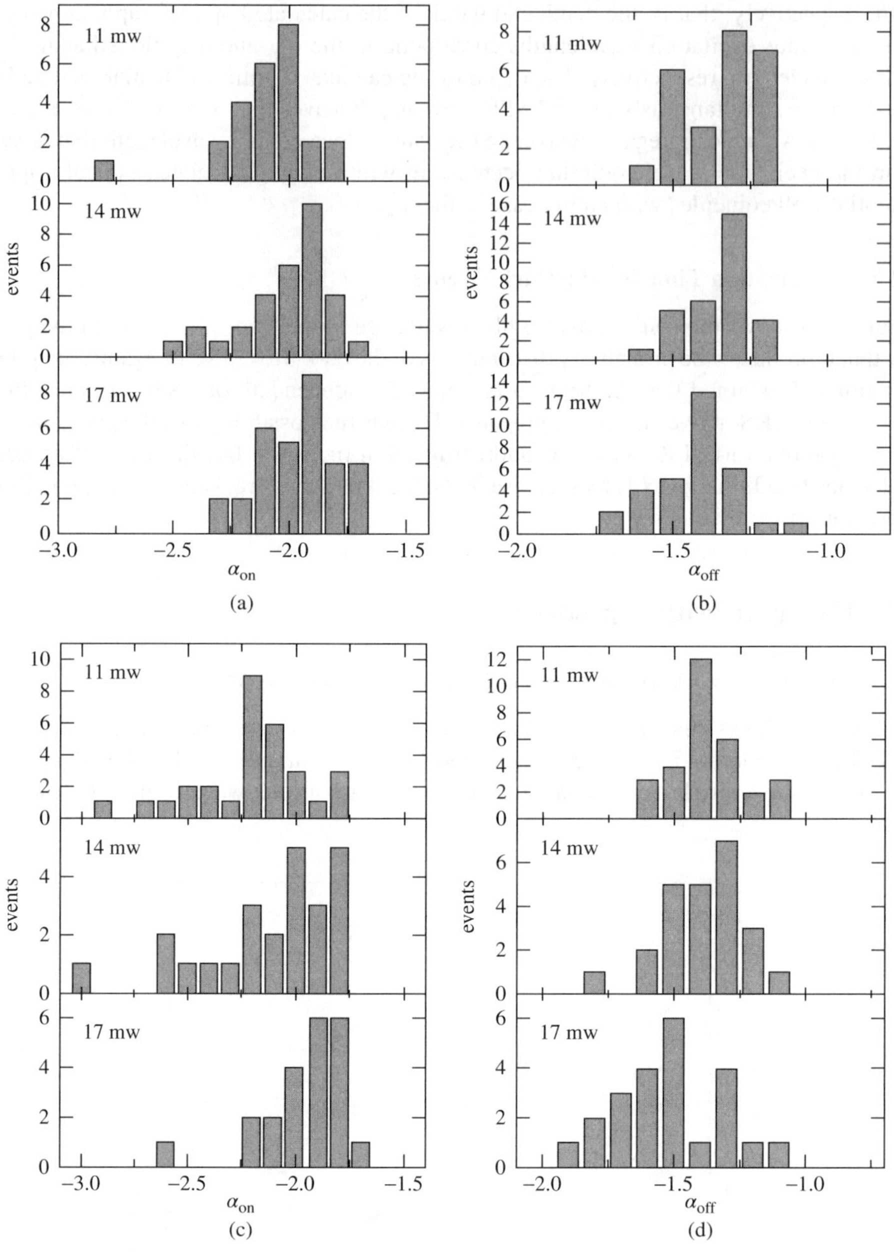

Figure 6.14 *Histograms for exponents of the power law that reproduces the probability distributions of bright/dark SERS events by excitation at (a,b) 514 and (c,d) 568 nm versus their duration times for TCC adsorbed on single Ag nanoaggregates excited under laser intensities of 11, 14, and 17 mW, respectively. Reproduced from [19] by permission of The Royal Society of Chemistry*

intensity increases (11, 14, and 17 mW, corresponding to power densities of 110, 140, and 170 W cm^{-2}, respectively), the maxima in the histograms of the power law exponents for the bright SERS increase from $\alpha_{on} \sim -2.0$ to -1.8 and those for the dark SERS decrease from $\alpha_{off} \sim -1.3$ to -1.5.

The maxima in the histograms of the power law exponents for the bright and dark SERS approach -1.5 as the laser intensity increases. The power laws can be explained by the distribution of the first-passage time required for the random walker to return to its starting point for a one- or two-dimensional random walk (Figure 6.2) [55]. For the one-dimensional random walk model, the power law exponent is derived to be -1.5, and many experimental power law exponents in the bright and dark fluorescence from a single QD are consistent with -1.5 [55–57]. For a QD, a photo-excited electron in a conduction band, which recombines with a hole in a valence band and then emits fluorescence, is restrained by Coulomb attraction. The restraint may be the reason why the diffusion of the photo-excited electron on a single QD is approximated to a one-dimensional random walk model as a function of the distance from the hole, rather than a two-dimensional random walk model. This situation may be similar to the SERS-active Ag nanoaggregate on which an adsorbate molecule is restrained by a surface-plasmon-enhanced optical trapping potential well around the junction [21, 66]. The optical trapping force due to the surface-plasmon-enhanced EM field is strengthened as the original EM field of the excitation light (the laser intensity) increases. The adsorbed molecule is then restrained by the surface-plasmon-enhanced optical trapping force. Thus, the power law exponents near -1.5 for the bright and dark SERS may originate from the strong surface-plasmon-enhanced optical trapping force. We confirmed also that the power law exponents are close to -1.5 as the LSPR wavelength of the Ag nanoaggregate with an adsorbed dye molecule approaches the excitation wavelength, as shown in Figure 6.10 [19].

In the SERS under the low laser intensity, the deviation from a value of -1.5 likely indicates some diffusion models other than a one-dimensional random walk model. The diffusion of a molecule adsorbed on an Ag surface that does not have a surface-plasmon-enhanced optical trapping potential, and where no SERS is emitted, is more likely to be expressed by a two-dimensional random walk model rather than a one-dimensional random walk model. From the two-dimensional random walk model, a power law exponent of -1 is derived [55]. This result may be consistent with the exponents in the dark SERS under the low laser intensity of 11 mW, corresponding to the low power density of 110 W cm^{-2} ($\alpha_{off} \sim -1.3$). The exponents in the dark SERS excited at a wavelength that is away from the LSPR wavelength also approach $\alpha_{off} \sim -1.1$, as shown in Figure 6.10 [19].

6.5.2 Truncation Time for the Dark Events

Figure 6.15 shows histograms of the truncation times in the power law that reproduces the probability distributions of the dark SERS events against their duration times under various laser intensities (11, 14, and 17 mW, corresponding to power densities of 110, 140, and 170 W cm^{-2}, respectively) [18]. The truncation times were almost constant despite changing the excitation laser intensity from 11 to 17 mW. Note that this differs from the truncation times for a single QD, which are shortened by photo-induced diffusion of the excited electron due to the higher excitation laser intensity [56, 58]. As the laser intensity, namely, the original EM field of excitation light (E_0 in Figure 6.8) increases, the energy barrier due to the periodic surface-plasmon-enhanced EM field becomes higher, which allows truncation

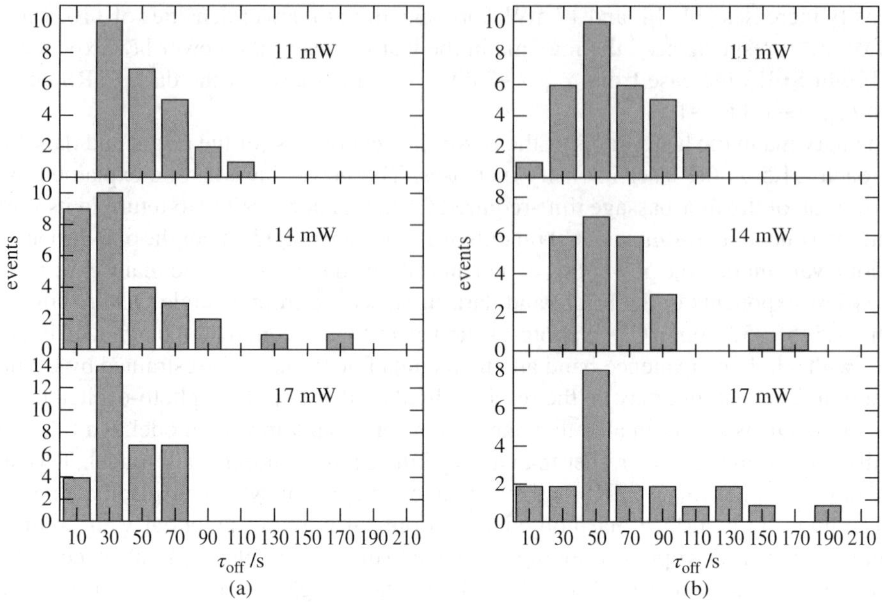

Figure 6.15 *Histograms for truncation times of the power law with exponential function that reproduces probability distributions of (a) 514 nm- and (b) 568 nm-excited bright SERS events versus their duration times for TCC adsorbed on single Ag nanoaggregates excited under laser intensities of 11, 14, and 17 mW. Reproduced from [18] by permission of The Royal Society of Chemistry*

to occur in the shorter duration time. However, diffusion of an adsorbate molecule is likely suppressed by the optical trapping potential due to the surface-plasmon-enhanced EM field. The longer diffusion time Γ (slower diffusion) probably cancels out the higher energy barrier E_a with the higher laser intensity. Thus, the truncation times ($\tau \propto \Gamma/E_a$) for SERS are then constant, as represented in Equation 6.4. If the blinking SERS was not affected by the photo-dependent energy barrier, the truncation times were changed by the excitation laser intensity through photo-induced diffusion of adsorbate in a similar way as for a single QD [56, 58]. Thus, the observed constant truncation times under various laser intensities suggest an energy barrier due to the periodic pattern of the surface-plasmon-enhanced EM field.

The maxima in the histograms of the truncation times were observed to be ~30 s at 514 nm and ~50 s at 568 nm, as shown in Figure 6.15a,b, respectively. By inducing excitation at a shorter wavelength than an absorption band of TCC dimer at 505 nm and that of the monomer at 550 nm [76, 77], photo-induced desorption may be accelerated, thus shortening the truncation time in a different manner to the excitation wavelength dependence of blinking fluorescence from a single QD, in which the truncation times were similar except for near-ultraviolet excitation [57].

6.5.3 Comparison with Other Analysis

In the case of a single QD [56, 58], the truncation times were shortened by an increase in the laser intensity in a different way from that of the blinking SERS of TCC [18]. For blinking

SERS peaks of Fe-protoporphyrin IX by excitation at 514 nm, the exponents were independent of the laser intensity (≤1 mW, corresponding to a power density of ≤130 kW cm^{-2}), and thus the possibility that the constant α_{off} could be due to irreversible photo-induced processes was ruled out [16]. The independence of the exponents may be attributed to the strong laser power density compared with the study for TCC (110–170 W cm^{-2}).

Under various intensities of the excitation laser (0–100 W cm^{-2}), blinking SERS of Rhodamine 6G (R6G) excited at 532 nm was evaluated based on the rates calculated by integrating the averaged autocorrelation functions [14]. The rates increased with increase in the laser intensity [14], which indicated that the phenomenon of blinking SERS is a photo-induced mechanism. In the blinking SERS of TCC [18], the number of bright SERS events increased with increase in the laser intensity, as represented in Figure 6.16, which shows histograms for the number of bright SERS events from each of the single Ag nanoaggregates. This may be similar to the increase in the rates by integrating the averaged autocorrelation functions.

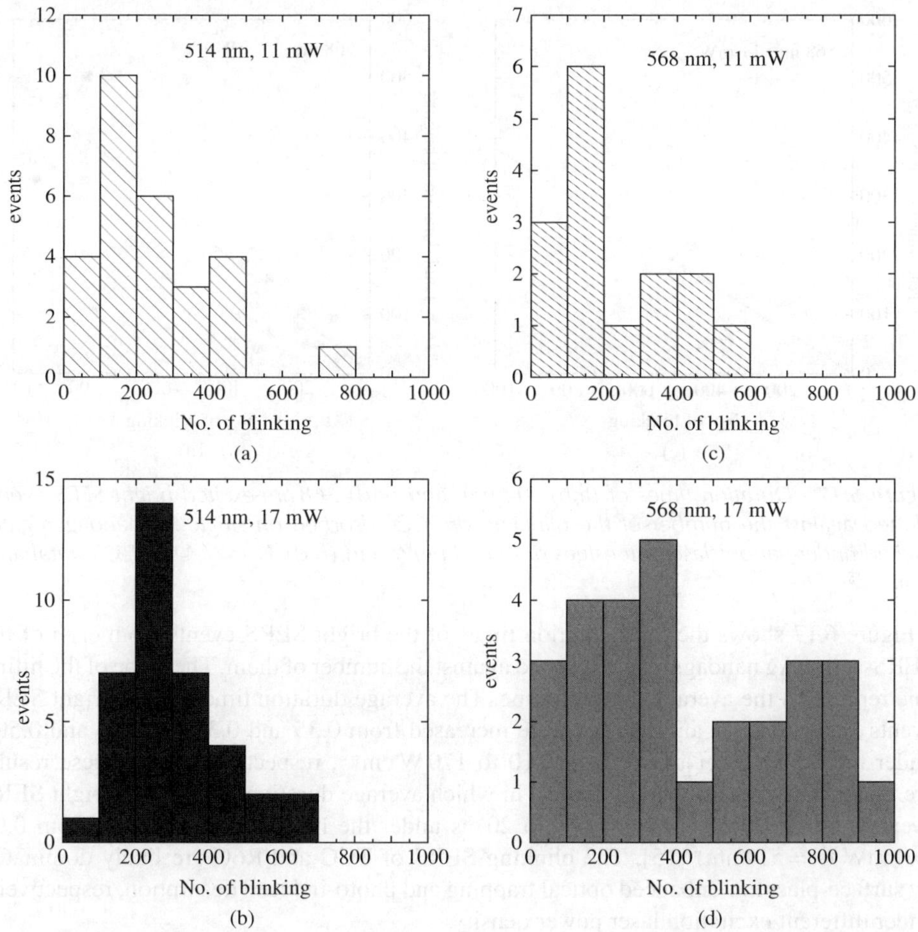

Figure 6.16 Histograms for the number of (a,b) 514 nm- and (c,d) 568 nm-excited bright SERS events for TCC adsorbed on single Ag nanoaggregates excited under laser intensities of (a,c) 11 mW and (b,d) 17 mW. © 2013, Y. Kitahama et al.

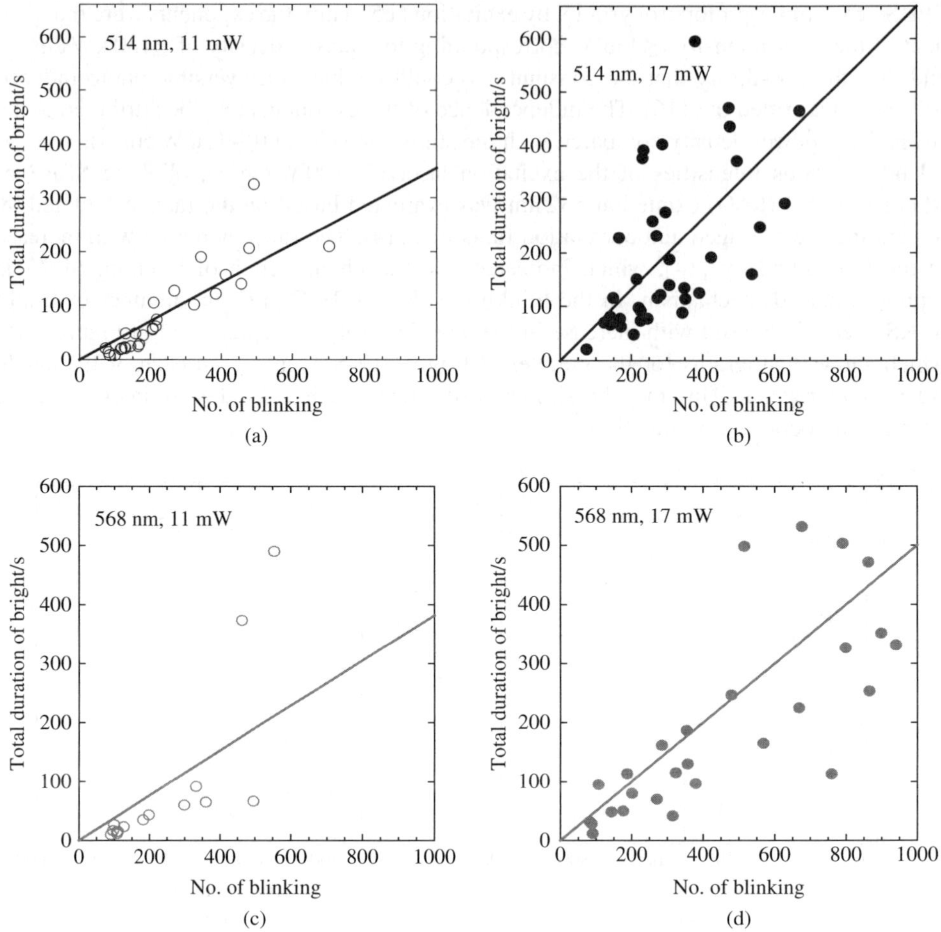

Figure 6.17 *Duration times of (a,b) 514 nm- and (c,d) 568 nm-excited bright SERS events plotted against the number of the blinking for TCC adsorbed on single Ag nanoaggregates excited under various laser intensities of (a,c) 11 mW and (b,d) 17 mW. © 2013, Y. Kitahama et al.*

Figure 6.17 shows the total duration times of the bright SERS events from each of the SERS-active Ag nanoaggregates plotted against the number of them. The slope of the fitting line represents the average duration time. The average duration times of the bright SERS events excited at 514 and 568 nm were increased from 0.35 and 0.37 s to 0.66 and 0.50 s under the higher laser power from 110 to 170 W cm^{-2}, respectively [18]. These results are in contrast to the previous studies, in which average duration time of the bright SERS event of R6G decreased from ~7 s to 20 ms under the higher laser intensity from 0.01 to 1 mW ($\lambda = 514$ nm) [15]. The blinking SERS of TCC and R6G are likely dominated by surface-plasmon-enhanced optical trapping and photo-induced desorption, respectively, under different excitation laser power density.

Incidentally, the total duration times of the dark SERS events from each SERS-active Ag nanoaggregate were shortened as the number of them increased, which is opposite to the results in Figure 6.17 for the bright SERS events. Thus, the blinking statistics of the dark SERS events should be evaluated not by an average, but by a power law with an exponential function.

6.6 Temperature Dependence

6.6.1 Power Law Exponents for the Bright and Dark Events

It is thought that change in temperature affects only diffusion of the molecules on the metal. At various temperatures from 50 to 8 °C, we observed SERS from a single Ag nanoaggregate with a TCC molecule, which was prepared by mixing a stock aqueous solution of TCC (4 µM), a NaCl aqueous solution (5 mM), and an Ag colloidal solution in a volume ratio of 1 : 1 : 2, and excited at 568 nm (the excitation laser intensity of 11 mW, corresponding to the power density of 110 W cm^{-2}). Then the blinking was analyzed by a power law with an exponential function. In the histograms of the power law exponents at various temperatures (Figure 6.18), the α_{on} and α_{off} seem to increase and decrease, respectively, at lower temperature.

6.6.2 Truncation Time for the Dark Events

For a single QD at higher temperature, truncation times in bright fluorescence are shortened due to thermo-activated diffusion, although the power law exponents are constant. In the

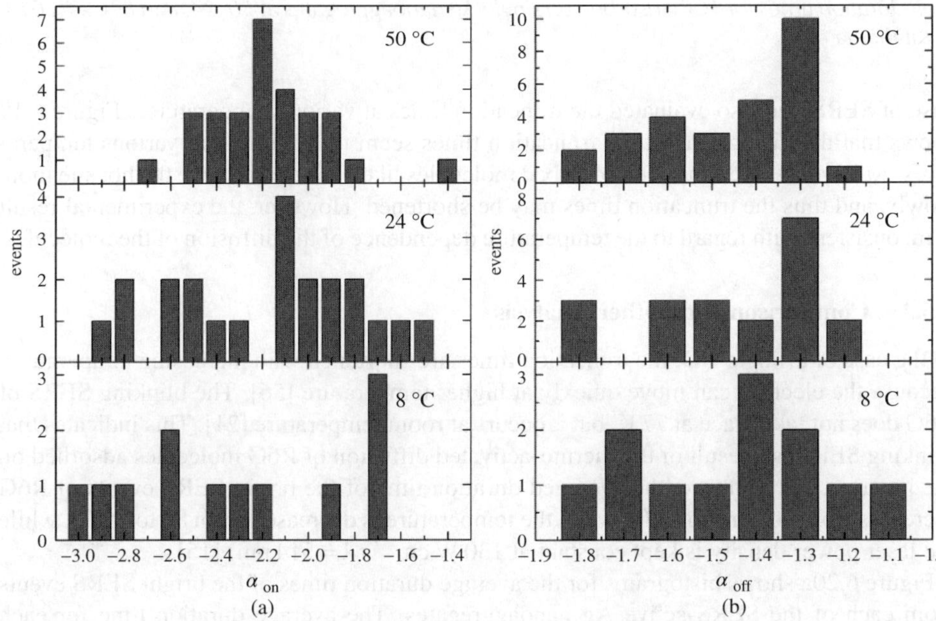

Figure 6.18 *Histograms for exponents of the power law that reproduces the probability distributions of 568 nm-excited (a) bright and (b) dark SERS events versus their duration times for TCC adsorbed on single Ag nanoaggregates at 50, 24, and 8 °C. © 2013, Y. Kitahama et al.*

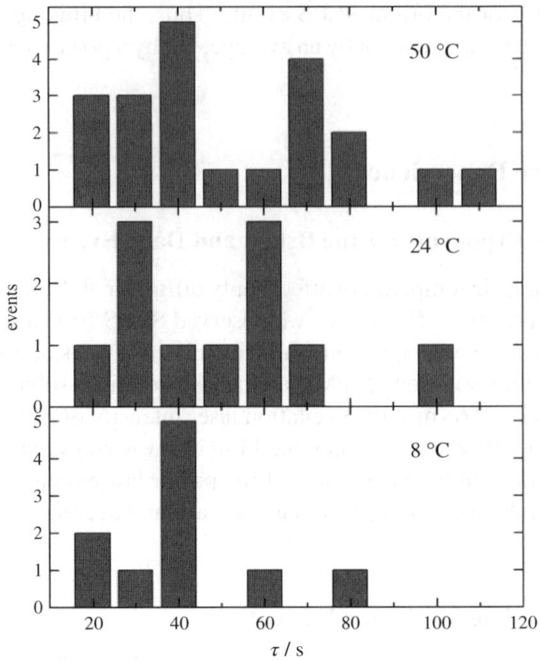

Figure 6.19 *Histograms for the truncation times of the power law with exponential function that reproduces probability distributions of 568 nm-excited bright and dark SERS events against their duration times for TCC adsorbed on single Ag nanoaggregates at 50, 24, and 8 °C. © 2013, Y. Kitahama et al.*

case of SERS, we also evaluated the truncation times at various temperatures. Figure 6.19 shows that the histograms of the truncation times seem to be similar at various temperatures. At lower temperature, the adsorbed molecules likely enter and leave the hot site more slowly, and thus the truncation times may be shortened. However, the experimental result is inconsistent with regard to the temperature dependence of the diffusion of the molecules.

6.6.3 Comparison with Other Analysis

In the case of a single QD, the truncation times are shortened with increasing temperature, because the electron can move quickly at higher temperature [56]. The blinking SERS of R6G does not take place at 77 K, but it occurs at room temperature [21]. This indicates that blinking SERS is a result of the thermo-activated diffusion of R6G molecules adsorbed on the junction. Furthermore, the averaged duration time of the bright SERS events of R6G increases from ~15 ms to ~10 s when the temperature is decreased from 80 to 25 °C, while the laser power density is kept constant at 130 W cm^{-2} ($\lambda = 514$ nm) [15].

Figure 6.20a shows histograms for the average duration times of the bright SERS events from each of the SERS-active Ag nanoaggregates. The average duration time for each shows a downward trend at lower temperature. Figure 6.20b shows the total duration times of the bright SERS events plotted against the number of them. By the slopes of the fitting lines in Figure 6.20b, the average of the whole was derived to be 0.86, 0.63, and 0.56 s at

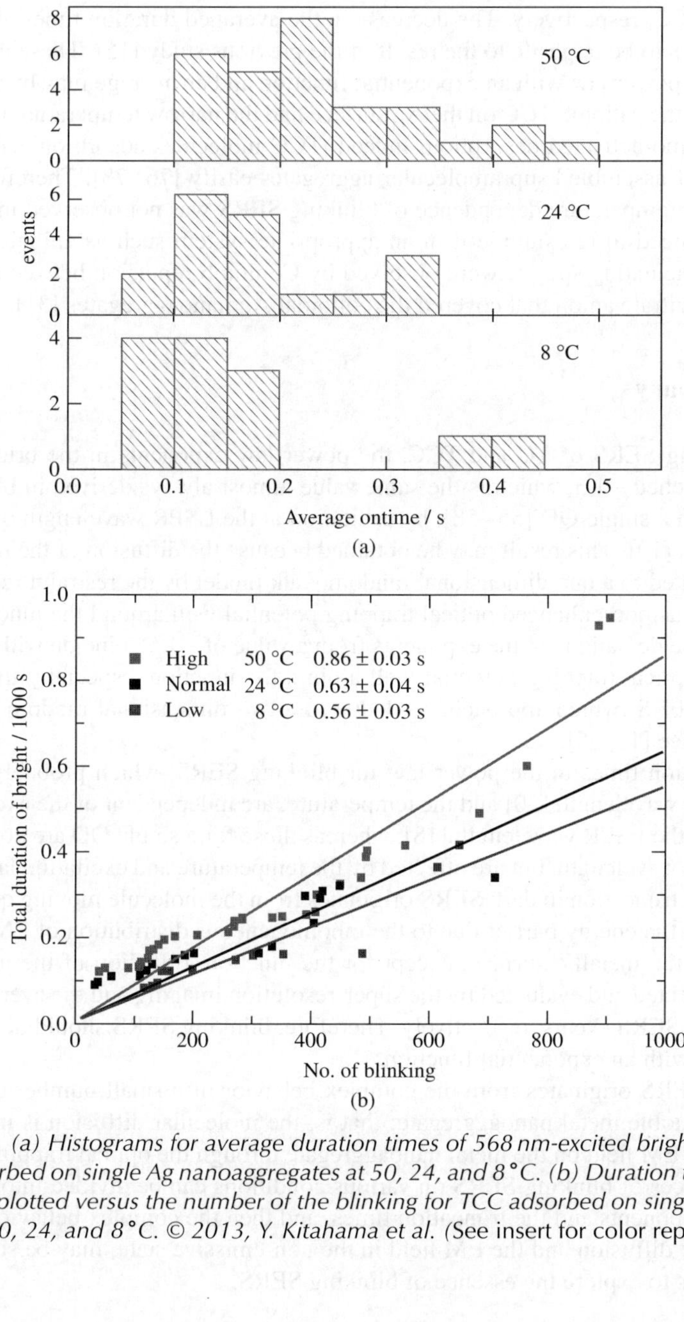

Figure 6.20 *(a) Histograms for average duration times of 568 nm-excited bright SERS events for TCC adsorbed on single Ag nanoaggregates at 50, 24, and 8 °C. (b) Duration times of bright SERS events plotted versus the number of the blinking for TCC adsorbed on single Ag nanoaggregates at 50, 24, and 8 °C. © 2013, Y. Kitahama et al. (See insert for color representation of the figure.)*

50, 24, and 8 °C, respectively. The decrease in the averaged duration times at lower temperature seems to be opposite to the result of the previous study [15]. These unreasonable analyses by a power law with an exponential function and an average may be due to stable adsorption of the cationic TCC on the Ag surface and the narrow temperature range, which suppress thermo-activated desorption. Indeed, TCC molecules adsorb on each other and then form self-assembled supramolecular aggregates easily [76–78]. Therefore, we conclude that the temperature dependence of blinking SERS was not observed in this system and thus we need to re-examine it in an appropriate system such as anionic TC, whose temporally fluctuating spectra were observed by Coulomb repulsion between the anionic dyes and the citrate anion that covered the colloidal Ag nanoaggregates [31].

6.7 Summary

In the blinking SERS of TC and TCC, the power law exponents in the bright and dark SERS approached −1.5, which is the same value almost always derived in blinking fluorescence from a single QD [55–58], by excitation at the LSPR wavelength or under high laser intensity [19]. This result may be obtained because the diffusion of the molecule can be approximated to a one-dimensional random walk model by the restraint resulting from the surface-plasmon-enhanced optical trapping potential well around the junction. On the other hand, the deviations of the exponents from a value of −1.5 coincide with the shallow and narrow optical trapping potential well around the junction, especially the exponents in the dark SERS, which approached −1 due to a two-dimensional random walk on the metallic surface [19, 55].

The truncation times in the power law for blinking SERS, which probably depend on the excitation wavelength [20] and the temperature, are independent of the excitation laser intensity and the LSPR wavelength [18], whereas those for a single QD are not affected by the excitation wavelength, but are affected by the temperature and excitation laser intensity [56–58]. The truncation in dark SERS originates from the molecule moving quickly to the "hot-site" via the energy barrier due to the inhomogeneous distribution of EM around the junction. On the metallic surface, except for the junction, diffusion of the molecule can hardly be detected and evaluated by the super-resolution imaging and the average duration times of dark SERS events, respectively. Therefore, blinking SERS should be analyzed by a power law with an exponential function.

Blinking SERS originates from the complex behavior of a small number of molecules and a single noble metal nanoaggregate; that is, the molecular diffusion is influenced by the enhanced EM field on the metal nanoaggregate through the optical trapping. However, the dependences of blinking SERS on various conditions can be divided into those of the power law exponents and the truncation times, and then the complex behavior, especially the molecular diffusion and the EM field in the non-emissive state, may be separated into simple factors to explore the essence of blinking SERS.

Acknowledgments

We would like to express appreciation to Prof. N. Tamai of Kwansei Gakuin University for enlightening us on the blinking fluorescence from a single QD. The authors are grateful to Dr T. Itoh of the National Institute of Advanced Industrial Science and Technology (AIST) for fruitful discussion. We also thank Mr Y. Tanaka, Mr T. Nagahiro, and Miss A. Enogaki, who are students of Kwansei Gakuin University, for their contributions to these studies. This work was supported by the Collaborative Development of Innovative Seeds of the Japan Science and Technology Agency (JST), KAKENHI (Grant-in-Aid for Scientific Research) in the priority area of "Strong Photon-Molecule Coupling Fields (No. 470, 21020034)" and for Young Scientists B (No. 23750025), and a Support Project to Assist Private Universities in Developing Bases for Research (Research Center for Single Molecule Vibrational Spectroscopy, 2009–2013) from the Ministry of Education, Culture, Sports, Science and Technology of Japan (MEXT).

References

1. Aroca, R. (2006) *Surface-Enhanced Vibrational Spectroscopy*, John Wiley & Sons, Ltd., Chichester.
2. Kneipp, K., Moskovits, M. and Kneipp, H. (2006) *Surface-Enhanced Raman Scattering–Physics and Applications*, Springer-Verlag, Heidelberg, Berlin.
3. Le Ru, E.C. and Etchegoin, P.G. (2009) *Principles of Surface-Enhanced Raman Spectroscopy and Related Plasmonic Effects*, Elsevier, Amsterdam.
4. Itoh, T., Sujith, A., and Ozaki, Y. (2009) Surface-enhanced Raman scattering. in *Frontiers of Molecular Spectroscopy*, ed. J. Laane, Chapter 10, pp 289–320, Elsevier, Amsterdam.
5. Kitahama, Y., Itoh, T., Pienpinijtham, P., Ekgasit, S., Han X. X., and Ozaki, Y. (2013) Biological applications of SERS using functional nanoparticles. in *Functional Nanoparticles for Bioanalysis, Nanomedicine and Bioelectronic Devices*, eds M. Hepel and C-J. Zhong, Vol. 2, Chapter 9, pp 181–234, ACS Symposium Series, Vol. **1113**, American Chemical Society, Washington, DC.
6. Kitahama, Y., Hossain, M. K., Itoh, T., Sujith, A., Han, X. X., and Ozaki, Y. (2010) Surface-enhanced Raman scattering imaging: application and experimental approach by far-field with conventional set-up. in *Raman, Infrared, and Near-Infrared Chemical Imaging*, eds S. Sacic and Y. Ozaki, Chapter 15, pp 285–300, John Wiley & Sons, Inc., Hoboken, NJ.
7. Kneipp, K., Wang, Y., Kneipp, H. *et al.* (1997) Single molecule detection using surface-enhanced Raman scattering (SERS). *Physical Review Letters*, **78**, 1667.
8. Nie, S. and Emory, S.R. (1997) Probing single molecules and single nanoparticles by surface-enhanced Raman scattering. *Science*, **275**, 1102.
9. Qian, X.-M. and Nie, S.M. (2008) Single-molecule and single-nanoparticle SERS: from fundamental mechanisms to biomedical applications. *Chemical Society Reviews*, **37**, 912.
10. Pieczonka, N.P.W. and Aroca, R.F. (2008) Single molecule analysis by surfaced-enhanced Raman scattering. *Chemical Society Reviews*, **37**, 946.
11. Kneipp, J., Kneipp, H. and Kneipp, K. (2008) SERS – a single-molecule and nanoscale tool for bioanalytics. *Chemical Society Reviews*, **37**, 1052.
12. Etchegoin, P.G. and Le Ru, E.C. (2008) A perspective on single molecule SERS: current status and future challenges. *Physical Chemistry Chemical Physics*, **10**, 6079.

13. Habuchi, S., Cotlet, M., Gronheid, R. *et al.* (2003) Single-molecule surface enhanced resonance Raman spectroscopy of the enhanced green fluorescent protein. *Journal of the American Chemical Society*, **125**, 8446.
14. Weiss, A. and Haran, G. (2001) Time-dependent single-molecule Raman scattering as a probe of surface dynamics. *The Journal of Physical Chemistry B*, **105**, 12348.
15. Emory, S.R., Jensen, R.A., Wenda, T. *et al.* (2006) Re-examining the origins of spectral blinking in single-molecule and single-nanoparticle SERS. *Faraday Discussions*, **132**, 249.
16. Bizzarri, A.R. and Cannistraro, S. (2005) Levy statistics of vibrational mode fluctuations of single molecules from surface-enhanced Raman scattering. *Physical Review Letters*, **94**, 068303.
17. Kitahama, Y., Tanaka, Y., Itoh, T. and Ozaki, Y. (2010) Power-law statistics in blinking SERS of thiacyanine adsorbed on a single silver nanoaggregate. *Physical Chemistry Chemical Physics*, **12**, 7457.
18. Kitahama, Y., Tanaka, Y., Itoh, T. and Ozaki, Y. (2011) Analysis of excitation laser intensity dependence of blinking SERRS of thiacarbocyanine adsorbed on single silver nanoaggregates by using a power law with an exponential function. *Chemical Communications*, **47**, 3888.
19. Kitahama, Y., Tanaka, Y., Itoh, T. and Ozaki, Y. (2011) Power-law analysis of surface-plasmon-enhanced electromagnetic field dependence of blinking SERS of thiacyanine or thiacarbocyanine adsorbed on single silver nanoaggregates. *Physical Chemistry Chemical Physics*, **13**, 7439.
20. Kitahama, Y., Enogaki, A., Tanaka, Y. *et al.* (2013) Truncated power law analysis of blinking SERS of thiacyanine molecules adsorbed on single silver nanoaggregates by excitation at various wavelengths. *Journal of Physical Chemistry C*, **117**, 9397.
21. Maruyama, Y., Ishikawa, M. and Futamata, M. (2004) Thermal activation of blinking in SERS signals. *The Journal of Physical Chemistry B*, **108**, 673.
22. Itoh, T., Hashimoto, K., Biju, V. *et al.* (2006) Elucidation of interaction between metal-free tetraphenylporphine and surface Ag atoms through temporal fluctuation of surface-enhanced resonance Raman scattering and background-light emission. *The Journal of Physical Chemistry B*, **110**, 9579.
23. Lombardi, J.R., Birke, R.L. and Haran, G. (2011) Single molecule SERS spectral blinking and vibronic coupling. *Journal of Physical Chemistry C*, **115**, 4540.
24. Šašić, S., Itoh, T. and Ozaki, Y. (2005) Surface enhanced Raman spectroscopy. *Journal of Raman Spectroscopy*, **36**, 593.
25. Ruan, C., Wang, W. and Gu, B. (2007) Single-molecule detection of thionine on aggregated gold nanoparticles by surface enhanced Raman scattering. *Journal of Raman Spectroscopy*, **38**, 568.
26. Doering, W.E. and Nie, S. (2002) Single-molecule and single-nanoparticle SERS: examining the roles of surface active sites and chemical enhancement. *The Journal of Physical Chemistry B*, **106**, 311.
27. Bizzarri, A.R. and Cannistraro, S. (2005) SERS and tunnelling spectroscopy investigation of iron-protoporphyrin IX adsorbed on a silver tip. *The Journal of Physical Chemistry B*, **109**, 16571.
28. Bizzarri, A.R. and Cannistraro, S. (2007) Statistical analysis of intensity fluctuations in single molecule SERS spectra. *Physical Chemistry Chemical Physics*, **9**, 5315.

29. Le Ru, E.C., Etchegoin, P.G. and Meyer, M. (2006) Enhancement factor distribution around a single surface-enhanced Raman scattering hot spot and its relation to single molecule detection. *The Journal of Chemical Physics*, **125**, 204701.
30. Etchegoin, P.G., Lacharmoise, P.D. and Le Ru, E.C. (2009) Influence of photostability on single-molecule surface enhanced Raman scattering enhancement factors. *Analytical Chemistry*, **81**, 682.
31. Kitahama, Y., Tanaka, Y., Itoh, T. *et al.* (2009) Identification of thiacyanine J-aggregates adsorbed on single silver nanoaggregates by surface-enhanced Raman scattering and emission spectroscopy. *Bulletin of the Chemical Society of Japan*, **82**, 1126.
32. Shegai, T.O. and Haran, G. (2006) Probing the Raman scattering tensors of individual molecules. *The Journal of Physical Chemistry B*, **110**, 2459.
33. Wang, Z. and Rothberg, L.J. (2005) Origins of blinking in single-molecule Raman spectroscopy. *The Journal of Physical Chemistry B*, **109**, 3387.
34. Peyser-Capadona, L., Zheng, J., González, J.I. *et al.* (2005) Nanoparticle-free single molecule anti-stokes Raman spectroscopy. *Physical Review Letters*, **94**, 058301.
35. Michaels, A.M., Nirmal, M. and Brus, L.E. (1999) Surface enhanced Raman spectroscopy of individual Rhodamine 6G molecules on large Ag nanocrystals. *Journal of the American Chemical Society*, **121**, 9932.
36. Michaels, A.M., Jiang, J. and Brus, L.E. (2000) Ag nanocrystal junctions as the site for surface-enhanced Raman scattering of single Rhodamine 6 G molecules. *The Journal of Physical Chemistry B*, **104**, 11965.
37. Bosnick, K.A., Jiang, J. and Brus, L.E. (2002) Fluctuations and local symmetry in single-molecule Rhodamine 6G Raman scattering on silver nanocrystal aggregates. *The Journal of Physical Chemistry B*, **106**, 8096.
38. Itoh, T., Iga, M., Tamaru, H. *et al.* (2012) Quantitative evaluation of blinking in surface enhanced resonance Raman scattering and fluorescence by electromagnetic mechanism. *The Journal of Chemical Physics*, **136**, 024703.
39. Stranahan, S.M. and Willets, K.A. (2010) Super-resolution optical imaging of single-molecule SERS hot spots. *Nano Letters*, **10**, 3777.
40. Weber, M.L., Litz, J.P., Masiello, D.J. and Willets, K.A. (2012) Characterization of the electron- and photon-driven plasmonic excitations of metal nanorods. *ACS Nano*, **6**, 1839.
41. Le Ru, E.C., Meyer, M. and Etchegoin, P.G. (2006) Resolving single molecules in surface-enhanced Raman scattering within the inhomogeneous broadening of Raman peaks. *The Journal of Physical Chemistry B*, **110**, 1944.
42. Dieringer, J.A., Lettan, R.B., Scheidt, K.A. and Van Duyne, R.P. (2007) A frequency domain existence proof of single-molecule surface enhanced Raman spectroscopy. *Journal of the American Chemical Society*, **129**, 16249.
43. Goulet, P.J.G., Pieczonka, N.P.W. and Aroca, R.F. (2005) Mapping single-molecule SERRS from Langmuir-Blodgett monolayers on nanostructured silver island films. *Journal of Raman Spectroscopy*, **36**, 574.
44. Sawai, Y., Takimoto, B., Nabika, H. *et al.* (2007) Observation of a small number of molecules at a metal nanogap arrayed on a solid surface using surface-enhanced Raman scattering. *Journal of the American Chemical Society*, **129**, 1658.

45. Takase, M., Sawai, Y., Nabika, H. and Murakoshi, K. (2011) Detection of adsorption sites at the gap of a hetero-metal nano-dimer at the single molecule level. *Journal of Photochemistry and Photobiology, A: Chemistry*, **221**, 169.
46. Nagasawa, F., Takase, M., Nabika, H. and Murakoshi, K. (2011) Polarization characteristics of surface-enhanced Raman scattering from a small number of molecules at the gap of a metal nano-dimer. *Chemical Communications*, **47**, 4514.
47. Kudelski, A. and Pettinger, B. (2000) SERS on carbon chain segments: monitoring locally surface chemistry. *Chemical Physics Letters*, **321**, 356.
48. Kudelski, A. (2006) Raman spectroscopy studies for diagnosis of cancers in human uterine cervix. *Vibrational Spectroscopy*, **41**, 83.
49. Domke, K.F., Zhang, D. and Pettinger, B. (2007) Enhanced Raman spectroscopy: single molecules or carbon? *Journal of Physical Chemistry C*, **111**, 8611.
50. Itoh, K., Kudryashov, I., Yamagata, J. *et al.* (2005) Raman microspectroscopic study on polymerization and degradation processes of a diacetylene derivative at surface enhanced Raman scattering active substrates. 2. Confocal Raman microscopic observation of polydiacetylene adsorbed on active sites. *The Journal of Physical Chemistry B*, **109**, 271.
51. Szeghalmi, A., Kaminskyj, S., Rösch, P. *et al.* (2007) Time-fluctuations and imaging in the SERS spectra of fungal hyphae grown on nanostructured substrates. *The Journal of Physical Chemistry B*, **111**, 12916.
52. Chaigneau, M., Picardi, G. and Ossikovski, R. (2010) Tip enhanced Raman imaging of amorphous carbon contamination on gold surfaces. *Surface Science*, **604**, 701.
53. Krichevsky, O. and Bonnet, G. (2002) Fluorescence-correlation spectroscopy: the technique and its applications. *Reports on Progress in Physics*, **65**, 251.
54. Hess, S.T., Huang, S., Heikal, A.A. and Webb, W.W. (2002) Biological and chemical applications of fluorescence correlation spectroscopy: a review. *Biochemistry*, **41**, 697.
55. Cichos, F., von Borczyskowski, C. and Orrit, M. (2007) Power-law intermittency of single emitters. *Current Opinion in Colloid & Interface Science*, **12**, 272.
56. Shimizu, K.T., Neuhauser, R.G., Leatherdale, C.A. *et al.* (2001) Blinking statistics in single semiconductor nanocrystal quantum dots. *Physical Review B*, **63**, 205316.
57. Crouch, C.H., Mohr, R., Emmons, T. *et al.* (2009) Excitation energy dependence of fluorescence intermittency in CdSe/ZnS core-shell nanocrystals. *Journal of Physical Chemistry C*, **113**, 12059.
58. Peterson, J.J. and Nesbitt, D.J. (2009) Modified power law behavior in quantum dot blinking: a novel role for biexcitons and auger ionization. *Nano Letters*, **9**, 338.
59. Tang, J. and Marcus, R.A. (2005) Mechanisms of fluorescence blinking in semiconductor nanocrystal quantum dots. *The Journal of Chemical Physics*, **123**, 054704.
60. Schroeder, M. (1991) *Fractals, Chaos, Power Laws*, W. H. Freeman & Co., New York.
61. Lee, P.C. and Meisel, D.J. (1982) Adsorption and surface-enhanced Raman of dyes on silver and gold sols. *The Journal of Physical Chemistry*, **86**, 3391.
62. Galloway, C.M., Etchegoin, P.G. and Le Ru, E.C. (2009) Ultrafast nonradiative decay rates on metallic surfaces by comparing surface-enhanced Raman and fluorescence signals of single molecules. *Physical Review Letters*, **103**, 063003.
63. Yoshida, K., Itoh, T., Tamaru, H. *et al.* (2010) Quantitative evaluation of electromagnetic enhancement in surface-enhanced resonance Raman scattering from plasmonic

properties and morphologies of individual Ag nanostructures. *Physical Review B*, **81**, 115406.
64. Itoh, T., Yoshida, K., Biju, V. *et al.* (2007) Second enhancement in surface-enhanced resonance Raman scattering revealed by an analysis of anti-Stokes and Stokes Raman spectra. *Physical Review B*, **76**, 085405.
65. Itoh, T., Biju, V., Ishikawa, M. *et al.* (2006) Surface-enhanced resonance Raman scattering and background light emission coupled with plasmon of single Ag nanoaggregates. *The Journal of Chemical Physics*, **124**, 134708.
66. Xu, H. and Käll, M. (2002) Surface-plasmon enhanced optical forces in silver nanoaggregates. *Physical Review Letters*, **89**, 246802.
67. Luo, X. and Ishihara, T. (2004) Surface plasmon resonant interference nanolithography technique. *Applied Physics Letters*, **84**, 4780.
68. García-Vidal, F.J. and Pendry, J.B. (1996) Collective theory for surface enhanced Raman scattering. *Physical Review Letters*, **77**, 1163.
69. Futamata, M. and Maruyama, Y. (2008) LSP spectral changes correlating with SERS activation and quenching for R6G on immobilized Ag nanoparticles. *Applied Physics B: Lasers and Optics*, **93**, 117.
70. Li, K., Stockman, M.I. and Bergman, D.J. (2003) Self-similar chain of metal nanospheres as an efficient nanolens. *Physical Review Letters*, **91**, 227402.
71. Sasaki, K., Tanaka, Y., and Fujiwara, H. (2012) Proceedings of Yamada Conference LXVI, OP08.
72. Tanaka, Y., Sanada, A. and Sasaki, K. (2012) Nanoscale interference patterns of gap-mode multipolar plasmonic fields. *Scientific Reports*, **2**, 764.
73. Yoshida, K., Itoh, T., Biju, V. *et al.* (2009) Experimental evaluation of the twofold electromagnetic enhancement theory of surface-enhanced resonance Raman scattering. *Physical Review B*, **79**, 085419.
74. Biju, V., Makita, Y., Nagase, T. *et al.* (2005) Subsecond luminescence intensity fluctuations of single CdSe quantum dots. *The Journal of Physical Chemistry B*, **109**, 14350.
75. Osborne, M.A., Balasubramanian, S., Furey, W.S. and Klenerman, D. (1998) Optically biased diffusion of single molecules studied by confocal fluorescence microscopy. *The Journal of Physical Chemistry B*, **102**, 3160.
76. Chibisov, A.K. and Görner, H. (2002) Photophysics of aggregated 9-methylthia carbocyanine bound to polyanions. *Chemical Physics Letters*, **357**, 434.
77. Slavnova, T.D., Chibisov, A.K. and Görner, H. (2002) Photoprocesses of thiacarbocyanine monomers, dimers, and aggregates bound to polyanions. *The Journal of Physical Chemistry A*, **106**, 10985.
78. Kitahama, Y., Ogawa, A., Tanaka, Y. *et al.* (2010) Difference in time dependence of surface-enhanced Raman scattering spectra of thiacarbocyanine J- and H-aggregates adsorbed on single silver nanoaggregates. *Chemical Physics Letters*, **493**, 309.

7

Tip-Enhanced Raman Spectroscopy (TERS) for Nanoscale Imaging and Analysis

Taka-aki Yano[1] *and Satoshi Kawata*[2]

[1]*Department of Electronic Chemistry, Tokyo Institute of Technology, Japan*
[2]*Department of Applied Physics, Osaka University, Japan*

7.1 Crucial Difference between TERS and SERS

Tip-enhanced Raman spectroscopy (TERS) is an evolved version of surface-enhanced Raman spectroscopy (SERS). A metallic nano-tip plays an important role in confining and enhancing the optical field in the vicinity of the tip apex [1–3]. This is due to local excitation of surface plasmon polaritons (SPPs), which is known as the primary mechanism of electromagnetic enhancement in SERS. In TERS measurement, the metallic nano-tip acts as a SERS probe, and is brought into close proximity to sample molecules. The tip is laterally scanned over the sample surface, as shown in Figure 7.1a. This approach enables one to perform SERS imaging at a spatial resolution far beyond the diffraction limit of light. On the other hand, in SERS measurement, sample molecules are put on immobilized metallic nanostructures (Figure 7.1b), and then the molecules near the metallic surface are detected without nano-resolution.

Differently from SERS, the metal–molecule distance (Δd) is controllable in TERS measurement. When Δd is ~10 nm, only the electromagnetic effect occurs (Figure 7.2a). Raman scattering intensity increases plasmonically as the distance decreases. When Δd is less than 1 nm, a metal–molecule complex is formed (Figure 7.2b). Due to charge

Frontiers of Surface-Enhanced Raman Scattering: Single Nanoparticles and Single Cells, First Edition.
Edited by Yukihiro Ozaki, Katrin Kneipp and Ricardo Aroca.

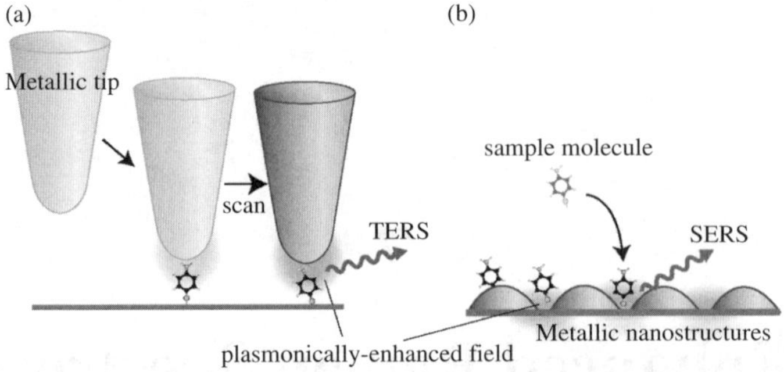

Figure 7.1 Conceptional difference between (a) TERS and (b) SERS

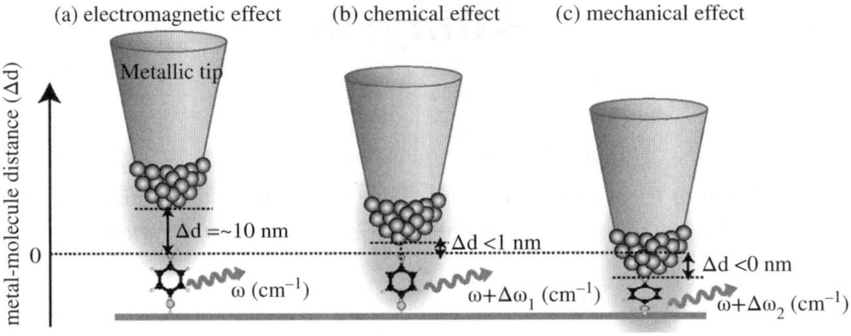

Figure 7.2 TERS-specific effects as a function of tip–sample distance (Δd): (a) electromagnetic, (b) chemical, and (c) mechanical effects

transfer between the metal and the molecule, both the Raman scattering intensity and frequency of the adsorbed molecule change. This is known as the chemical effect, which is observed also in SERS [4–6]. Subnanometric distance regulation in TERS can control the degree of the chemical effect. When the distance is further reduced, pressure can be applied by the tip onto the sample molecule. The pressure deforms the molecule locally under the tip, which results in a negative value of Δd (Figure 7.2c). The Raman intensity and frequency of the pressurized molecule are further modified by molecular deformation. This is a unique function of TERS, and is known as the mechanical effect. Details of the TERS-specific effects will be discussed in Section 7.2 with some experimental results.

The origin of TERS goes back to 1992 when Kawata and Inouye invented an apertureless near-field scanning optical microscope (NSOM). An illuminated metallic scanning tunneling microscope (STM) tip under total internal reflection was scanned over the sample to obtain an optical image. This result was published in 1994 [7]. This was the first proposal and the first demonstration of apertureless (scattering-type) NSOM with a probe to scatter the light. Bachelot also reported the same concept in 1995, where he used a bent conical tungsten tip in the tapping mode of atomic force microscopy (AFM) [8]. Keillmann and Knoll extended apertureless NSOM to work at radio [9] and microwave [10] frequencies, and in the mid-infrared [11]. There were no involvements of SPP because metals are not plasmonic but perfect conductors in the microwave frequency region. In 1999, Xie utilized apertureless NSOM to locally excite two-photon fluorescence of sample molecules using a gold tip, and obtained images with spatial resolution of 30 nm [12]. In the same year, Kawata *et al.* published the first tip-enhanced Raman study [13]. Zenobi also published in 2000 [14].

7.2 TERS-Specific Spectral Change as a Function of Tip–Sample Distance

Watanabe *et al.* investigated the chemical effect between adenine molecules and a silver tip (Figure 7.3a). Figure 7.3b shows two possible adenine–silver complexes (Ad-N7 and Ad-N3) with different adsorbed sites. The ring breathing modes (RBMs) of the Ad-N7 and Ad-N3 complexes were calculated by density functional theory (DFT). The calculated frequencies are plotted as a function of bond distance (*d*) between the silver atom and the nitrogen of adenine in Figure 7.3c [15, 16]. The Ad-N3 complex exhibits a significant shift toward a higher frequency by as much as 20 cm^{-1} with decreasing bond distance while the Ad-N7 exhibits a frequency shift of only a few cm^{-1}. Figure 7.3d shows the binding energies of both the Ad-N3 and the Ad-N7 complexes calculated by changing the bond distance. The equilibrated bond distances of both complexes were calculated to be around 2.5 Å. A bond distance shorter than 2.5 Å indicates contraction of the bond, and hence the contracted bond involves the mechanical effect in addition to the chemical effect. Inclusion of the mechanical effect in TERS measurement provides a clear difference in frequency between the Ad-N3 and the Ad-N7 complexes, which enables us to determine molecular orientation in the vicinity of the metallic tip [16]. The mechanical effect is a unique feature only observed in TERS and not in SERS.

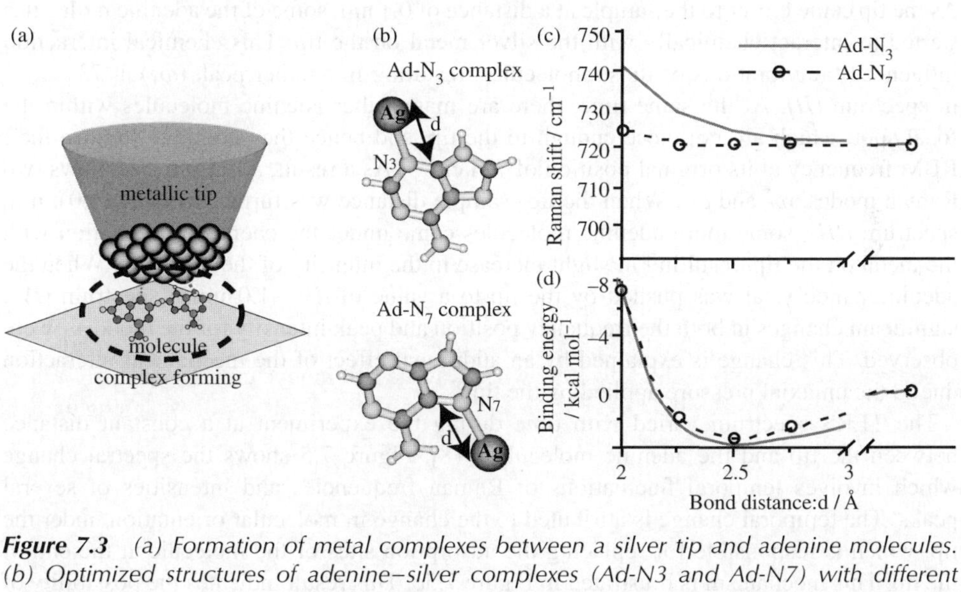

Figure 7.3 (a) Formation of metal complexes between a silver tip and adenine molecules. (b) Optimized structures of adenine–silver complexes (Ad-N3 and Ad-N7) with different adsorbed sites. (c) The calculated frequency shift of the Ad-N3 and Ad-N7 complexes, and (d) the calculated binding energies as a function of the bond distance for the Ag–N linkage. Reproduced with permission from [15]. © 2004 The American Physical Society

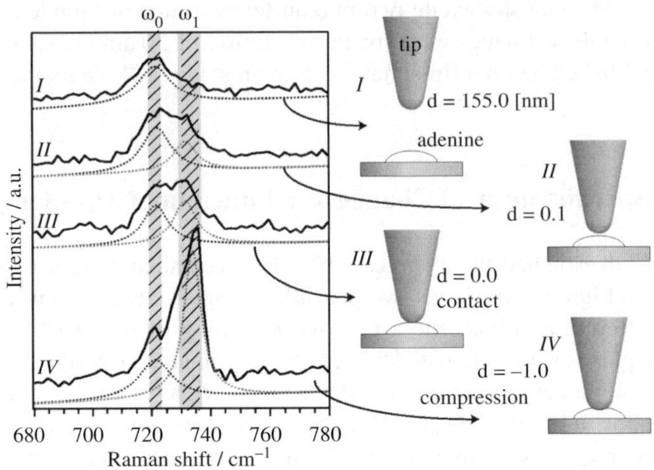

Figure 7.4 *Raman spectra of an adenine nanocrystal at tip–sample distances indicated in the corresponding illustrations. The peaks marked by ω_0 and ω_1 represent the unperturbed RBM and the frequency-shifted RBM, respectively. Deconvoluted Lorentzian peaks for each spectrum are also shown by the dotted curves. Reproduced with permission from [17].*

Figure 7.4 shows TERS spectra of an adenine crystal detected with a silver-coated tip [17]. The spectra were measured at silver–adenine distances between −1.0 and 155 nm. When the distance is 155 nm, a peak (ω_0) of RBM is observed at 721 cm^{-1} in spectrum *(I)*. As the tip came closer to the sample at a distance of 0.1 nm, some of the adenine molecules started to interact chemically with the silver metal on the tip. This chemical interaction influenced the vibrations of those molecules, resulting in another peak (ω_1) at 731 cm^{-1} in spectrum *(II)*. At the same time, there are many other adenine molecules within the focal spot, which are not close enough to the tip, and hence they continue to have their RBM frequency at its original position of 721 cm^{-1}. As a result, spectrum *(II)* shows two Raman modes, ω_0 and ω_1. When the tip–sample distance was further reduced to 0 nm in spectrum *(III)*, some more adenine molecules came under the chemical interaction with the metal on the tip, resulting in slight increase in the intensity of the ω_1 mode. When the adenine nanocrystal was pushed by the tip to a value of $d = -1.0$ nm in spectrum *(IV)*, significant changes in both the frequency position and peak intensity for the mode ω_1 were observed. This change is explained by an additional effect of the mechanical interaction due to the uniaxial pressure applied by the tip.

The TERS spectrum varied with time during the experiment at a constant distance between the tip and the adenine molecules [18]. Figure 7.5 shows the spectral change which involves temporal fluctuations of Raman frequencies and intensities of several peaks. The temporal change is attributed to the change in molecular orientation under the tip, which is responsible for replacing the adsorption sites of the molecule at the end of the tip. This mechanism is illustrated in Figure 7.5c. This result indicates the possibility of the single-molecule-detection sensitivity of TERS spectroscopy of DNA base molecules.

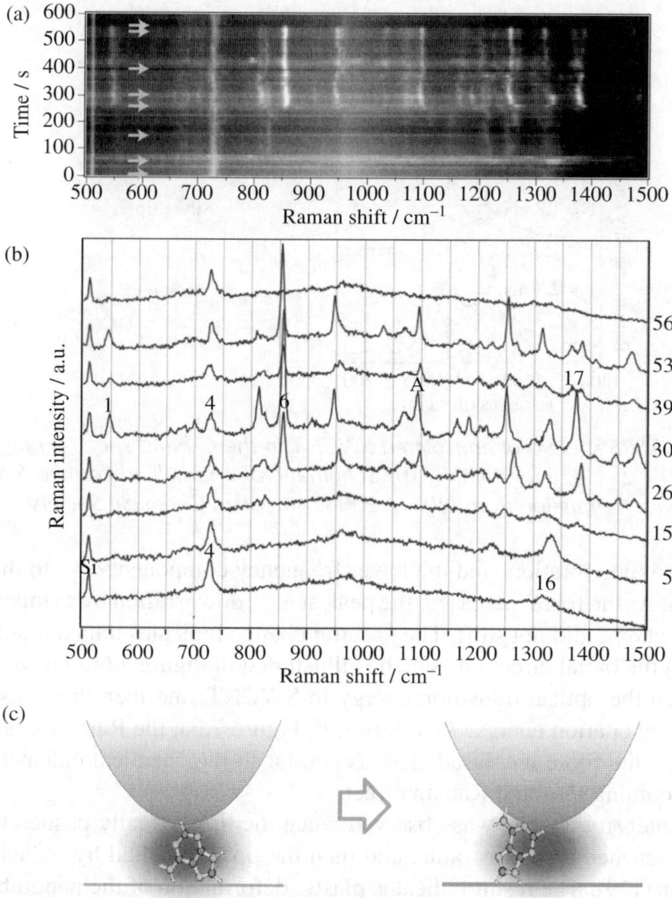

Figure 7.5 *(a) Waterfall plot of a time evolution of TERS spectra of an adenine nanocrystal. (b) Selected characteristic TERS spectra taken at different times indicated by arrows in (a). (c) Schematic of orientation change of an adenine molecule under the silver tip. Reprinted with permission from [18]. © 2007, American Chemical Society*

7.3 Mechanical Effect in TERS

When a metallic probe tip comes into contact with a sample molecule, and then force is applied to the sample by the tip, the molecule of the sample will be deformed. The deformed molecule will exhibit an interesting TERS spectral change in both peak-position and intensity [16, 17, 19–21]. The force can be controllable in the sub-nano newton scale if AFM is used.

Figure 7.6a shows the spectra of a single-walled carbon nanotube (SWCNT) [20]. The spectra were measured under a force of 0, 1.2, 1.8, and 2.4 nN in the G-band (tangential stretching mode of C–C bonds). In the spectra, the higher frequency component ω_{G+} is

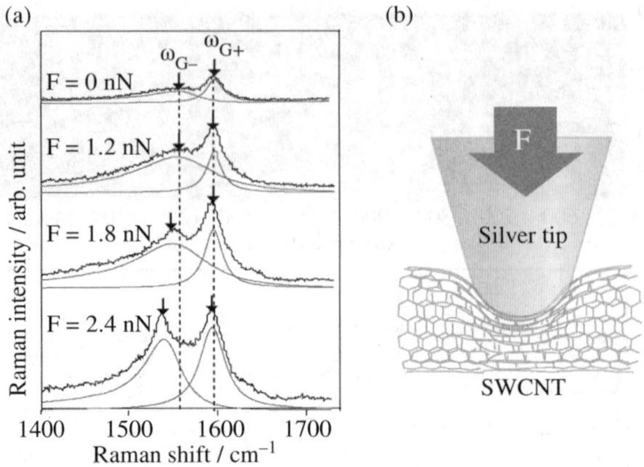

Figure 7.6 *(a) TERS spectra of an isolated SWCNT in the G-band spectral range at four different forces (0, 1.2, 1.8, and 2.4 nN). (b) Schematic of a radially-deformed SWCNT by the tip. Reprinted with permission from [20]. © 2006, American Chemical Society*

assigned to the axial vibration, and the lower frequency component ω_{G-} to the circumferential vibration. As the force increased, the peak at ω_{G-} downshifted by as much as 18 cm^{-1} while the peak at ω_{G+} did not shift. The spectral change indicates uniaxial deformation of the nanotube in the radial direction, which is illustrated in Figure 7.6b. The uniaxial deformation modified the optical transition energy in SWCNT, and then the transition energy got close to the excitation energy. As a result, in Figure 7.6a, the Raman scattering intensity increased as the force increased. This is similar to the chemical enhancement which involves the incoming resonant Raman effect.

Irreversible intensity change was observed when the tip gradually pushed the nanotube until the force reached a certain value, and then the tip was pulled by reducing the force (data not shown) [19]. The result indicates plastic deformation of the nanotube during the push–pull cycle. The irreversible spectral response enables us to determine the threshold value of the force for plasmatic deformation, which is a sample-dependent nanoelastic property.

C60 molecules were also measured under the controlled force (Figure 7.7a) [21]. TERS spectra were measured with increasing the force (F) from 0 to 8.7 nN by the silver tip. In Figure 7.7b, the peak of the radial breathing Ag mode was observed at 496 cm^{-1} without applying the force ($F = 0$ nN). The peak position shifted to higher frequency by 1.5 cm^{-1} as the force increased to 8.7 nN. The spectral shift is predicted from the DFT calculation, indicating the uniaxial deformation of the C60 molecules by 0.1 Å. The Hg mode was also measured in the same experiment (Figure 7.7c). No peak shifts were observed for the Hg mode with increasing force, but the peak gradually broadened. This result indicates the degeneracy of the Hg mode due to symmetry breaking of the C60 molecule.

7.4 Application to Analytical Nano-Imaging

TERS has been applied to analytical nano-imaging of materials, such as nano-carbon materials [14, 21–29], semiconductors [30–40], and biomolecules [18, 41–47].

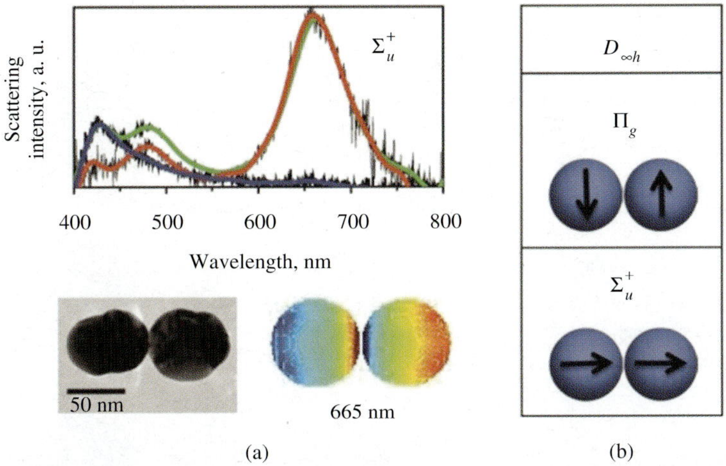

Figure 3.1 *Plasmonic dimer. (a) Plasmon scattering spectrum. Thin black lines – raw experimental data; other lines – low pass filter applied. The clusters are excited by either non-polarized light (green) or light linearly polarized parallel (red) and perpendicular (blue) to dimer axis. The polarized spectra are normalized to the corresponding peaks of the non-polarized spectra. Lower panel – TEM image of the dimer and numerically simulated surface charge distribution (the scale bar is 50 nm). (b) Schematic presentation of bonding plasmon modes. The symbols of the corresponding irreducible representations are shown. Adapted with permission from: Ref. [36]. Copyright (2011) American Chemical Society.*

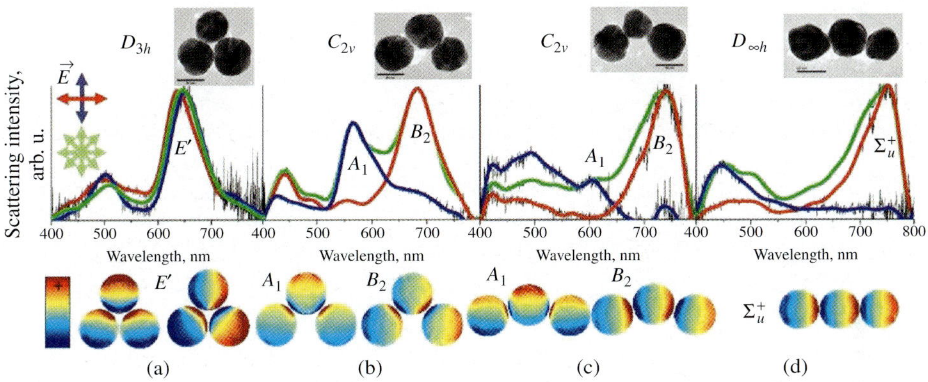

Figure 3.2 *(a–d) Scattering spectra of silver nanoparticle trimers showing mode splitting due to gradual opening of the vertex angle. Thin black lines, raw experimental data; color lines, low pass filter applied. The clusters are excited by either non-polarized light (green) or light linearly polarized along the longitudinal (red) and transverse (blue) directions (which are indicated by arrows of matching colors). The polarized spectra are normalized to the corresponding peaks of the non-polarized spectra. Plasmon mode symmetries of the bonding modes are marked. TEM images of the clusters are shown in the insets (the bars are 50 nm). Calculated surface charge distributions of the selectively excited bonding modes are shown below the spectra. Adapted with permission from: Ref. [35]. Copyright (2011) American Chemical Society.*

Frontiers of Surface-Enhanced Raman Scattering: Single Nanoparticles and Single Cells, First Edition.
Edited by Yukihiro Ozaki, Katrin Kneipp and Ricardo Aroca.

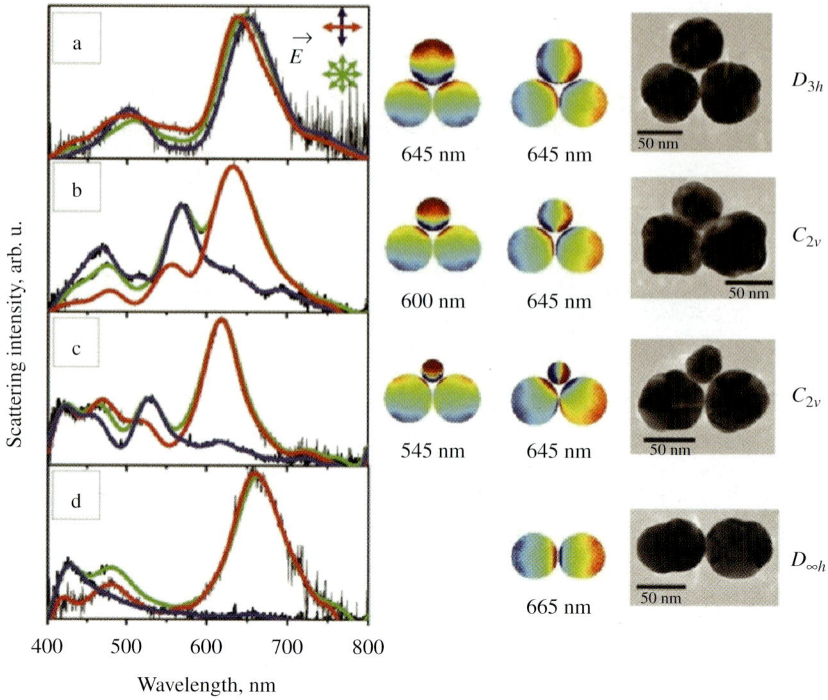

Figure 3.4 *Experimental scattering spectra, numerical simulations of surface charge distribution, and TEM images (scale bar is 50 nm) of the plasmon spectra of symmetry-broken trimers for different sizes of the vertex nanoparticle. Green lines denote excitation by non-polarized light, red lines excitation by linearly polarized light along the trimer's base, and blue lines excitation by light polarized perpendicular to the trimer's base. Polarized spectra are normalized to the corresponding peaks of the non-polarized spectra. The sizes of vertex nanoparticle (R_1) used in the simulations are 32, 24, 16, and 0 nm corresponding to R_1/R_2 values of 1, 0.75, 0.5, and 0 respectively. The corresponding symmetry point groups are also indicated. Adapted with permission from: Ref. [36]. Copyright (2011) American Chemical Society.*

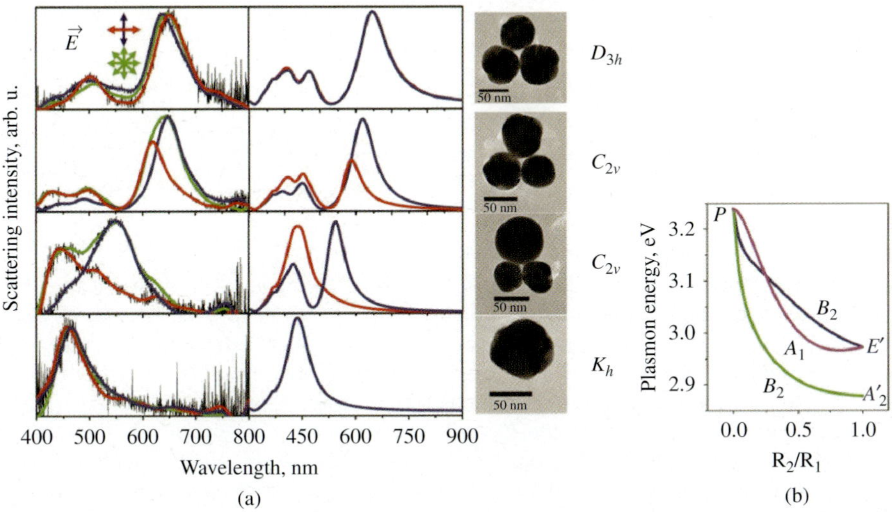

Figure 3.6 *(a) Experimental scattering spectra, corresponding TEM images (scale bar is 50 nm), and numerical simulations of the plasmon spectra of symmetry-broken trimers for varying sizes of the base nanoparticles. Green lines represent excitation by the non-polarized light, red lines excitation by linearly polarized light along the trimer's base, and blue lines excitation by light polarized perpendicularly to the trimer's base. The sizes of the base nanoparticles* R_2 *used in the simulations are 32, 24, 8, and 0 nm corresponding to* R_2/R_1 *ratios of 1, 0.75, 0.25, and 0 respectively. Polarized spectra are normalized to the corresponding peaks of the non-polarized spectra. The corresponding symmetry point groups are indicated. (b) Plasmon energy diagram of the in-plane bonding dipolar eigenmodes calculated by plasmon hybridization theory. The interparticle separation is 1 nm. The symmetry symbols of the plasmon modes are marked in agreement with the symmetry group correlation tables:* $R_2 = 1$ *corresponds to* D_{3h}*,* $0 < R_2/R_1 < 1$ *corresponds to* C_{2v}*, and* $R_2/R_1 = 0$ *corresponds to* K_h*. Adapted with permission from: Ref. [36]. Copyright (2011) American Chemical Society.*

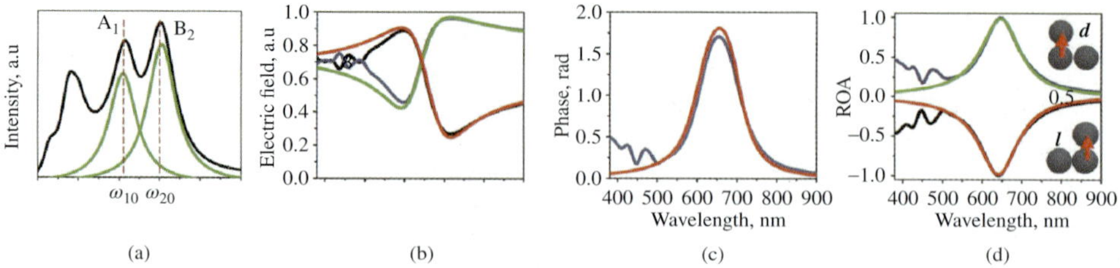

Figure 3.9 *SMSERs from a symmetry-broken silver trimer ($\phi = 90°$, $R_1 = R_2 = R_3 = 32$ nm, interparticle distance of 1 nm). (a) Emission spectrum with the symmetries of the excited plasmon modes marked. Green lines show Lorentzian fits of the spectral peaks. (b) Relative magnitudes of the electric field components of Raman-scattered light obtained from an electromagnetic simulation of a trimer of* d *configuration (E_z is negligible) and from the harmonic oscillators model based on the parameters of the Lorentzian fit (see(a)). Blue and black lines – simulated results for the E_x and E_y components, respectively, green and red lines – harmonic oscillator model results. (c) Relative phase of the electric field components from (b). Blue line – simulation and red line – model. (d) ROA of the two trimers with* d *and* l *configurations. Black and blue lines – electromagnetic simulation, red and green lines – harmonic oscillator model.*

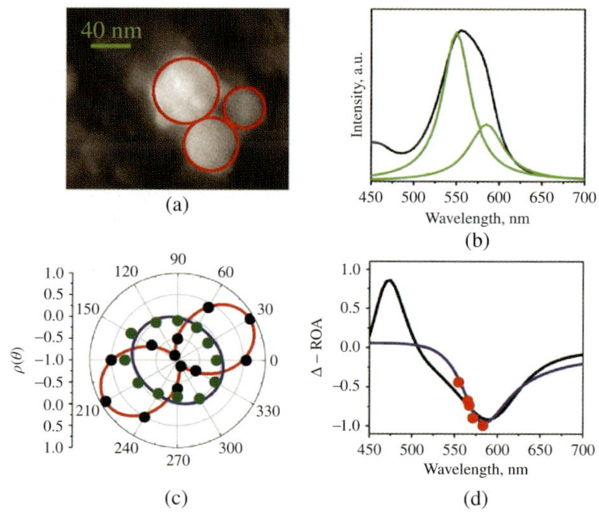

Figure 3.10 *Experimental observation of the effect of a symmetry broken trimer on SERS polarization. (a) SEM image of a trimer. Estimated nanoparticle sizes are $R_1 = 20$ nm, $R_2 = 27$, and $R_3 = 32$ nm, and the distances between nanoparticles are 1 nm (for particles 1–2 and 2–3), and 8 nm (for particles 1–3). (b) Simulated emission spectrum with the nanoparticle dimensions and gap sizes estimated from the image (see text). The molecule is assumed to be located between particles 1 and 2. (c) Experimental polarization functions of the Raman-scattered light measured at 555 nm (black dots) and 583 nm (green dots). Red and blue lines show a fit to Equation 3.3. (d) ROA calculated from the experimental data (dots), from the electromagnetic simulation (black line), and from the harmonic oscillator model (blue line).*

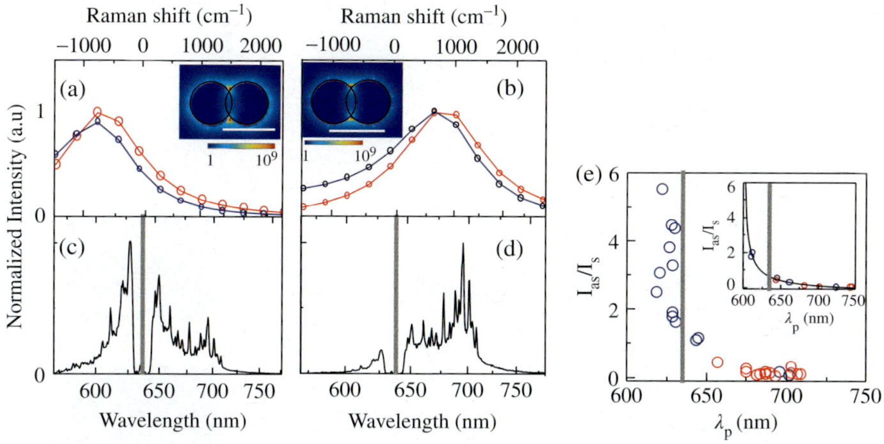

Figure 4.3 *Calculated spectra of plasmon resonance (red lines) and SERS enhancement factors (blue lines) at (a)* $n_{eff} = 1.3$ *and (b)* $n_{eff} = 1.5$. *Insets: spatial distribution of SERS enhancement factors at plasmon resonance maxima. Scale bars are 50 nm. Calculations of SERS spectra at (c)* $n_{eff} = 1.3$ *and (d)* $n_{eff} = 1.5$. *(e) Experiments of* λ_p *dependence of intensity ratios of anti-Stokes to Stokes SERS bands at 619 and 707 nm. Insets: calculations. Blue and red circles show the ratios of anti-Stokes to Stokes SERS bands at 619 and 707 nm without and with covering oil, respectively. Reproduced with permission from Ref. [40]. Copyright © 2009, American Institute of Physics.*

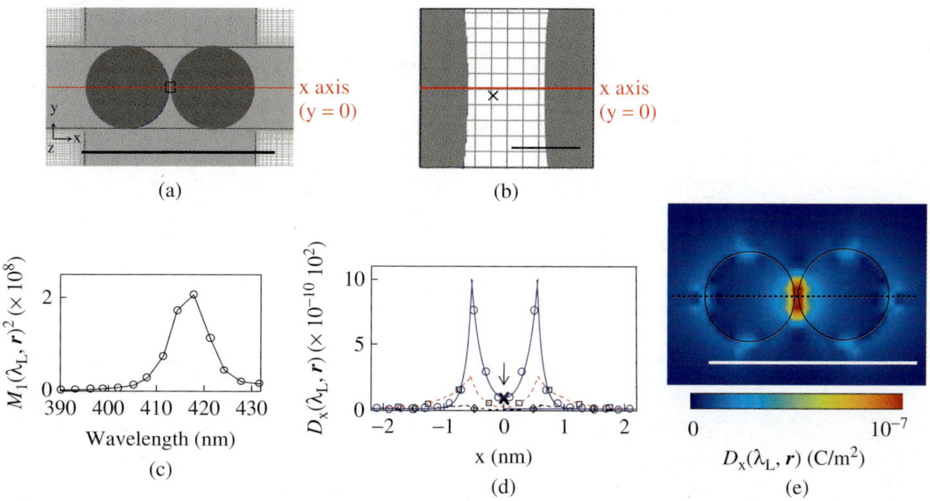

Figure 4.6 *(a) Modeled structure of an Ag NP dimer in the FDTD calculation. (b) Magnified view of the vicinity of the gap in the modeled structure. (c) Calculation spectrum of the twofold EM enhancement factor at the cross in (b). (d) Calculated cross-section of spatial distribution of the electric flux density (diamonds, squares, and circles; mesh size is 1.5, 0.5, and 0.2 nm respectively). (e) Spatial distribution of the electric flux density* D_x. *The cross indicated by the arrow in (d) indicates the calculated enhancement factor in Figure 2A of Ref. [24]. Note that the cross section in (d) indicates the electric flux density along the dashed line in (e). Scale bars in (a, b, and e) are 50, 1.0, and 50 nm, respectively. Reproduced with permission from Ref. [26]. Copyright © 2010, The American Physical Society.*

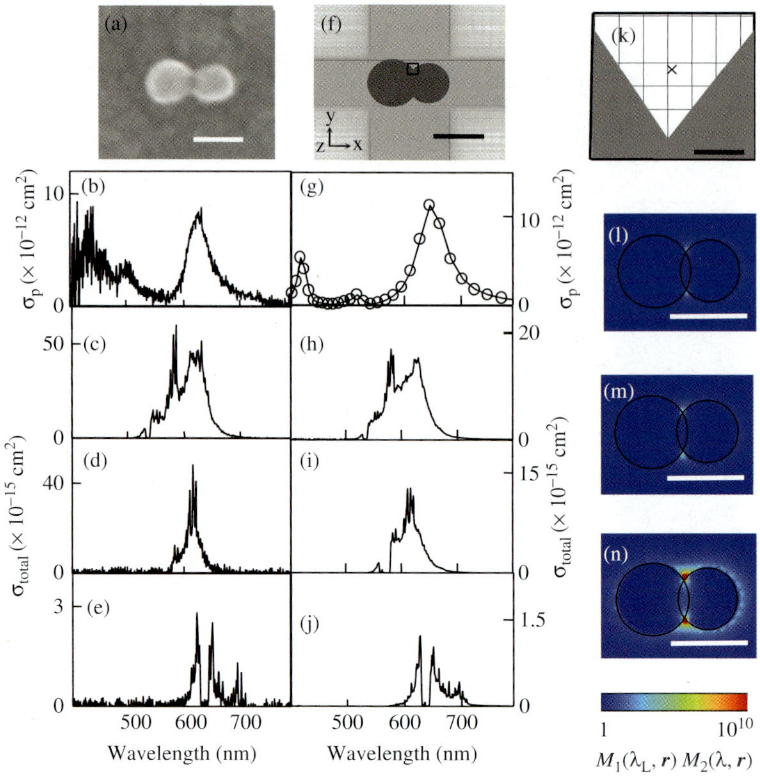

Figure 4.7 *(a) SEM image of an Ag NP dimer, (b) experimental plasmon resonance spectrum, and experimental SERS spectra excited at (c) 532 nm, (d) 561 nm, and (e) 633 nm. (f) Modeled structure of an Ag NP dimer in the FDTD calculation, (g) calculated plasmon resonance spectrum and calculated SERS spectra excited at (h) 532 nm, (i) 561 nm, and (j) 633 nm. (k) Magnified view in the vicinity of the crevice of the model structure in (f). Spatial distribution of calculated twofold EM enhancement factor excited at (l) 532 nm, (m) 561 nm, and (n) 633 nm. Experimental plasmon resonance maximum in (b) was ~620 nm, its FWHM was ~60 nm, and its cross section was ~1×10^{-11} cm^2. Circularly-polarized incident light (532, 561, and 633 nm), the mesh size 0.2 nm, and the refractive index 1.3 of the surrounding medium were selected in the FDTD calculation. Scale bars in (a, f, l, m) are 50 nm. Scale bar in (k) is 0.5 nm. Reproduced with permission from Ref. [26]. Copyright © 2010, The American Physical Society.*

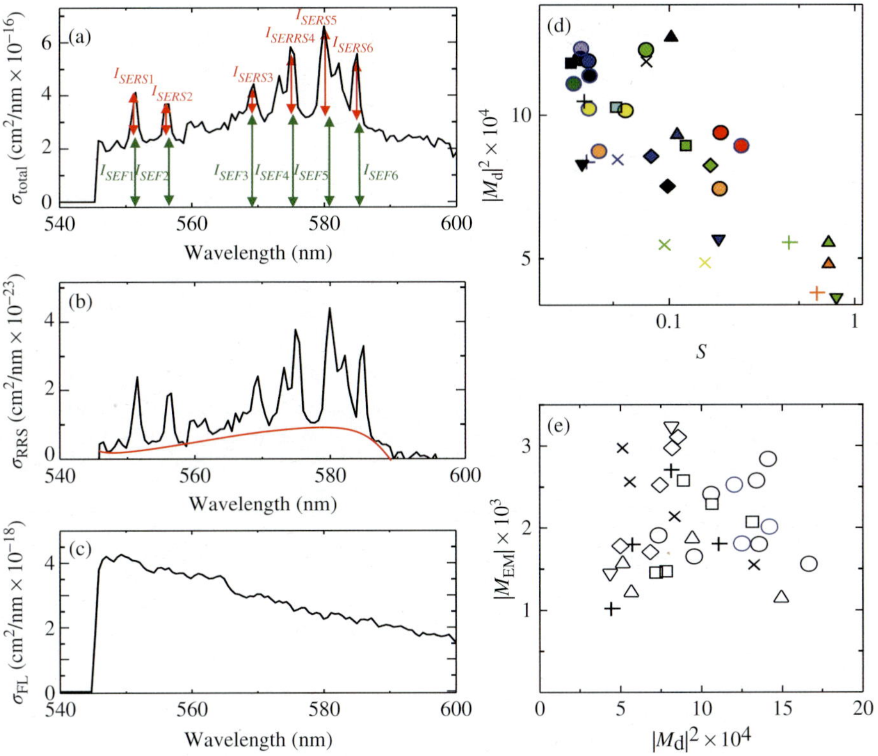

Figure 4.10 *(a) Evaluating I_{SERS}/I_{SEF} for panels (d). Total SERS intensity I_{SERS} and SEF intensity I_{SEF} were averaged over I_{SERS1} to I_{SERS6} (upper arrows) and that of I_{SEF1} to I_{SEF6} (lower arrows), respectively. (b) σ_{RS} of spectrum R6G in an aqueous solution. Raman intensity is defined as an upper part of the red curve that is considered as back ground fluorescence. The unit of the y-axis is converted into a cross section ($cm^2\ nm^{-1}$). (c) σ_{FL} spectrum of R6G in an aqueous solution detected with the same experimental set-up. The unit of the y-axis is converted into a cross section ($cm^2\ nm^{-1}$). (d) Normalized standard deviation scores S dependence of $|M_d|^2$ in $0 < S < 0.6$ for eight dimers denoted by eight symbols (○, ○, □, Δ, ∇, ◇, +, and ×). Filled circles are taken from the Ag NP dimer used for this figure. Number of irradiated laser pulses is indicated by colors. Gray color indicates before laser pulse irradiation, and faint blue, green, orange, and red indicates after first, second, third, and fourth, laser pulse irradiation. (e) $|M_{EM}|$ ($= (I_{SERS}/\sigma_{RS})^{1/2}$) dependence of $|M_d|^2$ ($= I_{SERS}/I_{SEF} \times \sigma_{RS}/\sigma_{FL}$) for the same eight dimers as those in (d). Reproduced with permission from Ref. [50]. Copyright © 2012, American Institute of Physics.*

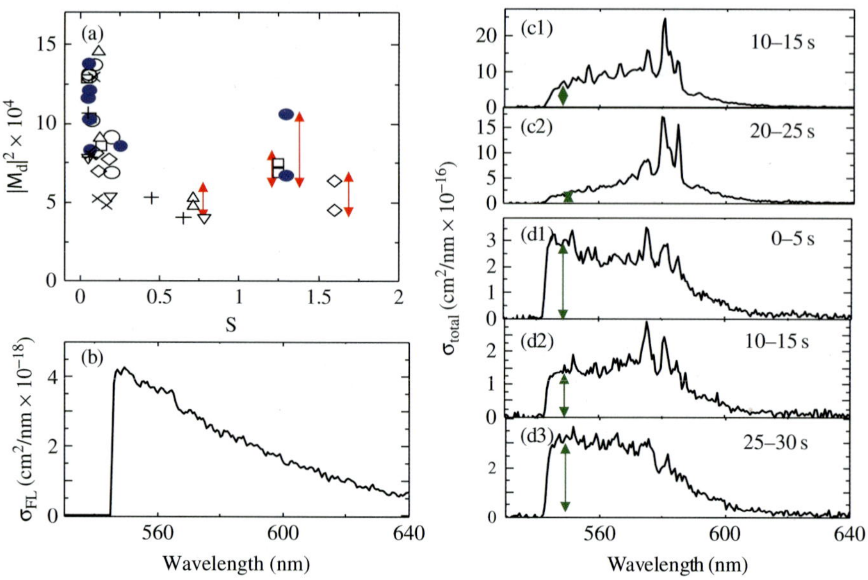

Figure 4.11 *(a) S dependence of $|M_d|^2$ in $0 < S < 2.0$ for the same eight Ag NP dimers as in Figure 4.10. Fluctuation in $|M_d|^2$ is highlighted by binary points (arrows). (b) σ_{FL} spectrum of R6G in an aqueous solution detected with the same experimental set-up. (c1,c2, d1–d3) Intensity fluctuation in SERS and SEF spectra from two Ag NP dimers just after irradiating six and four pulses without irradiating further pulse, respectively. Detection periods are indicated in each figure. The fluorescence intensity that dominantly fluctuated is highlighted by arrows. Reproduced with permission from Ref. [50]. Copyright © 2012, American Institute of Physics.*

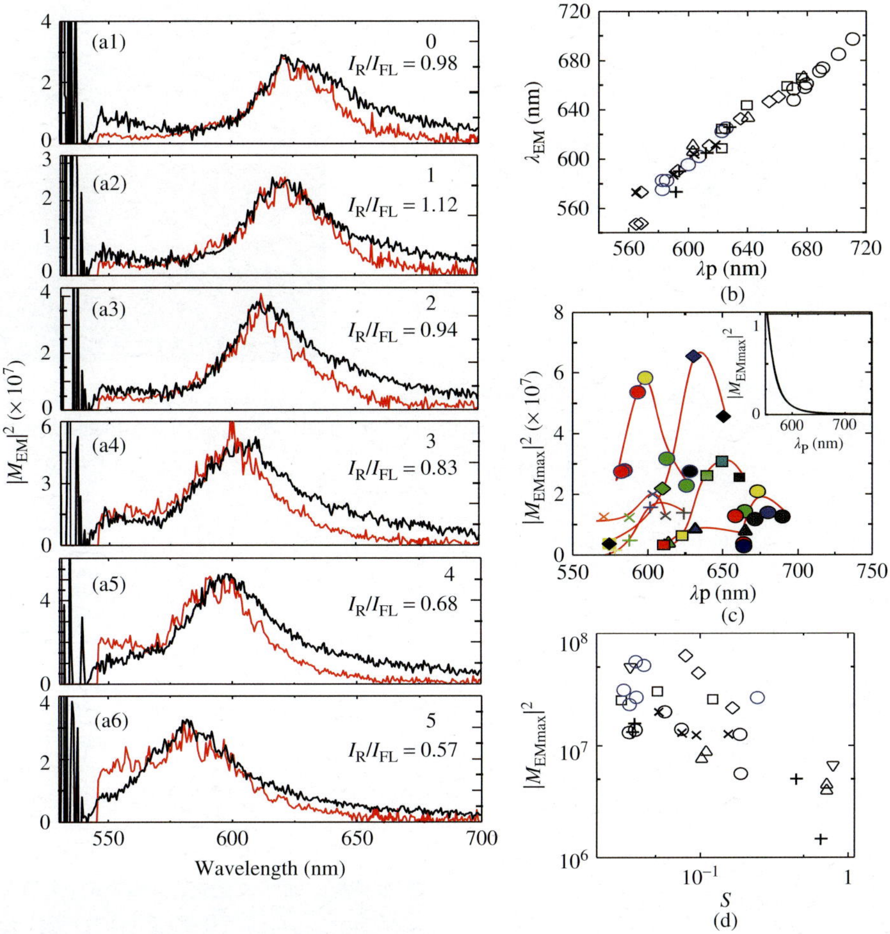

Figure 4.13 *(a1–a6) Spectral changes in plasmon resonance (black lines) and* $|M_{EM}|^2$ *(red lines) from an Ag dimer measured after irradiating with1064-nm laser pulses. The number of the NIR pulses and* I_{SERS}/I_{SEF} *are indicated in each figure.* I_{SERS}/I_{SEF} *is obtained in the same manner as that for Figure 4.11a. Spectra of* σ_{total} *in Figure 4.9b1–b4 are used to obtain* I_{SERS}/I_{SEF}*. (b) Maximum wavelength* λ_p *of plasmon resonance dependence of maximum wavelength* λ_M *of* $|M_{EM}|^2$ *for eight Ag NP dimers denoted by eight symbols (○, ○, □, Δ, ∇, ◇, +, and ×). (c)* λ_p *dependence of maximum value* $|M_{EMmax}|^2$ *of* $|M_{EM}|^2$ *for the eight Ag NP dimers. Inset is calculated* $|M_{EMmax}|^2$ *against* λ_p *assuming a Lorentzian line shape for plasmon resonance. The lines are fitted curves. Number of irradiated laser pulses is indicated by colors. Gray color indicates before NIR laser pulse irradiation, and faint blue, green, orange, red, brown, and moss green indicate after first, second, third, fourth, fifth, and sixth NIR laser pulse irradiation. (d) S dependence of* $|M_{EMmax}|^2$ *for the eight dimers. Filled circles in (b–d) are taken from (a1–a6). Reproduced with permission from Ref. [50]. Copyright © 2012, American Institute of Physics.*

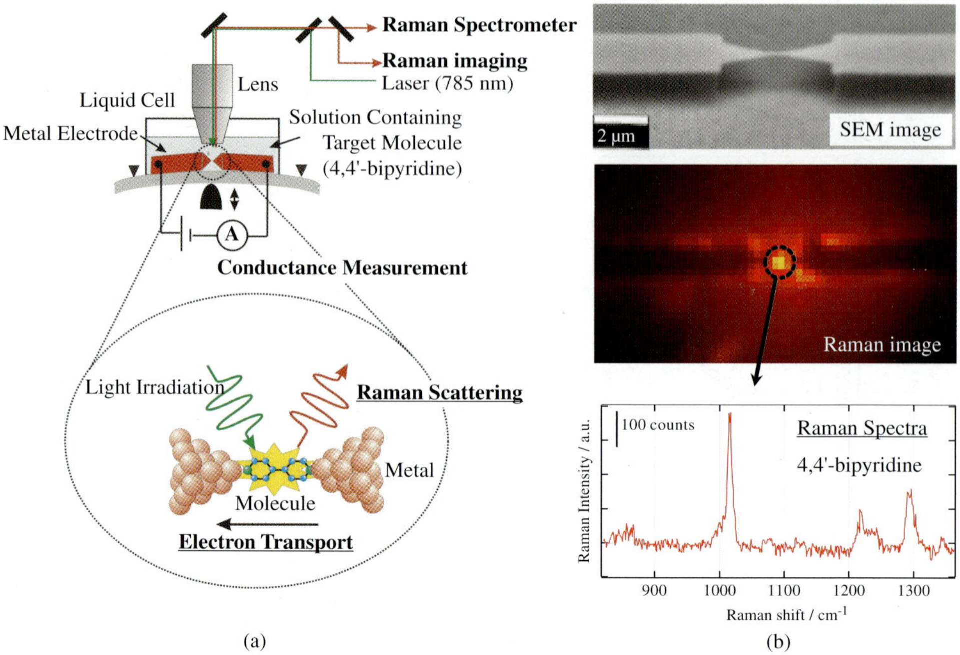

Figure 5.1 *(a) Schematic view of the MCBJ-SERS system. (b) SEM image of an Au nanobridge in the MCBJ sample and Raman imaging of the Au nanogap observed at 1015 cm*$^{-1}$

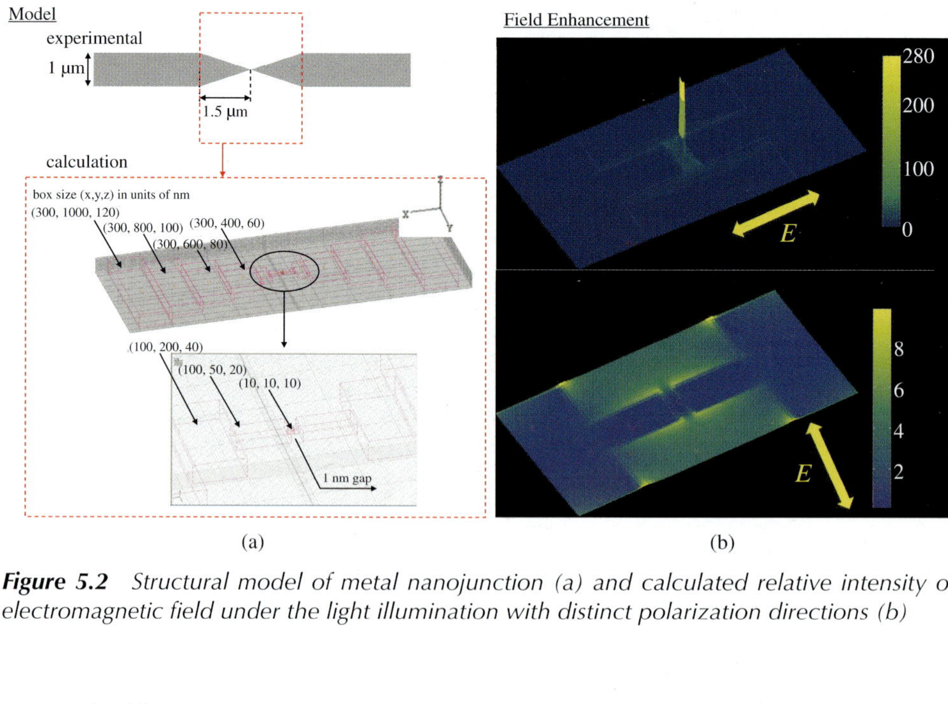

Figure 5.2 *Structural model of metal nanojunction (a) and calculated relative intensity of electromagnetic field under the light illumination with distinct polarization directions (b)*

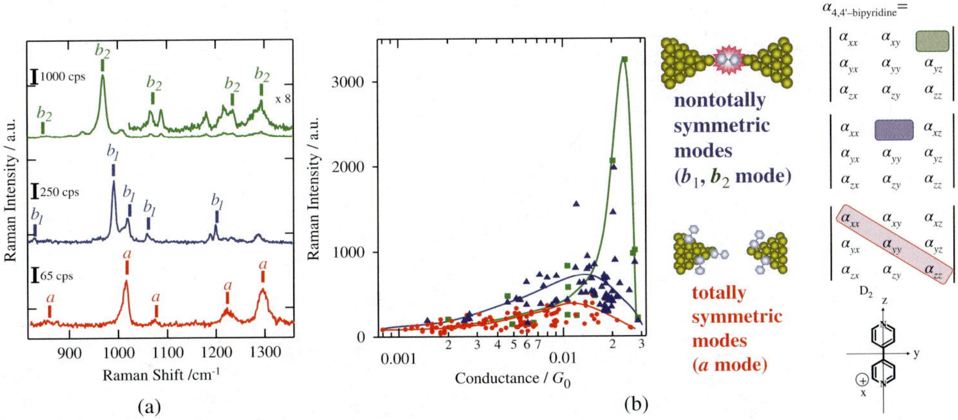

Figure 5.3 *(a) Three types of SERS spectra of a single 4,4′ bipyridine molecule junction. Assignment of the totally symmetric* a *mode , and the nontotally symmetric* b_1 *mode and* b_2 *mode are shown as dotted lines. (b) SERS intensity dependence of* a *(circles),* b_1 *(triangles), and* b_2 *(squares) modes on the conductance of the junction. Illustrations on the right-hand side depict possible structures of the junction and Raman tensors*

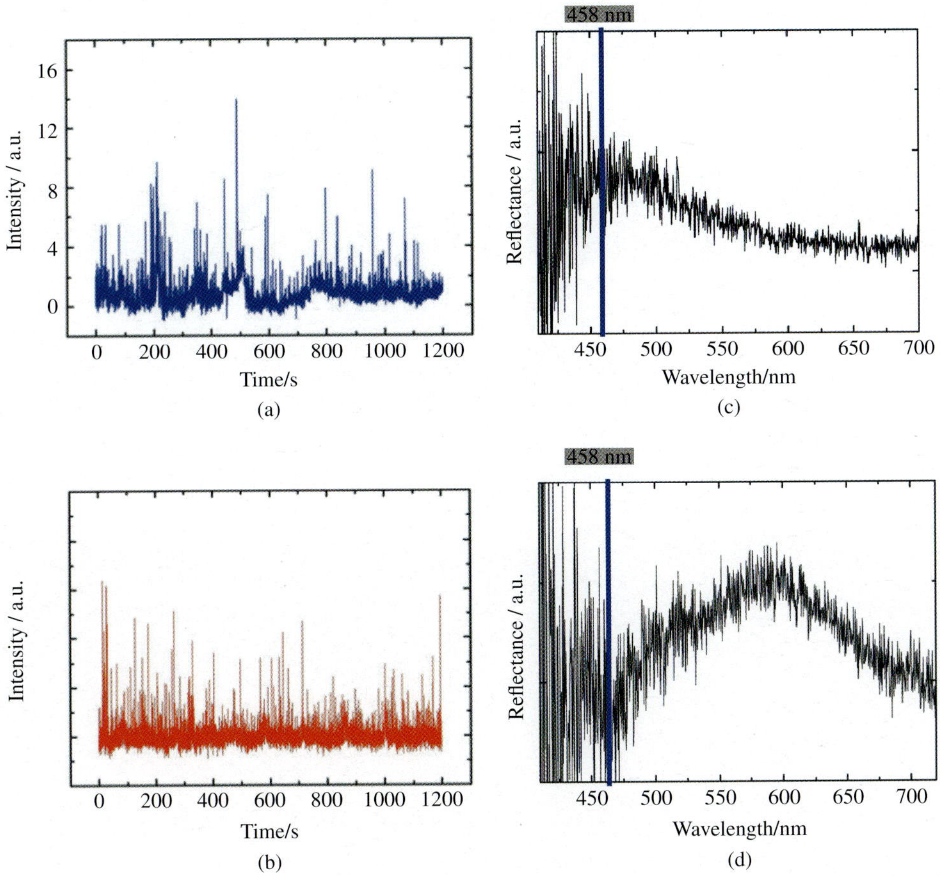

Figure 6.4 *Time-profiles of 458 nm-excited SERS intensities of TC adsorbed on single Ag nanoaggregates with LSPR at (a) 460 nm and (b) 590 nm [19]. (c,d) The LSPR spectra from the single Ag nanoaggregates which show the time-profiles in (a,b), respectively. Reproduced from [19] by permission of The Royal Society of Chemistry.*

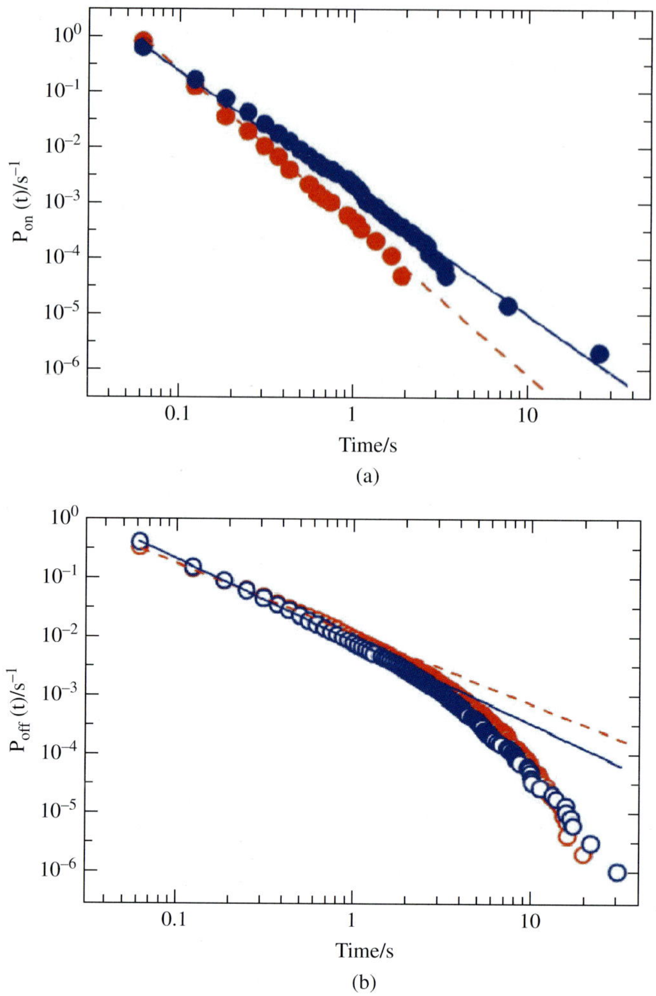

Figure 6.5 *Probability distributions of the 458 nm-excited (a) bright and (b) dark SERS events versus duration times for TC adsorbed on single Ag nanoaggregates with LSPR at 460 nm (solid line) and 590 nm (broken). Reproduced from [19] by permission of The Royal Society of Chemistry.*

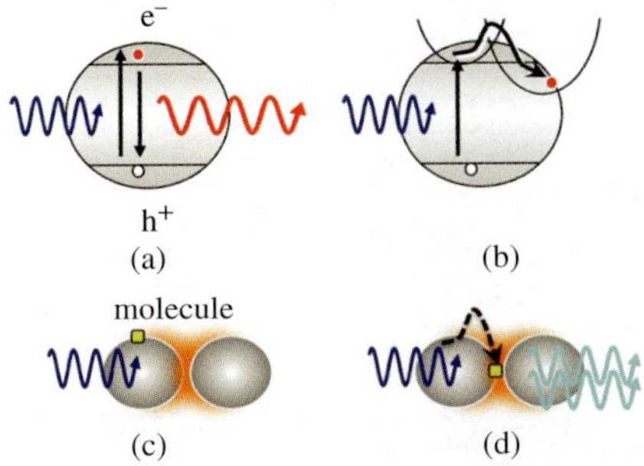

Figure 6.7 Schematic illustrations of (a) the bright and (b) the dark fluorescence events for a single quantum dot, and (c) the dark and (d) the bright SERS events for a single molecule adsorbed on an Ag nanoaggregate. Reproduced from [17] by permission of The Royal Society of Chemistry.

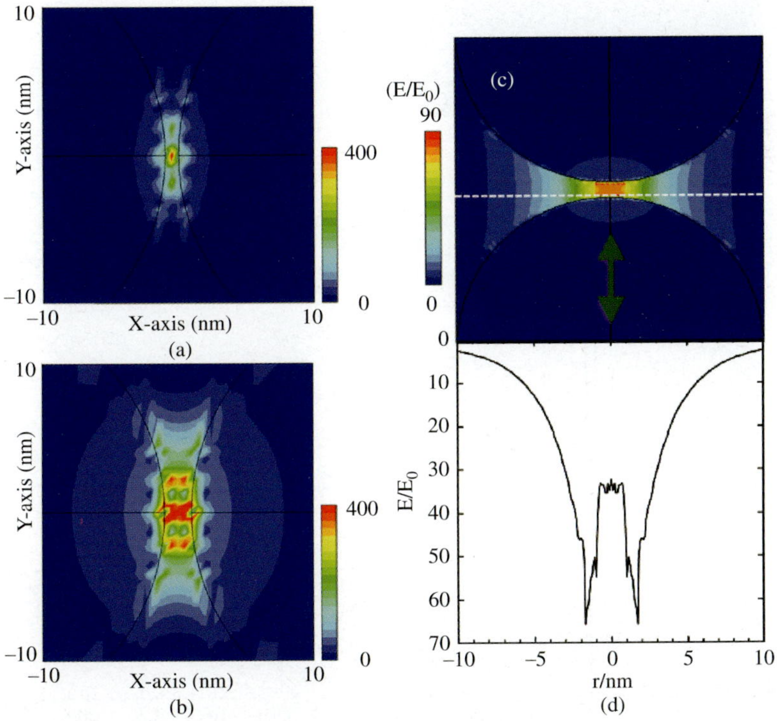

Figure 6.8 Calculated spatial distribution of the EM field around the gap of Ag nanoparticles. Linearly arrayed particles, 16 nm in radii with gap size (a) 1 nm and (b) 2 nm were used. The excitation wavelength was 470 nm [69]. (c) The Ag nanodimer with a diameter of 20 nm and a gap of 1 nm in 0.1 nm mesh was used. The 514-nm polarization along the long axis was used. (d) Calculated intensity of the EM field along a tangential line on the surface at the closest distance to the gap, shown as a white broken line in (c). Reproduced from [18] by permission of The Royal Society of Chemistry.

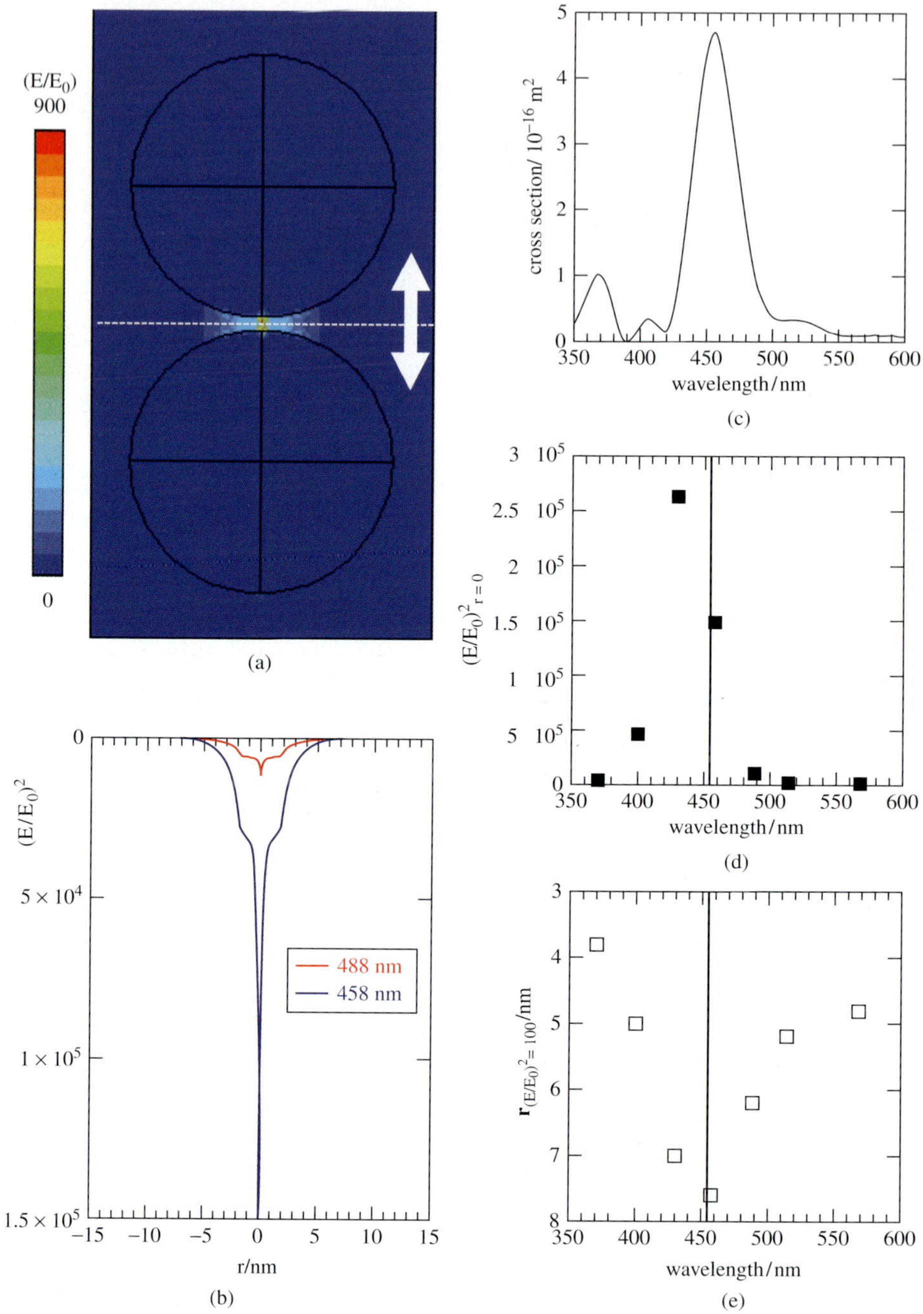

Figure 6.12 *(a) Calculated spatial distribution of the EM field around the gap (1 nm) in Ag nanoparticles with a diameter of 20 nm excited by vertical polarizations at 430 nm. (b) Calculated intensity of the optical trapping potential well along the horizontal broken line in (a). (c) Calculated far-field scattering efficiency spectrum for the Ag nanoparticles referred to in (a). (d) Calculated depths and (e) calculated half-widths where there is a 100-fold enhancement of optical trapping potential wells, like that in (b), at various excitation wavelengths. Reproduced from [19] by permission of The Royal Society of Chemistry.*

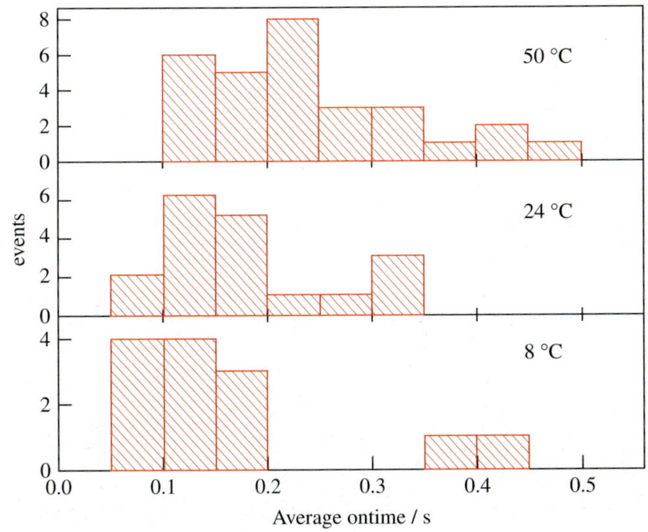

(a)

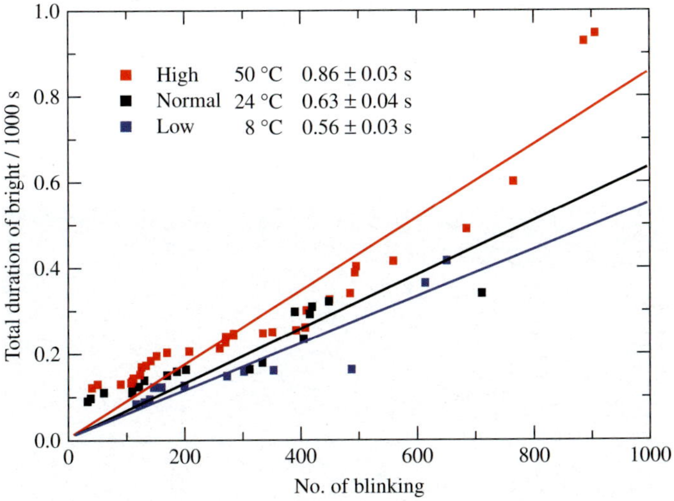

(b)

Figure 6.20 *(a) Histograms for average duration times of 568 nm excited bright SERS events for TCC adsorbed on single Ag nanoaggregates at 50, 24, and 8 °C. (b) Duration times of bright SERS events plotted versus the number of the blinking for TCC adsorbed on single Ag nanoaggregates at 50, 24, and 8 °C. © 2013, Y. Kitahama et al.*

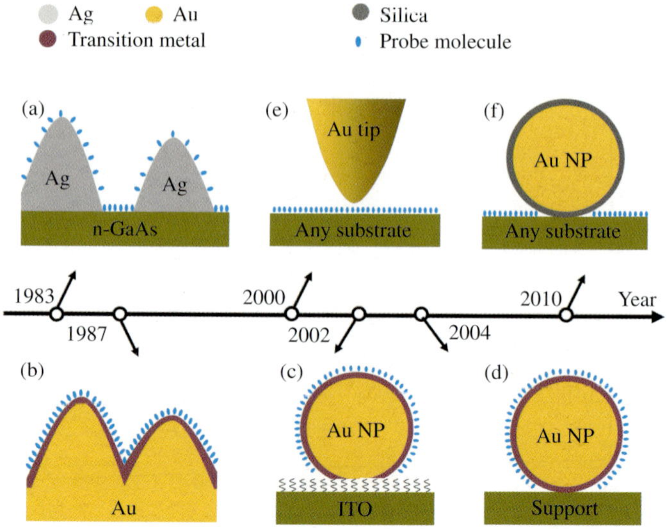

Figure 8.1 *Illustration of the development over the past three decades of the "borrowing SERS activity" methods utilizing different nanostructures in order to expand the materials and morphology generalities.*

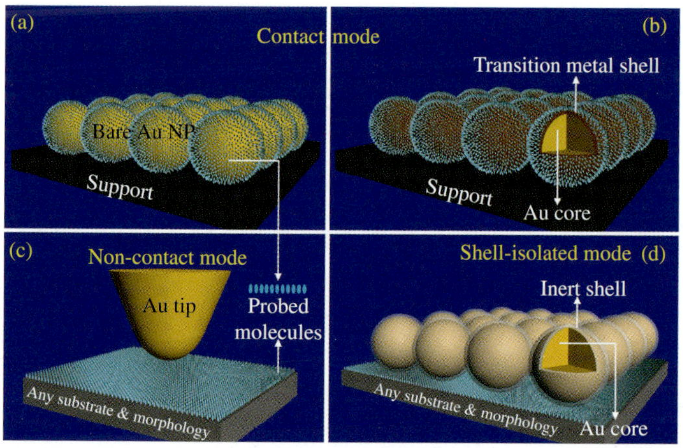

Figure 8.2 *Three working principles of conventional SERS, TERS, and SHINERS. (a) Bare Au nanoparticles, contacted mode; (b) Au core–transition metal shell nanoparticles, contacted mode; (c) tip-enhanced Raman spectroscopy, non-contact mode; and (d) shell-isolated nanoparticle-enhanced Raman spectroscopy, shell-isolated mode. Reprinted by permission from [45], Macmillan Publishers Ltd. Copyright © 2010.*

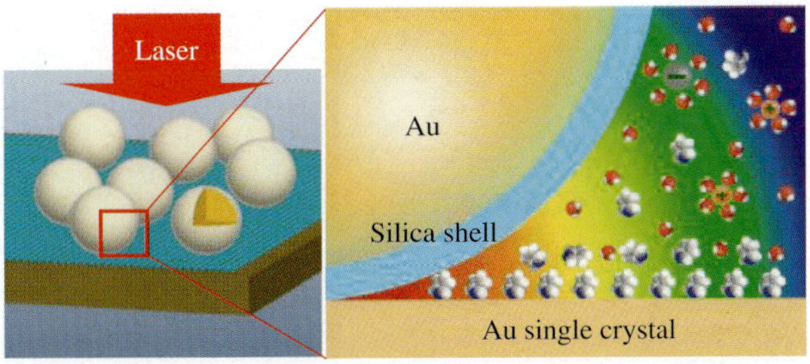

Figure 8.3 Schematic diagrams of the SHINERS experiment. Reproduced with permission from [52]. Copyright © 2011, American Chemical Society.

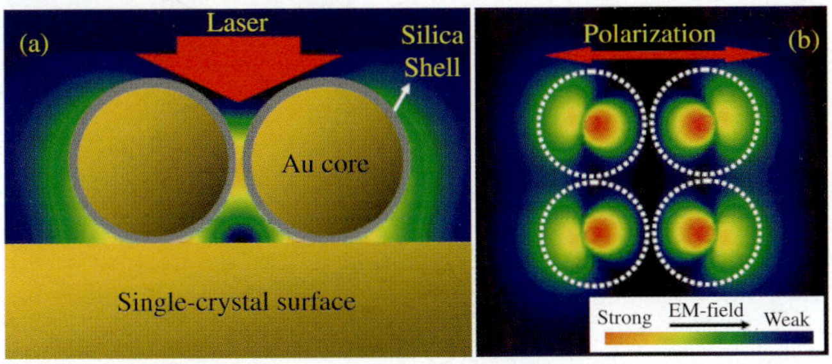

Figure 8.4 The FDTD simulation of the optical electric field distribution of a 2×2 array of 55 nm Au@4 nm SiO_2 NPs placed on a single-crystal gold surface. (a) Side-view and (b) top-view of the optical electric field distribution at the gap between the particle and the surface. The shell to shell distance is 4 nm. The polarization of the 632.8 nm laser is indicated in (b). Reprinted by permission from [45], Macmillan Publishers Ltd. Copyright © 2010.

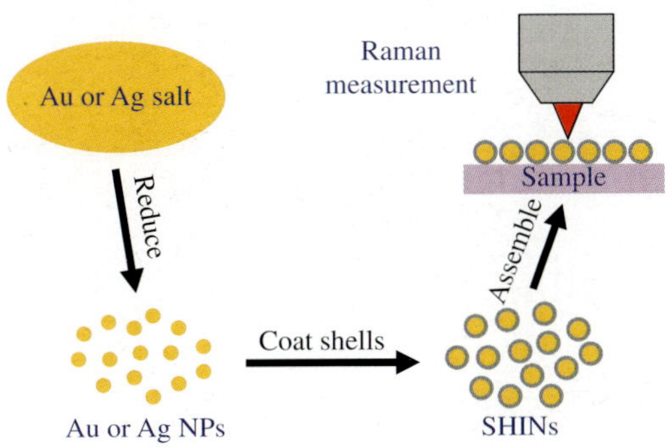

Figure 8.5 Schematic illustration of the SHINERS experiment.

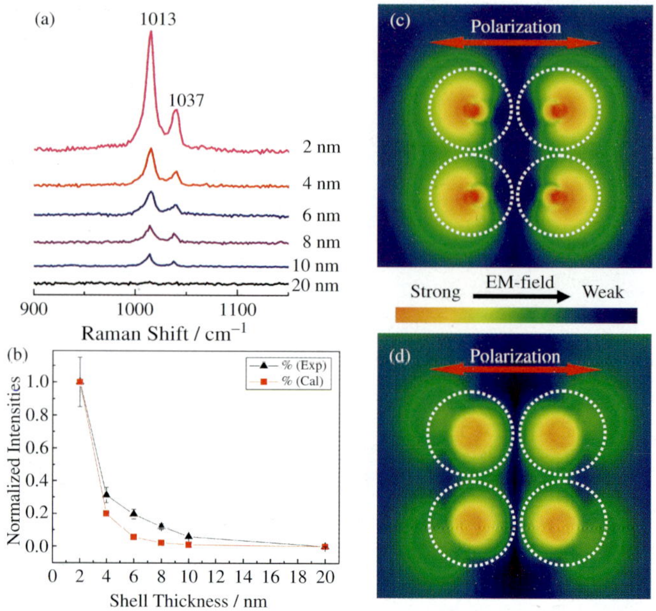

Figure 8.7 *(a) SHINERS spectra of Py adsorbed on a smooth Au surface modified by 55 nm Au@SiO_2 NPs with different silica shell thicknesses. (b) The shell thickness dependence of the integrated SHINERS intensity of Py (triangle) and the corresponding 3D-FDTD calculation result (square). (c,d) Top-view of the FDTD simulation of the optical electric field distribution at the plane between the smooth Au surface and 2 × 2 array of Au@SiO_2 NPs with different shell thickness of 2 nm (c) and 8 nm (d).*

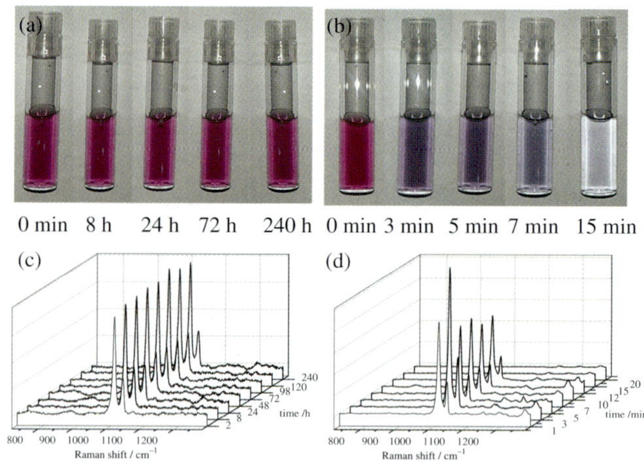

Figure 8.20 *Stability comparison of SHINs and bare Au NPs. Photos of (a) SHINs and (b) bare Au NPs in 0.01 M pyridine solution at different times. Corresponding SERS signals from Py obtained with (c) SHINs and (d) bare Au NPs. Reprinted by permission from [47], Macmillan Publishers Ltd. Copyright © 2010.*

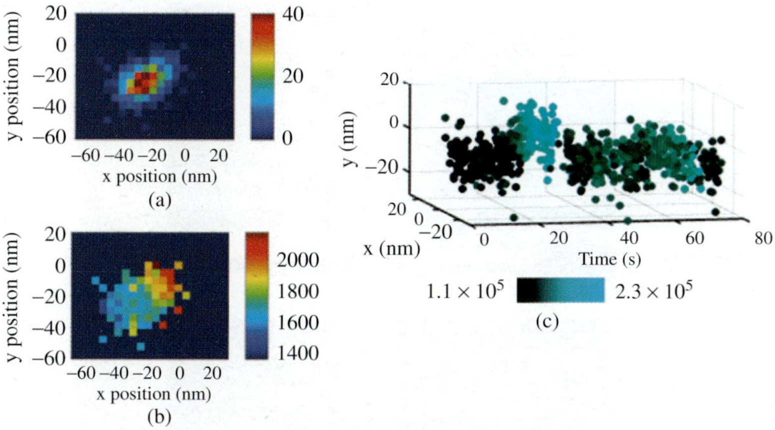

Figure 9.8 *(a) Frequency histogram showing the number of SERS centroids occurring in each bin. (b) Intensity histogram showing the average intensity of SERS centroids occurring in each bin (4.6 nm bin size). (c) 3D wander plot showing diffusion trajectory of SERS centroid (shade-bar indicates emission intensity). Reprinted with permission from Ref. [6]. Copyright © 2010, American Chemical Society.*

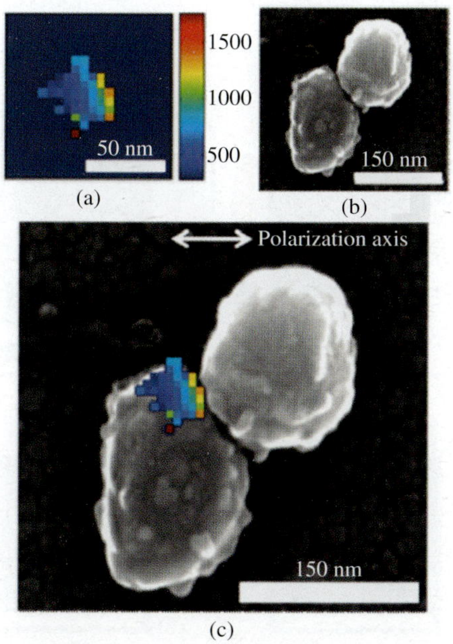

Figure 9.9 *(a) Intensity histogram of SM-SERS data collected on aggregate, with (b) SEM image of the same aggregate. (c) Intensity histogram scaled and qualitatively overlaid on SEM image of aggregate. Reprinted with permission from Ref. [22]. Copyright © 2011, American Chemical Society.*

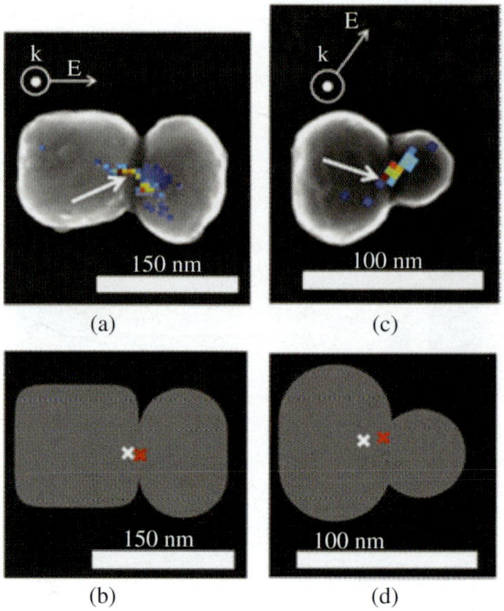

Figure 9.10 *(a) and (c) SEM images of two nanoparticle dimers with SERS intensity histogram overlaid on image (white arrow indicates luminescence centroid fit). (b) and (d) show corresponding theoretical* ε^2 *(white cross) and* ε^4 *(dark cross) centroid fits, modeled based on the structure of (a) and (c). Reprinted with permission from Ref. [24]. Copyright © 2012, American Chemical Society.*

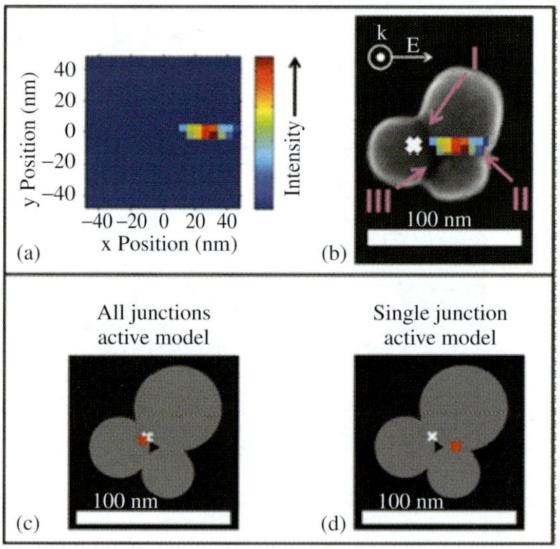

Figure 9.11 *(a) SERS intensity histogram and (b) SERS intensity histogram overlaid on SEM image of the same SM-SERS-active aggregate (white X denotes luminescence centroid). (c) Predicted* ε^2 *(white cross) and* ε^4 *(dark cross) centroid fits assuming that* ε^4 *samples all possible SM-SERS active junctions. (d) Same as (c), but* ε^4 *is calculated based only on junction II in (b). Reprinted with permission from Ref. [24]. Copyright © 2012, American Chemical Society.*

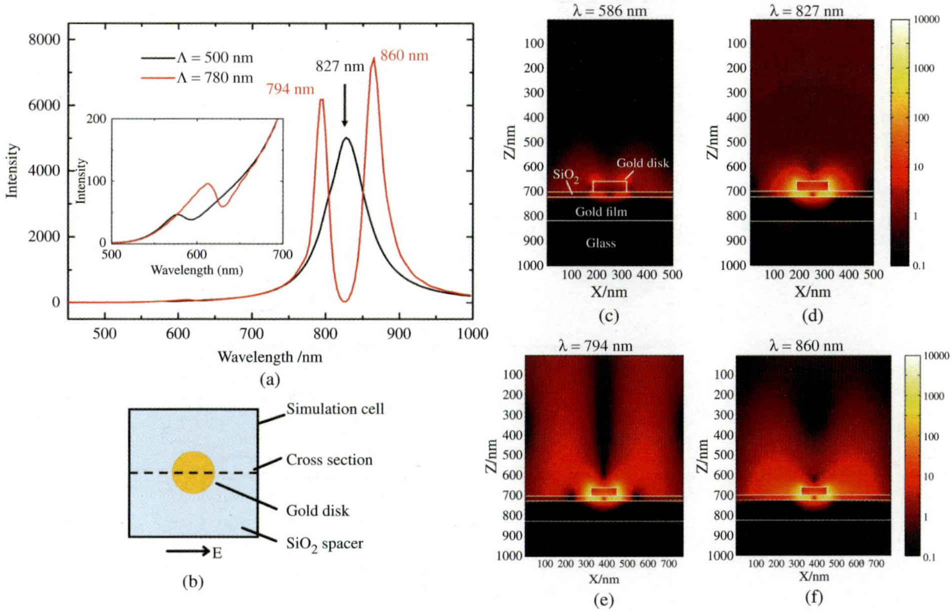

Figure 10.4 *(a) Simulated near-field intensity spectra. Gold disk diameter: 133 nm. SiO_2 spacer thickness: 25 nm. Black: device with array period of 500 nm. Red: device with array period of 780 nm. Inset: expanded view of near-field spectra over the range $\lambda = 500-700$ nm. (b) Top view of the structure in the simulation. (c–f) Intensity ($|E_s|^2$) distributions of scattered field on the cross section perpendicular to substrate for illumination at (c) $\lambda = 586$ nm and at (d) $\lambda = 827$ nm for the 500 nm period array, at (e) $\lambda = 794$ nm and at (f) $\lambda = 860$ nm for the 780 nm period array.*

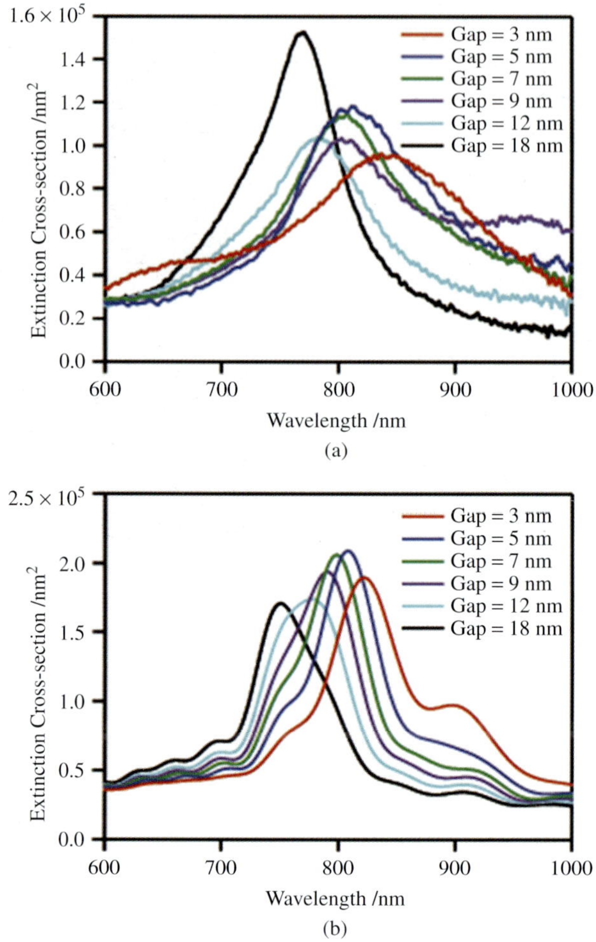

Figure 10.7 *Extinction cross-sections for dimer arrays with different gap sizes. (a) Measurements and (b) simulations.*

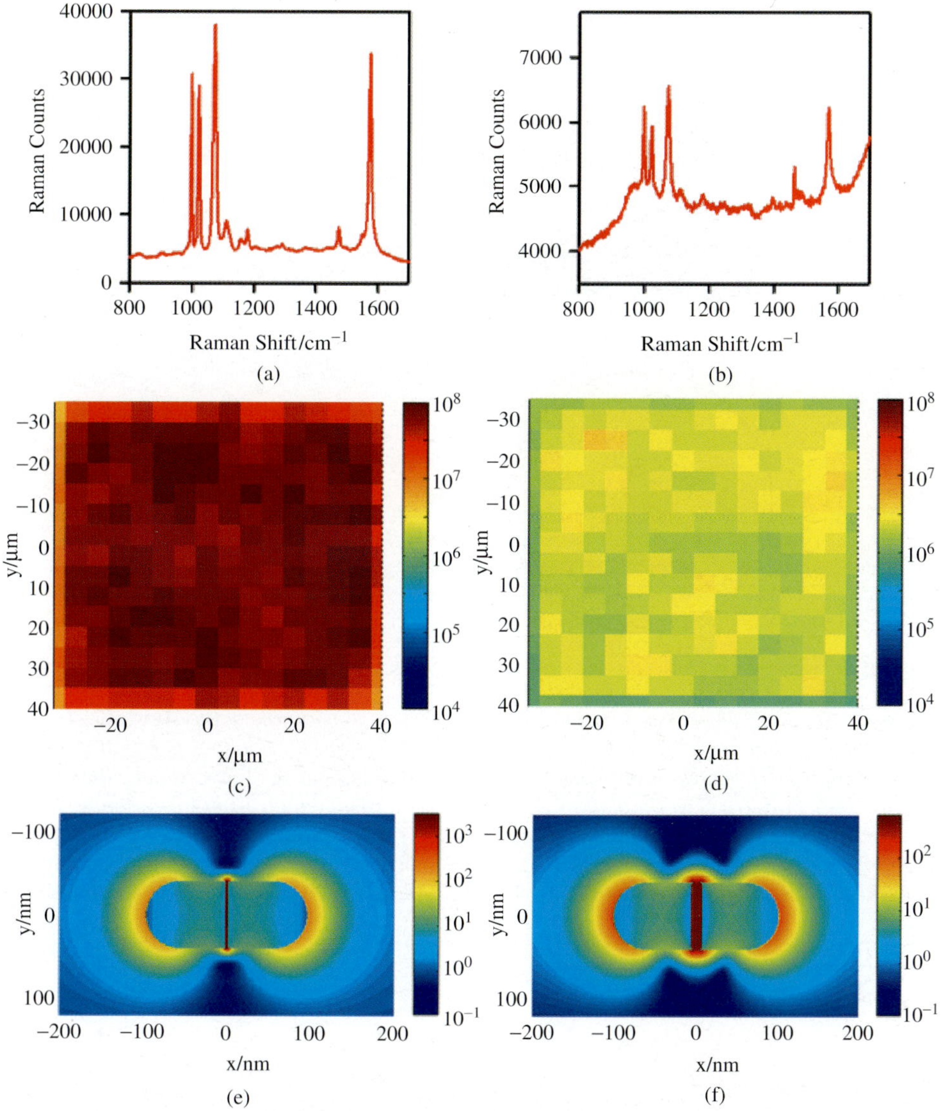

Figure 10.8 *(a). Benzenethiol SERS spectrum measured from 3 nm gap dimers. (b) Benzenethiol SERS spectrum measured from 12 nm gap dimers. (c) EF versus position on array with 3 nm gap dimers. (d) EF versus position on array with 12 nm gap dimers. (e) Simulated field intensity distribution ($|E|^2$) for 3 nm gap dimers, for plane wave illumination ($\lambda = 821$ nm) polarized along antenna axes. Field intensity distribution is shown over center plane of dimer, 15 nm above substrate. (f) Simulated field intensity distribution ($|E|^2$) for 3 nm gap dimers, for plane wave illumination ($\lambda = 776$ nm) polarized along antenna axes. Field intensity distribution is shown over center plane of dimer, 15 nm above substrate*

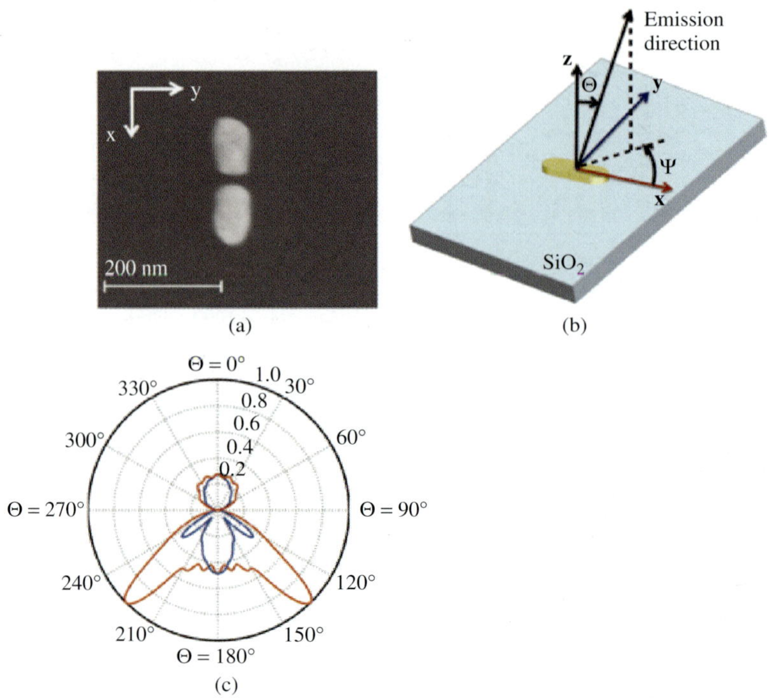

Figure 10.10 *Raman emission pattern from dimer antennas on dielectric substrate. (a) SEM image of dimer antenna. (b) Cartesian* xyz*-coordinate system that defines geometry of dimer structure and air/dielectric interface (*xy*-plane); and spherical* Θ, Ψ*-coordinate that defines emission direction. (c) FEM simulation of emission pattern, in which arbitrarily oriented Raman dipole is placed at gap center of dimer antenna. We specify plane of emission and allow* Θ *to range from 0° to 360°. Region below line from* $\Theta = 90°$ *to* $\Theta = 270°$ *represents dielectric substrate. Red: emission pattern in* yz*-plane (*$\Psi = 90°$*); blue: emission pattern in* xz*-plane (*$\Psi = 0°$*)*

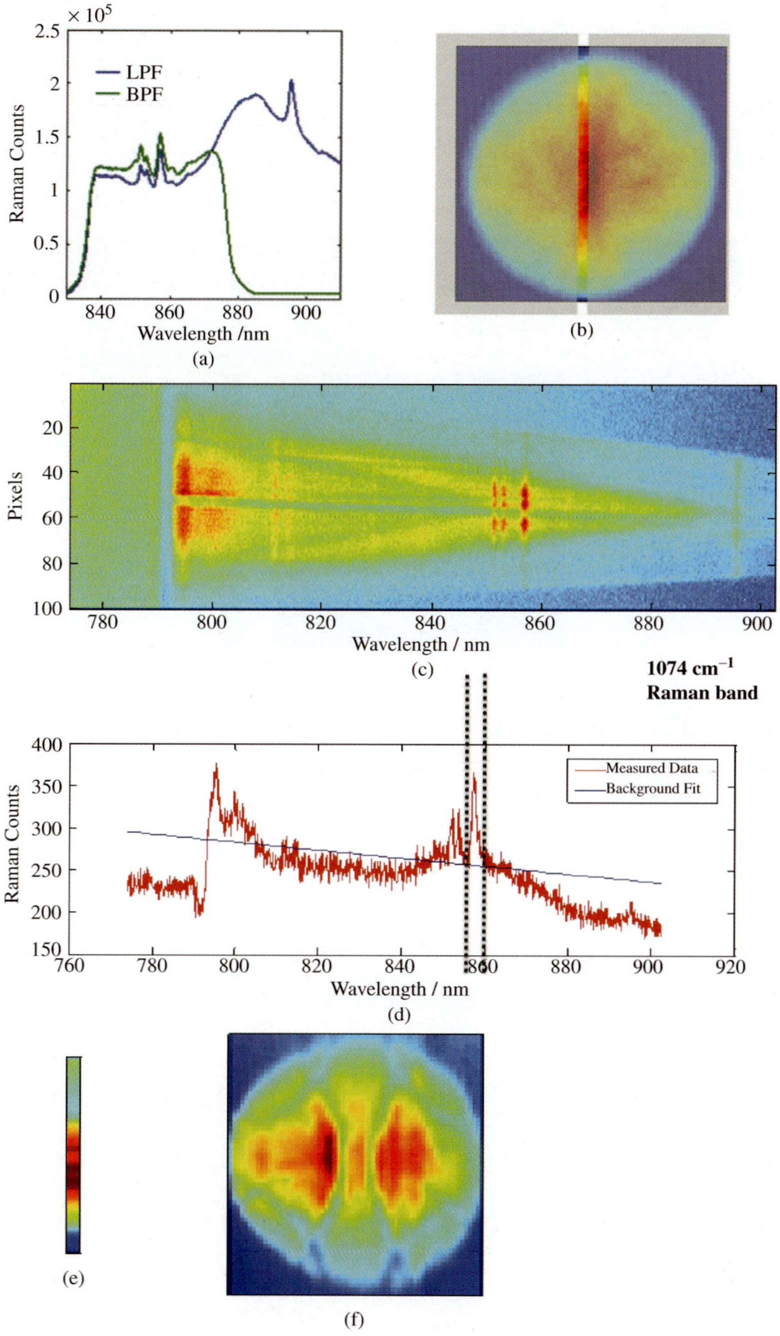

Figure 10.12 *Measuring the emission patterns of Raman lines by EMS. (a) SERS spectra obtained with LPF and BPF. (b) Second back-focal-plane image plane formed on the entrance slit of the spectrometer, using LPF and Bertrand lens. (c) Spectral map, as recorded by spectrometer CCD. (d) Raman spectrum of one row of pixels of (c). (e) Intensity of 1074 cm^{-1} Raman line at slit. (f) Reconstructed emission pattern of 1074 cm^{-1} Raman line*

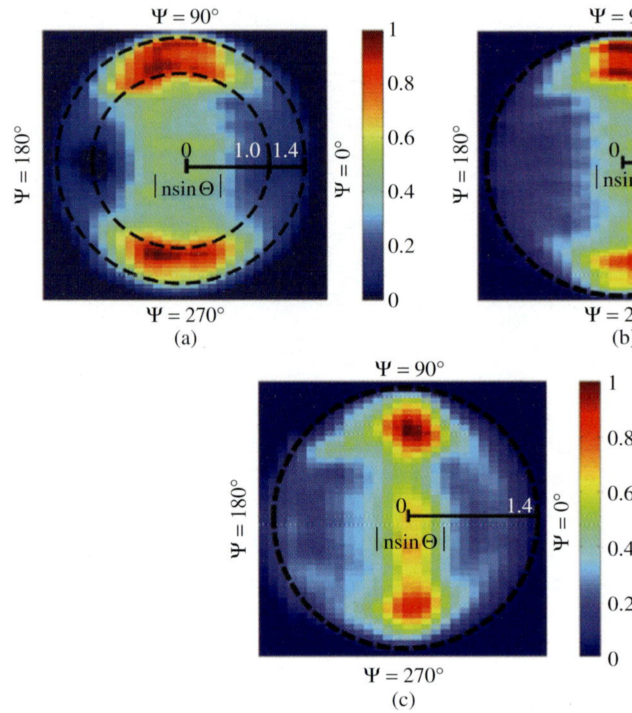

Figure 10.13 *(a) Measured emission pattern of thiophenol 1074 cm^{-1} Raman line retrieved from EMS measurements. Map represents the normalized scattering intensity. Outer black dashed circle denotes NA of objective lens. Measured emission patterns of (b) 415 cm^{-1} Raman line and (c) 1586 cm^{-1} Raman line*

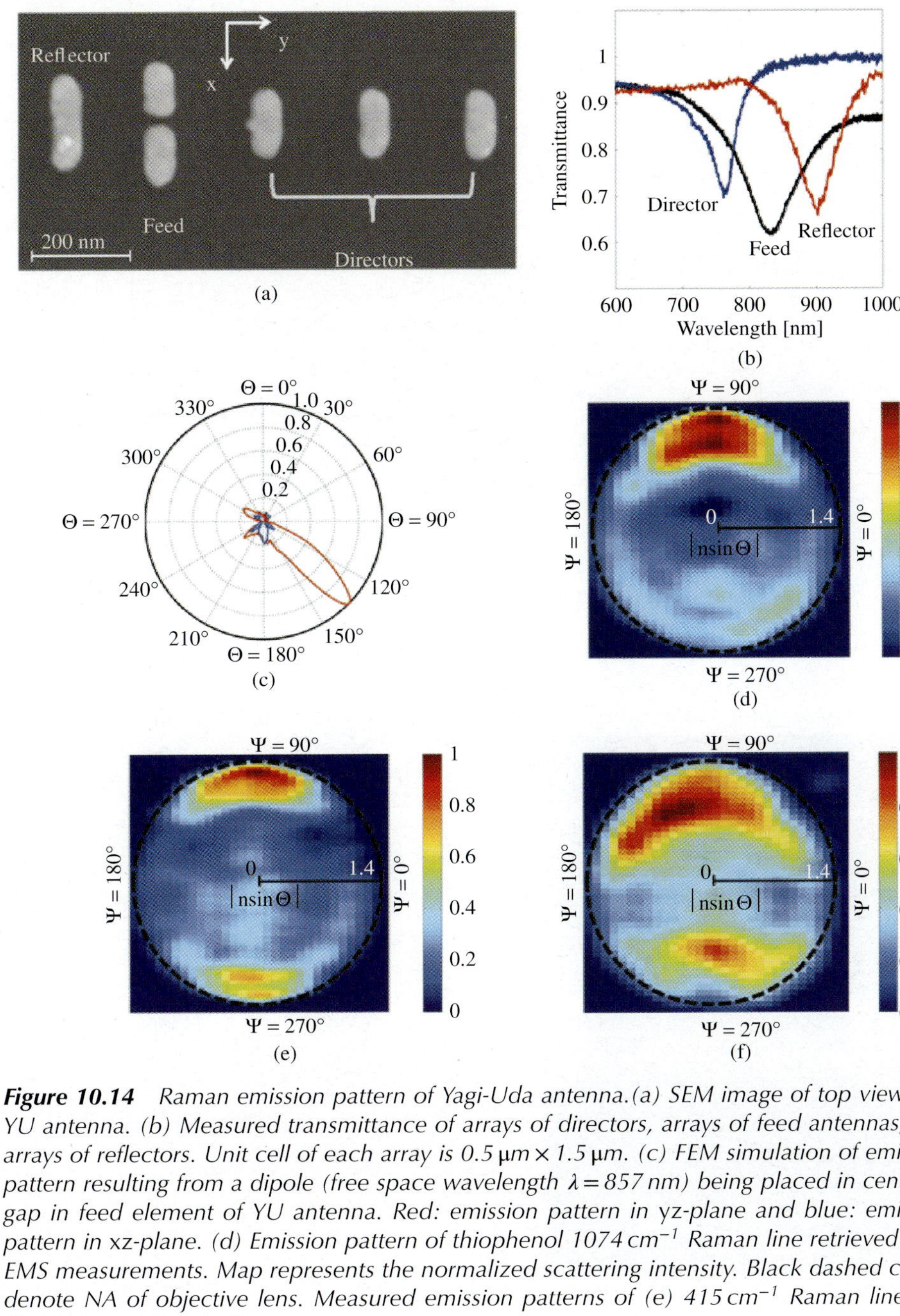

Figure 10.14 *Raman emission pattern of Yagi-Uda antenna.(a) SEM image of top view of a YU antenna. (b) Measured transmittance of arrays of directors, arrays of feed antennas, and arrays of reflectors. Unit cell of each array is 0.5 μm × 1.5 μm. (c) FEM simulation of emission pattern resulting from a dipole (free space wavelength $\lambda = 857$ nm) being placed in center of gap in feed element of YU antenna. Red: emission pattern in yz-plane and blue: emission pattern in xz-plane. (d) Emission pattern of thiophenol 1074 cm^{-1} Raman line retrieved from EMS measurements. Map represents the normalized scattering intensity. Black dashed circles denote NA of objective lens. Measured emission patterns of (e) 415 cm^{-1} Raman line and (f) 1586 cm^{-1} Raman line*

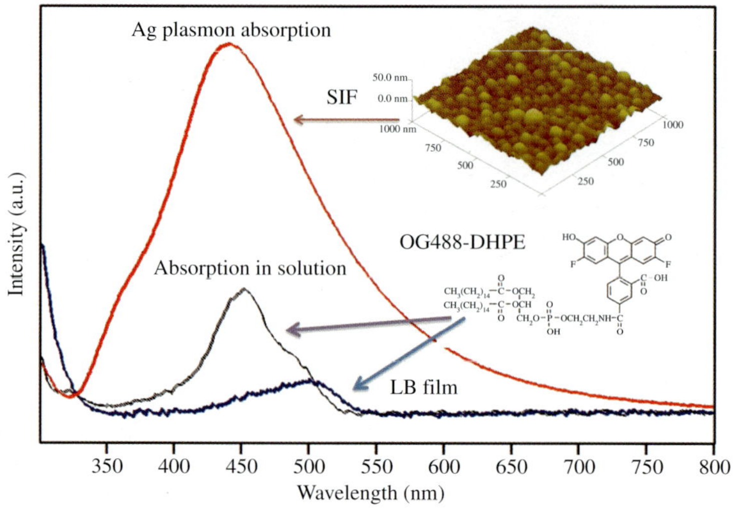

Figure 11.3 *Characterization of the Ag island film and OG488®-DHPE. Extinction spectrum and AFM of Ag island film, as well as the absorbance (1 cm optical path length) profiles for OG488®-DHPE in chloroform solution and monolayer LB.*

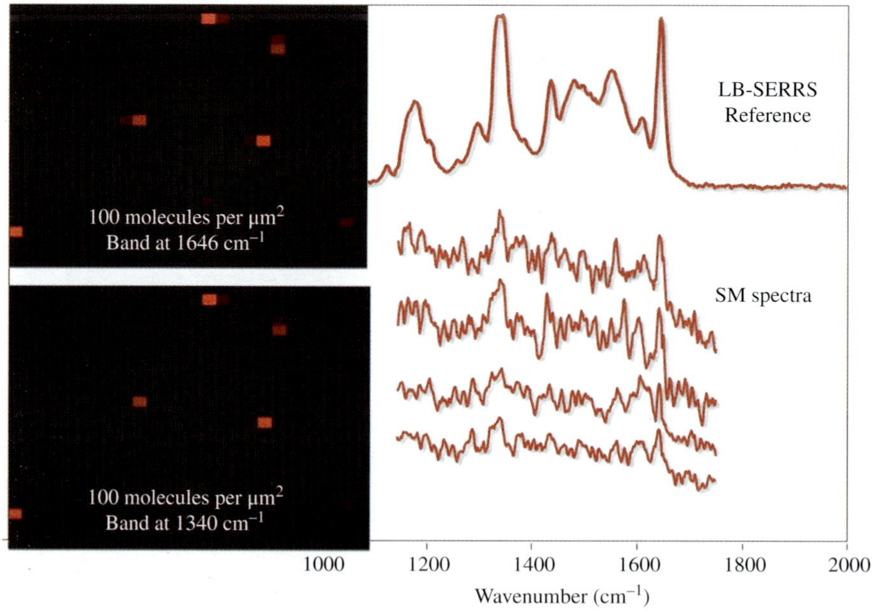

Figure 11.6 *Raman images (area mapping) for a concentration of 100 molecules of tag-phospholipid per μm² constructed using the intensity of 1340 and 1646 cm⁻¹ fundamental modes of OG488®-DHPE as a function of the x–y area. Each pixel in the figure corresponds to a single recorded spectrum from a total of 676 spectra.*

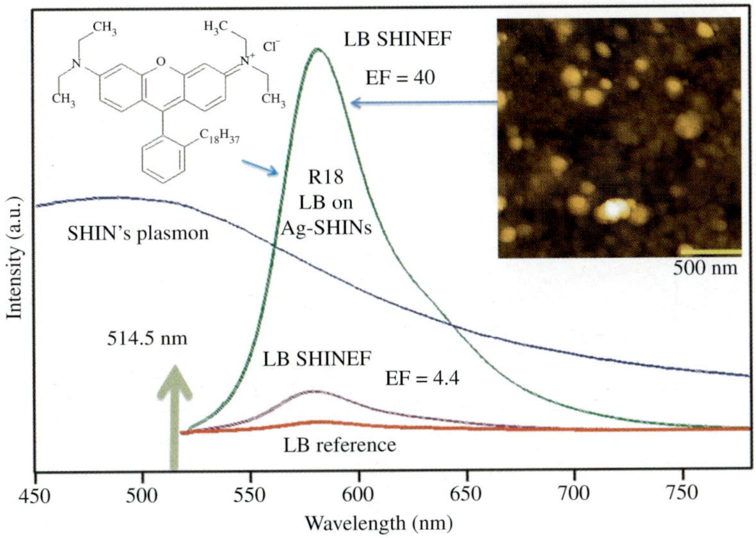

Figure 11.7 SHINEF spectra of the R18 monolayer on Ag-SHIN substrate after 6 and 16 h.

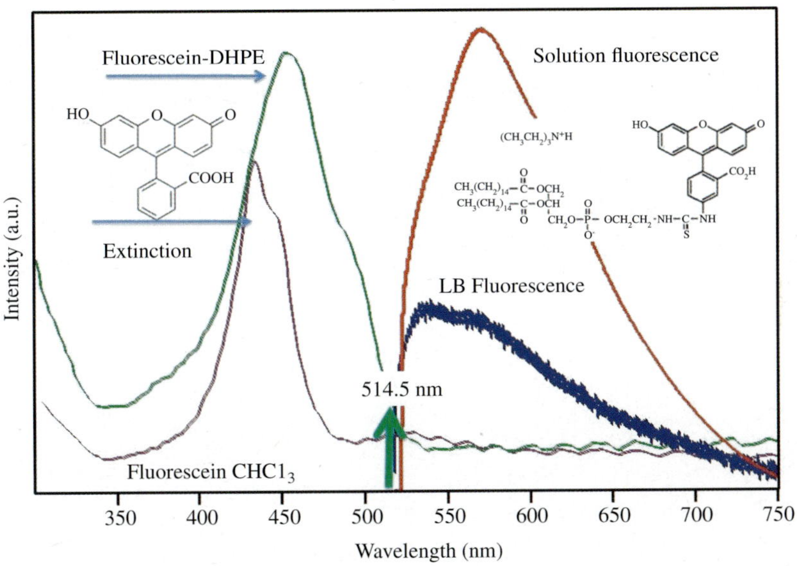

Figure 11.8 Extinction and emission spectra of fluorescein-DHPE. The extinction spectrum of fluorescein is included for comparison.

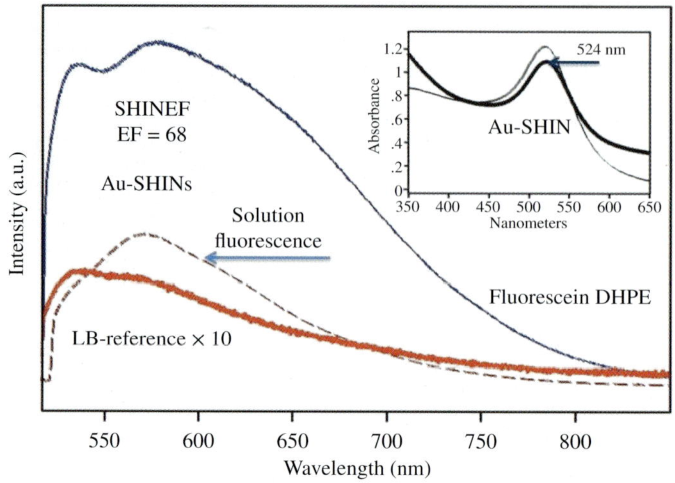

Figure 11.9 *Plasmon-enhanced fluorescence for fluorescein-DHPE with gold-SHINs. LSPR of Au-SHIN (bold line) and gold colloid (inset).*

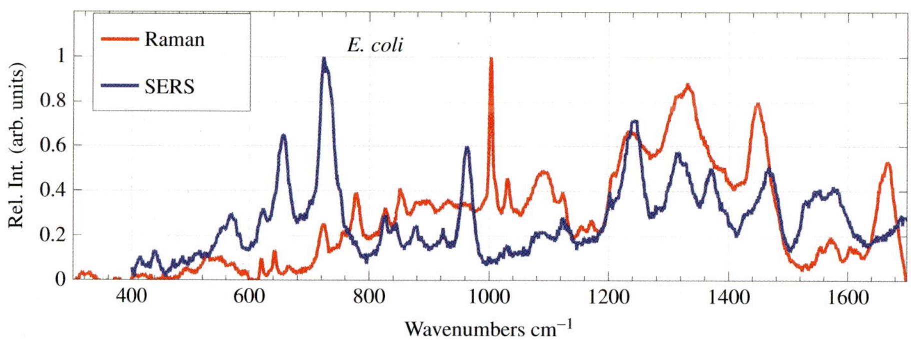

Figure 12.3 *Comparison of SERS and normal Raman spectrum of the same* E. coli *strain excited at 785 nm. Each spectrum has been normalized to the maximum band intensity. There is no correspondence between the bands that appear in the "normal" (non-SERS) and SERS spectrum.*

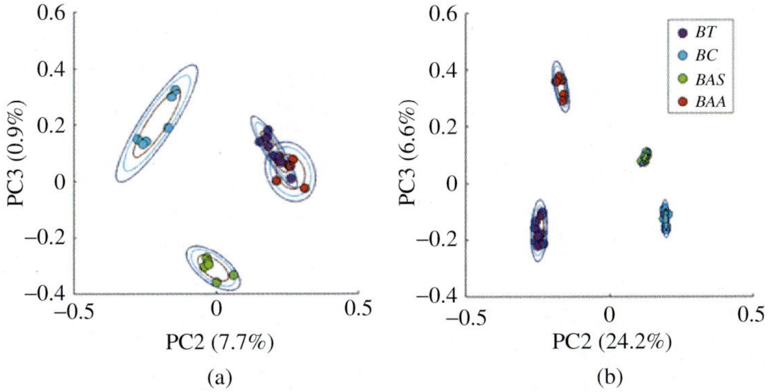

Figure 12.5 *Comparison of PCA plots (PC3 vs. PC2) for SERS spectra of four closely related* Bacillus *organisms (*B. thuringiensis, B. cereus, *B.* anthracis *Sterne,* B. anthracis *Ames). When the observed SERS spectra are the inputs to the PCA, the contour plot shown in (a) results. By contrast, when the input to the PCA are the barcode representations of the observed SERS spectra, the contour plot shown in (b) results. The enhancement in specificity resulting from the barcode spectra is clearly evident by this comparison. © L. Ziegler.*

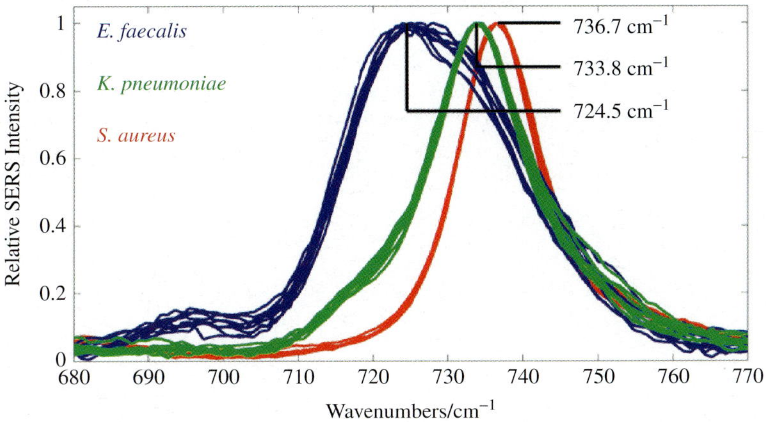

Figure 12.6 *Ten SERS spectra of* E. faecalis, K. pneumoniae, *and* S. aureus *bacterial strains in the 680–770 cm^{-1} region demonstrating the reproducibility of spectral shapes in this important part of the SERS spectrum. The Raman shift of the spectral peak in this region is indicated for each species.*

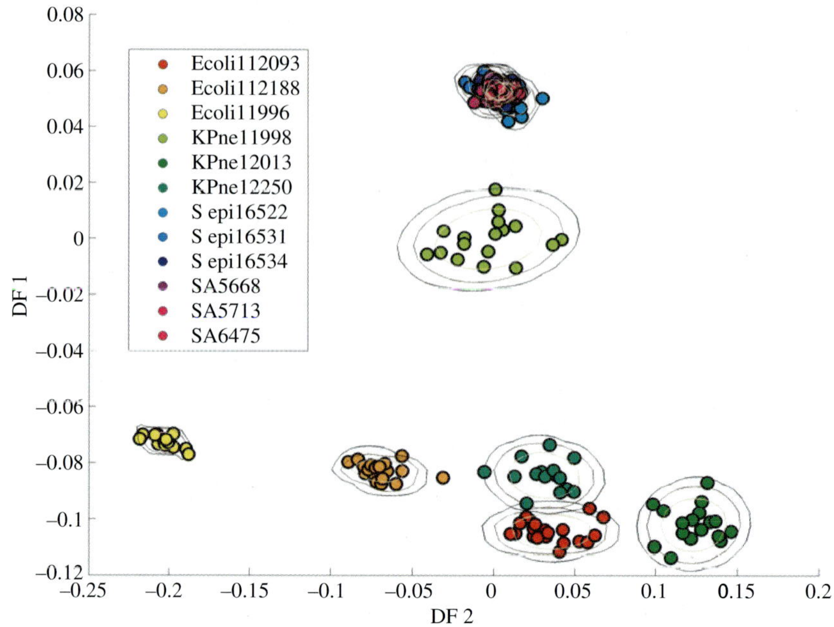

Figure 12.7 *Barcode-based discriminant function analysis (DFA) plot derived from twelve SERS spectra for three strains for each of four different species (144 SERS spectra);* E. coli, K. pneumoniae, S. epidermidis, *and* S. aureus. *All six staphylococcus strains are found to cluster together whereas the other six strains cluster separately indicating the biological significance of these SERS spectra. The six staphylococcus spectra separate in PCA treatments when analyzed separately.*

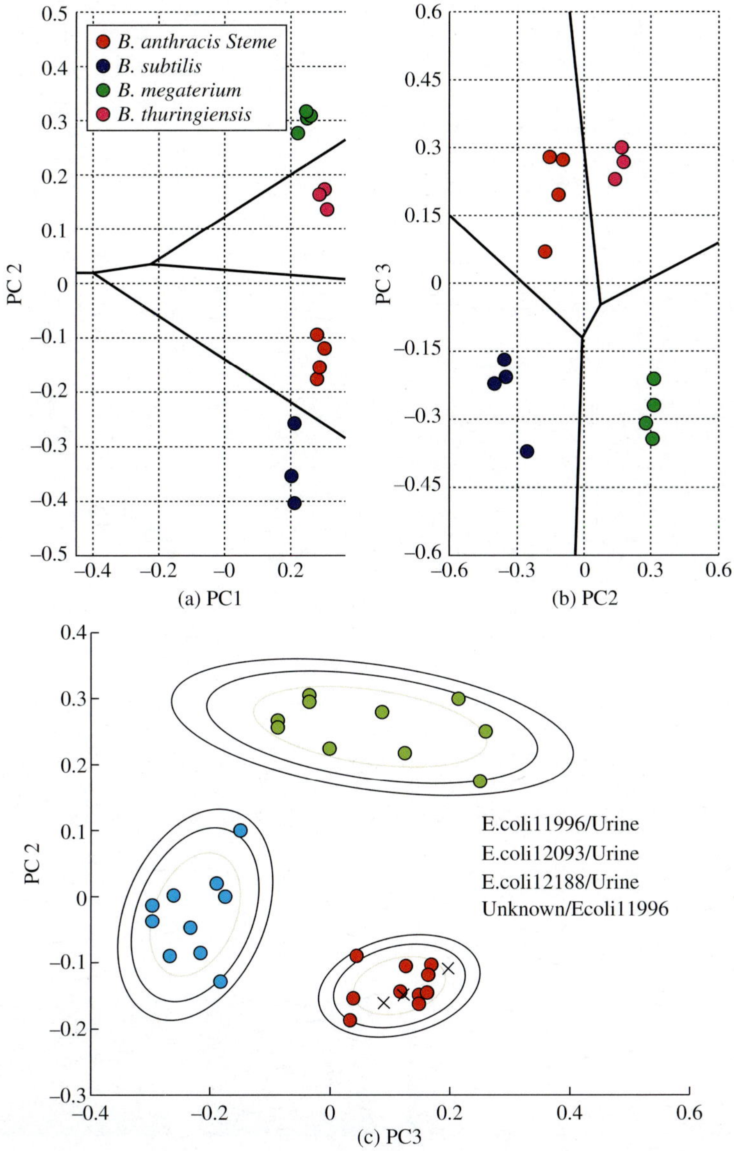

Figure 12.8 *(a) Barcode PCA plots for four Bacillus bacteria that have been spiked into human blood and then recovered, demonstrating that specificity is maintained via SERS after blood exposure. Black lines are equidistant to the mean of each cluster (Voroni diagram). (b) Barcode PCA plot for three* E. coli *strains spiked into urine at* 10^5*/ml and recovered by centrifugation. "X" corresponds to use of this limited library to identify the "unknown" in a test of the use of this limited library following enrichment of this bacterium from a spiked urine sample.*

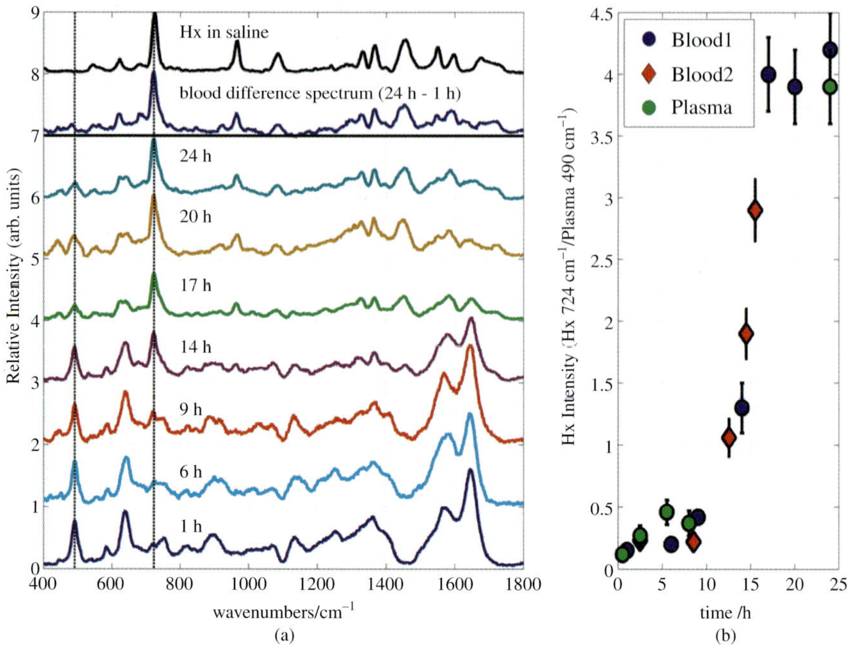

Figure 12.9 *SERS spectrum of whole blood as a function of storage time (4 °C), 24 h − 1 h SERS blood difference spectrum and the SERS spectrum of hypoxanthine (Hx) in normal saline solution are shown in (a). The relative intensities of the time dependent whole blood SERS spectra have been normalized by the observed signal size intensity of the largest feature in the 24 h spectrum at 724 cm^{-1}. The SERS intensity of the 724 cm^{-1} band of Hx relative to the 490 cm^{-1} band of whole blood, and the 724 cm^{-1} band of Hx relative to the 490 cm^{-1} band of a blood plasma sample as a function of storage time are shown in (b). Reprinted with permission from [50]. Copyright © 2012, American Chemical Society.*

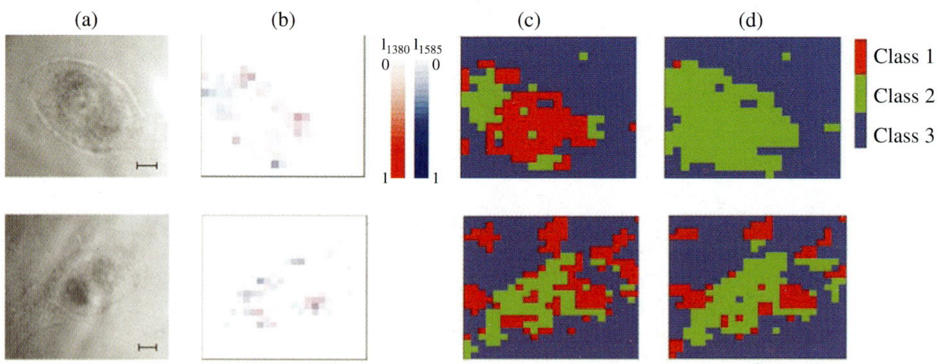

Figure 13.1 *Duplex imaging using two different SERS hybrid labels in single 3T3 fibroblast cells. (a) Photomicrograph of the cell (scale bar: 4 μm), (b) chemical image, based on one intensity from each of the labels (red scale: 2-NAT, 1380 cm^{-1}; blue scale: pABT 1585 cm^{-1}), and (c,d) images obtained using hierarchical cluster analysis and K-means clustering, respectively, with the spectral information in the region 300–1700 cm^{-1}, based on three spectral classes. Reprinted with permission from [15]. Copyright © 2010, American Chemical Society.*

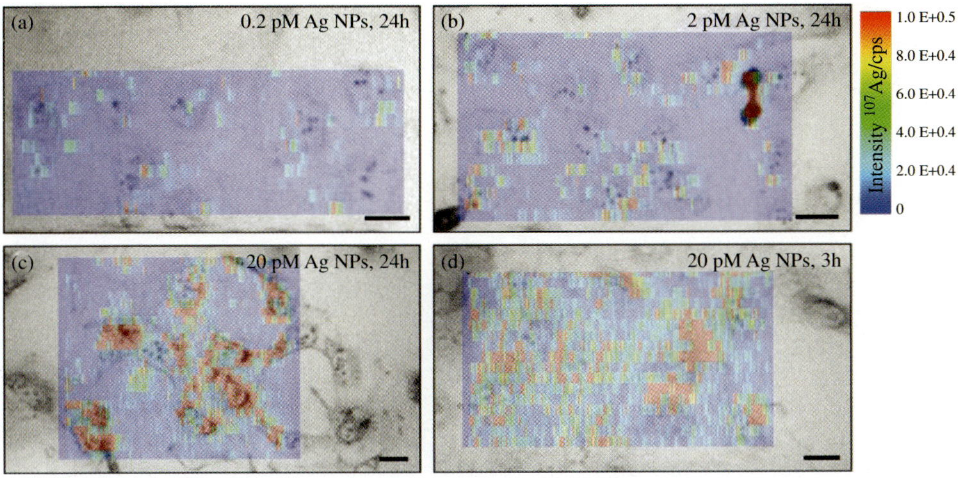

Figure 13.5 LA-ICP-MS images of the ^{107}Ag intensity distribution and the corresponding bright field images of fixed 3T3 fibroblast cells, incubated with silver nanoparticles (0.2/2/20 pM) or 3 hours (d) or 24 hours (a–c). Scale bars represent 25 μm. Parameters: laser spot size, 8 μm; scan speed, 8 $\mu m\, s^{-1}$; repetition rate, 5 Hz; pixel size, $1.2 \times 8\, \mu m^2$; fluence, 1.5 $J\, cm^{-2}$. Reproduced with permission from [2]. Copyright © 2009, American Chemical Society.

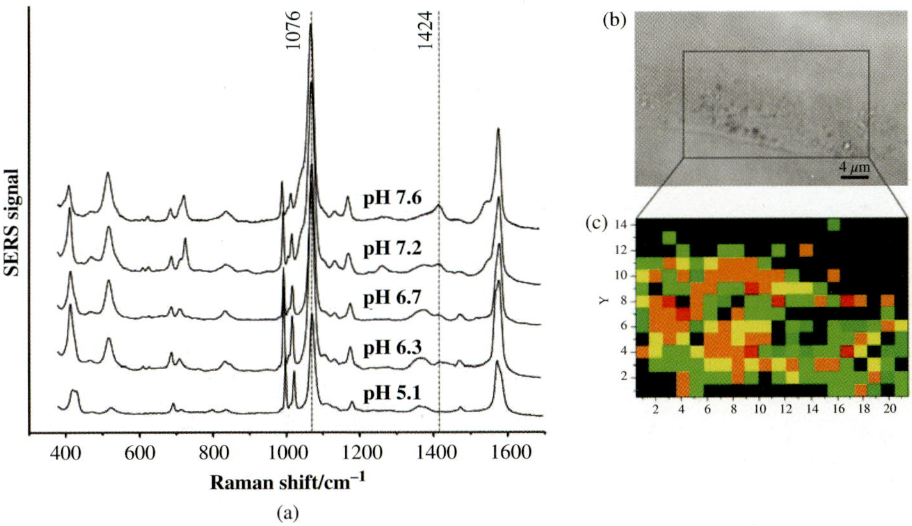

Figure 13.9 SERS investigation of the intraendosomal pH value. (a) Selected SERS spectra of the pH nanosensor, composed of para-mercaptobenzoic acid (pMBA) coated gold nanoparticles, in live cells. Mouse fibroblast cells were exposed to pMBA gold nanoparticles for 5 h (pMBA concentration 1 mM). (b) Bright field image of an individual fibroblast cell. (c) Raman chemical map indicating the local distribution of the intracellular pH value in vitro (intensity ratio of the pMBA Raman bands at 1076–1424 cm^{-1}). An excitation wavelength of 785 nm, a laser intensity of $8 \times 10^4\, W\, cm^{-2}$ and an acquisition time of 1 s were applied [89].

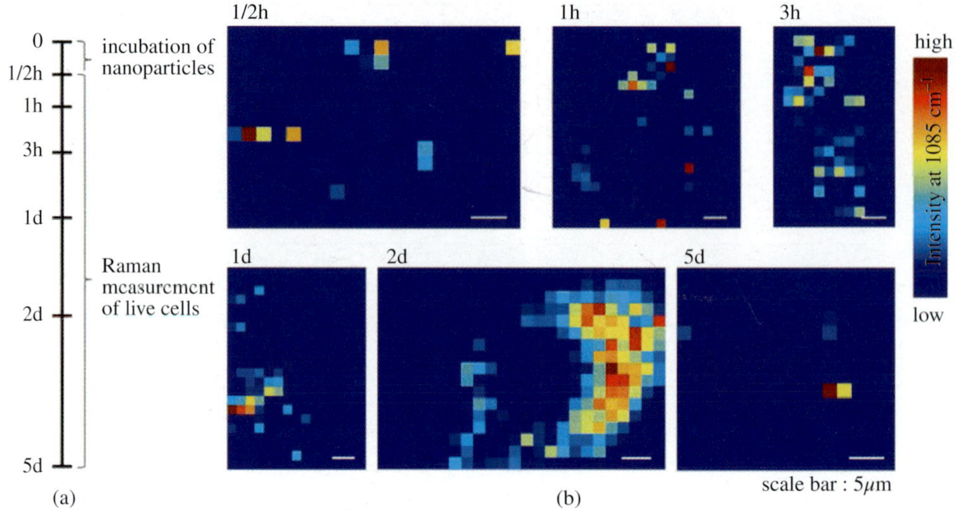

Figure 13.10 *SERS nanoprobes for time-dependent analysis of nanoparticle uptake and their intracellular distribution in live cells. (a) Experimental design: Fibroblast cells were incubated with* para*-aminothiophenol coated gold nanoparticles (*p*ABT, 100 nM) for 30 min. SERS measurements were conducted after different time points. (b) Raman chemical maps illustrate the nanoparticle distribution inside fibroblast cells for different stages of incubation. The intensity of the C-S stretching vibration of* p*ABT at 1085 cm^{-1} was used for imaging.*

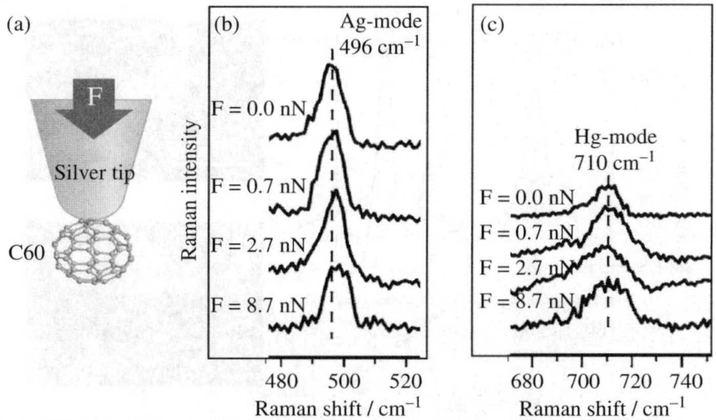

Figure 7.7 *(a) Schematic of a tip-pressurized C60 molecule. TERS spectra of the Ag-mode (b) and the Hg-mode (c) were measured with increasing the force from 0 to 8.7 nN. Reproduced with permission from [21]. © 2006 The American Physical Society*

SWCNT has been analyzed because the structures (chirality and diameter) and the electric properties (semiconducting or metallic) of SWCNT are spectroscopically probed by TERS. Figure 7.8a shows an example of TERS spectrum of a SWCNT bundle [24]. Several peaks in the spectrum represent the radial breathing mode (RBM). Since the frequency of the RBM is inversely proportional to the diameter of the SWCNT, the presence of peaks in the spectrum indicates the presence of SWCNTs with different diameters in the bundle, as illustrated in Figure 7.8b. Figure 7.8c–e show the images measured at three RBM frequencies marked in Figure 7.8a. The tube diameters are 0.85, 0.97, and 1.23 nm for (c–e), respectively. These images reveal how the nanotubes with different diameters are spatially distributed within the bundled sample. Such information cannot be obtained by conventional confocal Raman spectroscopy due to both the diffraction limit of the probing light and the weak Raman efficiency of nano-sized samples.

Isolated individual SWCNTs have also been analyzed using the TERS technique [22, 23, 25]. Figure 7.9a shows that the RBM frequency varies along an isolated SWCNT [25]. The frequency variation indicates the local change in chirality along the nanotube. The spatial resolution was a few tens of nanometers. The spectral change of the G-band along the nanotube showed that the electronic properties of the nanotube changed from semiconducting to metallic. Local defects in the SWCNT were also detected by measuring the TERS intensity ratio between the D-band and the G-band [29].

Figure 7.9b shows TERS spectra of organic-molecule-encapsulated SWCNTs together with the AFM image [26]. The spectra *a–e* were measured at the sample positions in the AFM image. The spectra include the vibrational modes of both $\nu 1$ (1523 cm^{-1} conjugated C=C mode) from β-carotene and the G-band (1592 cm^{-1} C–C stretching mode) from the nanotube. The intensity ratio $I_{\nu 1}/I_{G}$ represents the local filling rate of β-carotene molecule inside the nanotube, which is highest at the position c. The result indicates that the organic molecules were not uniformly distributed inside the nanotube.

Graphene has been attracting a lot of attention from scientists due to its unique electronic properties. TERS allowed us to detect nanometric defects created on single-layered

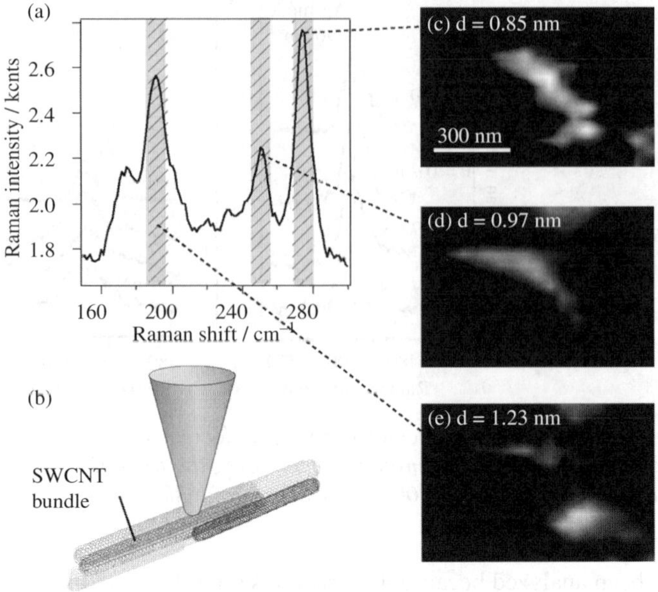

Figure 7.8 *(a) TERS spectrum of a SWCNT bundle in the RBM frequency region. Various distinct peaks indicate the existence of SWCNTs with different diameters. (b) Schematic of the sample with bundled nanotubes. TERS images in (c–e) were constructed from the three shaded peaks in the TERS spectrum, which correspond to the SWCNTs with diameter 0.85, 0.97, and 1.23 nm, respectively. Reprinted with permission from [24]. Copyright © 2006, American Institute of Physics*

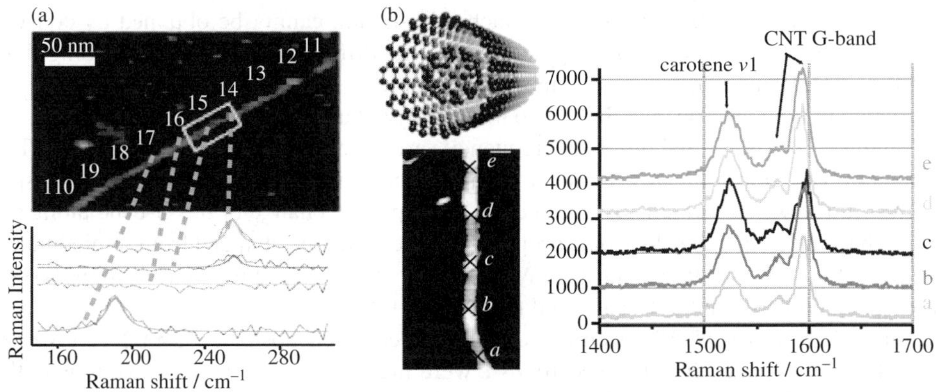

Figure 7.9 *TERS spectral mapping of (a) an isolated SWCNT. Reprinted with permission from [25]. © 2007, American Chemical Society. and (b) carotene-encapsulated SWCNT. Reprinted with permission from [26]. Copyright ©2006 The Japan Society of Applied Physics*

graphene by measuring the intensity of the D-band [27]. Hydrogen-terminated areas and contaminated areas on the single-layered graphene were also locally imaged. For a multi-layered sample, the edge boundaries of multi-layered graphene were locally imaged [28]. Local stress distribution due to interlayer interactions was measured through the intensity fluctuation of the 2D-band (the second-order overtone of the D-band). The spatial resolution was about 10 nm.

Semiconducting materials of nano-electronic devices have also been analyzed [30–34]. Figure 7.10a shows a TERS spectrum of a thin silicon layer, which includes stress induced by lattice mismatch with the underlying Ge-doped silicon buffer layer [40]. The higher frequency peak originates from the strained Si layer. The amount of strain can be quantitatively evaluated from the frequency shift. The lower frequency peak originates from the underlying SiGe layer. Figure 7.10b shows a TERS image of the strained Si surface, which was constructed by the frequency of the strained silicon. The image clearly addresses the local distribution of strain. The spatial resolution was 25 nm. On the other hand, the far-field Raman frequency image in Figure 7.10c does not show a distinct frequency shift because the far-field Raman shift values show only the average over the diffraction-limited focusing spot.

Figure 7.11a shows an AFM image of a ferroelectric ($BaTiO_3$) nanocrystal [37]. Figure 7.11b shows the TERS image of the nanocrystal, measured with the integrated intensity of the A_1 LO and A_1 TO phonon modes. The large intensity of the A_1 TO mode was

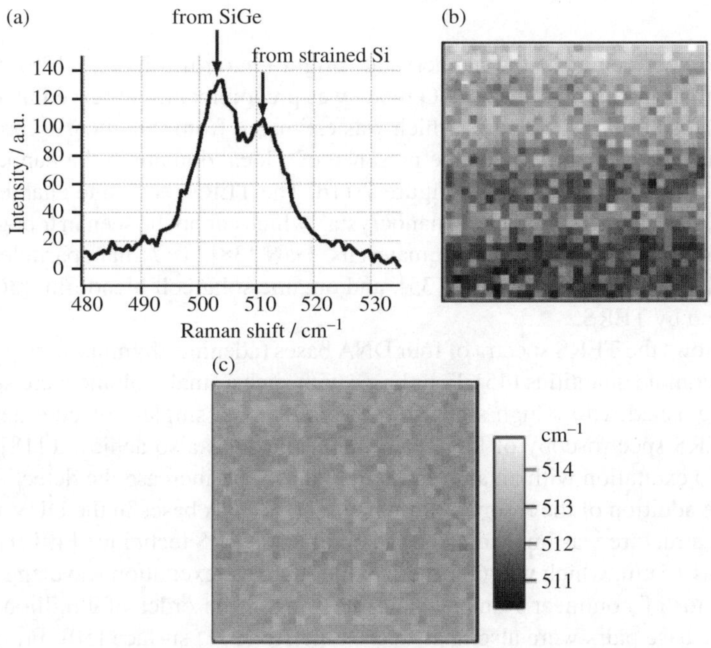

Figure 7.10 *(a) TERS spectra of strained Si underlaid with a Ge-doped silicon buffer layer. (b) TERS and (c) far-field Raman images of the strained Si surface constructed by the frequency shift of the LO phonon mode. Reproduced with permission from [40]. © 2008 John Wiley & Sons, Ltd*

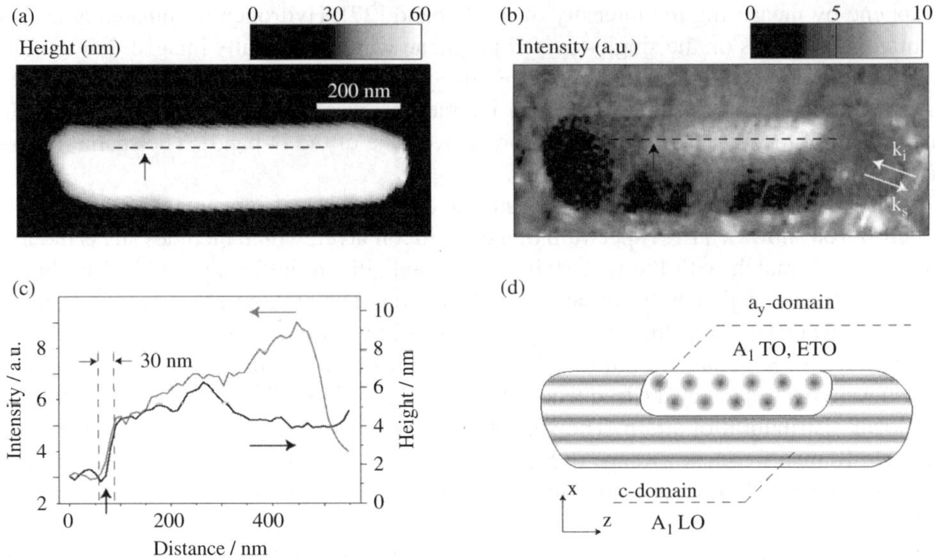

Figure 7.11 *(a) Topography of a $BaTiO_3$ nanorod. (b) TERS intensity image obtained simultaneously with the topographic image. (c) Lateral cross-section along the dashed lines in (a,b). (d) Domain assignment of the $BaTiO_3$ nanorod. Reproduced with permission from [37]. Copyright © 2009, Rights Managed by Nature Publishing Group*

observed on the upper center of the nanocrystal (bright region in Figure 7.11b). On the other hand, the small intensity of the A_1 LO mode was predominantly observed in another area. The spatial resolution was 30 nm, which was estimated from the intensity profile in Figure 7.11c. The TERS image reveals the presence of a local domain in the nanocrystal. The domain assignment is illustrated in Figure 7.11d. The TERS technique enables us to identify the local domain in a ferroelectric nanocrystal, which cannot be seen in topographic measurement. In addition to ferroelectric materials, GaN [38], TiO_2 nanoparticles [39], hybrid polymer blend [48], Ge nanowire [35], and organic solar cell blend film [36] have been characterized by TERS.

Figure 7.12a shows the TERS spectra of four DNA bases (adenine, thymine, cytosine, and guanine) with picomole quantities [45]. Four bases with such a small volume were spectroscopically distinguished with a high signal-to-noise ratio. The single-molecule-detection sensitivity of TERS spectroscopy of DNA base molecules was also achieved [18]. Deep ultraviolet (DUV) excitation with an aluminum tip [41] would increase the detection sensitivity due to the addition of the resonant Raman effect of DNA bases in the DUV region. A DNA-network structure was imaged using the nonlinear TERS technique [49]. The spatial resolution was 15 nm, which was 60 times smaller than the excitation wavelength. The enhancement factor of nonlinear Raman scattering was on the order of 1 million times. Adenine-thymine base pairs were also analyzed on the Au (111) surface [50]. From careful spectral analysis, the base pair was found to adsorb at the surface with the adenine molecule close to the substrate surface and the thymine molecule away from the surface. Besides the DNA samples, a single RNA strand was measured [43]. Figure 7.12b shows the spectra measured along the strand. Slight changes in peak-position and peak-intensity

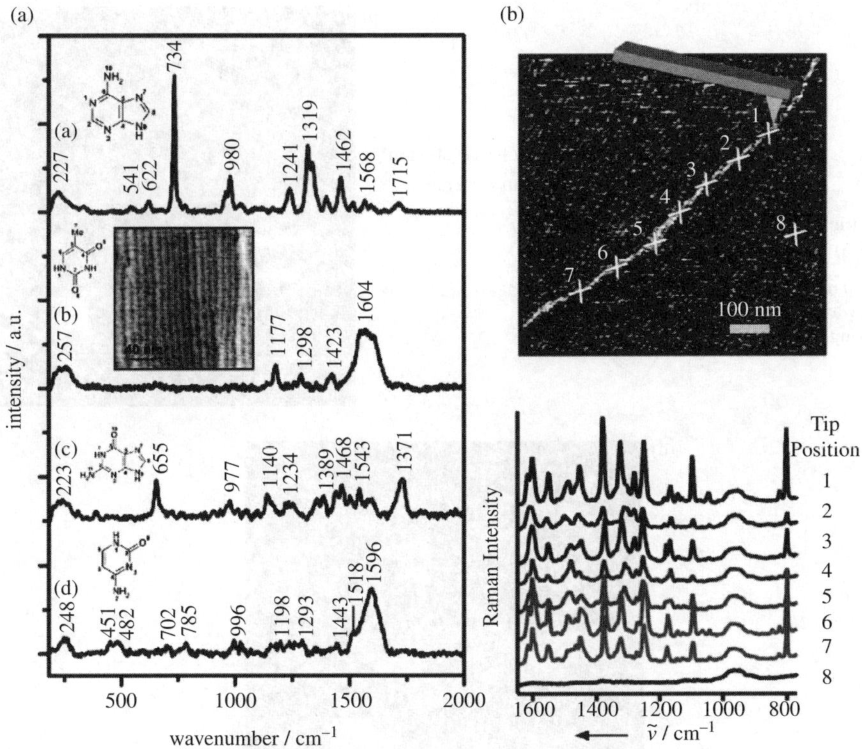

Figure 7.12 *(a) TERS spectra of DNA nucleobases. Reproduced with permission from [45]. Copyright © 2007, American Chemical Society. (b) TERS spectral mapping of a single RNA strand. Reproduced with permission from [43]. Copyright © 2008 WILEY-VCH Verlag GmbH & Co. KGaA, Weinheim*

were observed. These results open up a promising possibility of label-free DNA and RNA sequencings with base-to-base resolution.

Complex biological samples, such as immobilized cells, cell organelles, single viruses, and bacteria have also been measured [42, 47, 51–56]. However, analysis of complicated TERS spectra of complex biological mixtures is still a challenging issue. Further improvement of spatial resolution would help to simplify the TERS spectra of bio-samples due to further reduction of detection volume.

7.5 Metallic Probe Tip: Design and Fabrication

It has been an issue for TERS scientists that the yield of efficient tips in production has been low [57]. Huge effort has been devoted to improving the yield. The finite-difference time-domain method (FDTD) [58–67] is one of the most popular methods to optimize the design. The finite-element method (FEM) [68–70] and multiple multipole method (MMP) [71] have also been used for simulating experimental results. Field enhancement and plasmon resonant wavelength have been obtained through electromagnetic field simulation.

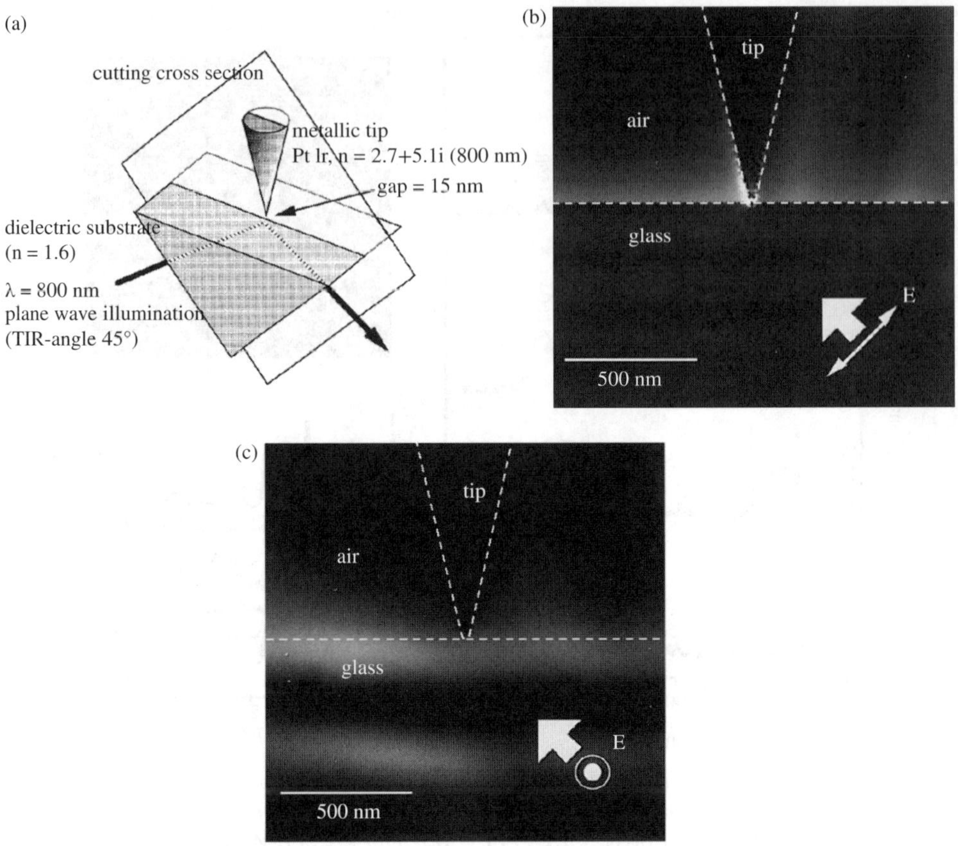

Figure 7.13 *(a) Simulation model. Sectional field distribution including a probe (platinum iridium) and a sample (flat glass substrate), for (b) p-polarized illumination, and (c) s-polarized illumination. Reproduced with permission from [67]. Copyright © 1998, Elsevier*

Polarization of incident excitation light is one of the critical factors for field enhancement [67, 72]. Incident light polarized in parallel to the tip axis (p-polarized) is predominantly enhanced while s-polarized light is not much enhanced, as shown in Figure 7.13 [67]. The cone angle [58, 73] and radius of the apex [63–65, 74] determine the factor of field enhancement.

When a dielectric under layer with metal coating is used, the plasmon resonant wavelength of the metal-coated dielectric tip is determined by the refractive index of the underlying dielectric material. Figure 7.14a shows plasmon spectra of both SiO_2 and Si tips with silver coating, calculated by the FDTD method [75]. The Ag-coated SiO_2 tip exhibits plasmon resonance in the visible wavelength region while the plasmon resonance of the Ag-coated Si tip is out of the visible. This is due to the large difference in refractive index between Si and SiO_2. In Figure 7.14a, field enhancement of the Ag-coated SiO_2 tip at the resonant wavelength is higher than that of the Ag-coated Si tip because SiO_2 is less absorptive than Si. Figure 7.14b shows the spectra of an Ag-coated dielectric tip as a function of the refractive index of the dielectric material [76]. This result represents a red shift of the

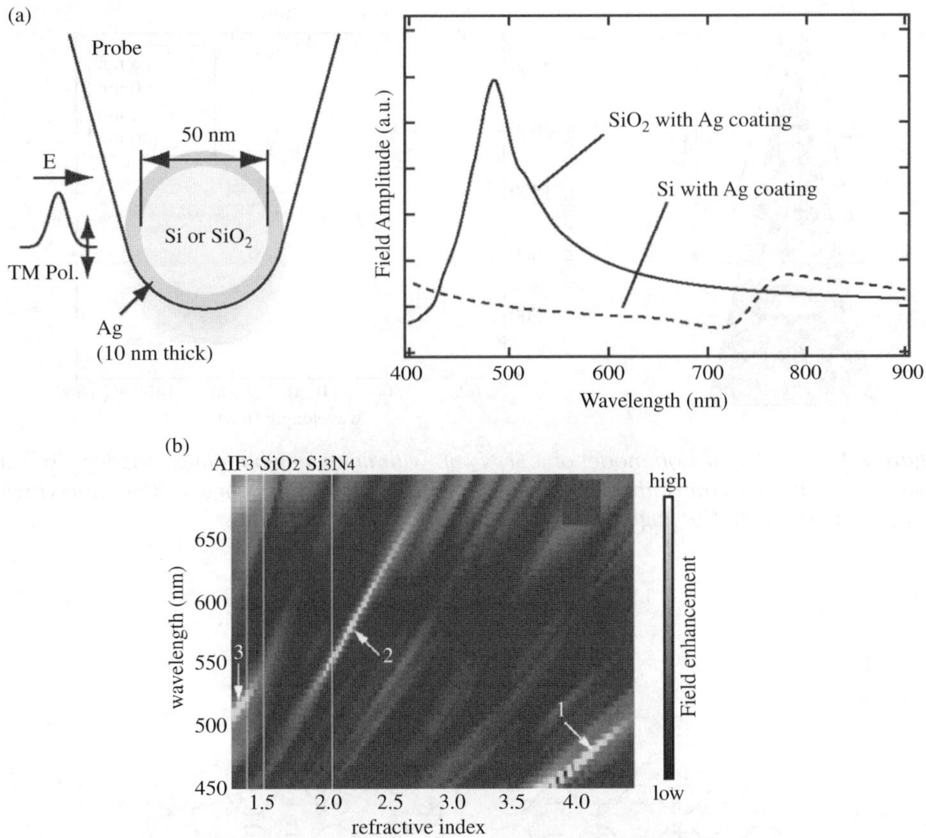

Figure 7.14 *(a) Calculated plasmon spectra of Ag-coated Si and SiO_2 tips [75]. (b) Calculated plasmon spectra of an Ag-coated dielectric tip with variable refractive index of the underlying dielectric material. Reproduced from [75] and [76] with permission from The Optical Society*

plasmon resonant wavelength with increasing refractive index of the underlying dielectric material.

When the tip surface consists of metallic nano-grains (i.e., the surface is rough), each grain works as a cavity. Huge TERS enhancement occurs at the grain attached to the end of the tip. On the other hand, when the surface is smooth and flat, SPPs propagate over the surface without any cavity resonances. As a result, the smooth surface of the metallic tip gives lower TERS enhancement than the rough one, which also occurs in SERS. However, when the length of the smooth tip is short enough, the finite-length smooth tip also works as a cavity [74, 77, 78]. Figure 7.15a shows the calculation model of an Au-coated tip with finite length (*L*). Figure 7.15b shows the plasmon spectra, calculated with four lengths between 188 and 240 nm [77]. With increasing length, the plasmon resonance shifts to longer wavelength. This is an intrinsic feature of antenna resonance.

When the substrate surface is metallic, gap-mode field enhancement occurs between the metallic tip and the substrate. The field enhancement is about two order of magnitude stronger than with a non-metallic substrate [79]. The resonant wavelength is red-shifted

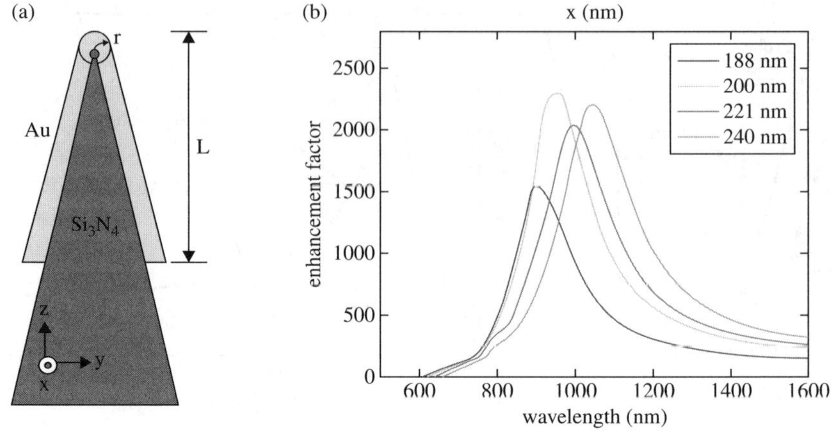

Figure 7.15 *(a) Simulation model of a Si_3N_4 tip with finite-length metallic coating. (b) Calculated plasmon spectra of the Au-coated Si_3N_4 tip with varying lengths. Reproduced with permission from [77]. Copyright © 2009, American Institute of Physics*

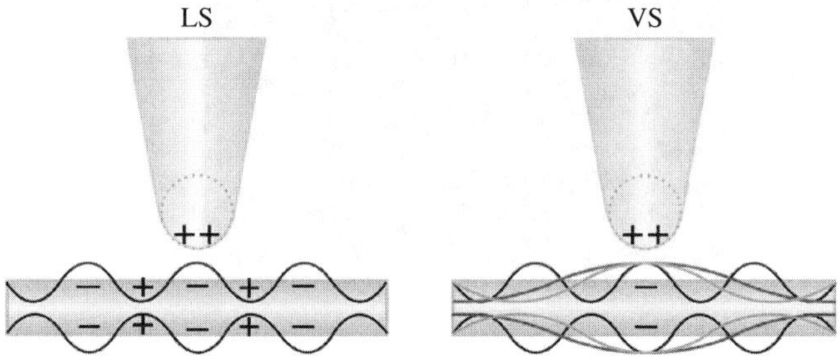

Figure 7.16 *Hybridized plasmon polaritons induced between a metallic tip and a thin metal substrate. Reproduced from [80] by permission of The Royal Society of Chemistry*

as the distance decreases. However, when the thickness of the substrate is thin enough, hybridized plasmon resonances occur due to interaction between the localized plasmon resonance of the tip and the delocalized SPPs of the thin substrate (Figure 7.16) [80]. The lower energy resonance is highly tunable, and decreases with reduced substrate thickness. On the other hand, the higher energy resonance is not sensitive to substrate thickness due to the energy mismatch between the localized plasmon of the tip and the substrate plasmon.

Silver and gold are the most commonly used plasmonic materials in visible TERS experiments. This is because they exhibit a very small value in the imaginary part of the dielectric function with a negative real part throughout the visible wavelength. Aluminum is less attractive at visible wavelengths because of the relatively large imaginary part of the dielectric function. However, in the DUV region, the dielectric function of aluminum shows a reasonably small imaginary part with a negative real part, and hence aluminum is suitable for TERS experiments in the DUV region [41]. The choice of a suitable plasmonic material is important for effective tip-enhancement.

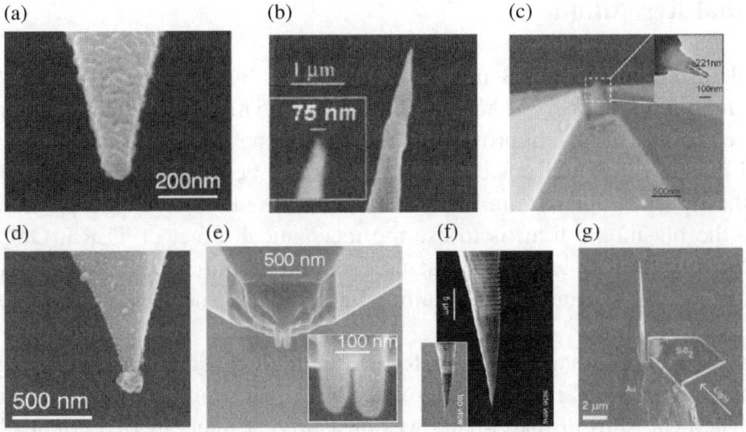

Figure 7.17 *SEM images of metallic tips fabricated by (a) thermal evaporation. Reproduced with permission from [75]. © Optical Society of American, 2009. (b) electrochemical etching. Reproduced with permission from [82] Copyright © 2004, American Institute of Physics, (c) FIB lithography. Reproduced with permission from [77]. Copyright © 2009, American Institute of Physics, (d) photoreduction. Reprinted with permission from [84]. Copyright ©2012 The Japan Society of Applied Physics, and (e) EB lithography [85, 86]. Tips with grating (f). Reproduced with permission from [87]. Copyright © 2007, American Chemical Society and prism (g). Reproduced with permission from [88]. Copyright © 2002, American Institute of Physics were also fabricated by nanolithography techniques*

There are a variety of fabrication methods for TERS-active tips. The thermal evaporation method is commonly used to fabricate a metal-coated dielectric tip. A metal film with controlled thickness is deposited on an AFM cantilever probe tip (Figure 7.17a). The surface roughness of the deposited metal surface is well controlled by the evaporation parameters. The chemical reduction method with mirror reaction is also used to fabricate roughened metal-coated tips. The size of the metallic nano-grains is controlled by the reaction time and reactant concentration [81]. Metallic tips with a smooth surface are prepared by electrochemical etching of a metallic wire (Figure 7.17b) [82]. A single-grained tip is fabricated by attaching a gold nanoparticle onto an end of a pointed optical fiber [83]. Finite-length metallic tips with length varying from 150 to 250 nm are fabricated by focused ion beam (FIB) milling after coating Si_3N_4 cantilever tips with gold (Figure 7.17c) [77]. Photoreduction is also utilized to control the size of a metallic nanostructure at the apex of the dielectric tip. The exposure time in a photochemical reaction is a crucial factor to control the size (Figure 7.17d) [84]. Recent progress in nanolithography techniques enables us to fabricate a bow-tie nano-antenna at the tip apex (Figure 7.17e) [85, 86]. Electron beam (EB) and FIB nanolithography techniques are also utilized to fabricate a grating [87] and a prism [88] on the shaft (i.e., side surface) of the tip, as shown in Figure 7.17f and g, respectively. These structured surfaces efficiently couple with far-field excitation light, and then the excited SPPs propagate along the shaft toward the tip apex where they are re-radiated into the far-field.

7.6 Spatial Resolution

The spatial resolution of TERS microscopy is given by the size of the tip apex. The resolution reported in papers is mostly around 10–15 nm. One would assume that the resolution could be simply improved by decreasing the size of the tip apex. However, the spatial resolution would not be improved further because a large number of metal atoms at the tip apex are required for field enhancement. One of the possible ways to go beyond the plasmonic limit is to use the mechanical effect of TERS. Only a single, or at most a very few, molecules in a sample can be deformed with an optimal pressure (force) to the sample (Figure 7.18), resulting in the achievement of single-molecule-level resolution [19].

Force-assisted TERS imaging was performed under a constant force [19]. Figure 7.19a shows a topographic AFM image of an isolated SWCNT, overlaid with a schematic of tip-scanning. The peak shift of the G-band was measured at intervals of 1 nm, as the tip was raster-scanned along the dashed line crossing over the SWCNT under a constant force of 2.4 nN. The peak shift is shown in Figure 7.19b. As the tip reaches close to the SWCNT, the amount of the peak shift starts to increase, reachinging a maximum value of ~10 cm^{-1} when the tip is exactly at the center of the SWCNT, and then decreases as the tip moves away from the SWCNT. The spatial resolution is 4 nm, which is estimated from the FWHM of this scanning profile. We calculated the contact area between the tip and the nanotube. The result indicates that the total contact area is less than 1 nm^2, confirming that only a very small part of the SWCNT is under compression.

A thin and flat adenine nanocrystal was also measured by the force–assisted imaging technique. Figure 7.19c shows an AFM image of the crystal. The peak position of the RBM was measured while the tip was scanned across the edge of the crystal under a constant force of 0.3 nN. The line profile is shown in Figure 7.19d. The peak position shifted from ~730 to 723 cm^{-1} at the edge. The steep edge response shows a spatial resolution of about 4 nm. The mechanical interaction between tip and sample enables us to go beyond the plasmonic limit of the spatial resolution which is a few tens of nanometers.

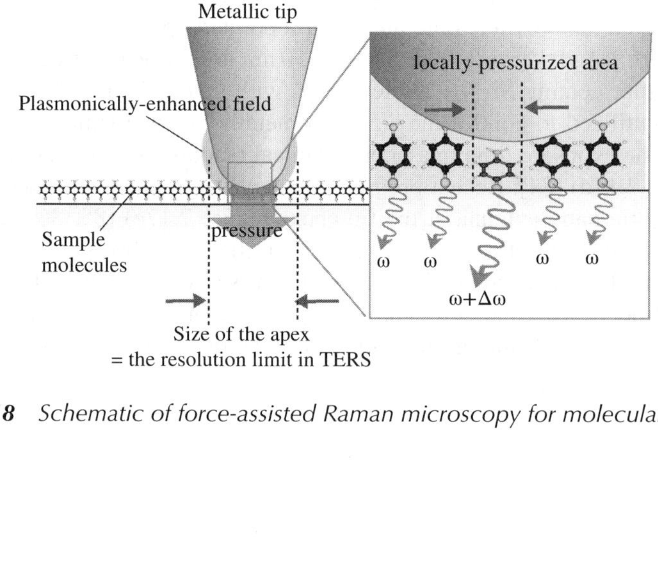

Figure 7.18 *Schematic of force-assisted Raman microscopy for molecular resolution*

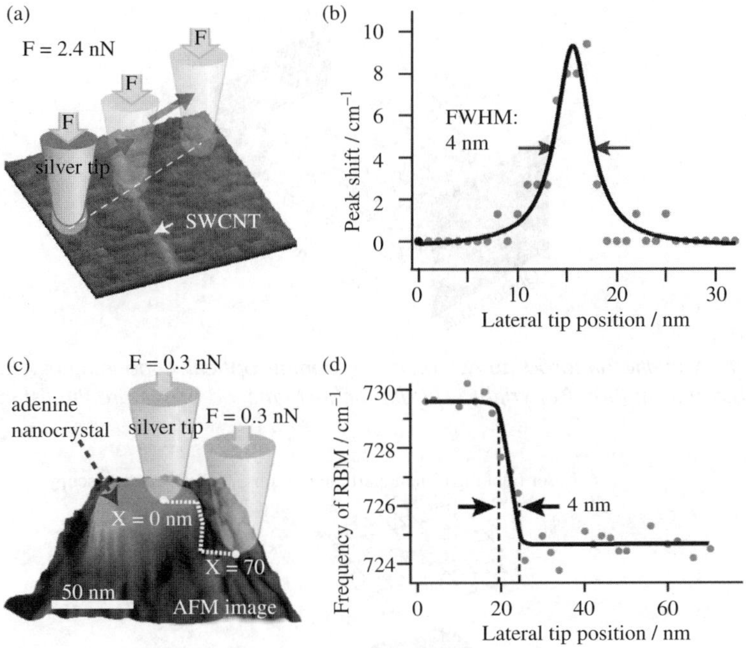

Figure 7.19 *(a) Topographic AFM image of an isolated SWCNT, overlaid with an illustration of tip scanning across the SWCNT. The tip was scanned along the dotted line at steps of 1 nm at a constant tip-applied force of 2.4 nN. (b) Experimental data for peak shift of the G-mode as a function of the lateral position of the tip. The best fitting curve shows a spatial resolution of 4 nm in this one-dimensional imaging. (c) AFM image of an adenine nanocrystal, overlaid with an illustration of tip scanning across the edge of the crystal. (d) Frequency of the RBM peak as a function of the lateral position of the tip. The steep edge of the line profile shows a spatial resolution of about 4 nm. Reproduced with permission from [19]. Copyright © 2009, Rights Managed by Nature Publishing Group*

7.7 Real-Time and 3D Imaging: Perspectives

In TERS imaging, a metallic tip is required to be raster-scanned over the sample surface. The scanning process prolongs the measurement time. One of the most promising ways to overcome the limit of temporal resolution is to use a nanolens [89, 90]. The nanolens is made of silver nanorods arranged in a stacked layered configuration, as illustrated in Figure 7.20. It provides broad resonance for achieving color imaging. It also provides long-distance image transfer without significant loss. A tapered arrangement of the nanorods within the stacked layers enables magnification of the image. The non-scanning imaging technique has the possibility to perform real-time TERS imaging.

TERS is limited to two-dimensional surface imaging. Three-dimensional nano-imaging is an important issue for biological application, such as intracellular nano-analysis. This issue is solved by performing three-dimensional tracking of a plasmonic nanoparticle together with SERS measurement. The dynamic SERS imaging technique is shown in

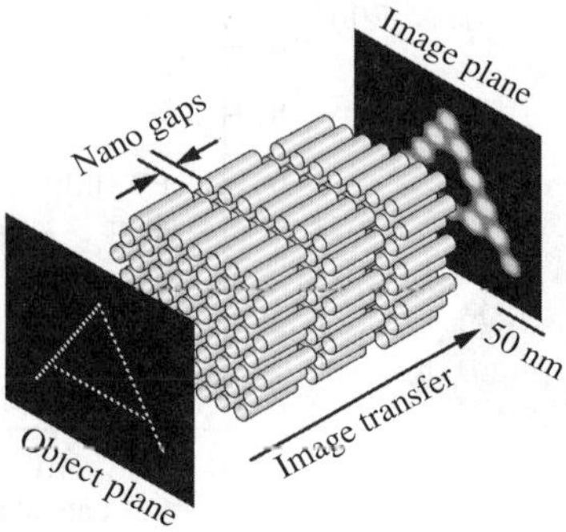

Figure 7.20 Metallic nanorod arrays for non-scanning optical nano-imaging. Reproduced with permission from [90]. Copyright © 2008, Rights Managed by Nature Publishing Group

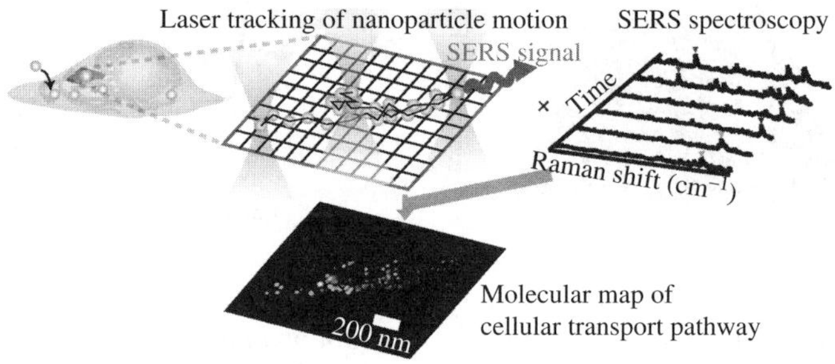

Figure 7.21 Dynamic SERS imaging of cellular transport pathways with endocytosed gold nanoparticles. Reproduced with permission from [91]. Copyright © 2011, American Chemical Society

Figure 7.21 [91]. This technique allows us to perform molecular mapping not of the whole cell but of cellular transport pathways inside a living cell at 65 nm spatial resolution and 50 ms temporal resolution.

References

1. Kawata, S. and Shalaev, V.M. (2007) *Tip Enhancement*, Elsevier, Amsterdam, Boston, MA.
2. Kawata, S. (2001) *Near-Field Optics and Surface Plasmon Polaritons*, Springer, Berlin, New York.
3. Kawata, S., Inouye, Y. and Verma, P. (2009) Plasmonics for near-field nano-imaging and superlensing. *Nature Photonics*, **3**, 388–394.
4. Jeanmaire, D.L. and Vanduyne, R.P. (1977) Surface Raman spectroelectrochemistry. 1. Heterocyclic, aromatic, and aliphatic-amines adsorbed on anodized silver electrode. *Journal of Electroanalytical Chemistry*, **84**, 1–20.
5. Pettinger, B. and Wetzel, H. (1981) Surface enhanced Raman-spectroscopy of pyridine on Ag electrodes – surface complex-formation. *Chemical Physics Letters*, **78**, 398–403.

6. Adrian, F.J. (1982) Charge-transfer effects in surface-enhanced Raman-scattering. *Journal of Chemical Physics*, **77**, 5302–5314.
7. Inouye, Y. and Kawata, S. (1994) Near-field scanning optical microscope with a metallic probe tip. *Optics Letters*, **19**, 159–161.
8. Gleyzes, P., Boccara, A.C. and Bachelot, R. (1995) Near-field optical microscopy using a metallic vibrating tip. *Ultramicroscopy*, **57**, 318–322.
9. Keilmann, F., vander Weide, D.W., Eickelkamp, T. *et al.* (1996) Extreme subwavelength resolution with a scanning radio-frequency transmission microscope. *Optics Communications*, **129**, 15–18.
10. Knoll, B., Keilmann, F., Kramer, A. and Guckenberger, R. (1997) Contrast of microwave near-field microscopy. *Applied Physics Letters*, **70**, 2667–2669.
11. Knoll, B. and Keilmann, F. (1998) Scanning microscopy by mid-infrared near-field scattering. *Applied Physics A: Materials Science and Processing*, **66**, 477–481.
12. Sanchez, E.J., Novotny, L. and Xie, X.S. (1999) Near-field fluorescence microscopy based on two-photon excitation with metal tips. *Physical Review Letters*, **82**, 4014–4017.
13. Inouye, Y., Hayazawa, N., Hayashi, K. *et al.* (1999) Near-field scanning optical microscope using a metallized cantilever tip for nanospectroscopy. *Proceedings of SPIE*, **3791**, 40–48.
14. Stockle, R.M., Suh, Y.D., Deckert, V. and Zenobi, R. (2000) Nanoscale chemical analysis by tip-enhanced Raman spectroscopy. *Chemical Physics Letters*, **318**, 131–136.
15. Watanabe, H., Ishida, Y., Hayazawa, N. *et al.* (2004) Tip-enhanced near-field Raman analysis of tip-pressurized adenine molecule. *Physical Review B*, **69**, 155418.
16. Hayazawa, N., Watanabe, H., Saito, Y. and Kawata, S. (2006) Towards atomic site-selective sensitivity in tip-enhanced Raman spectroscopy. *Journal of Chemical Physics*, **125**, 244705.
17. Ichimura, T., Fujii, S., Verma, P. *et al.* (2009) Subnanometric near-field Raman investigation in the vicinity of a metallic nanostructure. *Physical Review Letters*, **102**, 186101.
18. Ichimura, T., Watanabe, H., Morita, Y. *et al.* (2007) Temporal fluctuation of tip-enhanced Raman spectra of adenine molecules. *Journal of Physical Chemistry C*, **111**, 9460–9464.
19. Yano, T., Verma, P., Saito, Y. *et al.* (2009) Pressure-assisted tip-enhanced Raman imaging at a resolution of a few nanometres. *Nature Photonics*, **3**, 473–477.
20. Yano, T.A., Inouye, Y. and Kawata, S. (2006) Nanoscale uniaxial pressure effect of a carbon nanotube bundle on tip-enhanced near-field Raman spectra. *Nano Letters*, **6**, 1269–1273.
21. Verma, P., Yamada, K., Watanabe, H. *et al.* (2006) Near-field Raman scattering investigation of tip effects on C-60 molecules. *Physical Review B*, **73**, 045416.
22. Hayazawa, N., Yano, T., Watanabe, H. *et al.* (2003) Detection of an individual single-wall carbon nanotube by tip-enhanced near-field Raman spectroscopy. *Chemical Physics Letters*, **376**, 174–180.
23. Hartschuh, A., Sanchez, E.J., Xie, X.S. and Novotny, L. (2003) High-resolution near-field Raman microscopy of single-walled carbon nanotubes. *Physical Review Letters*, **90**, 095503.
24. Yano, T., Verma, P., Kawata, S. and Inouye, Y. (2006) Diameter-selective near-field Raman analysis and imaging of isolated carbon nanotube bundles. *Applied Physics Letters*, **88**, 093125.

25. Anderson, N., Hartschuh, A. and Novotny, L. (2007) Chirality changes in carbon nanotubes studied with near-field Raman spectroscopy. *Nano Letters*, **7**, 577–582.
26. Saito, Y., Yanagi, K., Hayazawa, N. *et al.* (2006) Vibrational analysis of organic molecules encapsulated in carbon nanotubes by tip-enhanced Raman spectroscopy. *Japanese Journal of Applied Physics Part 1: Regular Papers Brief Communications and Review Papers*, **45**, 9286–9289.
27. Stadler, J., Schmid, T. and Zenobi, R. (2011) Nanoscale chemical imaging of single-layer graphene. *ACS Nano*, **5**, 8442–8448.
28. Saito, Y., Verma, P., Masui, K. *et al.* (2009) Nano-scale analysis of graphene layers by tip-enhanced near-field Raman spectroscopy. *Journal of Raman Spectroscopy*, **40**, 1434–1440.
29. Georgi, C. and Hartschuh, A. (2010) Tip-enhanced Raman spectroscopic imaging of localized defects in carbon nanotubes. *Applied Physics Letters*, **97**.
30. Saito, Y., Motohashi, M., Hayazawa, N. *et al.* (2006) Nanoscale characterization of strained silicon by tip-enhanced Raman spectroscopy in reflection mode. *Applied Physics Letters*, **88**, 143109.
31. Zhu, L., Georgi, C., Hecker, M. *et al.* (2007) Nano-Raman spectroscopy with metallized atomic force microscopy tips on strained silicon structures. *Journal of Applied Physics*, **101**, 104305.
32. Lee, N., Hartschuh, R.D., Mehtani, D. *et al.* (2007) High contrast scanning nano-Raman spectroscopy of silicon. *Journal of Raman Spectroscopy*, **38**, 789–796.
33. Tarun, A., Hayazawa, N. and Kawata, S. (2009) Tip-enhanced Raman spectroscopy for nanoscale strain characterization. *Analytical and Bioanalytical Chemistry*, **394**, 1775–1785.
34. Hayazawa, N., Motohashi, M., Saito, Y. *et al.* (2007) Visualization of localized strain of a crystalline thin layer at the nanoscale by tip-enhanced Raman spectroscopy and microscopy. *Journal of Raman Spectroscopy*, **38**, 684–696.
35. Ogawa, Y., Yuasa, Y., Minami, F. and Oda, S. (2011) Tip-enhanced Raman mapping of a single Ge nanowire. *Applied Physics Letters*, **99**, 053112.
36. Zhang, D., Wang, X., Braun, K. *et al.* (2009) Parabolic mirror-assisted tip-enhanced spectroscopic imaging for non-transparent materials. *Journal of Raman Spectroscopy*, **40**, 1371–1376.
37. Berweger, S., Neacsu, C.C., Mao, Y.B. *et al.* (2009) Optical nanocrystallography with tip-enhanced phonon Raman spectroscopy. *Nature Nanotechnology*, **4**, 496–499.
38. Matsui, R., Verma, P., Ichimura, T. *et al.* (2007) Nanoanalysis of crystalline properties of GaN thin film using tip-enhanced Raman spectroscopy. *Applied Physics Letters*, **90**, 061906.
39. Pan, D.H., Klymyshyn, N., Hu, D.H. and Lu, H.P. (2006) Tip-enhanced near-field Raman spectroscopy probing single dye-sensitized TiO2 nanoparticles. *Applied Physics Letters*, **88**, 093121.
40. Saito, Y., Motohashi, M., Hayazawa, N. and Kawata, S. (2008) Stress imagining of semiconductor surface by tip-enhanced Raman spectroscopy. *Journal of Microscopy-Oxford*, **229**, 217–222.
41. Taguchi, A., Hayazawa, N., Furusawa, K. *et al.* (2009) Deep-UV tip-enhanced Raman scattering. *Journal of Raman Spectroscopy*, **40**, 1324–1330.

42. Budich, C., Neugebauer, U., Popp, J. and Deckert, V. (2008) Cell wall investigations utilizing tip-enhanced Raman scattering. *Journal of Microscopy-Oxford*, **229**, 533–539.
43. Bailo, E. and Deckert, V. (2008) Tip-enhanced Raman spectroscopy of single RNA strands: towards a novel direct-sequencing method. *Angewandte Chemie International Edition*, **47**, 1658–1661.
44. Neugebauer, U., Schmid, U., Baumann, K. *et al.* (2007) Towards a detailed understanding of bacterial metabolism – spectroscopic characterization of staphylococcus epidermidis. *ChemPhysChem*, **8**, 124–137.
45. Domke, K.F., Zhang, D. and Pettinger, B. (2007) Tip-enhanced Raman spectra of picomole quantities of DNA nucleobases at Au(111). *Journal of the American Chemical Society*, **129**, 6708–6709.
46. Cialla, D., Deckert-Gaudig, T., Budich, C. *et al.* (2009) Raman to the limit: tip-enhanced Raman spectroscopic investigations of a single tobacco mosaic virus. *Journal of Raman Spectroscopy*, **40**, 240–243.
47. Bohme, R., Richter, M., Cialla, D. *et al.* (2009) Towards a specific characterisation of components on a cell surface – combined TERS-investigations of lipids and human cells. *Journal of Raman Spectroscopy*, **40**, 1452–1457.
48. Yeo, B.S., Amstad, E., Schmid, T. *et al.* (2009) Nanoscale probing of a polymer-blend thin film with tip-enhanced Raman spectroscopy. *Small*, **5**, 952–960.
49. Ichimura, T., Hayazawa, N., Hashimoto, M. *et al.* (2004) Tip-enhanced coherent anti-stokes Raman scattering for vibrational nanoimaging. *Physical Review Letters*, **92**, 220801.
50. Zhang, D., Domke, K.F. and Pettinger, B. (2010) Tip-enhanced Raman spectroscopic studies of the hydrogen bonding between adenine and thymine adsorbed on Au (111). *ChemPhysChem*, **11**, 1662–1665.
51. Yeo, B.S., Madler, S., Schmid, T. *et al.* (2008) Tip-enhanced Raman spectroscopy can see more: the case of cytochrome C. *Journal of Physical Chemistry C*, **112**, 4867–4873.
52. Bohme, R., Cialla, D., Richter, M. *et al.* (2010) Biochemical imaging below the diffraction limit – probing cellular membrane related structures by tip-enhanced Raman spectroscopy (TERS). *Journal of Biophotonics*, **3**, 455–461.
53. Bohme, R., Mkandawire, M., Krause-Buchholz, U. *et al.* (2011) Characterizing cytochrome c states – TERS studies of whole mitochondria. *Chemical Communications*, **47**, 11453–11455.
54. Hermann, P., Hermelink, A., Lausch, V. *et al.* (2011) Evaluation of tip-enhanced Raman spectroscopy for characterizing different virus strains. *Analyst*, **136**, 1148–1152.
55. Opilik, L., Bauer, T., Schmid, T. *et al.* (2011) Nanoscale chemical imaging of segregated lipid domains using tip-enhanced Raman spectroscopy. *Physical Chemistry Chemical Physics*, **13**, 9978–9981.
56. Richter, M., Hedegaard, M., Deckert-Gaudig, T. *et al.* (2011) Laterally resolved and direct spectroscopic evidence of nanometer-sized lipid and protein domains on a single cell. *Small*, **7**, 209–214.
57. Hayazawa, N., Yano, T. and Kawata, S. (2012) Highly reproducible tip-enhanced Raman scattering using an oxidized and metallized silicon cantilever tip as a tool for everyone. *Journal of Raman Spectroscopy*, **43**, 1177–1182.

58. Goncharenko, A.V., Chang, H.C. and Wang, J.K. (2007) Electric near-field enhancing properties of a finite-size metal conical nano-tip. *Ultramicroscopy*, **107**, 151–157.
59. Baida, F.I. and Belkhir, A. (2009) Superfocusing and light confinement by surface Plasmon excitation through radially polarized beam. *Plasmonics*, **4**, 51–59.
60. Wang, Z.B., Luk'yanchuk, B.S., Li, L. *et al.* (2007) Optical near-field distribution in an asymmetrically illuminated tip-sample system for laser/STM nanopatterning. *Applied Physics A: Materials Science & Processing*, **89**, 363–368.
61. Yang, Z.L., Aizpurua, J. and Xu, H.X. (2009) Electromagnetic field enhancement in TERS configurations. *Journal of Raman Spectroscopy*, **40**, 1343–1348.
62. Sukharev, M. and Seideman, T. (2009) Optical properties of metal tips for tip-enhanced spectroscopies. *Journal of Physical Chemistry A*, **113**, 7508–7513.
63. Behr, N. and Raschke, M.B. (2008) Optical antenna properties of scanning probe tips: plasmonic light scattering, tip-sample coupling, and near-field enhancement. *Journal of Physical Chemistry C*, **112**, 3766–3773.
64. Roth, R.M., Panoiu, N.C., Adams, M.M. *et al.* (2006) Resonant-Plasmon field enhancement from asymmetrically illuminated conical metallic-probe tips. *Optics Express*, **14**, 2921–2931.
65. Demming, A.L., Festy, F. and Richards, D. (2005) Plasmon resonances on metal tips: understanding tip-enhanced Raman scattering. *Journal of Chemical Physics*, **122**, 184716.
66. Goncharenko, A.V., Dvoynenko, M.M., Chang, H.C. and Wang, J.K. (2006) Electric field enhancement by a nanometer-scaled conical metal tip in the context of scattering-type near-field optical microscopy. *Applied Physics Letters*, **88**.
67. Furukawa, H. and Kawata, S. (1998) Local field enhancement with an apertureless near-field-microscope probe. *Optics Communications*, **148**, 221–224.
68. Downes, A., Salter, D. and Elfick, A. (2006) Finite element simulations of tip-enhanced Raman and fluorescence spectroscopy. *Journal of Physical Chemistry B*, **110**, 6692–6698.
69. Downes, A., Salter, D. and Elfick, A. (2008) Simulations of tip-enhanced optical microscopy reveal atomic resolution. *Journal of Microscopy-Oxford*, **229**, 184–188.
70. Notingher, I. and Elfick, A. (2005) Effect of sample and substrate electric properties on the electric field enhancement at the apex of SPM nanotips. *Journal of Physical Chemistry B*, **109**, 15699–15706.
71. Esteban, R., Vogelgesang, R. and Kern, K. (2007) Tip-substrate interaction in optical near-field microscopy. *Physical Review B*, **75**, 195410.
72. Martin, O.J.F. and Girard, C. (1997) Controlling and tuning strong optical field gradients at a local probe microscope tip apex. *Applied Physics Letters*, **70**, 705–707.
73. Goncharenko, A.V., Wang, J.K. and Chang, Y.C. (2006) Electric near-field enhancement of a sharp semi-infinite conical probe: material and cone angle dependence. *Physical Review B*, **74**, 235442.
74. Zhang, W.H., Cui, X.D. and Martin, O.J.F. (2009) Local field enhancement of an infinite conical metal tip illuminated by a focused beam. *Journal of Raman Spectroscopy*, **40**, 1338–1342.
75. Taguchi, A., Hayazawa, N., Saito, Y. *et al.* (2009) Controlling the Plasmon resonance wavelength in metal-coated probe using refractive index modification. *Optics Express*, **17**, 6509–6518.

76. Cui, X.D., Zhang, W.H., Yeo, B.S. *et al.* (2007) Tuning the resonance frequency of Ag-coated dielectric tips. *Optics Express*, **15**, 8309–8316.
77. Zou, Y.S., Steinvurzel, P., Yang, T. and Crozier, K.B. (2009) Surface Plasmon resonances of optical antenna atomic force microscope tips. *Applied Physics Letters*, **94**, 171107.
78. Krug, J.T., Sanchez, E.J. and Xie, X.S. (2002) Design of near-field optical probes with optimal field enhancement by finite difference time domain electromagnetic simulation. *Journal of Chemical Physics*, **116**, 10895–10901.
79. Yano, T., Ichimura, T., Taguchi, A., Hayazawa, N., Verma, P., Inouye, Y. and Kawata, S. (2007) Confinement of enhanced field investigated by tip-sample gap regulation in tapping-mode tip-enhanced Raman microscopy. *Applied Physics Letters*, **91**, 121101.
80. Uetsuki, K., Verma, P., Nordlander, P. and Kawata, S. (2012) Tunable Plasmon resonances in a metallic nanotip-film system. *Nanoscale*, **4**, 5391–5395.
81. Saito, Y., Wang, J.J., Batchelder, D.N. and Smith, D.A. (2003) Simple chemical method for forming silver surfaces with controlled grain sizes for surface Plasmon experiments. *Langmuir*, **19**, 6857–6861.
82. Ren, B., Picardi, G. and Pettinger, B. (2004) Preparation of gold tips suitable for tip-enhanced Raman spectroscopy and light emission by electrochemical etching. *Review of Scientific Instruments*, **75**, 837–841.
83. Anger, P., Bharadwaj, P. and Novotny, L. (2006) Enhancement and quenching of single-molecule fluorescence. *Physical Review Letters*, **96**, 113002.
84. Umakoshi, T., Yano, T.A., Saito, Y. and Verma, P. (2012) Fabrication of near-field plasmonic tip by photoreduction for strong enhancement in tip-enhanced Raman spectroscopy. *Applied Physics Express*, **5**, 052001.
85. Farahani, J.N., Eisler, H.J., Pohl, D.W. *et al.* (2007) Bow-tie optical antenna probes for single-emitter scanning near-field optical microscopy. *Nanotechnology*, **18**, 125506.
86. Farahani, J.N., Pohl, D.W., Eisler, H.J. and Hecht, B. (2005) Single quantum dot coupled to a scanning optical antenna: a tunable superemitter. *Physical Review Letters*, **95**, 017402.
87. Ropers, C., Neacsu, C.C., Elsaesser, T. *et al.* (2007) Grating-coupling of surface plasmons onto metallic tips: a nanoconfined light source. *Nano Letters*, **7**, 2784–2788.
88. Sanchez, E.J., Krug, J.T. and Xie, X.S. (2002) Ion and electron beam assisted growth of nanometric SimOn structures for near-field microscopy. *Review of Scientific Instruments*, **73**, 3901–3907.
89. Ono, A., Kato, J. and Kawata, S. (2005) Subwavelength optical imaging through a metallic nanorod array. *Physical Review Letters*, **95**, 267407.
90. Kawata, S., Ono, A. and Verma, P. (2008) Subwavelength colour imaging with a metallic nanolens. *Nature Photonics*, **2**, 438–442.
91. Ando, J., Fujita, K., Smith, N.I. and Kawata, S. (2011) Dynamic SERS imaging of cellular transport pathways with endocytosed gold nanoparticles. *Nano Letters*, **11**, 5344–5348.

8

Shell-Isolated Nanoparticle-Enhanced Raman Spectroscopy (SHINERS)

Jian-Feng Li and Zhong-Qun Tian
State Key Laboratory of Physical Chemistry of Solid Surfaces and College of Chemistry and Chemical Engineering, Xiamen University, China

8.1 Introduction

Surface-enhanced Raman scattering [1–5] (SERS) is a nanoscale phenomenon because the metal nanoparticles (NPs) or surface nanostructures responsible for the much enhanced Raman signal depend critically on the size, shape, and interparticle spacing [6–15]. This nanostructure-based molecular vibrational spectroscopy can provide non-destructive and ultra-sensitive characterization down to single molecule level, comparable to single molecule fluorescence spectroscopy [16–31]. However, SERS has not been adopted as a general tool due to lack of substrate (materials) generality and surface (morphology) generality. Only a few metals, for example, Ag, Au, and Cu, provide a huge SERS effect under the condition that the metal surface exhibits roughness or the colloid size must be at the scale of several tens of nanometers [12, 13, 32, 33]. This condition has severely limited the breadth of practical applications of SERS to other materials widely used in materials, energy, medical, and life sciences. Moreover, atomically flat surfaces of various single-crystals, which are commonly used in surface science and semiconductor technology, were almost excluded.

To overcome these two problems several groups have developed a strategy of "borrowing SERS activity" to extend the SERS study to non-traditional SERS substrates [33–47]. This approach employs the long-range effect of the enormous electromagnetic (EM) field created by the high SERS-active Au or Ag nanostructures nearby. The Raman signals

Frontiers of Surface-Enhanced Raman Scattering: Single Nanoparticles and Single Cells, First Edition.
Edited by Yukihiro Ozaki, Katrin Kneipp and Ricardo Aroca.

of probed molecules on non-traditional SERS substrates, spatially separated from the SERS-active substrate, can be enhanced to some extent. The essential notion rests on the expectation from EM theories that the enhancement effect should occur, albeit in attenuated form, for molecules located close to (within 1–10 nm) SERS-active NPs or nanostructured surfaces [48–51]. It should be noted that the strong EM field will be attenuated exponentially with increasing thickness of the "spacer" film, therefore the film has to be ultra-thin, normally a few atomic layers.

In 1983, the approach of "borrowing SERS activity" was first proposed by Van Duyne and coworkers. They obtained enhanced Raman signals of molecular species adsorbed on non-SERS-active materials, such as n-GaAs electrodes, by electrodepositing a discontinuous SERS-active overlayer of Ag [34]. The long-range effect of the very strong EM field created by the SERS-active Ag islands can enhance the Raman scattering of adsorbates at the semiconductor nearby. As illustrated in Figure 8.1a, this configuration has a limitation for studying surface adsorption, as it requires that the probed species should be adsorbed selectively at the non-SERS-active substrate but not the Ag islands. In fact, most molecules studied adsorb preferentially at Ag or Au as the SERS-active substrates.

In 1987, the groups of Fleischmann [35, 36] and Weaver [37–39] developed a more effective and feasible way to coat an ultra-thin film of transition metals (Group VIIIB elements) onto SERS-active Au and Ag electrodes, respectively (Figure 8.1b). With the aid of the long-range effect of the strong EM field created by the SERS-active Au or Ag underneath, Raman signals of molecules on the transition metal surfaces can be enormously enhanced. However, it was very difficult to cover completely the very irregular roughened surfaces with so thin a film. In 2002, the Weaver group successfully coated an ultra-thin (pinhole-free) film of Pt-group metal onto Au NPs immobilized on an

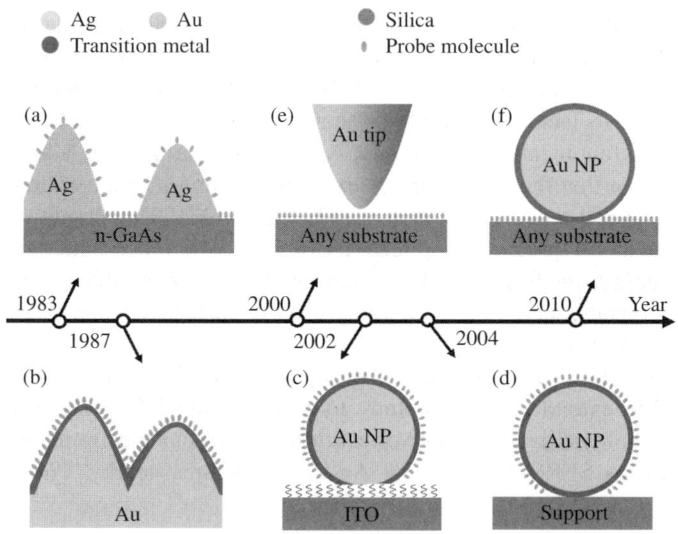

Figure 8.1 *Illustration of the development over the past three decades of the "borrowing SERS activity" methods utilizing different nanostructures in order to expand the materials and morphology generalities.* (See insert for color representation of the figure.)

ITO electrode, by under-potential deposition (UPD) and the chemical redox replacement method (Figure 8.1c) [40], which then showed much more intense SERS signals. However, the whole preparation procedure, involving surface functionalization, electrochemical UPD, and chemical reaction, is quite complicated. It is therefore highly desirable to develop a simpler and straightforward method to prepare transition metal coated Au NPs to improve the substrate generality.

In 2004, our group developed a simpler and more straightforward strategy to synthesize transition metal coated Au NPs by a seed-mediated growth method (Figure 8.1d) [33, 41]. In this approach Au or Ag NPs or nanostructures are first prepared and then coated with ultra-thin shells (1–10 atomic layers) of various metals. Boosted by the long-range effect of the enhanced EM field generated by the highly SERS-active Au core, the inherently low surface enhancement of the shell material can be substantially boosted to obtain total enhancement factors (EFs) up to 10^4–10^5, for Pt, Pd, Ru, Rh, Ni, and Co, respectively [33, 41]. However, for many other materials, such as oxides, polymers, insulators, or biological membranes, it is very difficult, if not impossible, to coat them as uniform ultra-thin shells onto Au NPs. It is therefore highly desirable to explore an innovative approach to make SERS universally applicable.

As we look carefully over all conventional SERS substrates, including the abovementioned approaches, we can find a similarity, that is, the probed molecules (substance) are either adsorbed at (in contact with) the surface of SERS-active NPs (nanostructures) or the over-coated layer. We name these approaches as the contact mode, see Figure 8.2a,b. The NP acts as a Raman signal amplifier and a host of the probed molecules simultaneously, causing the limitation of the materials and morphology generalities.

In 2000, tip-enhanced Raman spectroscopy (TERS) [42–44] was invented (Figure 8.1e), leading to a breakthrough in the limitation of substrate and surface generalities for SERS,

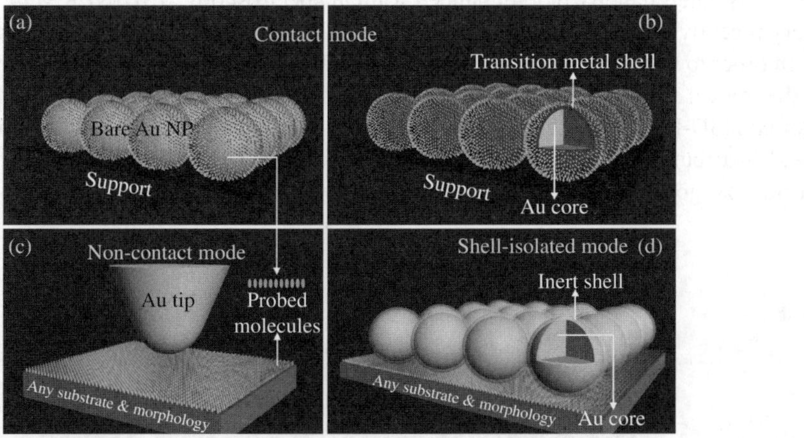

***Figure 8.2** Three working principles of conventional SERS, TERS, and SHINERS. (a) Bare Au nanoparticles, contacted mode; (b) Au core–transition metal shell nanoparticles, contacted mode; (c) tip-enhanced Raman spectroscopy, non-contact mode; and (d) shell-isolated nanoparticle-enhanced Raman spectroscopy, shell-isolated mode. Reprinted by permission from [45], Macmillan Publishers Ltd. Copyright © 2010. (*See insert for color representation of the figure.*)*

which is different from the previous approaches. TERS utilizes a new non-contact working mode (Figure 8.2c) that separates the probed substance (molecule or materials surface) from an Au tip that acts as the Raman signal amplifier. The largely enhanced EM field generated on the probe tip under the excitation of a suitable laser can extend to the sample that is in close proximity to the tip apex. In principle, the Raman signals from any substrate, regardless of material and surface smoothness, can be enhanced by the tip apex with high spatial resolution [43]. However, the total Raman scattering signal from the tip area (about 20–50 nm in diameter) is rather weak, thus TERS studies have been largely limited to molecules having large Raman cross sections, which prohibits its application to a range of molecules and materials. If TERS is applied to study electrochemical systems, the tip has to be immersed in the solution, thus some solution species could adsorb at the bare tip. The Raman spectra recorded could be dominated by molecules adsorbed at the tip with the highest SERS activity, which may provide misleading information.

In 2010, we invented a new method, at least to a large extent, to circumvent all the shortcomings described above. This approach is to replace the tip of TERS by a monolayer of SERS-active NPs (Figure 8.1f), each coated with an ultra-thin and inert shell of silica (here denoted Au@SiO_2 NPs). Each NP acts as the TERS tip, and thus simultaneously brings thousands of TERS tips to the substrate surface to be probed. One can obtain the jointly enhanced Raman signal contributed by all of these NPs (Figure 8.3), which is two to three orders of magnitude higher than that for a single TERS tip. Thereby, the physical property (strong EM field) of Au NPs is transferred to enhance the chemical signal, for example, Raman vibrational bands, of the adsorbed molecules on the materials we are interested in. Moreover, the use of a chemically inert shell coating around the Au NP protects the SERS-active nanostructure from contact with what is being probed. The main virtue of such a shell-isolated mode (Figure 8.2d) is its much higher detection sensitivity and vast practical applications to various materials with diverse morphologies. We call this new technique "shell-isolated nanoparticle-enhanced Raman spectroscopy" (SHINERS) [45–47], which very recently has been applied successfully to several systems. [45–47, 52–63].

In order to understand the mechanism of SHINERS and calculate the field enhancement, a theoretical simulation was carried out using the three-dimensional finite-difference time-domain (3D-FDTD) method [64, 65]. Figure 8.4a,b illustrate the calculated optical electric-field distribution on which a 2×2 array of Au@SiO_2 NPs is placed on a single-crystal surface of gold. As the Au NPs are brought very close to an atomically flat surface with

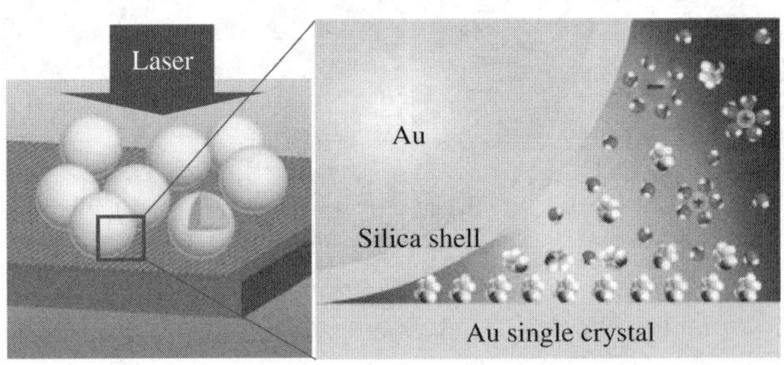

Figure 8.3 *Schematic diagrams of the SHINERS experiment. Reproduced with permission from [52]. Copyright © 2011, American Chemical Society.* (See insert for color representation of the figure.)

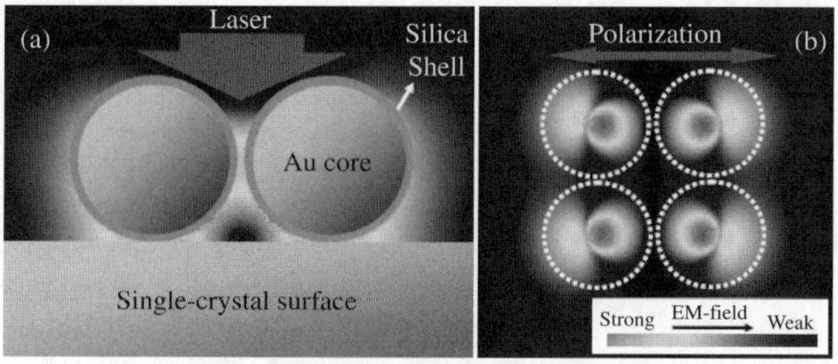

Figure 8.4 *The FDTD simulation of the optical electric field distribution of a 2×2 array of 55 nm Au@4 nm SiO_2 NPs placed on a single-crystal gold surface. (a) Side-view and (b) top-view of the optical electric field distribution at the gap between the particle and the surface. The shell to shell distance is 4 nm. The polarization of the 632.8 nm laser is indicated in (b). Reprinted by permission from [45], Macmillan Publishers Ltd. Copyright © 2010.* (See insert for color representation of the figure.)

controllable spacing by using the glass shell, localized surface plasmons can be excited in the well-controlled nanogap, producing a large EM field enhancement and a tremendous increase in the Raman intensity for molecules adsorbed on the material's surface beneath the particles, thus allowing the detection of monolayer adsorbed species on a practically important substrate of general interest. The calculated results reveal that the magnitude of the maximally enhanced electric field is about 85 and 142 times higher than that of the incident light for the 4 nm and 2 nm shells, respectively; and the highest enhancement of Raman scattering appears at the junction between the particle and the substrate as hot spots with about 5×10^7- and 4×10^8-fold magnitude, respectively [45]. The 3D-FDTD simulation also shows that the interparticle distance does not affect in any appreciable way the intensity distribution of the electric field at the particle–substrate region.

8.2 Synthesis of Various Shell-Isolated Nanoparticles (SHINs)

An overview of the experimental procedure is provided in Figure 8.5 [47]. Different kinds of SHINs (shell-isolated nanoparticles) are synthesized to meet different analysis requirements.

We use a 55 nm Au NP core in most cases, because these type of SHINs are quick and easy to prepare (Figure 8.6a) [45] and they can also provide enough Raman enhancement for most samples. If higher enhancement is required, a 120 nm Au NP core will be used (Figure 8.6b) [63]. The 3D-FDTD simulations have shown that the 120 nm core size can generate the strongest SERS activity when a 633 nm excitation wavelength is employed. Nanocube and nanorod SHINs (Figure 8.6c,d) [56] were developed for their unique and tunable surface plasmon resonance (SPR) properties. The SPR absorption band maximum can be tuned to the near-infrared by adjusting the nanorod aspect ratio, which is essential for biosensor applications. SHINs with a mostly Ag core instead of an Au core were synthesized (Figure 8.6e) [62] because of the greater surface plasmon efficiency on Ag rather than on Au, and the SPR of Ag nanostructures can be forced to any wavelength in the visible region of the spectrum.

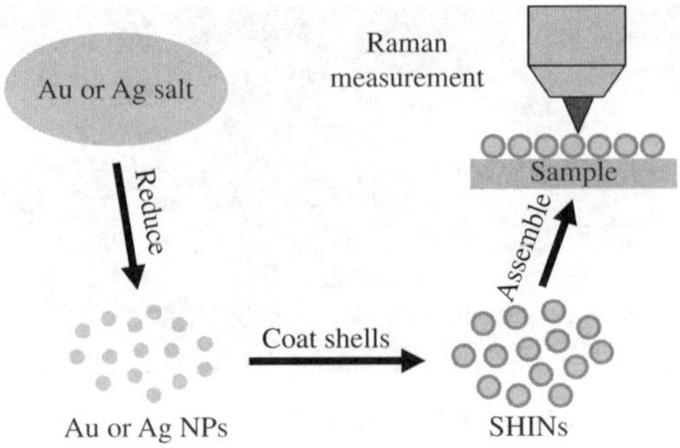

Figure 8.5 *Schematic illustration of the SHINERS experiment.* (See insert for color representation of the figure.)

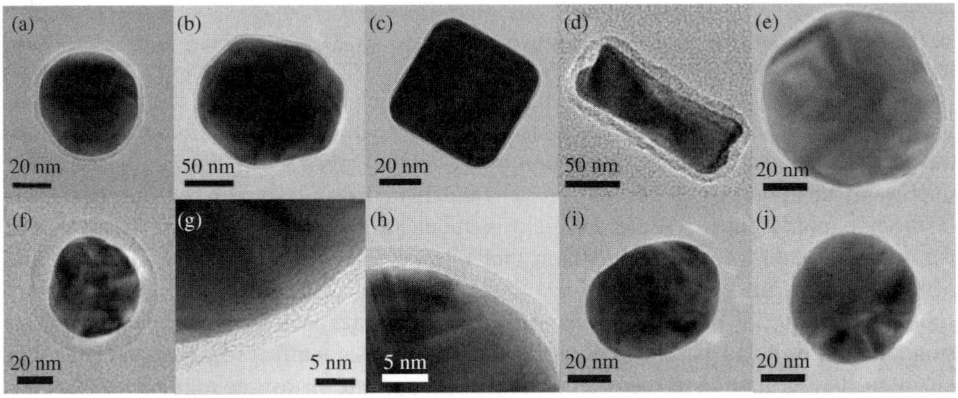

Figure 8.6 *TEM images of various kinds of SHINs. (a) 55 nm, (b) 120 nm Au@SiO$_2$ NPs, (c) cube, (d) rod Au@SiO$_2$ NPs, (e) Ag@SiO$_2$ NP, and (f) 55 nm Au@SiO$_2$ NP with 20 nm silica shell, Au SHINs shell with different materials: (g) SiO$_2$, (h) Al$_2$O$_3$, (i) MnO$_2$, and (j) Ag$_2$S*

The shell must be less than 5 nm thick to ensure sufficient EM enhancement. A thicker shell (Figure 8.6f) may not generate the desired Raman signal strength. We have also synthesized shells consisting of different materials, such as SiO_2 (Figure 8.6g), Al_2O_3 (Figure 8.6h) [45], MnO_2 (Figure 8.6i) [60], and Ag_2S (Figure 8.6j), to meet the demands of various chemical environments (e.g., stability in different pH solutions).

We take 55 nm Au@SiO_2 SHINs as a typical example to describe the detailed experimental procedure of making the core–shell NPs. The 55 nm Au NPs were prepared by using a standard sodium citrate reduction method [66]. Briefly, 200 ml of chloroauric acid (0.01 wt%) were placed in a round-bottom flask and brought to the boil. Next, 1.4 ml of sodium citrate (1.0 wt%) were added quickly to the boiling solution. The mixture was refluxed for 40 min and then allowed to cool to room temperature.

Au@SiO_2 SHINs were prepared [45, 47] by placing 30 ml of the 55 nm Au NPs solution into a round-bottom flask, adding 0.4 ml of 1 mM (3-aminopropyl) trimethoxysilane, and stirring for 15 min at room temperature. A 27% solution of sodium silicate was diluted to

0.54% and adjusted to pH ~10.3 using hydrochloric acid. 3.2 ml of the diluted and acidified sodium silicate solution were added to the reaction mixture, and it was stirred for a further 3 min at room temperature. The solution was then transferred to a 90 °C bath and stirred for a period of time. The shell thickness could be tuned from a few nanometers to tens of nanometers by simply controlling the reaction time, pH, temperature, and concentration [67–69]. For example, 20 min, 1 , 2 , 4 , 8 , or 20 h of heating will result in 2 , 4 , 6 , 8 , 10 , or 20 nm shell thicknesses, respectively, if the other parameters are fixed as described above [45].

The hot mixture was then transferred into some test tubes (1.5 ml volume), and cooled in an ice bath, which slowed or stopped the reaction dramatically. These test tubes were centrifuged at 5500 rpm for 15 min and the supernatant was removed. Then the concentrated SHINs at the bottom of each test tube were diluted, and the SHINs were centrifuged again. The supernatant was removed a second time, and clean concentrated SHINs were obtained from the bottom of the test tubes.

After several rounds of cleaning and concentrating by centrifugation, $Au@SiO_2$ NPs were diluted with Milli-Q water, dropped onto sample surfaces by a micropipette, and dried in Ar for further electrochemical or Raman measurement. The SHINs form a 2D-submonolayer on a substrate nearly without 3D aggregates thanks to isolation by the inert silica shell. It should be mentioned that neither too many nor too few SHINs should be spread on the sample surface. A monolayer of SHINs may generate the strongest Raman signal.

8.3 Characterizations of SHINs

8.3.1 Correlation of the SHINERS Intensity and Shell Thickness

Since the SPR-based (physical) enhancement of Raman scattering depends critically on the interparticle spacing of the NPs (nanostructures), we performed a combined experimental and theoretical study on the dependence of SHINERS intensity on the silica shell thickness with a fixed Au core of 55 nm in diameter. The shell thickness was controllably changed from 2 to 20 nm in sequence, which was confirmed by high resolution transmission electron microscopy (HRTEM).

Figure 8.7a presents a set of the SHINERS spectra of pyridine (Py) adsorbed on a smooth Au surface with different silica shell thickness [45]. As expected, the SHINERS intensity of Py decreases significantly with the increasing shell thickness. In other words, the thinner the silica shell, the higher the SHINERS intensity. To investigate the correlation of the shell thickness and the SHINERS intensity more quantitatively, we plotted the integrated intensity (1013 cm^{-1}, ν_1 mode) against the shell thickness in Figure 8.7b (black triangle). The integrated intensities were normalized by the signal obtained at 2 nm thickness with error bars of 15%. It can be seen that the SHINERS signal decreases exponentially with increasing shell thickness.

We also carried out a 3D-FDTD simulation of the shell thickness dependence using a 2×2 array of $Au@SiO_2$ NPs coated at an Au single-crystal surface as the model. Figure 8.7c,d show the optical electric field distribution at the gap between the Au surface and NPs with 2 and 8 nm silica shells, respectively. It can be clearly seen that the optical electric field enhancement is considerably higher for the thinner shell. Based on the fact

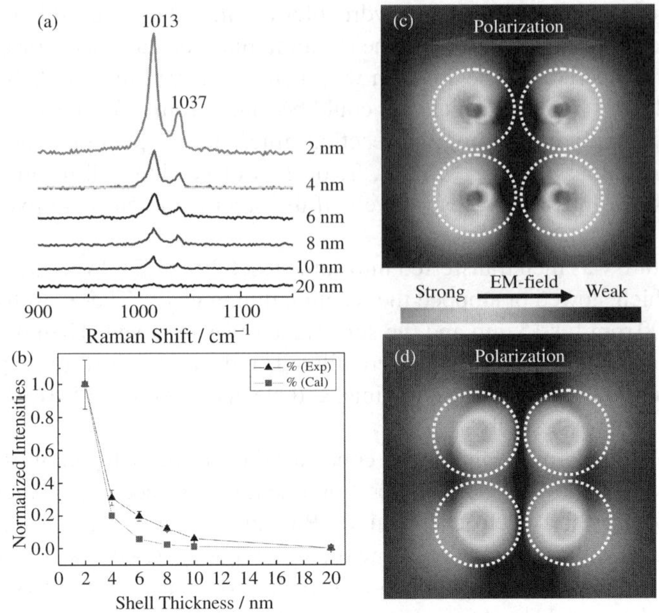

Figure 8.7 *(a) SHINERS spectra of Py adsorbed on a smooth Au surface modified by 55 nm Au@SiO$_2$ NPs with different silica shell thicknesses. (b) The shell thickness dependence of the integrated SHINERS intensity of Py (triangle) and the corresponding 3D-FDTD calculation result (square). (c,d) Top-view of the FDTD simulation of the optical electric field distribution at the plane between the smooth Au surface and 2×2 array of Au@SiO$_2$ NPs with different shell thickness of 2 nm (c) and 8 nm (d). (*See insert for color representation of the figure.*)*

that the surface Raman scattering intensity is enhanced by a factor of approximately proportional to the fourth power of the enhancement of the local incident near-field, that is, $|E_L|^4$, we can correlate the field strength to the SHINERS intensity.

Since the field distribution is not uniform (see Figure 8.7c,d), we calculated the averaged SHINERS intensity of the Au(111) coated with Au@SiO$_2$ NPs by integrating the whole sampled area with consideration of the area with different field strength, including the hot spot and other low enhanced field areas. The FDTD data (squares) are normalized with the thinnest shell system of 2 nm. The plot shows essentially the same trend as the experimental result (triangles), as shown in Figure 8.7b. Both the experimental and theoretical results confirm that the high sensitivity achieved in SHINERS is attributed to the long-range effect of the enhanced field generated by the encapsulated Au cores.

Regarding the thickness-dependent EF, our FDTD simulation also indicates that, for 55 nm Au@4 nm SiO$_2$ NPs, the electric field EF drops from 85 for the case of a 2×2 particle matrix to 55 for a single particle case, and the corresponding enhancement of Raman signal drops from 5×10^7 to 9×10^6, which is only a factor of 5.7. Therefore, even for the isolated particle case, the EF is still rather large. Accordingly, for many applications, it is not necessary to require that the spread NPs are highly ordered or density packed, as long as the spread particles form about a monolayer (multiple layers will probably block the incident laser light to the probed substrate surface). This is a unique advantage of SHINERS, making it easy to use.

8.3.2 Characterization of the Ultra-Thin Uniform Silica Shell

The most important but difficult part for SHINERS is how to effectively transfer the strong EM field from the surface of the Au core to the probed surface. It is essential to fabricate the SHINs with an ultra-thin inert shell. However, the pinhole effect should be very carefully considered, because the pinholes in the shell will expose some tiny areas of the Au core. In this case, the comparably strong SERS signals from molecules directly adsorbed on the Au cores through the pinholes may severely interfere with the SHINERS signals from the molecules adsorbed on the surface of interest, which may then mislead the interpretation of the SERS spectrum. Therefore, if the SHINERS strategy is to be applied to a wide range of chemical systems under various conditions, the shells of SHINs must be suitably thin, uniform, and ideally free of pinholes.

We use three methods to ensure that the synthesized Au@SiO_2 core–shell NPs with ultra-thin silica shells are really pinhole-free [45]. First, the uniformity of the coated silica shell can be directly checked by HRTEM. Secondly, an electrochemical method (cyclic voltammetry, CV) was used to test the isolation of the Au cores. As shown in Figure 8.8, the dotted line reveals the typical feature of 55 nm Au NPs with a characteristic reduction peak at ~0.9 V in 0.1 M H_2SO_4. On the other hand, the solid line for 55 nm Au@4 nm SiO_2 NPs is without the characteristic reduction peak, indicating the shell-isolated property of the SHINs. In addition to electrochemical examination, we used an SERS method employing pyridine as a typical probed molecule that is more sensitive to small numbers of pinholes to demonstrate the pinhole-free character.

Here, we spread SHINs over a Si wafer, and dried under vacuum. Then one drop of Py solution (10 mM) was added and a quartz window was placed on top. Py molecules are adsorbed on the Au NPs surfaces, but are not strongly adsorbed on the silica shell or Si wafer. The SHINs with pinholes generated strong Raman signals at 1011 cm^{-1} during the experiments (see curve a in Figure 8.9), indicating the Py molecules adsorbed at the Au cores through the pinholes. While using pinhole-free SHINs we did not observe any signal related to Py for several hours (see curve b).

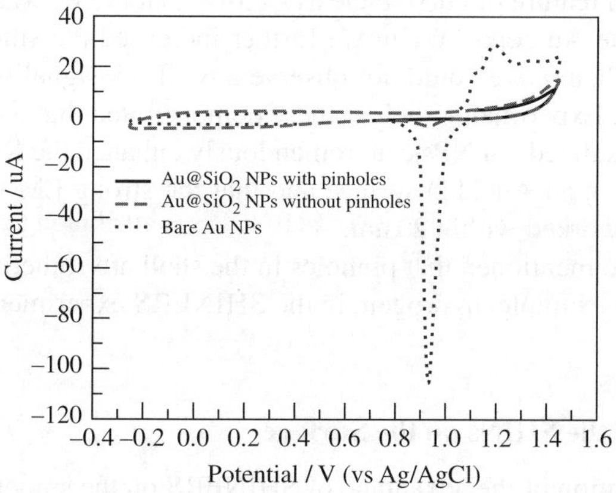

Figure 8.8 *Cyclic voltammograms of 55 nm Au (dotted line) and 55 nm Au@4 nm SiO_2 NPs with (dashed line) and without (solid line) pinholes on glassy carbon (GC) electrodes in 0.1 M H_2SO_4 solution. Scan rate: 10 mV/s*

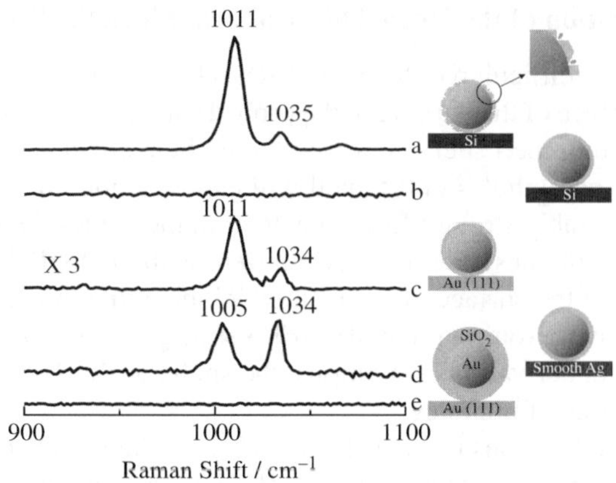

Figure 8.9 *SHINERS spectra of (a) 55 nm Au@1 nm SiO_2 NPs with pinholes on a Si wafer, 55 nm Au@4 nm SiO_2 NPs without pinholes on a (b) Si wafer, (c) Au(111) single-crystal electrode, (d) smooth Ag electrode, and (e) 55 nm Au@20 nm SiO_2 NPs on Au(111) single-crystal electrode*

When we spread the same pinhole-free SHINs over a freshly prepared Au(111) single-crystal electrode surface, we obtained a strong Raman peak related to the adsorbed Py on Au (see curve c). The signals should come from the Au(111) single-crystal surface underneath but not from the capsulated Au cores. To further improve this, we carried out this experiment on a smooth Ag electrode surface. Two strong Raman peaks at 1005 and 1034 cm^{-1} were observed (see curve d). The Ag substrate was used because the SERS spectral features of Py on an Ag surface are quite different from Py on other metal surfaces, such as Au, Pt, Ni, and so on. Py interacts with Ag considerably more weakly than it interacts with those metals [41]. Thus, the intensities of the ring breathing band (ν_1) at 1005 cm^{-1} and the symmetric triangular ring deformation band (ν_{12}) at 1034 cm^{-1} are about the same for Py on an Ag surface, while the ν_1 mode is substantially stronger than ν_{12} for Py on the other metals. The spectrum feature of curve d clearly shows that the Py was adsorbed on the Ag surface but not on the Au cores. When we further increased the silica shell thickness of the SHINs to about 20 nm, we could not observe any SERS signal of Py on the Au(111) surface (see curve e). Experiments such as these demonstrated that the enormous EM field generated by the capsulated Au NPs can tremendously enhance the Raman intensity of the molecules adsorbed on an Au(111) surface and that the strong EM field enhancement of 55 nm Au NPs only worked within 20 nm.

Finally, it should be mentioned that pinholes in the shell are sometimes acceptable if the probed molecule, for example, hydrogen, in the SHINERS experiment does not adsorb on the core material.

8.3.3 Influence of the SHINs on the Surface

Apart from consideration of the advantage of SHINERS on the smooth electrode surfaces, it is also necessary to consider the signal perturbation in an electrochemical environment where the local chemistry may change. Differential capacitance is a widely used technique to study the potential dependent charge of the electric double-layer at an electrode in

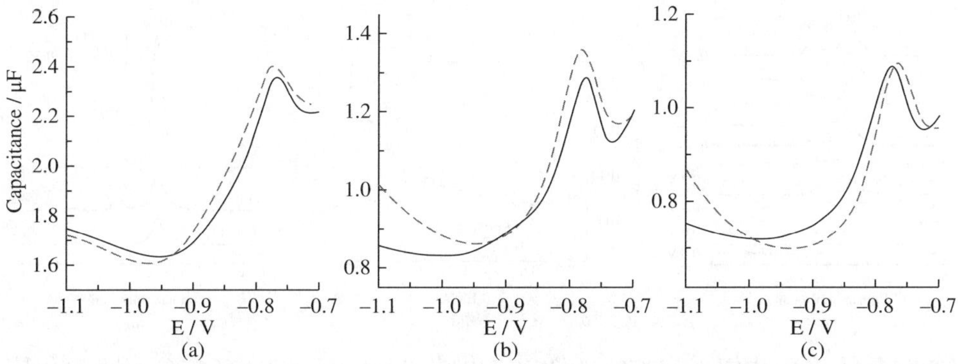

Figure 8.10 *Differential capacitance curves of the smooth silver electrodes covered by (a) 0%, (b) 10%, and (c) 100% of SHINs. Solution: 0.5 mM pyridine + 0.05 M* $NaClO_4$*, 18 Hz, solid line: negative sweep; dash line: positive sweep*

solution. This technique can reveal information about the interfacial structure and give insight into the perturbation brought about by introduction of Au@SiO_2 NPs into the system. Here we compare a smooth silver electrode with varying amounts of SHINs on its surface [55].

Figure 8.10 displays the typical differential capacitance curve of a smooth silver electrode in 0.5 mM Py and 0.05 M $NaClO_4$ solution. In the negative and positive sweeps, a minimum of the capacitance appears at about −0.95 V (Figure 8.10a) which is assigned to the potential of zero total charge (PZTC). It is quite similar to that on silver single-crystal electrode surfaces described in the literature [70–72]. When the SHINs were introduced onto the smooth electrode (Figure 8.10b), the capacitance decreased a little when compared to the bare electrode. This is because of the insulating effect from the SiO_2 shells and the reduction of effective electrode area when the SHINs occupied the electrode surface. As the NP concentration is increased from 10 to 100% (Figure 8.10c), the features of the capacitance are almost the same. The comparative study on difference capacitance infers that the coated NPs do not considerably alter the surface charge and thus adsorption behavior of the Py on Ag, that is, the SHINs on the electrode are stable and do not significantly influence the electrochemical system. This is important when considering the properties of an ideal system probe species, that is, minimum system perturbation.

8.4 Applications of SHINERS

8.4.1 Single-Crystal Electrode Surface

SERS, as a nanostructure-based molecular vibrational spectroscopy, is in principle extremely difficult to use to directly study atomically flat single-crystal surfaces that cannot effectively support the strong SPR. The SERS of atomically flat single-crystal surfaces, which are commonly used in surface electrochemistry, if possible, will be very helpful for studying the adsorption phenomena unambiguously and for comparing surface selection rules of conventional Raman spectroscopy and SERS.

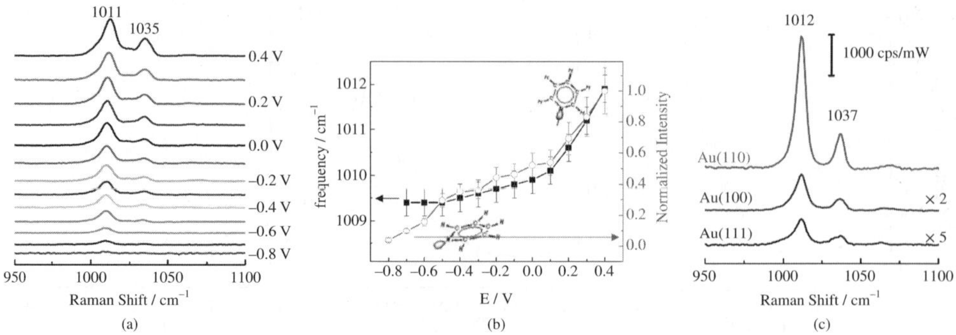

Figure 8.11 *(a) SHINER spectra of Py adsorbed on Au(111) surfaces from −0.8 to 0.4 V. (b) Raman frequencies (black squares) and normalized Raman intensities (circles) of the ring breathing mode (1011 cm^{-1}) dependent on applied potentials. (c) SHINERS spectra of Py adsorbed on Au(111), Au(100), and Au(110) at 0.00 V. Solution: 0.01 M pyridine + 0.1 M $NaClO_4$*

Here, the SHINERS method was introduced to solve this problem and probe the Au(*hkl*) single-crystal electrode surface using Py as the probed molecule, which is an ideal compound for model studies on adsorption, orientation, and surface coordination of molecules at electrified solid/liquid interfaces. *In situ* SHINERS spectra of Py adsorbed on an Au(111) single-crystal electrode are plotted in Figure 8.11a. The strongest peaks at 1011 and 1035 cm^{-1} are assigned to the ring breathing mode (ν_1) and the symmetric triangular ring deformation (ν_{12}) of the Py molecule [73, 74]. The intensity and frequency shifts of the ring breathing mode (ν_1) are sensitive to potential-dependent structure changes of the Py adlayer electrostatic effects, σ-donation and π-back donation, and so on [75, 76].

Figure 8.11b displays the potential dependence of Raman frequencies and the integrated intensities of the ν_1 mode. The frequencies (black squares) change little in the negative potential region, which points to a slight deviation from a fully planar orientation of the π-system. A continuous increase is observed at $E \geq 0.1$ V, with a stark tuning rate ($\delta\nu/\delta E$) of ~5.6 cm^{-1} V^{-1}. This value is rather typical for Py molecules adsorbed at electrochemical interfaces with an upright orientation [41, 77]. Comparing the Raman intensities of the ν_1 mode with the results from chronocoulometry and voltammetry [78], we find that the ν_1 mode appears with the onset of Py adsorption at $E \geq -0.7$ V and grows until −0.4 V, where a first plateau is reached (circles). A second increase in intensity is observed at $E \geq 0.1$ V, simultaneously with the increase in frequency. The above SHINERS data demonstrate unambiguously that Py is adsorbed in a flat but most probably slightly tilted orientation with respect to the plane surface at a negatively charged Au(111) surface, and reorients into a fully upright orientation upon passing the potential of zero charge at $E > 0.10$ V [78], which is in good agreement with previous electrochemical studies [78–80].

Comparative experiments of Py adsorbed on the three low-index Au(*hkl*) single-crystal surfaces show that SHINERS intensity is strongly dependent on the surface crystallographic orientation [52]. Figure 8.11c illustrates the trend at a substrate potential of 0.0 V. The Raman signal from Au(110) is about 8 times stronger than that obtained from Au(100), and about 30 times stronger than that obtained from Au(111). The same qualitative trend in SHINERS intensity was also found on the low-index Pt(*hkl*) surfaces. Based on our DFT calculations and theoretical analysis, we conclude that the dependent SHINERS intensities observed for Py on Au(*hkl*) and Pt(*hkl*) are dominated by dielectric properties of the single-crystal surfaces, which leads to a considerably stronger EM field coupling between our SHINERS NPs and the (110) surface compared to other low-index surfaces [52]. This facet-dependent Raman effect should, in principle, also be observed for the high-index single-crystal surfaces.

We have also applied SHINERS to examine hydrogen adsorption on the Pt(111) surface, which is a very important system in surface science, chemistry, and industry, including fuel cell applications. To our knowledge, no Raman spectra of any hydrogen adsorbed at single-crystal surfaces have been reported because of its extremely low Raman cross section. The 55 nm Au@1 nm SiO_2 NPs were spread over a Pt(111) single-crystal electrode in 0.1 M $NaClO_4$ and the electrode potential was applied in the region of the hydrogen evolution reaction. As can be seen in Figure 8.12 from curves a, b, and c, a broad band at about 2023 cm^{-1} was observed, being attributed to the Pt-H stretching vibration mode according to the previous reports on highly roughened Pt surfaces [81].

For comparison, we collected Raman signals on Pt(111) under the same condition but without SHINs. No Raman peak in this frequency region can be observed (curve d), which indicates that the Au@SiO_2 NPs are necessary to obtain the signals. Moreover, when we spread these SHINs on a smooth Au electrode, the Raman signal also disappeared (curve e), illustrating that the band at around 2023 cm^{-1} truly came from the hydrogen adsorbed on Pt(111). The reason that we did not obtain the Au-H band may simply be because of the weak interaction of hydrogen with Au.

8.4.2 Non-Metallic Material Surfaces

To examine the capability of SHINERS on non-metallic materials, we extended the study to atomically flat silicon (Si) wafer, such as that which is used widely in integrated circuit (IC) technology and microelectronics with potential use as optical sensors and possible

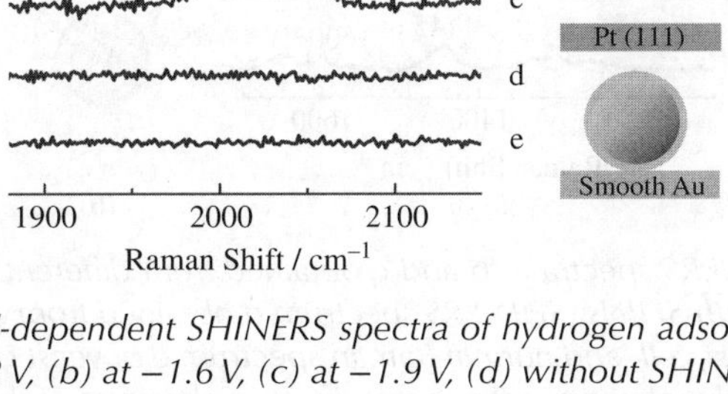

Figure 8.12 *Potential-dependent SHINERS spectra of hydrogen adsorbed on a Pt(111) surface. Curve (a) at −1.2 V, (b) at −1.6 V, (c) at −1.9 V, (d) without SHINs, and (e) on a smooth Au electrode at −1.9 V. Reprinted by permission from [45], Macmillan Publishers Ltd. Copyright © 2010*

data storage. We have studied the Si-H system on the Si(111) surface using the SHINs under different procedures applied in the IT industry. A Si wafer was first treated with sulfuric acid (98%), no Raman band related to Si-H vibration in this region can be detected. A SERS spectrum was obtained from the Si wafer when treated with HF solution. The broad band at about 2145 cm^{-1} can be attributed to the Si-H stretching vibration mode [82]. For comparison, we collected Raman signals on Si(111) under the same conditions but without SHINs. No Raman peak in this frequency region can be observed, which indicates that the SHINs are necessary to obtain the signals. After the last treatment with O_2 plasma, the Si-H band vanished. This confirms that the Si-H bond is only formed in the HF solution. It is important to mention here that there is no SERS or TERS study on the Si-H system on smooth Si surfaces, although there have been some Raman studies of hydrogen bonded porous Si [82].

Another important SHINERS application is the study of biological systems, such as living cells. Yeast cells were selected as they are one of the most extensively studied model eukaryotic organisms from genetics to biochemistry. Importantly, in yeast the cell walls are of considerable interest due to their sensitivity toward the different biological functions of the cell. First, living yeast cells were incubated with SHINs solution for 30 min at room temperature, 30 °C. Then an aliquot of an incubated cell solution was dropped onto a slide glass and covered with a cover slip for SERS detection. Figure 8.13a–c shows the SERS spectra from different spots. They are very different from the normal Raman spectra (Figure 8.13e) of yeast cells, but similar to the SERS spectra of mannoproteins [83], which were considered to be the main components of the yeast cell wall. There are some other peaks which are attributed to amide, protein backbone, and amino acid vibrations, related to the bioactivity of living cells, such as protein secretion, movement, and so on. These results demonstrate the potential of SHINERS as a safe and convenient technique for *in situ*

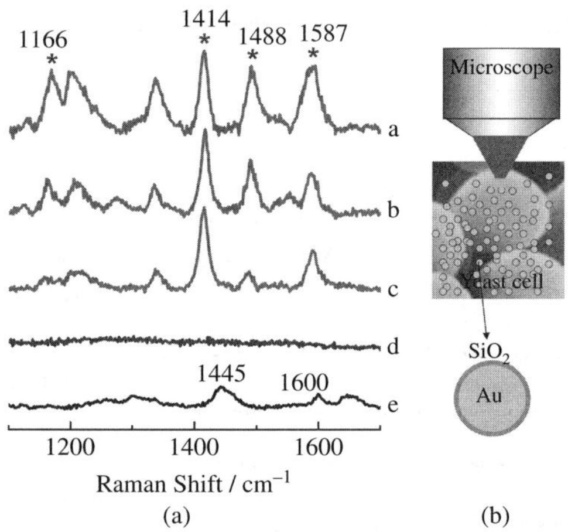

Figure 8.13 *(a) SHINERS spectra a, b and c obtained from different spots at the wall of an yeast cell incubated with SHINs. SHINERS spectrum d obtained from a substrate coated with SHINs but without yeast cell, and normal Raman spectrum e of yeast cells. The peaks marked with asterisks are closely related to mannoprotein. (b) Schematic of the apparatus. Reprinted by permission from [45], Macmillan Publishers Ltd. Copyright © 2010*

detection of cell wall proteins and as a possible probe of the dynamics and mechanisms of living systems.

Given the availability of using a portable Raman spectrometer, SHINERS could be applied even widely as an accurate and rapid simple-to-use tool for food safety inspection, drug and explosives screening, and environment protection. To inspect pesticide residues on fruit we spread SHINs as "smart dust" over the probed orange surface and then took *in situ* measurements. Figure 8.14a shows that normal Raman spectra recorded on a fresh orange with clean pericarps (curve a) or contaminated by parathion (curve b) only show two bands at about 1165 and 1526 cm^{-1}, attributed to carotenoid molecules contained in citrus fruits [84]. SHINs were then applied and two additional peaks at around 1109 and 1340 cm^{-1} (curve c), attributed to the characteristic bands of parathion residues [85], were clearly detected. The same result can also be obtained by using a confocal Raman spectrometer.

On the other hand, testing for multiple pesticides at the same time is also possible with SHINERS. Figure 8.14b curves d and e show typical SHINERS spectra for cabbage

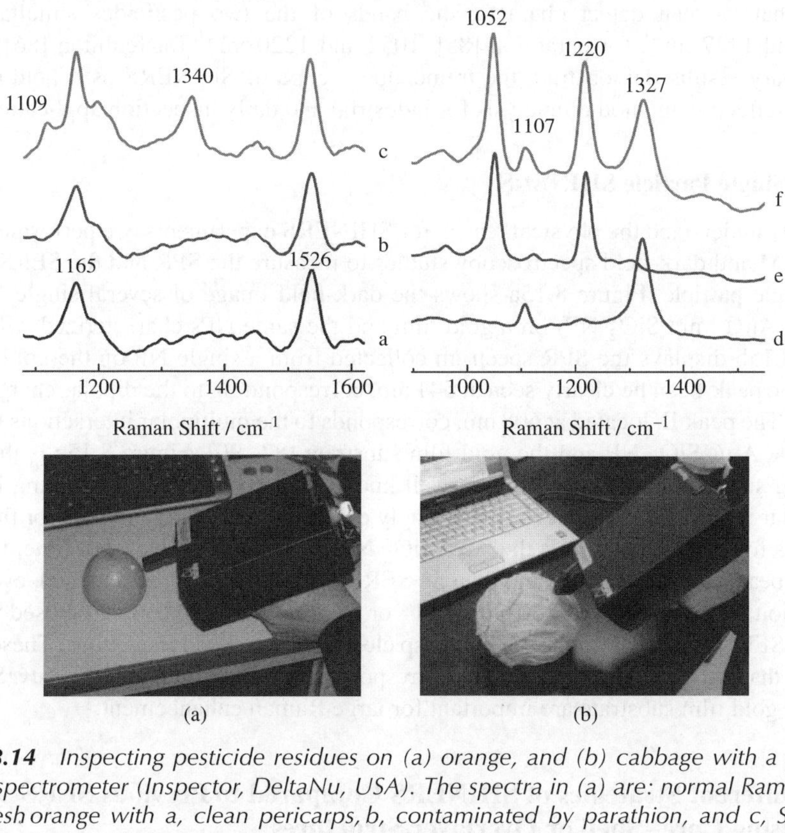

Figure 8.14 *Inspecting pesticide residues on (a) orange, and (b) cabbage with a portable Raman spectrometer (Inspector, DeltaNu, USA). The spectra in (a) are: normal Raman spectra of fresh orange with a, clean pericarps, b, contaminated by parathion, and c, SHINERS spectrum of contaminated orange modified by SHINs. The spectra in (b) are: SHINERS spectra of cabbage contaminated by d, parathion, e, fenthion, and f, mixed pesticides. The excitation line was 785 nm, laser power on the sample was 30 mW, and the integration acquisition time, was 30 s*

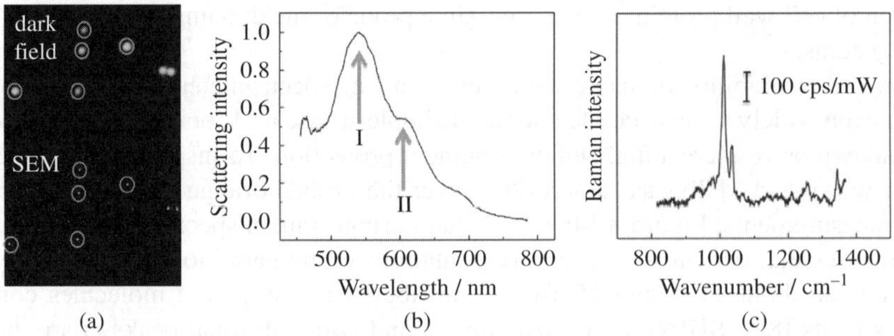

Figure 8.15 *Single particle SERS of Py on a gold film and the corresponding SPR spectrum of the particle. (a) Correlated scattering (top), and SEM (bottom) images of the same 120 nm Au@1 nm SiO_2 NPs; (b) single NP scattering spectra; and (c) corresponding single NP SERS spectrum of Py. Reprinted with permission from [63]. Copyright © 2013, John Wiley & Sons, Ltd*

leaves contaminated by parathion and fenthion pesticides, respectively. Curve f clearly shows that we can detect characteristic bands of the two pesticides simultaneously (1107 and 1327 cm^{-1} for parathion [85], 1052 and 1220 cm^{-1} for fenthion [86]). These preliminary results demonstrate the tremendous scope of SHINERS as a field-portable and cost-effective method of analysis for industrial and daily inspection applications.

8.4.3 Single Particle SHINERS

To deeply understand the physical reason for SHINERS experiments, we performed correlated SEM and dark-field spectroscopy studies to measure the SPR and the SERS spectra on a single particle. Figure 8.15a shows the dark-field image of several single particles (120 nm Au@1 nm SiO_2 NP) on a gold film and the same NPs characterized with SEM. Figure 8.15b displays the SPR spectrum collected from a single NP on the gold film, in which the peak I can be clearly seen at 541 nm, corresponding to the dipolar charge distributions. The peak II, located at 608 nm, corresponds to the multipolar interactions between the single Au@SiO_2 NP and the gold film substrate [87–90]. Figure 8.15c is the corresponding single NP SERS of Py. It is well known that peak II of the coupling between multipolar modes is important for the strongly enhanced near-field and thus for the SERS hot spots formed in the gap. If the Au@SiO_2 NPs are placed on a Si substrate, there are no SPR peaks around 608 nm, and then no SERS signal of Py can be obtained, even if the acquisition time is increased 20-fold. A 514 or 532 nm laser line cannot be used with Au NPs for SERS due to the dominance of d–sp electronic interband transitions. These results show us that the multipolar interactions corresponding to the coupling of the Au@SiO_2 NP with the gold film substrate are important for large Raman enhancement.

8.5 Different Strategies of SHINERS Compared to Previous SERS Works Using Core–Shell or Overlayer Structures

Our strategy of using Au@SiO_2 NPs is totally different from that of other groups using a much thicker silica shell (generally from 20 to 50 nm) [67–69, 91–93] or an ALD (atomic layer deposition) alumina-layer over AgFON (silver film-over-nanosphere) [12, 94–96] for the SERS studies. In the cases using the thicker shell, a Raman marker (preferably molecules with a resonant Raman effect) is adsorbed on the very first layer of the SERS-active Au NP surfaces to produce a strong Raman signal of the marker [69, 91–93]. Then the Raman marker tagged NPs were modified with (protected by) a thick SiO_2, polyelectrolyte, or polymer shell to provide the mechanical and chemical stability of the tagged molecules and NPs. The shell surface may provide various functional groups for binding with specific molecules for further immunoassay, as exemplified in Figure 8.16a. In this configuration, its function is the same as the SiO_2-coated quantum dots, which works for the image purpose. The Raman signal detected is from the Raman marker on the NPs that are linked to the target molecules [69, 91, 92].

In the case of AgFON@Al_2O_3, an ultrathin alumina layer (from several angstroms to several nanometers, fabricated by using ALD) is coated over AgFON that is fixed at a substrate such as glass [12, 94–96]. The strategy is to use the alumina layer to stabilize the SERS activity of the underlying Ag film (because Ag is easily oxidized) and to add the alumina layer to imparting a new chemical functionality, for example, coating a specific catalysts, to interact (link) with analytes [12, 94–96], as presented in Figure 8.16b.

Although the structure of AgFON@Al_2O_3 looks quite similar to SHINs, it in fact works in a different way. The analytes closely contact either directly to the Al_2O_3 overlayer or indirectly through the catalyst deposited at the Al_2O_3 overlayer (Figure 8.16b). As a consequence, the preparation of the AgFON@Al_2O_3 nanostructure fixed in a large substrate is very complicated and its fixed structure limits the wide and flexible applications. Therefore SHINs act just as "smart dust" that can freely spread over any substrates thus amplifying the Raman signal of the analyte or substance underneath, as shown in Figure 8.16e.

The unique feature of SHINERS is that the SHINs are totally free from the analyte or any probed substance. One only needs to find a way to coat a limited number of materials

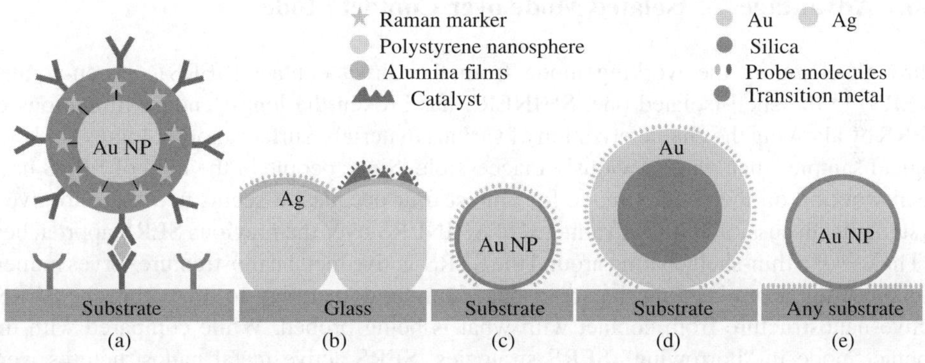

Figure 8.16 *Schematic illustration of five core–shell NPs with different strategies. (a) Au NP coated with a thick silica shell with the Raman marker adsorbed directly on the core surface and the inert shell modified with a functional group that can bind to the target molecules. (b) ALD alumina-modified AgFON substrates stabilize the SERS activity of the underlying silver and detect the analyte adsorbed at the alumina layer. (c) Silica-core–Au-shell NPs. (d) Au-core–transition metal shell NP. (e) Au NP coated with ultra-thin and chemically inert silica shell. Reprinted by permission from [45], Macmillan Publishers Ltd. Copyright © 2010*

(such as SiO_2, Al_2O_3, or silicon nitrides) over Au NPs as the isolated shell, then thousands of different materials (substrates) can be studied flexibly.

It should be noted that there are two other types of core–shell NPs widely used in SERS studies, which also work in the contact mode. The silica-core–Au-shell NPs show a great potential in tuning plasmonic structure by simply changing the aspect ratio of the Au nanoshell, the ratio of inner to outer radius, and the shell thickness. These NPs can serve as constituents in a new class of materials that are capable of uniquely controlling radiation in the visible and infrared spectral regions. The Raman enhancement can be designed to follow the plasmon resonance of the individual NPs [97–99]. Au-core–transition-metal-shell NPs have been widely used in electrochemical SERS studies. The inherently low surface enhancement of the transition metals as the shell material can be substantially boosted [33, 41]. However, the SERS studies are limited to those analytes that must adsorb directly at the Au shell (Figure 8.16c) or transition metal shells (Figure 8.16d), which is again due to that fact that these two nanostructures work based on the contact mode.

Overall, conceptually the strategy of SHINERS is completely different from that of the other four core–shell or overlayer nanostructures that have been used successfully for various purposes, that is, the Au-core–silica-shell SERS marker, ALD alumina-modified AgFON, silica-core–Au-shell, and Au-core–transition-metal-shell NPs. This is because all these systems enhance the Raman signal of analytes that contact directly or indirectly to the nanostructures as the signal enhancer. Whereas in SHINERS, the SHINs as the enhancer are separated freely from the target analytes that can be adsorbed species at single-crystal surfaces or the surface components of living cells or other materials (Figure 8.16e). The key feature of SHINERS is that its shell is de-functionalized and chemically inert, simply acting as a spacer and isolator, working in the isolated mode. This makes SHINERS unprecedentedly applicable for diverse samples and molecules with large flexibility (see Figure 8.16e).

8.6 Advantages of Isolated Mode over Contact Mode

Through changing the working mode from the direct contact (SERS) or non-contact (TERS) to the shell-isolated one, SHINERS has broken the long-standing limitations of SERS by allowing the characterization of various materials, surface morphologies, and biological samples that were previously inaccessible. Since people in the field of SERS have been working on the contact mode for almost four decades, it seems necessary to give a systematic discussion of the advantages of SHINERS over the previous SERS approaches.

The use of a thin-shell coating around the SERS-active metal nanostructure serves numerous purposes, from keeping the smart dust from agglomerating to protecting the SERS-active nanostructure from contact with what is being probed. While compared with the contact mode in "borrowing" SERS strategies, SERS-active metal nanostructures were directly deposited onto a surface that was not SERS-active, or a SERS-inactive material was deposited onto a SERS-active one. For instance, bare Ag or Au NPs constitute the simplest and strongest Raman signal amplifier, and have been used by some groups [100, 101], however, certain problems may arise when a chemically inert dielectric shell is absent:

Contact with the chemical environment. The use of bare NPs is inapplicable to liquid (for biology and electrochemistry) or gas (for catalysis) containing the same probe

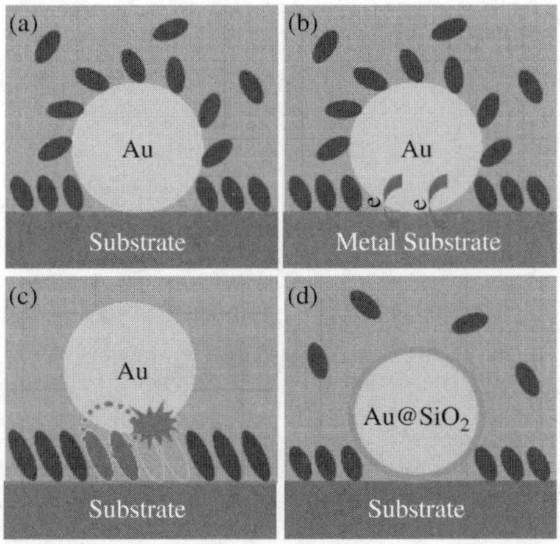

Figure 8.17 *Schematic illustrations of four different surface configurations for SERS and SHINERS measurements. (a) Contact with the chemical environment, may lead to nearby molecules adsorption. (b) Contact with the probe molecule, may lead to photocatalytic degradation. (c) Electrical contact with the surface, may lead to charge transfer. (d) All these problems can be avoided by use of our SHINs*

molecules, other molecules, or ions capable of adsorbing on the surfaces of bare NPs (see Figure 8.17a). Even in the case of liquid or air containing no such molecules, the molecules adsorbed on the substrate are likely to diffuse to the probed substrate to bare NPs. The SERS signal generated by the molecules that are adsorbed on the NP can interfere with or overshadow the SERS signal generated by the molecules adsorbed on the probed substrate. Since bare NPs are more favorable to SPR than the flat metal or other substrate, the signal from the NPs is likely to be dominant [32, 48, 102].

Contact with the probe molecule. In a self-assembled monolayer (SAM) system, since bare NPs lie in direct contact with one face of the SAM, the molecules located between the NPs and the substrate will most probably adopt a two-ended adsorption geometry instead of a one-ended adsorption geometry (Figure 8.17b). This direct contact is likely to change the electron density distribution in the molecule and its adsorption behavior, possibly resulting in a considerable distortion or drastic change in the SERS spectral feature. This will cause a misinterpretation of the recorded data. For some systems, photocatalytic reactions can be initiated in SAM molecules by the directly contacted bare NPs, resulting in a change in molecular structure or even new products (Figure 8.17b) [103]. The PATP (*p*-aminothiophenol) example illustrated below is an example of such a case.

Figure 8.18a shows the SERS or SHINERS spectra of two sandwich configurations (Figure 8.18b) with PATP as the probe molecule. When bare Au NPs were used, trace a shows the recorded spectrum of PATP adsorbed on flat Au, the spectrum shows misleading spectral features due to direct contact of the amino group of PATP with the Au NP, which leads to possible photocatalytic dimerization of PATP into an aromatic azo compound (Figure 8.18c), as evidenced by the new SERS bands observed at 1139, 1387, and 1433 cm^{-1} [103, 104]. Whereas when the same experiment was carried out using

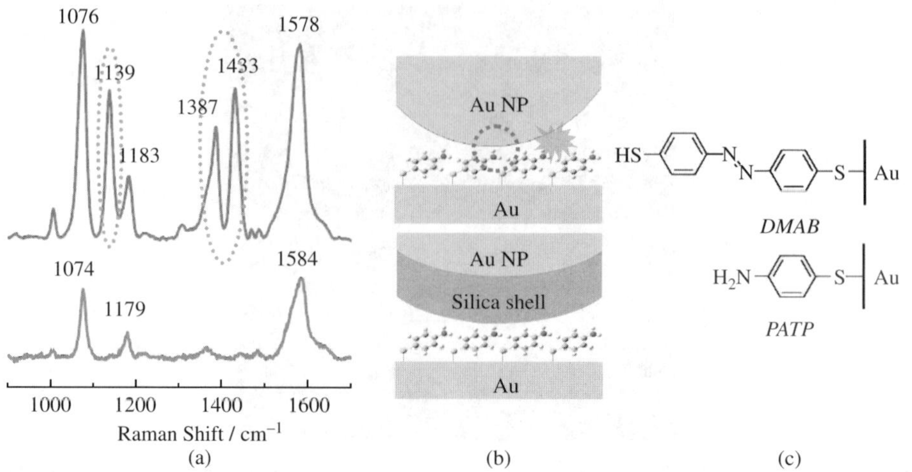

Figure 8.18 (a) SERS or SHINERS spectra of PATP molecules interconnected in the two different sandwich configurations, Au/PATP/Au NPs (upper curve) and Au/PATP/Au@SiO_2 NPs (lower curve). (b) Schematic illustration of the two sandwich configurations. (c) The possible photocatalytic dimerization

Au@SiO_2 NPs, trace b, only the spectrum for the PATP adsorbed on Au is obtained. Therefore, SHINERS has minimized the interference of the Au NP probe in molecular identification. In this study, the as-synthesized Au@SiO_2 NPs were carefully examined to ensure that the surface coating was free of pinholes, to prevent direct contact of the Au core with the detected molecules or the sample surface.

Electrical contact with the surface. If this methodology is extrapolated to non-coinage metals and semiconductors, or to non-SAM molecule systems (see Figure 8.17c), the Au NPs could be in direct contact with the surface of the material. The difference in the Fermi levels of Au and the substrate may result in a charge transfer (CT) and the creation of a contact potential that, as a consequence, can significantly affect the electronic structure of the probe molecules located below the Au NPs [105, 106].

Here we used 55 nm Au NPs to probe CO adsorbed at a single-crystal Pt(111) surface as an example. It can be seen from Figure 8.19 that there are two strong Raman bands located at 2060 and 2130 cm^{-1}, respectively. The former originates from the stretching vibration of CO adsorbed on Pt, while the latter is for CO adsorbed on Au. The strong and broad CO band related to the bare Au NPs could give a misleading interpretation of CO adsorption on Pt. More importantly, if we compare the CO(Pt) stretching frequency in the SERS and SHINERS spectra, we will see it shifts remarkably from 2072 to 2060 cm^{-1}. This shift is mainly due to CT from bare Au NPs to the Pt surface because the work functions of these two metals are quite different (Au: 5.1 eV, Pt: 5.65 eV). This CT will increase the local electron density at the localized Pt surface in contact with Au NPs [105, 106], leading to the large shift of the CO stretching frequency. However, SHINERS, using shell-isolated Au NPs, can provide correct and meaningful vibrational information regarding the surface adsorbates.

Stability. SHINs are much more stable than bare NPs because they are coated with an inert shell [47]. We placed 0.5 ml of Py solution (0.01 M) into each of two cuvettes,

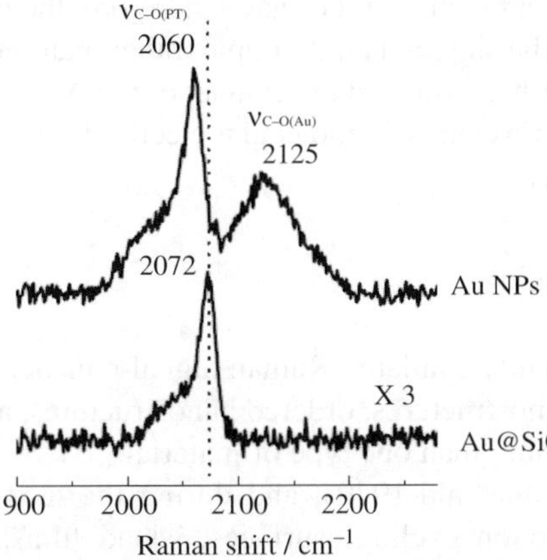

Figure 8.19 *SERS and SHINERS spectra of CO adsorption, using Au NPs or Au@SiO$_2$ NPs spread over a Pt(111) electrode surface in a solution of 0.1 M HClO$_4$ saturated by CO gas*

then added 10 μl of concentrated SHINs (~1.07 nM) to one of the cuvettes and 10 μl of concentrated bare Au NPs (~1.07 nM) to the other. Photographs and Raman spectra of these mixtures at different times are provided in Figure 8.20. During the 10-day stability test, the SHIN solution did not change color (Figure 8.20a) and gave a stable Raman signal (Figure 8.20c). The bare Au NP solution changed color in just 3 min and showed complete aggregation after 15 min (Figure 8.20b,d).

SHINERS with its unique isolated mode can overcome the three problems discussed above (Figure 8.17d). By coating the Au NPs with a thin (chemically and electronically) inert shell, it completely isolates the Au particles from the probe molecules regardless

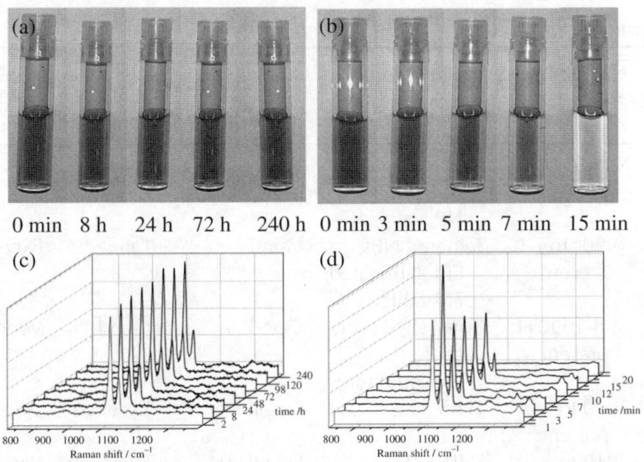

Figure 8.20 *Stability comparison of SHINs and bare Au NPs. Photos of (a) SHINs and (b) bare Au NPs in 0.01 M pyridine solution at different times. Corresponding SERS signals from Py obtained with (c) SHINs and (d) bare Au NPs. Reprinted by permission from [47], Macmillan Publishers Ltd. Copyright © 2010. (See insert for color representation of the figure.)*

of whether the measurement is in liquid or gas. Therefore, the measured Raman signal comes from nowhere but the single-crystal sample (other materials) to be measured. The selection of the inert layer has insulated the contact of the Au NPs and the substrate surface, thus, there is no CT, no change in molecular electronic structure, no photocatalysis reaction, and high stability.

8.7 Concluding Discussion

In Table 8.1 we group currently available Raman signal-enhancing structures into three broad categories: random nanostructures, ordered nanostructures, and complex nanostructures that are composed of more than one type of material.

Work on SERS began in the mid-1970s, and during the next 25 years or so, ORC-roughened (oxidation–reduction cycling) surfaces, island films, and surfaces covered with disordered NP aggregates dominated the literature. All these substrates possessed random features over which limited control could be exercised, so morphological generality is described as poor. Because they were composed of Au, Ag, or Cu, materials generality is also given as poor. The EFs obtained were high enough that a wide variety of probe/analyte molecules could be detected easily, and hence we describe molecule generality as very good.

Ordered nanostructures revitalized SERS research and may be considered the second generation of SERS substrates. Examples include NPs tethered to a solid surface in a uniformly packed monolayer, substrates fabricated by electron beam lithography (EBM), and substrates prepared by focused ion beam milling. Here, control over the substrate's morphology at the nanoscale may be considered very good but at the atomic level the morphology is only good as the surfaces are not structurally well-defined. This control has led to greater EFs and to greater reproducibility. Greater EFs mean that weakly enhancing

Table 8.1 *Enhancement strategies and their generality*

	Time frame	Examples	Morphological (surface) generality	Materials (substrate) generality	Molecule generality (sensitivity)	Spatial resolution
Random nanostructures	Mid-1970s to 2000s	ORC-generated, island films, NP aggregates	Poor	Poor	Very good	Poor
Ordered nanostructures	1990s to present	Tethered NPs EBL-fabricated, FIB-milled	Good	Fair	Excellent	Poor
More complex (multicomponent) nanostructures	Mid-1990s to present	Core–shell NPs, coated nanostructures	Good	Good	Very good	Poor
	2000 to present	TERS	Excellent	Excellent	Good	Excellent
	2010 to present	SHINERS	Excellent	Excellent	Very good	Poor

Morphological generality, materials generality, molecule generality, and spatial resolution are rated on a five-point scale as poor, fair, good, very good, or excellent. ORC, oxidation–reduction cycling; NP, nanoparticle; EBL, electron beam lithography; FIB, focused ion beam; TERS, tip-enhanced Raman spectroscopy; SHINERS, shell-isolated nanoparticle-enhanced Raman spectroscopy.

materials, such as the Group VIIIB transition metals can be used with more success. In practice, however, most of these second-generation substrates are still composed of Au, Ag, or Cu, and materials generality is only slightly improved over that of the first-generation substrates. Greater EFs also mean improved molecule generality when coinage metals are employed – molecules with lower Raman scattering cross sections can be, and are being, used in connection with these substrates.

The more complex nanostructures category is further divided into three subcategories. These include core-shell NPs and coated (ordered) nanostructures, TERS, and SHINERS.

As explained above, some weakly enhancing or non-enhancing material may be coated onto a coinage metal nanostructure as an ultra-thin shell to obtain strong SERS from a molecule on the surface, thus improving materials generality. Most procedures deposit coatings that follow the morphology of the underlying ordered nanostructure in a fairly uniform way, so morphological generality is described as good here, as it was for the uncoated ordered nanostructure. Field strength at the coating's surface is weaker than field strength at the surface of the underlying coinage metal, so EFs are lower here than they were for the uncoated ordered nanostructure and molecule generality is reduced from excellent for the uncoated ordered nanostructure to very good.

The TERS method was a conceptual breakthrough in that the signal amplifier was separated from the probed surface for the first time, and TERS may be used to examine a surface of any morphology and any material with excellent spatial resolution. The EF may be large, but the signal originates from a very small region of the sample so the absolute intensity of light detected is still weak. Therefore, only probe/analyte molecules with large Raman scattering cross sections are appropriate for use with TERS.

Our SHINERS technique is in one sense an extension of the core–shell and coatings strategy already described. However, the reason for the shell and the way in which the NPs are used differ significantly in SHINERS, and a considerable step forward in generality is the result. The material of interest is no longer added to the SERS-active coinage metal; rather, an isolating shell is added and the structure is then spread onto a surface of any morphology and any material. So, in another sense, SHINERS is an extension of what was accomplished for the first time with TERS: The signal amplifier is again separated from the probed surface. With the advent of SHINERS, SERS may be obtained from surfaces with the unlikely morphology of an atomically flat single-crystal and the unlikely material of an orange skin. The EFs achieved with SHINERS are comparable to those achieved with core–shell NPs and coated (ordered) nanostructures, so the molecule generality is also comparable.

8.8 Outlook

With the advent of SHINERS, we expect a growing emphasis on the direct detection and identification of analytes by SERS. SHINERS could be widely used to examine the surface of the skin, cells, and tissues after biopsy, or in surface chemistry as part of an industrial quality-control procedure. In addition to the direct detection and identification of analytes, we anticipate the in-depth characterization of probe–surface interactions for a wide variety of weakly enhancing and non-enhancing surfaces through detailed examination of their SHINERS spectra. They could include the probing of surface processes associated with catalysis and corrosion.

We also expect new structures to emerge. The SHIN surface could be modified with a trimethoxysilane-based linker, and then a biological receptor could be added, or the SHINs could be attached to a solid substrate in a uniformly packed monolayer. Any incompatibilities that may exist between certain chemical systems and the currently available shells, which include silica and alumina, will vanish as we learn how new materials can be deposited to form a shell that is ultrathin and, ideally, pinhole free. The semi-quantitative nature of the signal in SHINERS is also expected to be improved. If SHINs can be assembled on a surface in a uniform monolayer, Raman intensities will better correlate with the amount of the analyte present on the surface.

Finally, the concept of shell-isolated-nanoparticle enhancement may also be applicable to more general spectroscopy, such as infrared spectroscopy and sum frequency generation, as well as shell-isolated nanoparticle-enhanced fluorescence spectroscopy (SHINEFS) [107] that has been reported by Aroca and coworkers. We are optimistic that SHINERS will be developed into a simple, fast, non-destructive, flexible, and portable characterization tool, widely used in basic and applied research in natural sciences, industry, and even people's daily needs [108].

Acknowledgments

We are deeply indebted to Z. L. Yang, Z. L. Wang, T. Wandlowski, B. Ren, N. F. Zheng and J. R. Anema for their contributions and Y. F. Huang, X. D. Tian, Y. Ding, S. B. Li, X. D. Lin, Y. M. Zeng, and Q. Z. Chen for their experimental assistance during the development of SHINERS in the past six years. This work was supported by the Ministry of Science and Technology (MOST) of China (2011YQ030124, 2010IM040100 and 2009CB930703) and the National Natural Science Foundation of China (NSFC) (21033007 and 21021002).

References

1. McCreery, R.L. (2000) *Raman Spectroscopy for Chemical Analysis*, John Wiley & Sons, Inc., New York.
2. Long, D.A. (2002) *The Raman Effect: A Unified Treatment of the Theory of Raman Scattering by Molecules*, John Wiley & Sons, Ltd, Chichester.
3. Smith, E. and Dent, G. (2005) *Modern Raman Spectroscopy A Practical Approach*, John Wiley & Sons, Ltd, Chichester.
4. Chang, R.K. and Furtak, T.E. (1982) *Surface Enhanced Raman Scattering*, Plenum Press, New York.
5. Aroca, R. (2006) *Surface-Enhanced Vibrational Spectroscopy*, John Wiley & Sons, Inc., New York.
6. Le Ru, E.C. and Etchegoin, P.G. (2009) *Principles of Surface-Enhanced Raman Spectroscopy and Related Plasmonic Effects*, Elsevier, Amsterdam/Oxford.
7. Schlucker, S. (2011) *Surface Enhanced Raman Spectroscopy: Analytical, Biophysical and Life Science Applications*, Wiley-VCH Verlag GmbH, Weinheim.
8. Tian, Z.Q. ed. (2005) Surface enhanced Raman spectroscopy. *Journal of Raman Spectroscopy*, **36**, 465–747.

9. Kneipp, K., Moskovits, M. and Kneipp, H. (2006) *Surface-Enhanced Raman Scattering Physics and Applications*, Springer, Berlin/Heidelberg.
10. Brown, R.J.C., Cohen, L.F. and Milton, M.J.T. eds. (2006) Surface enhanced Raman spectroscopy. *Faraday Discussions*, **132**, 1–340.
11. Graham, D. and Goodacre, R. (eds) (2008) Surface enhanced Raman spectroscopy. *Chemical Society Reviews*, **37**, 873–1076.
12. Stiles, P.L., Dieringer, J.A., Shah, N.C. and Van Duyne, R.P. (2008) Surface-enhanced Raman spectroscopy. *Annual Review of Analytical Chemistry*, **1**, 601–626.
13. Moskovits, M. (1985) Surface-enhanced spectroscopy. *Reviews of Modern Physics*, **57**, 783–826.
14. Baumberg, J.J., Kelf, T.A., Sugawara, Y. *et al.* (2005) Angle-resolved surface-enhanced Raman scattering on metallic nanostructured plasmonic crystals. *Nano Letters*, **5**, 2262–2267.
15. Camden, J.P., Dieringer, J.A., Zhao, J. and Van Duyne, R.P. (2008) Controlled plasmonic nanostructures for surface-enhanced spectroscopy and sensing. *Accounts of Chemical Research*, **41**, 1653–1661.
16. Nie, S.M. and Emery, S.R. (1997) Probing single molecules and single nanoparticles by surface-enhanced Raman scattering. *Science*, **275**, 1102–1106.
17. Kneipp, K., Wang, Y., Kneipp, H. *et al.* (1997) Single molecule detection using Surface-Enhanced Raman Scattering (SERS). *Physical Review Letters*, **78**, 1667–1670.
18. Cao, Y.W.C., Jin, R.C. and Mirkin, C.A. (2002) Nanoparticles with Raman spectroscopic fingerprints for DNA and RNA detection. *Science*, **297**, 1536–1540.
19. Jackson, J.B. and Halas, N.J. (2004) Surface-enhanced Raman scattering on tunable plasmonic nanoparticle substrates. *Proceedings of the National Academy of Sciences of the United States of America*, **101**, 17930–17935.
20. Chen, Z., Tabakman, S.M., Goodwin, A.P. *et al.* (2008) Protein microarrays with carbon nanotubes as multicolor Raman labels. *Nature Biotechnology*, **26**, 1285–1292.
21. Jain, P.K., Huang, X.H., El-Sayed, I.H. and El-Sayed, M.A. (2008) Noble metals on the nanoscale: optical and photothermal properties and some applications in imaging, sensing, biology, and medicine. *Accounts of Chemical Research*, **41**, 1578–1586.
22. Graham, D., Thompson, D.G., Smith, W.E. and Faulds, K. (2008) Control of enhanced Raman scattering using a DNA-based assembly process of dye-coded nanoparticles. *Nature Nanotechnology*, **3**, 548–551.
23. Qian, X.M., Peng, X.H., Ansari, D.O. *et al.* (2008) In vivo tumor targeting and spectroscopic detection with surface-enhanced Raman nanoparticle tags. *Nature Biotechnology*, **26**, 83–90.
24. Anker, J.N., Hall, W.P., Lyandres, O. *et al.* (2008) Biosensing with plasmonic nanosensors. *Nature Materials*, **7**, 442–453.
25. Qin, L.D., Zou, S.L., Xue, C. *et al.* (2006) Designing, fabricating, and imaging Raman hot spots. *Proceedings of the National Academy of Sciences of the United States of America*, **103**, 13300–13303.
26. Nie, S.M. and Zare, R.N. (1997) Optical detection of single molecules. *Annual Review of Biophysics and Biomolecular Structure*, **26**, 567–596.

27. Michaels, A.M., Nirmal, M. and Brus, L.E. (1999) Surface enhanced Raman spectroscopy of individual rhodamine 6G molecules on large Ag nanocrystals. *Journal of the American Chemical Society*, **121**, 9932–9939.
28. Xu, H.X., Bjerneld, E.J., Kall, M. and Borjesson, L. (1999) Spectroscopy of single hemoglobin molecules by surface enhanced Raman scattering. *Physical Review Letters*, **83**, 4357–4360.
29. Willets, K.A. and Van Duyne, R.P. (2007) Localized surface plasmon resonance spectroscopy and sensing. *Annual Review of Physical Chemistry*, **58**, 267–297.
30. Dieringer, J.A., Wustholz, K.L., Masiello, D.J. *et al.* (2009) Surface-enhanced Raman excitation spectroscopy of a single rhodamine 6G molecule. *Journal of the American Chemical Society*, **131**, 849–854.
31. Han, X.X., Zhao, B. and Ozaki, Y. (2009) Surface-enhanced Raman scattering for protein detection. *Analytical and Bioanalytical Chemistry*, **394**, 1719–1727.
32. Moskovits, M. (2005) Surface-enhanced Raman spectroscopy: a brief retrospective. *Journal of Raman Specroscopy*, **36**, 485–496.
33. Tian, Z.Q., Ren, B., Li, J.F. and Yang, Z.L. (2007) Expanding generality of surface-enhanced Raman spectroscopy with borrowing SERS activity strategy. *Chemical Communications*, 3514–3534.
34. Vanduyne, R.P. and Haushalter, J.P. (1983) Surface-enhanced resonance Raman-spectroscopy of adsorbates on semiconductor electrode surfaces – Tris(Bipyridine) Ruthenium(II) adsorbed on silver-modified N-GaAs(100). *Journal of Physical Chemistry*, **87**, 2999–3003.
35. Fleischmann, M., Tian, Z.Q. and Li, L.J. (1987) Raman-spectroscopy of adsorbates on thin-film electrodes deposited on silver substrates. *Journal of Electroanalytical Chemistry*, **217**, 397–410.
36. Fleischmann, M. and Tian, Z.Q. (1987) The induction of SERS on smooth Ag by the deposition of Ni and Co. *Journal of Electroanalytical Chemistry*, **217**, 411–416.
37. Leung, L.W.H. and Weaver, M.J. (1987) Extending surface-enhanced Raman-spectroscopy to transition-metal surfaces – carbon-monoxide adsorption and electrooxidation on platinum-coated and palladium-coated gold electrodes. *Journal of the American Chemical Society*, **109**, 5113–5119.
38. Leung, L.W.H. and Weaver, M.J. (1987) Extending the metal interface generality of surface-enhanced Raman-spectroscopy – underpotential deposited layers of mercury, thallium, and lead on gold electrodes. *Journal of Electroanalytical Chemistry*, **217**, 367–384.
39. Zou, S.Z. and Weaver, M.J. (1998) Surface-enhanced Raman scattering an uniform transition metal films: toward a versatile adsorbate vibrational strategy for solid-non vacuum interfaces? *Analytical Chemistry*, **70**, 2387–2395.
40. Park, S., Yang, P.X., Corredor, P. and Weaver, M.J. (2002) Transition metal-coated nanoparticle films: vibrational characterization with surface-enhanced Raman scattering. *Journal of the American Chemical Society*, **124**, 2428–2429.
41. Wu, D.Y., Li, J.F., Ren, B. and Tian, Z.Q. (2008) Electrochemical surface-enhanced Raman spectroscopy of nanostructures. *Chemical Society Reviews*, **37**, 1025–1041.
42. Stockle, R.M., Suh, Y.D., Deckert, V. and Zenobi, R. (2000) Nanoscale chemical analysis by tip-enhanced Raman spectroscopy. *Chemical Physics Letters*, **318**, 131–136.

43. Pettinger, B., Ren, B., Picardi, G. *et al.* (2004) Nanoscale probing of adsorbed species by tip-enhanced Raman spectroscopy. *Physical Review Letters*, **92**, 096101.
44. Ren, B., Picardi, G., Pettinger, B. *et al.* (2005) Tip-enhanced Raman spectroscopy of benzenethiol adsorbed on Au and Pt single-crystal surfaces. *Angewandte Chemie International Edition*, **44**, 139–142.
45. Li, J.F., Huang, Y.F., Ding, Y. *et al.* (2010) Shell-isolated nanoparticle-enhanced Raman spectroscopy. *Nature*, **464**, 392–395.
46. Anema, J.R., Li, J.F., Yang, Z.L. *et al.* (2011) Shell-isolated nanoparticle-enhanced Raman spectroscopy: expanding the versatility of surface-enhanced Raman scattering. *Annual Review of Analytical Chemistry*, **4**, 129–150.
47. Li, J.F., Tian, X.D., Li, S.B. *et al.* (2013) Surface analysis using shell-isolated nanoparticle-enhanced Raman spectroscopy. *Nature Protocols*, **8**, 52–65.
48. Gersten, J. and Nitzan, A. (1980) Electromagnetic theory of enhanced Raman-scattering by molecules adsorbed on rough surfaces. *Journal of Chemical Physics*, **73**, 3023–3037.
49. Nitzan, A. and Brus, L.E. (1981) Theoretical-model for enhanced photochemistry on rough surfaces. *Journal of Chemical Physics*, **75**, 2205–2214.
50. Aravind, P.K., Nitzan, A. and Metiu, H. (1981) The interaction between electromagnetic resonances and its role in spectroscopic studies of molecules adsorbed on colloidal particles or metal spheres. *Surface Science*, **110**, 189–204.
51. Murray, C.A., Allara, D.L. and Rhinewine, M. (1981) Silver-molecule separation dependence of surface-enhanced Raman-scattering. *Physical Review Letters*, **46**, 57–60.
52. Li, J.F., Ding, S.Y., Yang, Z.L. *et al.* (2011) Extraordinary enhancement of Raman scattering from pyridine on single crystal Au and Pt electrodes by shell-isolated Au nanoparticles. *Journal of the American Chemical Society*, **133**, 15922–15925.
53. Li, J.F., Anema, J.R., Yu, Y.C. *et al.* (2011) Core-shell nanoparticle based sers from hydrogen adsorbed on a rhodium(111) electrode. *Chemical Communications*, **47**, 2023–2025.
54. Li, J.F., Li, S.B., Anema, J.R. *et al.* (2011) Synthesis and characterization of gold nanoparticles coated with ultrathin and chemically inert dielectric shells for SHINERS applications. *Applied Spectroscopy*, **65**, 620–626.
55. Huang, Y.F., Li, C.Y., Broadwell, I. *et al.* (2011) Shell-isolated nanoparticle-enhanced Raman spectroscopy of pyridine on smooth silver electrodes. *Electrochimica Acta*, **56**, 10652–10657.
56. Li, S.B., Li, L.M., Anema, J.R. *et al.* (2011) Shell-isolated nanoparticle-enhanced Raman spectroscopy (SHINERS) based on gold-core silica-shell nanorods. *Zeitschrift für Physikalische Chemie*, **225**, 775–783.
57. Liu, B., Blaszczyk, A., Mayor, M. and Wandlowski, T. (2011) Redox-switching in a viologen-type adlayer: an electrochemical shell-isolated nanoparticle enhanced Raman spectroscopy study on Au(111)-(1 X 1) single crystal electrodes. *ACS Nano*, **5**, 5662–5672.
58. Butcher, D.P., Boulos, S.P., Murphy, C.J. *et al.* (2012) Face-dependent shell-isolated nanoparticle enhanced Raman spectroscopy of 2,2 ’-bipyridine on Au(100) and Au(111). *Journal of Physical Chemistry C*, **116**, 5128–5140.

59. Honesty, N.R. and Gewirth, A.A. (2012) Shell-isolated nanoparticle enhanced Raman spectroscopy (SHINERS) investigation of benzotriazole film formation on Cu(100), Cu(111), and Cu(Poly). *Journal of Raman Specroscopy*, **43**, 46–50.
60. Lin, X.D., Uzayisenga, V., Li, J.F. *et al.* (2012) Synthesis of ultrathin and compact Au@MnO_2 nanoparticles for shell-isolated nanoparticle-enhanced Raman spectroscopy (SHINERS). *Journal of Raman Spectroscopy*, **43**, 40–45.
61. Qian, K., Liu, H.L., Yang, L.B. and Liu, J.H. (2012) Functionalized shell-isolated nanoparticle-enhanced Raman spectroscopy for selective detection of trinitrotoluene. *Analyst*, **137**, 4644–4646.
62. Uzayisenga, V., Lin, X.D., Li, L.M. *et al.* (2012) Synthesis, characterization, and 3D-FDTD simulation of Ag@SiO_2 nanoparticles for shell-isolated nanoparticle-enhanced Raman spectroscopy. *Langmuir*, **28**, 9140–9146.
63. Tian, X.D., Liu, B.J., Li, J.F. *et al.* (2013) SHINERS and plasmonic properties of Au-core–SiO_2-shell nanoparticles with optimal core size and shell thickness. *Journal of Raman Spectroscopy*, 44, 994–998.
64. Kunz, K.S. and Luebbers, R.J. (1993) *The Finite Difference Time Domain Method for Electromagnetics*, CRC Press, Boca Raton, FL.
65. Sherry, L.J., Chang, S.H., Schatz, G.C. *et al.* (2005) Localized surface plasmon resonance spectroscopy of single silver nanocubes. *Nano Letters*, **5**, 2034–2038.
66. Frens, G. (1973) Controlled nucleation for regulation of particle-size in monodisperse gold suspensions. *Nature Physical Science*, **241**, 20–22.
67. Liz-Marzan, L.M., Giersig, M. and Mulvaney, P. (1996) Synthesis of nanosized gold-silica core-shell particles. *Langmuir*, **12**, 4329–4335.
68. Lu, Y., Yin, Y.D., Li, Z.Y. and Xia, Y.N. (2002) Synthesis and self-assembly of Au@SiO_2 core-shell colloids. *Nano Letters*, **2**, 785–788.
69. Mulvaney, S.P., Musick, M.D., Keating, C.D. and Natan, M.J. (2003) Glass-coated, analyte-tagged nanoparticles: a new tagging system based on detection with surface-enhanced Raman scattering. *Langmuir*, **19**, 4784–4790.
70. Hamelin, A., Morin, S., Richer, J. and Lipkowski, J. (1989) Adsorption of pyridine on the (110) face of silver. *Journal of Electroanalytical Chemistry*, **272**, 241–252.
71. Hamelin, A., Morin, S., Richer, J. and Lipkowski, J. (1990) Adsorption of pyridine on the (311) face of silver. *Journal of Electroanalytical Chemistry*, **285**, 249–262.
72. Hamelin, A., Morin, S., Richer, J. and Lipkowski, J. (1991) Adsorption of pyridine on the (210) face of silver. *Journal of Electroanalytical Chemistry*, **304**, 195–209.
73. Wu, D.Y., Ren, B., Jiang, Y.X. *et al.* (2002) Density functional study and normal-mode analysis of the bindings and vibrational frequency shifts of the pyridine-M (M = Cu, Ag, Au, Cu+, Ag+, Au+, and Pt) complexes. *Journal of Physical Chemistry A*, **106**, 9042–9052.
74. Creighton, J.A., Blatchford, C.G. and Albrecht, M.G. (1979) Plasma resonance enhancement of Raman-scattering by pyridine adsorbed on silver or gold sol particles of size comparable to the excitation wavelength. *Journal of the Chemical Society, Faraday Transactions II*, **75**, 790–798.
75. Wu, D.Y., Liu, X.M., Duan, S. *et al.* (2008) Chemical enhancement effects in SERS spectra: a quantum chemical study of pyridine interacting with copper, silver, gold and platinum metals. *Journal of Physical Chemistry C*, **112**, 4195–4204.

76. Zhao, L.L., Jensen, L. and Schatz, G.C. (2006) Pyridine-Ag-20 cluster: a model system for studying surface-enhanced Raman scattering. *Journal of the American Chemical Society*, **128**, 2911–2919.
77. Tian, Z.Q., Ren, B. and Wu, D.Y. (2002) Surface-enhanced Raman scattering: from noble to transition metals and from rough surfaces to ordered nanostructures. *Journal of Physical Chemistry B*, **106**, 9463–9483.
78. Stolberg, L., Morin, S., Lipkowski, J. and Irish, D.E. (1991) Adsorption of pyridine at the gold (111)-solution interface. *Journal of Electroanalytical Chemistry*, **307**, 241–242.
79. Lipkowski, J. and Stolberg, L. (1992) Molecular adsorption at gold and silver electrodes, in *Adsorption of Molecules at Metal Electrodes* (eds J. Lipkowski and P.N. Ross), Wiley-VCH Verlag GmbH, New York, pp. 171–238.
80. Lipkowski, J., Stolberg, L., Yang, D.F. *et al.* (1994) Molecular adsorption at metal electrodes. *Electrochimica Acta*, **39**, 1045–1056.
81. Tian, Z.Q. and Ren, B. (2004) Adsorption and reaction at electrochemical interfaces as probed by surface-enhanced Raman spectroscopy. *Annual Review of Physical Chemistry*, **55**, 197–229.
82. Ren, B., Liu, F.M., Xie, J. *et al.* (1998) In situ monitoring of Raman scattering and photoluminescence from silicon surfaces in HF aqueous solutions. *Applied Physics Letters*, **72**, 933–935.
83. Sujith, A., Itoh, T., Abe, H. *et al.* (2008) Surface enhanced Raman scattering analyses of individual silver nanoaggregates on living single yeast cell wall. *Applied Physics Letters*, **92**, 103901.
84. Schulte, F., Mader, J., Kroh, L.W. *et al.* (2009) Characterization of pollen carotenoids with in situ and high-performance thin-layer chromatography supported resonant Raman spectroscopy. *Analytical Chemistry*, **81**, 8426–8433.
85. Lee, D., Lee, S., Seong, G.H. *et al.* (2006) Quantitative analysis of methyl parathion pesticides in a polydimethylsiloxane microfluidic channel using confocal surface-enhanced Raman spectroscopy. *Applied Spectroscopy*, **60**, 373–377.
86. Skoulika, S.G., Georgiou, C.A. and Polissiou, M.G. (1999) Quantitative determination of fenthion in pesticide formulations by FT-Raman spectroscopy. *Applied Spectroscopy*, **53**, 1470–1474.
87. Noguez, C. (2007) Surface plasmons on metal nanoparticles: the influence of shape and physical environment. *Journal of Physical Chemistry C*, **111**, 3806–3819.
88. Maier, S.A. and Atwater, H.A. (2005) Plasmonics: localization and guiding of electromagnetic energy in metal/dielectric structures. *Journal of Applied Physics*, **98**, 011101.
89. Halas, N.J., Lal, S., Chang, W.S. *et al.* (2011) Plasmons in strongly coupled metallic nanostructures. *Chemistry Review*, **111**, 3913–3961.
90. Gramotnev, D.K. and Bozhevolnyi, S.I. (2010) Plasmonics beyond the diffraction limit. *Nature Photonics*, **4**, 83–91.
91. Natan, M.J. (2006) Concluding remarks – surface enhanced Raman scattering. *Faraday Discussions*, **132**, 321–328.
92. Smith, W.E. (2008) Practical understanding and use of surface enhanced Raman scattering/surface enhanced resonance Raman scattering in chemical and biological analysis. *Chemical Society Reviews*, **37**, 955–964.

93. Doering, W.E. and Nie, S.M. (2003) Spectroscopic tags using dye-embedded nanoparticles and surface-enhanced Raman scattering. *Analytical Chemistry*, **75**, 6171–6176.
94. Zhang, X.Y., Zhao, J., Whitney, A.V. *et al.* (2006) Ultrastable substrates for surface-enhanced Raman spectroscopy: Al_2O_3 overlayers fabricated by atomic layer deposition yield improved anthrax biomarker detection. *Journal of the American Chemical Society*, **128**, 10304–10309.
95. Whitney, A.V., Elam, J.W., Zou, S.L. *et al.* (2005) Localized surface plasmon resonance nanosensor: a high-resolution distance-dependence study using atomic layer deposition. *Journal of Physical Chemistry B*, **109**, 20522–20528.
96. Dieringer, J.A., McFarland, A.D., Shah, N.C. *et al.* (2006) Surface enhanced Raman spectroscopy: new materials, concepts, characterization tools, and applications. *Faraday Discussions*, **132**, 9–26.
97. Oldenburg, S.J., Averitt, R.D., Westcott, S.L. and Halas, N.J. (1998) Nanoengineering of optical resonances. *Chemical Physics Letters*, **288**, 243–247.
98. West, J.L. and Halas, N.J. (2003) Engineered nanomaterials for biophotonics applications: improving sensing, imaging, and therapeutics. *Annual Review of Biomedical Engineering*, **5**, 285–292.
99. Jackson, J.B., Westcott, S.L., Hirsch, L.R. *et al.* (2003) Controlling the surface enhanced Raman effect via the nanoshell geometry. *Applied Physics Letters*, **82**, 257–259.
100. Zheng, J.W., Zhou, Y.G., Li, X.W. *et al.* (2003) Surface-enhanced Raman scattering of 4-aminothiophenol in assemblies of nanosized particles and the macroscopic surface of silver. *Langmuir*, **19**, 632–636.
101. Ikeda, K., Fujimoto, N., Uehara, H. and Uosaki, K. (2008) Raman scattering of aryl isocyanide monolayers on atomically at Au(111) single crystal surfaces enhanced by gap-mode plasmon excitation. *Chemical Physics Letters*, **460**, 205–208.
102. Mullins, D.R. and Campion, A. (1984) Unenhanced Raman-scattering from pyridine chemisorbed on a stepped silver surface – implications for proposed sers mechanisms. *Chemical Physics Letters*, **110**, 565–570.
103. Grirrane, A., Corma, A. and Garcia, H. (2008) Gold-catalyzed synthesis of aromatic Azo compounds from anilines and nitroaromatics. *Science*, **322**, 1661–1664.
104. Wu, D.Y., Liu, X.M., Huang, Y.F. *et al.* (2009) Surface catalytic coupling reaction of p-mercaptoaniline linking to silver nanostructures responsible for abnormal sers enhancement: a DFT study. *Journal of Physical Chemistry C*, **113**, 18212–18222.
105. Rodriguez, J.A. and Goodman, D.W. (1992) The nature of the metal-metal bond in bimetallic surfaces. *Science*, **257**, 897–903.
106. Zhang, J., Sasaki, K., Sutter, E. and Adzic, R.R. (2007) Stabilization of platinum oxygen-reduction electrocatalysts using gold clusters. *Science*, **315**, 220–222.
107. Guerrero, A.R. and Aroca, R.F. (2011) Surface-enhanced fluorescence with shell-isolated nanoparticles (SHINEF). *Angewandte Chemie International Edition*, **50**, 665–668.
108. Graham, D. (2010) The next generation of advanced spectroscopy: surface enhanced Raman scattering from metal nanoparticles. *Angewandte Chemie International Edition*, **49**, 9325–9327.

9

Applying Super-Resolution Imaging Techniques to Problems in Single-Molecule SERS

Eric J. Titus and Katherine A. Willets
Department of Chemistry and Biochemistry, The University of Texas at Austin, USA

9.1 Introduction

9.1.1 Single-Molecule Surface-Enhanced Raman Scattering (SM-SERS)

Since the first observation of single-molecule surface-enhanced Raman scattering (SM-SERS) in 1997, there have been many studies into the nature of this phenomenon [1–12]. Thus far, most SM-SERS substrates consist of randomly-assembled nanoparticle aggregates, which exhibit sizable electromagnetic field enhancements in gaps between adjacent nanoparticles to facilitate SM-SERS sensitivity. These regions of intense electromagnetic enhancement are known as "hot spots" and are predicted to be on the size scale of a single molecule [7, 13]. Due to their small size relative to the wavelength of light, hot spots are difficult to study using standard far-field optical microscopy techniques. Understanding the nature of hot spots is critical to the rational design of SM-SERS active substrates. In this chapter we present several examples showing that super-resolution optical microscopy techniques give important insight into the nature of SERS hot spots and how molecules interact with them.

One challenge when studying the optical properties of aggregated SM-SERS-active nanoparticles is that the aggregates are typically smaller than the diffraction limit of light. The diffraction-limited resolution of a standard optical microscope is $0.61\,\lambda/\mathrm{NA}$, where λ is

Frontiers of Surface-Enhanced Raman Scattering: Single Nanoparticles and Single Cells, First Edition.
Edited by Yukihiro Ozaki, Katrin Kneipp and Ricardo Aroca.

the wavelength of the emitter, and NA stands for the numerical aperture of the microscope objective. This limits the resolution of a microscope with a 1.3 NA objective to ~260 nm when imaging emission at 550 nm. Because most SM-SERS aggregates are on the order of 50–200 nm in size, they appear as diffraction-limited spots, as shown in Figure 9.1a, and contain no information about the shape and size of the underlying nanoparticle or the location of the molecule on the surface. The analysis is further complicated by the presence of background silver luminescence from the aggregate itself (Figure 9.1b). As a result, the point spread function (PSF) of the emission from the nanoparticle aggregate is the superposition of multiple, unresolved emitters [6, 14]. Representative diffraction-limited images of simultaneous SERS emission and nanoparticle luminescence, as well as silver luminescence alone can be seen in Figure 9.1a,b, respectively, showing that when these two emission sources are emitting at the same time, they cannot be resolved using standard far-field optical techniques.

9.1.2 Super-Resolution Imaging

In order to overcome the diffraction limit of light, super-resolution imaging has recently emerged as a far-field optical microscopy technique able to provide <5 nm

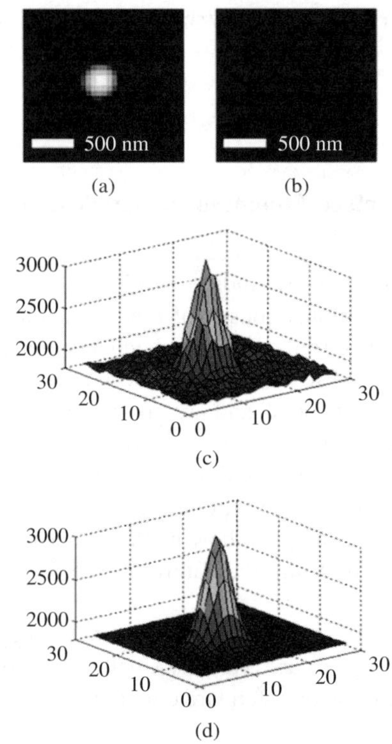

Figure 9.1 *Diffraction limited images of a nanoparticle aggregate exhibiting (a) SERS and (b) nanoparticle luminescence only. (c) 3D representation of SERS PSF and (d) the resulting Gaussian fit. Adapted with permission from Ref. [12]. Copyright © 2012, American Chemical Society*

resolution [15–18]. Super-resolution imaging is based upon two principles: (i) super-localization imaging and (ii) modulated emission [19–21]. In super-localization imaging, the PSF of an emitter is modeled by some fitting function – typically, a two-dimensional Gaussian – in order to extract the centroid position of the emitter [6, 22]. The equation for a two-dimensional Gaussian is:

$$I(x,y) = z_0 + I_0 \exp\left[-\frac{1}{2}\left[\left(\frac{x-x_0}{s_x}\right)^2 + \left(\frac{y-y_0}{s_y}\right)^2\right]\right] \tag{9.1}$$

In this equation, $I(x,y)$ is the spatially-varying intensity; z_0 is the background; x_0 and y_0 are the centroid coordinates; and $s_{x,y}$ is the standard deviation of the PSF Gaussian in the *x*- and *y*-dimensions. By fitting diffraction-limited images to this equation, the centroid of the emitter can be localized with an uncertainty of about 1 nm if the emitter has a high enough signal-to-noise ratio [20, 23]. A comparison of an experimental PSF and the corresponding two-dimensional Gaussian fit is shown in Figure 9.1c,d, respectively.

The second component of super-resolution imaging is signal modulation. In the case when multiple emitters are spaced by less than the diffraction limit of light, their PSFs will be superimposed and the centroid will be an intensity-weighted superposition of all emitting species (as in Figure 9.1a). However, if the emitters are switching between an emissive (on) and non-emissive (off) state, then each individual emitter can be localized when it is on and all others are off. In single-molecule fluorescence experiments, signal modulation is achieved by using photoswitchable fluorophores, which can be actively controlled by light [19–21]. On the other hand, SM-SERS signals fluctuate stochastically over time, providing an inherent signal modulation. This signal modulation allows us to deconvolve the background silver luminescence of the nanoparticle substrate from the SM-SERS emission, allowing localization of the SERS centroid separately from the nanoparticle luminescence background, as described in detail below [6, 24].

By combining super-localization imaging with signal modulation, it is possible to reconstruct images of nanoscale objects by mapping the centroids associated with individual emitters; this is the concept of super-resolution imaging. For SM-SERS, however, the situation is complicated by the fact that the SERS emission centroid is not reporting directly on the position of the molecule, as its emission is expected to couple out through the plasmon modes of the nanoparticle aggregate [6, 24–26]. As a result, super-resolution imaging of SERS-active substrates reflects the inherent coupling between an SERS analyte and the underlying nanoparticle substrate. In the remainder of this chapter, we will describe the experimental details of super-resolution SERS imaging, provide details on how to analyze SM-SERS super-resolution data, and then provide specific examples of super-resolution imaging of SM-SERS hot spots.

9.2 Experimental Considerations for Super-Resolution SM-SERS

9.2.1 Sample Preparation

Samples for super-resolution SM-SERS experiments require nanoparticle aggregates with an adsorbed SERS reporter molecule, some form of alignment marker to correct for any mechanical drift in the imaging system, and (for structure correlation studies) a gridded and/or functionalized coverslip. For the super-resolution SM-SERS experiments described herein, silver nanoparticles prepared by the Lee and Meisel method are used exclusively [27]. This preparation yields citrate capped silver nanoparticles with sizes ranging from roughly 20 to 150 nm. To create SM-SERS-active aggregates, an aliquot of nanoparticles is combined with a low concentration of an SERS reporter molecule or dye, typically Rhodamine 6G (R6G), and a salt solution (typically, sodium chloride or sodium bromide), which displaces the citrate capping agent and induces nanoparticle aggregation [6, 22, 24, 28, 29]. When using nanoparticles prepared with the Lee and Meisel method, the concentration of the dye used for the SM-SERS sample should be ~1 nM in the final nanoparticle/salt/dye mixture to obtain a roughly 1 : 1 nanoparticle:dye ratio [7, 9, 30, 31]. This is based on previous SM-SERS experiments and can also be calculated based on the amount of silver nitrate used in the synthesis, the average size of the particles, and the final volume of the nanoparticle synthesis.

The resulting solution is incubated for 1–18 h, and then drop cast onto an argon-plasma cleaned coverslip and dried with a stream of dry N_2. After drying the sample, the coverslip should be gently rinsed to remove excess salt, and dried again with N_2. Next, an aliquot of fluorescent beads (200–500 nm in diameter) are dropcast on the same sample to act as stationary markers to correct for mechanical drift. For experiments using green laser excitation, 500 nm Sky Blue beads from SpheroTech are used, as their low extinction, and therefore fluorescence, at 532 nm will not saturate the sensitive detectors needed for SM-SERS.

9.2.2 Instrument Set-up

For the studies described herein, the instrumental set-up is based on an inverted optical microscope with a 100×/1.3 variable NA oil immersion objective. 532 nm laser excitation is introduced in a wide-field, epi-illumination geometry to excite SM-SERS [32]. An overhead tungsten filament lamp and dark-field condenser are also available to measure scattering from the nanostructures of interest. The SM-SERS signal is collected by the objective and filtered with a dichroic beamsplitter and long pass filter to reject Rayleigh scattered excitation light. In order to probe hot spots of varying orientation in the sample plane, circularly polarized laser light is used for super-resolution experiments, but linear polarization can be used to selectively excite hot spots aligned along the polarization of the beam. A block-diagram of the microscope set-up is shown in Figure 9.2.

The SM-SERS signal is sent out of the side port of the microscope and through a non-magnifying relay lens and then to a non-polarizing cube beamsplitter, which splits the signal into two paths. The use of a non-polarizing cube beamsplitter is critical because SERS emission can be highly polarized, especially when originating from dimer nanostructures, which can distort the resulting PSFs [33, 34]. A plate beamsplitter that was mounted inside the microscope was used in earlier experiments [22, 24], but these beamsplitters can give rise to optical distortions and have a mild polarization dependence; as such, a non-polarizing cube beamsplitter is a better option. After passing through the beamsplitter, half the signal is sent through a magnifying relay lens and onto a Princeton Instruments ProEM 512 × 512 pixel electron multiplied charge-coupled device (EM-CCD) for imaging. The remaining signal is sent through a non-magnifying relay lens and onto a 2500i Acton SpectraPro spectrometer with a 1200 g/mm grating, equipped with a

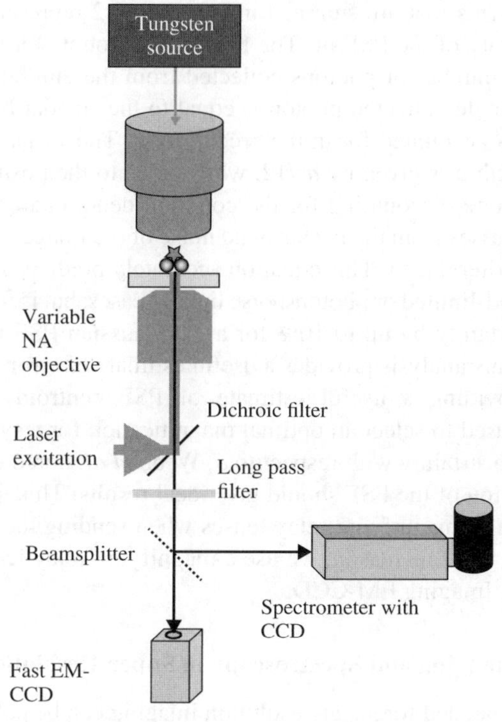

Figure 9.2 *Block diagram of the super-resolution SM-SERS optical microscope*

1340 × 400 pixel back-illuminated, liquid nitrogen cooled CCD camera. A USAF 1951 resolution test target is used to calibrate the EM-CCD pixel size.

9.2.3 Camera Pixels and Theoretical Uncertainties

An important consideration for super-resolution imaging is the relationship between the PSF and the pixel size of the imaging CCD [35]. If the pixel size is larger than the diffraction-limited spot size, then the image of the PSF will appear as a delta function on the CCD, with all photons falling into a single bin, which completely eliminates any spatial information. On the other hand, if the PSF is much larger than the pixel size of the camera, then the photon counts seen by each pixel are reduced, which lowers signal-to-noise levels across the image. In order to optimize the imaging system to obtain the lowest uncertainties when fitting PSFs, Equation 9.2 is used [35].

$$\sigma_{\mu_i} = \sqrt{\left(\frac{s_i^2}{N} + \frac{a^2/12}{N} + \frac{8\pi s_i^4 b^2}{a^2 N^2}\right)} \tag{9.2}$$

This equation gives the standard error, σ_{μ}, of the fitted centroid position, which is dependent on the width of the fitted Gaussian, s_i; the image pixel size of the EM-CCD in nanometers, a; the standard deviation of the background in the image, b; and the total photon count of the entire image, N.

Each of the three terms that are summed in Equation 9.2 represents a different contribution to the uncertainty of the PSF fit. The first term accounts for the uncertainty in the fit due to the limited number of photons collected from the emitter, that is, the position uncertainty for any single collected photon is equal to the standard deviation of the PSF. The pixilation error is accounted for in the second term. The variance of a top-hat distribution with edge length a is given by $a^2/12$, which adds to the position uncertainty from each photon in term one (accounting for the common denominator, N). Finally, the last term in Equation 9.2 arises from the background noise of the image and is derived from the least-squares fitting criteria [35]. This equation accurately predicts localization uncertainties in both background-limited or photon-noise limited cases, but is found to underestimate the localization uncertainty by up to 10% for a 2D Gaussian PSF in between these two cases [35]. Even so, this analysis provides a useful estimate of error in PSF fitting.

In addition to providing a useful estimate of PSF centroid position uncertainty, Equation 9.2 can be used to select an optimal magnification for a system. By minimizing the theoretical PSF uncertainty with respect to a, Webb *et al.* noted that a pixel size equal to the standard deviation of the PSF should give ideal results. This ideal pixel size can be achieved with the use of magnifying relay lenses when sending the collected light to the imaging camera [35]. For this reason, we use a magnifying relay lens before sending our SM-SERS data to our imaging EM-CCD.

9.2.4 Correlated Imaging and Spectroscopy in Super-Resolution SM-SERS

Although all the data needed for super-resolution imaging can be collected on an imaging EM-CCD, simultaneous spectral correlation provides a wealth of useful information when studying SM-SERS-active nanostructures. Spectral correlation helps discriminate fluctuating nanoparticle luminescence, SERS from contaminant molecules (like nanoparticle capping agents), and surface-enhanced fluorescence from SM-SERS signals [22, 24]. Additionally, with the use of a bi-analyte approach, spectroscopic data allow multiple molecules to be spectroscopically discriminated and individually localized on the surface of an SM-SERS-active aggregate [29]. Due to the complexity of this powerful technique, there are a few extra experimental considerations before it can be applied effectively, the most important of which are how collected photons are sent to both detectors and the timing errors of the correlated acquisitions.

One of the major problems that can be introduced when adding spectral correlation to the microscope is due to the beamsplitter. Any imaging artifacts introduced by this optic can strongly bias the PSF fits to the data, invalidating the calculated localization of the emitters. These can be introduced by interferences, that is, non-ideal anti-reflective coatings on plate beamsplitters, or by the polarization response of the optic, as discussed above.

A second important consideration when using correlated spectral imaging with super-resolution techniques is the timing registration between the two detectors. CCDs have a readout time associated with them, which limits how quickly sequential images can be acquired. For EM-CCDs, the readout times are <100 ms, based on the choice of analog-to-digital converter and the tolerated threshold in the read-noise. On the other hand, spectral acquisition often uses slower analog-to-digital converters to minimize read-noise; this leads to longer overall readout times. In addition to the readout time, the integration time of the detector is also based on acquiring high enough signal-to-noise data for analysis. For the

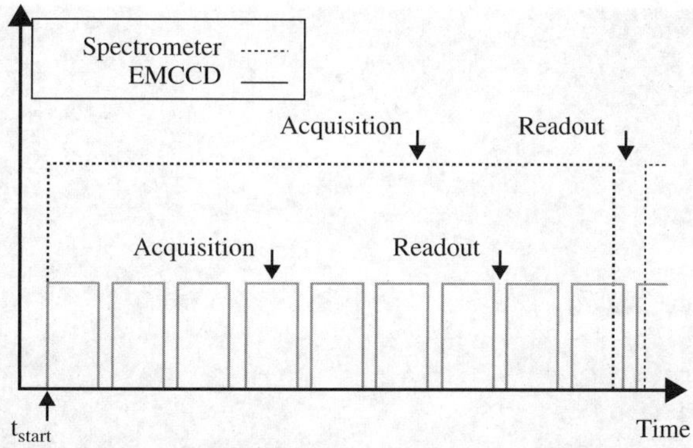

Figure 9.3 *Diagram of EM-CCD and spectrometer acquisition timing. High signal indicates acquisition, low signal indicates that the CCD is being read out*

EM-CCD data, integration times are typically on the order of 100–400 ms, while spectral acquisition often requires 1–4 s to acquire usable signals. Given the order of magnitude difference in timing speeds, there is an associated error when using spectral data to assign a diffraction-limited spot as originating from SERS of the analyte, nanoparticle luminescence, or other signals. Figure 9.3 illustrates these timing issues due to differences in both integration times (high signal) and readout times (low signal).

9.2.5 Correlated Optical and Structural Data

In order to correlate optical information from an SM-SERS aggregate with the nanoparticle structure, optical experiments are carried out on gridded coverslips for correlation with atomic force microscopy (AFM) or scanning electron microscopy (SEM). Gridded slides are prepared via a shadow deposition technique, in which copper transmission electron microscopy (TEM) reference grids are taped over a coverslip, and then ~30 nm of aluminum is thermally evaporated through the mask and deposited onto the coverslip. Removal of the mask leaves a grid pattern on the slide that can be used to register data between the optical microscope and the AFM/SEM. For SEM studies, the coverslip is coated with a thin layer of indium tin oxide (ITO), which is both conductive and optically transparent. Figure 9.4a shows an SEM image of a gridded ITO substrate, as well as a comparison of optical and SEM images of the same region of interest (Figure 9.4b,c). In Figure 9.4b,c, both an SERS-active nanoparticle aggregate and several fluorescent bead alignment markers are observed. The alphanumeric grid is used for coarse image registration between the optical data and SEM, while the location of the alignment markers allows us to definitively assign the optical signal to the nanostructure of interest.

If nanoparticle height information is required, AFM can also be used to correlate nanoparticle structure to optical activity. For this purpose, a combined AFM/total internal reflection optical microscope is used [36–38]. The optical microscopy system allows the user to locate the same area investigated in the super-resolution study, while the AFM gives sub-diffraction limited structural information on the nanoparticle aggregates. For this purpose, the same alphanumeric grid deposition procedure is used, but it is done on a glass coverslip instead of an ITO-coated substrate. When using AFM as a structure

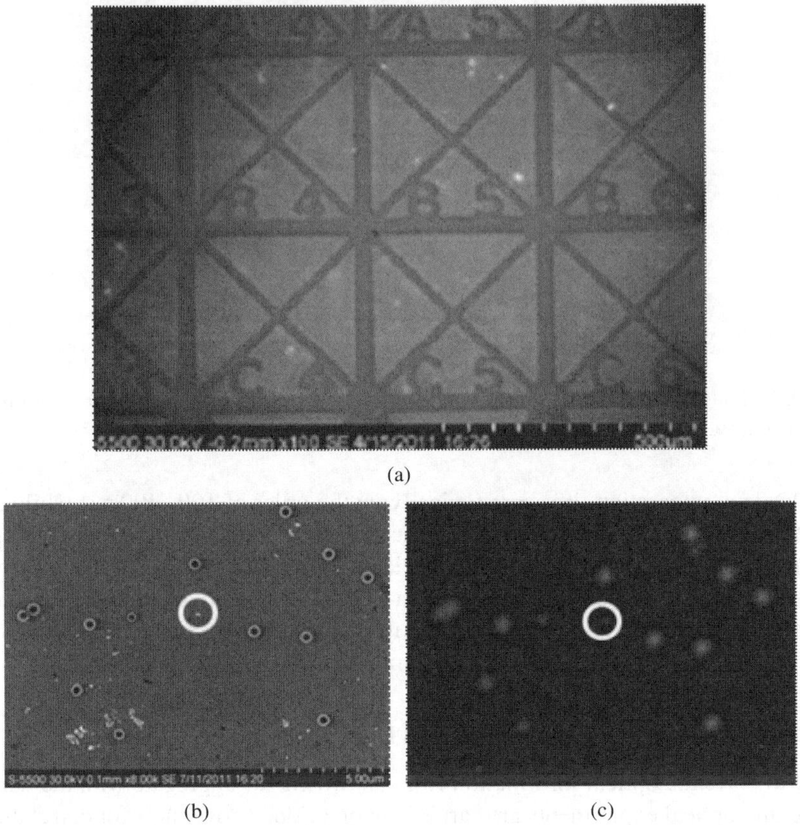

Figure 9.4 *(a) SEM image of a gridded ITO-coated coverslip with nanoparticles. (b) SEM image of nanoparticle aggregate (white circle) and fluorescent beads. (c) Image of the same region on optical microscope*

correlation technique, care must be taken that nanoparticle aggregates are not perturbed when interacting with the AFM probe. To prevent this, we functionalize glass coverslips with a layer of (3-aminopropyl) triethoxysilane (APTES) to anchor the nanoparticles to the substrate surface. In this procedure, argon plasma-cleaned coverslips are incubated in a 0.5–2% APTES–methanol solution for approximately 5 min, and then thoroughly rinsed with methanol to remove excess APTES [37]. This allows AFM characterization without disturbing aggregates on the coverslip surface.

9.3 Super-Resolution SM-SERS Analysis

The raw data from a super-resolution SM-SERS experiment consist of a stack of images from the EM-CCD with a corresponding stack of correlated spectra. Spectral analysis is usually carried out first, in order to confirm that the SERS signal is due to the R6G in the sample. Sample spectra are shown in Figure 9.5a. Once this has been confirmed, PSF analysis can be carried out on the image stack from the EM-CCD. For the data presented

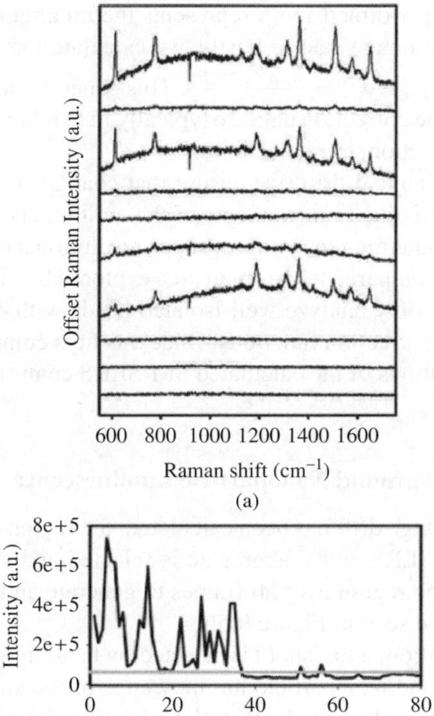

Figure 9.5 *(a) SERS spectra and (b) intensity–time trace of SM-SERS-active aggregate. Adapted with permission from Ref. [6]. Copyright © 2010, American Chemical Society*

here, all PSF fitting and analysis is carried out using algorithms written in MATLAB. Due to the nature of the fitting algorithm, it is important to note that initial guesses in the fit parameters can have strong effects on the quality and speed of fitting. For this reason, it is sometimes necessary to initially characterize PSF of emitters with a different method to obtain reasonable initial guesses.

9.3.1 Mechanical Drift Correction

Because super-resolution imaging offers resolution of <5 nm, it is critical that any mechanical drift in the experimental equipment be accounted for. The largest source of drift is from the sample mounted on a microscope translation stage; if the stage has any mechanical drift in the translation mechanism, this will appear as a shift in the calculated centroids. Thus, we use alignment markers, typically fluorescent beads, to account for and correct mechanical drift.

The first step in the PSF fitting analysis is to select two to three fluorescent alignment beads from the EM-CCD image, and localize the centroid of each bead using Equation 9.1 for each frame in the image stack. Next, we calculate how the *x*- and *y*-centroid coordinate of each individual bead changes over each frame (e.g., $x_{i,1} - x_{i,j}$ where x represents the centroid coordinate in the x-dimension, i represents the ith alignment marker and j represents the jth frame in the image stack). Lastly, we calculate the average shift in the bead position over all beads (i.e., $\Delta x(j) = \frac{\sum_{i=1}^{i}(x_{i,1}-x_{i,j})}{i}$). This generates a series of shifts in both x_0 and y_0 over all frames. The correction data are typically fit to a function, either a polynomial or a spline smoothing function, to reduce noise.

By correcting for mechanical drift, we ensure that changes in the centroid of the SM-SERS-active aggregate are due to movements of the emitter, and not just stage drift. This correction is critical for obtaining meaningful super-resolution data due to the large relative magnitude of stage drift compared to the small area explored by the R6G molecule. For this reason, it is important to only analyze well-isolated beads with defect-free PSFs in order to generate reliable drift correction functions. Once this fit is completed, the data are saved and used to shift the positions of the calculated SM-SERS centroids to offset the effects of stage drift.

9.3.2 Analysis of Background Nanoparticle Luminescence

Once the correction for stage drift has been calculated, the region of interest corresponding to a signal from an SM-SERS-active aggregate is selected and analyzed. The first step is to sum the intensity of the region over all frames to generate an intensity versus time plot (time trace), which can be seen in Figure 9.5b.

After time trace generation, a threshold is selected by hand in order to discriminate R6G SERS from the background nanoparticle luminescence of the silver particle. A threshold can often be drawn without the aid of spectral correlation due to the strongly fluctuating signals associated with SM-SERS (as in Figure 9.5b). SM-SERS from a molecule will be intense and will fluctuate, while background luminescence from the nanoparticle will be significantly less intense and relatively stable [6, 24]. Isolating frames that exhibit only nanoparticle luminescence is an important step because the PSF is a superposition of the PSFs from both the molecular emission and the nanoparticle luminescence when the SERS signal is on [15, 39]. Thus, in order to accurately localize the centroid position of the SM-SERS, the contribution of the nanoparticle luminescence to the overall PSF of the SM-SERS-active aggregate needs to be accounted for.

Once a threshold has been selected based on the time trace, frames with intensity that fall below the threshold are fit using Equation 9.1. This provides the centroid position corresponding to nanoparticle luminescence only. The stability of the nanoparticle luminescence has been previously confirmed by fitting the PSF of unlabeled silver nanoparticle aggregates, as shown in Figure 9.6. The time trace of unlabeled nanoparticle aggregate shows stable intensity (Figure 9.6c), and the resulting centroid coordinates show that the nanoparticle emission is symmetrically distributed (Figure 9.6d). The scatter in the data originates from noise in the measurement, and is predicted from Equation 9.2 above. However, we expect the centroid fits to be symmetrically distributed around the mean value, which is observed in the histograms in Figure 9.6d; this indicates that the origin of this emission is from a stationary emitter.

9.3.3 Calculating the SM-SERS Centroid Position

Next, we analyze all frames corresponding to SM-SERS emission (e.g., above the threshold defined in Figure 9.5b). Because the overall PSF of the SM-SERS-active

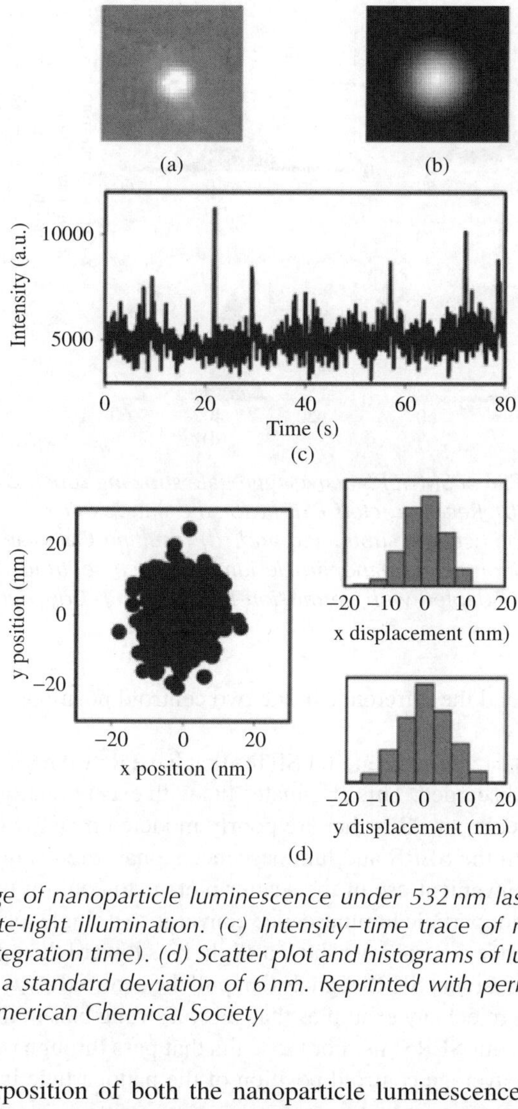

Figure 9.6 *(a) Image of nanoparticle luminescence under 532 nm laser excitation and (b) under dark-field white-light illumination. (c) Intensity–time trace of nanoparticle luminescence signal (0.1 s integration time). (d) Scatter plot and histograms of luminescence centroid localization showing a standard deviation of 6 nm. Reprinted with permission from Ref. [6]. Copyright © 2010, American Chemical Society*

aggregate is a superposition of both the nanoparticle luminescence and emission from the molecule [6, 24], the average PSF of the nanoparticle luminescence is subtracted from frames exhibiting SM-SERS in order to isolate the SM-SERS PSF. An example of this process is shown in Figure 9.7. In this example, the overall PSF of the ROI when SERS is "on" shows two superimposed PSFs (Figure 9.7a). Using the fit parameters associated with the luminescence PSF (as determined above), we reconstruct the PSF of the luminescence (Figure 9.7b). The reconstructed PSF of the nanoparticle luminescence is then subtracted from the overall PSF to give the PSF associated with the SERS only (i.e., Figure 9.7a − b = c). This subtracted PSF is then fit with Equation 9.1 to calculate the SERS centroid (Figure 9.7d). For the example shown in Figure 9.7, the SERS and luminescence emission sources are sufficiently far apart that we can see the superimposed PSFs by eye (as in Figure 9.7a,e); this example was specifically chosen to illustrate the subtraction process. However, for most examples, the two emitters are spaced by less than

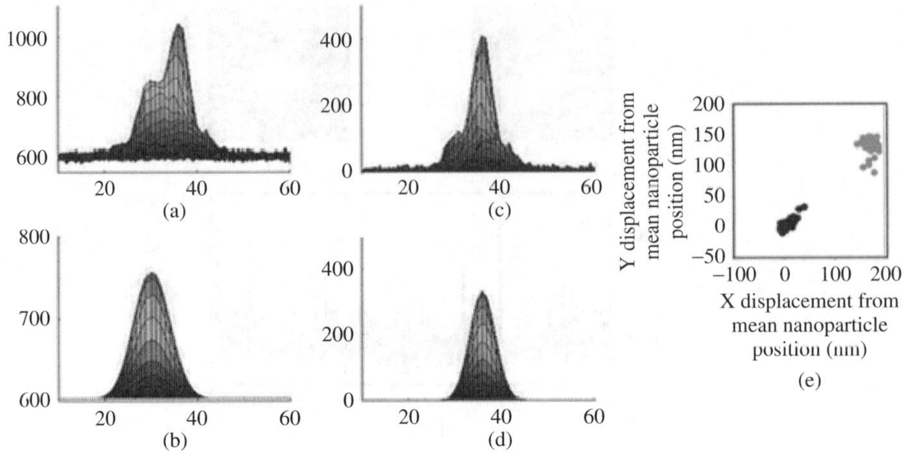

Figure 9.7 *(a) Raw PSF of SM-SERS-active aggregate showing superposition of the SERS and luminescence PSFs. (b) Reconstructed PSF fit for the nanoparticle luminescence. (c) Data from (a) with luminescence PSF subtracted and (d) resulting Gaussian fit of (c). (e) Scatter plot showing the position of the nanoparticle luminescence centroid (black) relative to the SERS centroid (gray). Adapted with permission from Ref. [6]. Copyright © 2010, American Chemical Society*

the diffraction limit, and the difference in the two centroid positions is only apparent after data processing.

Once all SM-SERS data have been fit, PSF fits that have R^2 values below a certain threshold (usually 0.8) are discarded. This eliminates fits with exceptionally low signal-to-noise, fits that did not converge, or PSFs that are poorly modeled by a 2D Gaussian. Next, centroid positions of both the SERS and luminescence signals are examined in a scatter plot (Figure 9.7e). One convenient use of the scatter plots is to confirm that fitted PSFs of the frames assigned as nanoparticle luminescence converge to a single location. As Figure 9.7e shows, the luminescence centroid fits show a slight bias toward the upper right corner, due to the contribution of residual SERS signals that pull the centroid away from the mean value in certain frames. We reject any examples that show a strong bias in the luminescence centroid, as this will bias our SERS fits. For those fits that pass through our screening process, we arbitrarily set the average centroid position of the nanoparticle luminescence at (0, 0) for convenience.

9.4 Super-Resolution SM-SERS Examples

9.4.1 Mapping SM-SERS Hot Spots

A major advantage of using super-resolution imaging to study SM-SERS is that we are able to map out the relationship between the SERS intensity and the spatial origin of the SERS signals with <5 nm resolution. For SM-SERS, where signals are believed to occur in junction regions between adjacent nanoparticles, super-resolution imaging offers a unique strategy for relating the SERS signals to the shape and size of the junction regions.

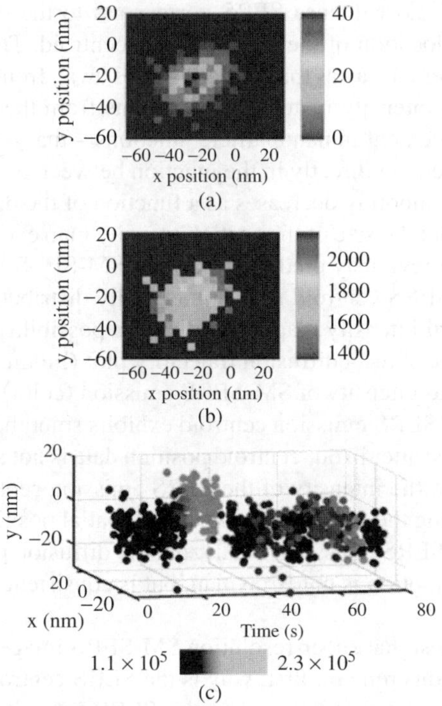

Figure 9.8 *(a) Frequency histogram showing the number of SERS centroids occurring in each bin. (b) Intensity histogram showing the average intensity of SERS centroids occurring in each bin (4.6 nm bin size). (c) 3D wander plot showing diffusion trajectory of SERS centroid (shade-bar indicates emission intensity). Reprinted with permission from Ref. [6]. Copyright © 2010, American Chemical Society. (*See insert for color representation of the figure.*)*

To analyze these data, we create histograms from our original SERS centroid data to visualize how the centroids are distributed in space.

We begin by calculating a frequency histogram, as shown in Figure 9.8a. In the frequency histogram, the region of interest is divided into 4.6 nm × 4.6 nm bins (corresponding to one-tenth of our imaging pixel size). Next, we count the number of SERS centroid events that fall into a particular spatial bin. For consistency, the histogram axes are set such that the nanoparticle luminescence centroid average is at (0,0). The frequency histogram in Figure 9.8a reveals several interesting features [6]. First, the centroids are asymmetrically distributed, with an ellipticity that is inconsistent with a stationary emitter, suggesting the possibility of a changing or mobile SERS centroid. Additionally, it shows that that the bulk of the SERS emission centroids are spatially offset from the nanoparticle luminescence centroid by >20 nm for this aggregate [6].

Next, we calculate a spatial intensity map, which represents how the SERS intensity is related to the spatial origin of the SERS emission [6]. In this histogram, we calculate the average SERS intensity of all points within each individual bin of the frequency histogram. This histogram shows the location of the SM-SERS centroids that are associated with the most intense emission. In the example shown in Figure 9.8c, the centroid positions exhibiting the most intense SERS correspond to the upper right region of the histogram, close to the location of the luminescence centroid. The SERS emission intensity also exhibits a gradient decay as the centroid moves away from the most intense region. The decay in the SERS intensity matches expectations from theory with respect to electromagnetic field enhancement in nanoparticle junctions – that is, we expect the strongest electromagnetic enhancement directly in the junction between two adjacent nanoparticles, while the enhancement smoothly decreases as a function of the distance from the junction [40, 41]. We also note that the size of the spatial intensity map extends over ~40 nm, which is much larger than the previously predicted size of a SM-SERS-active hot spot [6, 7, 30].

As noted earlier, the SERS centroid is asymmetrically distributed in both the frequency histogram and the spatial intensity map, indicating the possibility of a mobile SERS centroid. To explore this, we show a diffusion trajectory plot (Figure 9.8c), in which the centroid position (x_0, y_0) and intensity of SM-SERS emission (color) are plotted against time. This plot shows that the SERS emission centroid exhibits smooth, continuous motion over time, indicating that the scatter in our centroid position data is not simply due to noise in the measurement. Moreover, the intensity of the SERS emission centroid tracks the changing centroid position, showing the clear link between the spatial position of the SERS centroid and the intensity of the SERS emission. Analyzing the diffusion properties of the centroid shows that the centroid motion is not Brownian, but is consistent with constrained motion on the surface [42].

While these data suggest that super-resolution SM-SERS images are mapping out SERS hot spots, several questions remain: first, why is the SERS centroid changing and second, how do we explain the relationship between the SERS intensity and centroid position? These questions are especially challenging to answer because SERS is a highly coupled emission process, with the light scattered by the molecule coupled into the plasmon modes of the nanoparticle substrate. Calculations from Ausman and Schatz indicate that the position of a dipole emitter on the surface of a nanoparticle dimer will perturb the near-field scattering distribution of the emitted SERS [26]. Thus, we hypothesize that if the position of the molecule on the surface of the nanoparticle is changing, this will lead to a change in how the light is coupled out by the nanoparticle, and thus a shift in the SERS centroid. Moreover, as the molecule moves on the surface, it experiences different local electromagnetic field enhancements, leading to changes in the SERS intensity. Thus, the implication is that our hot spot maps are a function of both the local electromagnetic field enhancement (dictated by the structure of the nanoparticle aggregate) and the location of the molecule on the nanoparticle surface. In the following sections, we explore both of these ideas in more detail.

9.4.2 The Role of Plasmon-Enhanced Electromagnetic Fields: Structure Correlation Studies

In order to understand more about the nature of SM-SERS-active "hot spots," it is important to consider how the SERS spatial intensity maps relate to the actual structure of the SM-SERS-active nanoparticle aggregate. For these studies, we first calculate the SERS spatial intensity map for a given SERS-active nanostructure (Figure 9.9a), then use SEM to characterize the structure of the SM-SERS substrate (Figure 9.9b) [22]. In the example shown, the spatial intensity map shows a high intensity edge on the far right, accompanied by a gradient decay in intensity, similar to the data shown in Figure 9.8.

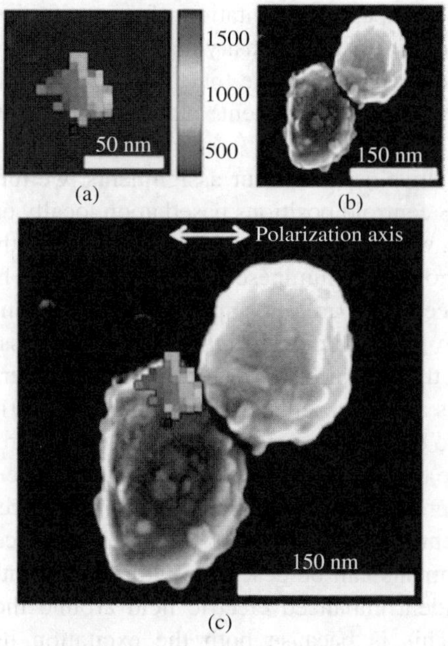

Figure 9.9 *(a) Intensity histogram of SM-SERS data collected on aggregate, with (b) SEM image of the same aggregate. (c) Intensity histogram scaled and qualitatively overlaid on SEM image of aggregate. Reprinted with permission from Ref. [22]. Copyright © 2011, American Chemical Society. (*See insert for color representation of the figure.*)*

Next, we scale and overlay the spatial intensity map over the image of the aggregate. In order to do this, assumptions must be made about the most probable location of the hot spot(s) on the aggregate. Given the straightforward dimer nanoparticle structure shown in Figure 9.9b, the region of highest enhancement is predicted to occur in the junction region between the two nanoparticles and is expected to decay as the SERS centroid moves away from the junction [40, 41]. Based on these assumptions, the spatial intensity map is overlaid on the particle image, as shown in Figure 9.9c, with the region of highest SERS intensity aligned with the junction between the two nanoparticles. The orientation of the high intensity edge of the spatial intensity map agrees with the alignment of the nanoparticle junction, and the intensity decays as the centroid moves away from the junction, as expected.

It is important to note that the overlay in Figure 9.9c is a qualitative assignment, based on the shape of the spatial intensity map and our chemical intuition of where the hot spot should be located on a dimer nanostructure. For the dimer shown here, the size of the spatial intensity map does not occupy the entire length of the junction region, but instead represents a subset of this junction; we have chosen to place the hot spot along the top portion of the junction, but it could just as easily lie closer to the middle or along the bottom of the edge. However, the important result is that the shape of the hot spot is consistent with our expectations from theory [40, 43]. The observed agreement between the shape of the spatial intensity map and the structure of the SERS-active nanoparticle aggregate has been observed across multiple nanoparticle examples [22]. In fact, even with nanoparticle trimers, where multiple junctions may be SERS-active [44], we find that the spatial intensity maps reflect the shape and orientation of a specific junction. Given the agreement between the shape of the spatial intensity maps and specific structural features of the nanoparticle substrates, these data indicate that the shape of the nanostructure (and thus the local electromagnetic field enhancement) plays a significant role in dictating SERS hot spots.

To improve the quantitative aspect of our assignments, we turned to theory to help us calculate predicted SERS centroid positions based upon locally enhanced electromagnetic fields [24]. For this work, we exploited our ability to measure both the SERS centroid position(s) as well as the nanoparticle luminescence. One trend we had previously noted was that the localized SERS centroids are often spatially distinct from the silver luminescence centroid average position [6]. While one might expect the emission centroid of both signals to be centered near the region of highest enhancement, our super-resolution studies showed that the centroids are often separated by distances of 10 nm or more. Thus, calculations would enable us to both assign centroid positions in a more quantitative manner and understand the difference between the SERS and luminescent centroid positions.

One basis for the difference between the silver luminescence centroid and the SERS centroid fits is the different roles that the electromagnetic enhancement plays in each case. SERS intensity enhancements can be described with an $|\varepsilon|^4$ enhancement factor, where ε is the spatially dependent enhanced electric field around the nanoparticle aggregate ($\varepsilon = |E_{\text{local}}|/|E_{\text{incident}}|$). This is because both the excitation intensity and the Raman scattered emission are enhanced by a factor of $|\varepsilon|^2$, yielding an overall Raman intensity enhancement as $|\varepsilon|^4$ [40, 45]. On the other hand, based on other luminescent processes, we predict an $|\varepsilon|^2$ enhancement for the nanoparticle luminescence [46]. With this in mind, correlated super-resolution SM-SERS experiments with SEM structural correlation were carried out, along with theoretical discrete-dipole approximation (DDA) calculations modeling the expected $|\varepsilon|^4$ and $|\varepsilon|^2$ centroid positions based on the nanoparticle structure [24].

To calculate the expected centroid positions in each case, the enhancement ε was calculated around an aggregate, using 532 nm plane wave excitation and a grid with 1 nm^3 resolution. The centroid position, $\langle X \rangle$, was calculated as a weighted-average over all possible locations using Equation 9.3.

$$\langle x \rangle(\omega) = \frac{\sum_i x_i f(\varepsilon(x_i, \omega))}{\sum_i f(\varepsilon(x_i, \omega))} \qquad \begin{aligned} f(\varepsilon)_{\text{luminescence}} &= \varepsilon^2 \\ f(\varepsilon)_{\text{SERS}} &= \varepsilon^4 \end{aligned} \tag{9.3}$$

In the case of the SERS centroid, the average centroid position was weighted by ε^4 while the luminescence centroid was weighted by ε^2. This analysis only accounts for the local electromagnetic field enhancement under plane wave excitation; there is no assumption about the location of the molecules on the nanoparticle surface.

For the initial investigation into this phenomenon, several nanoparticle dimers were investigated, as dimers have a single junction, which offers a more straightforward interpretation of experimental results. Two dimer examples with associated theoretical calculations are presented in Figure 9.10. In Figure 9.10a,c, the calculated spatial intensity maps are qualitatively overlaid on the SEM of the corresponding aggregate, and the measured average luminescence centroid position is indicated with a white arrow. As before,

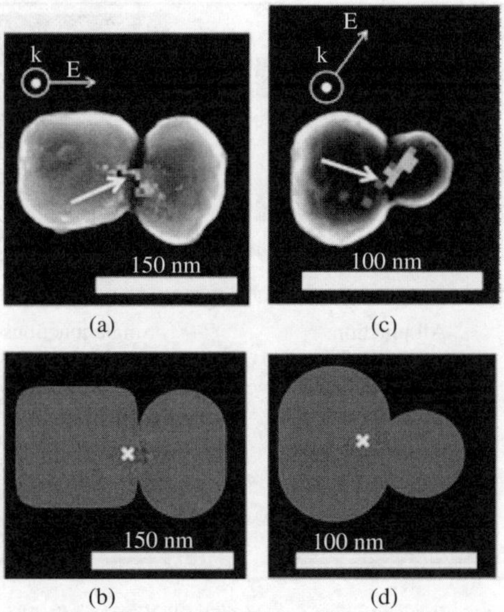

Figure 9.10 *(a) and (c) SEM images of two nanoparticle dimers with SERS intensity histogram overlaid on image (white arrow indicates luminescence centroid fit). (b) and (d) show corresponding theoretical ε^2 (white cross) and ε^4 (dark cross) centroid fits, modeled based on the structure of (a) and (c). Reprinted with permission from Ref. [24]. Copyright © 2012, American Chemical Society. (*See insert for color representation of the figure.*)*

the most intense SERS is placed on the junction region between the two nanoparticles, although we note that in Figure 9.10a, we have two regions of strong SERS, which we place on either side of the nanoparticle junction. In both cases, the qualitative overlay of the SERS spatial intensity maps leads the luminescence centroid to be slightly offset from the dimer junction. The corresponding centroid calculations are shown in Figure 9.10b,d, with the calculated $|\varepsilon|^2$ centroid position shown as a white X and the $|\varepsilon|^4$ centroid given as a dark X. We find that the offset between the experimental SERS and luminescence centroid positions correlate well with the predicted offset between the calculated ε^4 (SERS) and ε^2 (luminescence) centroids [24]. Moreover, the qualitative assignment of the position of the SERS and luminescence centroids in Figure 9.10a,c, agree remarkably well with the calculated centroids in Figure 9.10b,d, indicating that our intuitive assignments of SERS being localized near dimer junctions is a reasonable approximation.

To show that we can apply this same analysis to more complex nanostructures, we next examine a silver nanoparticle trimer, as shown in Figure 9.11. This structure has three potential SERS-active junctions, labeled I, II, and III. Based on the size and shape of the SERS intensity histogram (Figure 9.11a), it appears that the R6G molecule is exploring the single junction denoted II in Figure 9.11b, which places the silver luminescence centroid on the leftmost particle. We also note that the experimentally determined offset between the SERS and luminescence centroids is more than 30 nm [24]. Next, we compare the experimental results to the predicted ε^2 and ε^4 centroid positions, as calculated by Equation 9.3, and we observe that the calculated centroids nearly overlap (Figure 9.11c), in stark contrast to the experimental results.

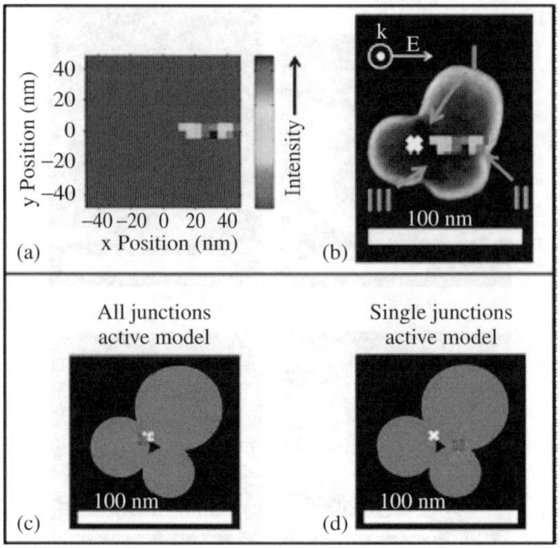

Figure 9.11 *(a) SERS intensity histogram and (b) SERS intensity histogram overlaid on SEM image of the same SM-SERS-active aggregate (white X denotes luminescence centroid). (c) Predicted ε^2 (white cross) and ε^4 (dark cross) centroid fits assuming that ε^4 samples all possible SM-SERS active junctions. (d) Same as (c), but ε^4 is calculated based only on junction II in (b). Reprinted with permission from Ref. [24]. Copyright © 2012, American Chemical Society. (*See insert for color representation of the figure.*)*

To explain this disagreement, we recognize that Equation 9.3 weights the centroid by the enhanced electromagnetic fields associated with all junctions in the calculation of the ε^2 and ε^4 centroid positions. While this is appropriate for the luminescence centroid, which originates from the bulk nanostructure, we expect the SERS centroid to be dominated by the enhanced electromagnetic field associated with the junction occupied by the SERS reporter molecule. Taking this into account, the ε^4 centroid is recalculated based only on the electromagnetic enhancement associated with junction II, and the results (Figure 9.11d) show much stronger agreement with the experimental data [24].

Taken together, the correlated super-resolution imaging and structural data indicate that the plasmon-enhanced electromagnetic fields play a strong role in dictating the spatial origin and associated intensity of the SERS signals. In the cases when multiple junctions are present in a nanostructure, the centroid is strongly associated with one junction, indicating that the SERS emission is coupling into local plasmon modes of the nanostructure. We conclude this based on both the agreement between the spatial intensity maps and the structure of the "hot" junctions, as well as our calculations of expected centroid locations based purely on electromagnetic enhancement. Having verified the important role played by plasmonic enhancement on the calculated hot spot maps, we next turn to the role of the molecule in dictating the specific location of the calculated SERS centroid.

9.4.3 The Role of the Molecule: Isotope-Edited Studies

While the results shown in the previous sections suggest that the measured SERS intensity depends on the plasmon-enhanced electric fields of the aggregate, it remains unclear as to how the position of the molecule affects the centroid position [6, 22, 24]. Our hypothesis is that the motion in the SM-SERS centroid is due partially to the motion of the SM-SERS analyte as it explores the nanostructure. However, others have suggested that the changing centroid could be due to reorientation of the SERS probe, nanoparticle damage, or polarization effects [26, 47]. To confirm the dependence of the SM-SERS centroid on the location of the SERS probe, we utilized a bi-analyte approach in order to measure centroid positions of multiple molecules on the nanoparticle surface [29]. Because two molecules cannot occupy the same physical space, we know that they will be in distinct spatial locations on the nanoparticle surface. By calculating the SERS centroids for each emitter, we can determine whether the centroids also appear in distinct spatial positions, indicating that the location of the molecule is responsible for the position of the SERS centroid.

The isotope-edited bi-analyte approach, in which two isotopologs of R6G are used as SERS tags, was first introduced by Van Duyne and coworkers as a method to confirm the presence of SM-SERS [7]. In this method, the protons on the phenyl ring of the R6G (denoted now as R6G-d0, for clarity) are replaced with four deuterium atoms (R6G-d4). Using isotopologs of the same species for a bi-analyte experiment is useful because the two analytes have the same Raman cross section and probability of adsorption to the nanoparticles, but will have distinct Raman spectra, allowing their signals to be discriminated [7]. In particular, we use the low energy Raman peaks at ~609 cm^{-1} (R6G-d0) and at ~597 cm^{-1} (R6G-d4) to identify the emitting species. Other than using equal amounts of each isotopolog at a slightly elevated concentration when preparing SERS samples, sample preparation for this experiment is identical to the previously described SM-SERS experiments [29].

Figure 9.12 shows the results of two different SERS-active aggregates labeled with multiple molecules. In Figure 9.12a, the spectra initially show the signature of R6G-d0 at 608 cm^{-1}, but after ~40 s, the R6G-d0 mode disappears and the spectral signature of R6G-d4 at 594 cm^{-1} is observed. Figure 9.12b shows a scatter plot of the associated SERS centroids as the identity of the emitter switches from R6G-d0 to R6G-d4. In this plot, the average centroid positions of the two scatterers are offset by ~2 nm, while the predicted uncertainty of the centroid fits (from Equation 9.2) are only ~0.2 nm (R6G-d4), and ~0.3 nm (R6G-d0) [35]. Because the centroid from each molecule is spatially distinct, we hypothesize that the plasmonic-enhancement from the nanostructure cannot be the sole determinant of the centroid location, and that the molecule must play a role in dictating the location of the measured centroid.

Figure 9.12 also shows a second example of SERS from multiple analytes on the same nanostructure, but in this case we observe emission from two different dye species simultaneously. Initially, emission is observed from both R6G species, as evidenced by SERS peaks at both 592 and 604 cm^{-1} in Figure 9.12c, but after ~48 s, the 592 cm^{-1} disappears, indicating that the R6G-d4 is no longer emitting. When the R6G-d4 stops emitting, a jump in the average position of the SM-SERS centroid of ~8.2 nm occurs, as shown in Figure 9.12d. Initially, the overall fitted centroid is an intensity-weighted superposition of both emission centroids, but when the R6G-d4 stops emitting, the centroid reflects emission from R6G-d0 only [39]. As before, the change in the centroid that accompanies a change in the identity of the emitter indicates that the molecule plays a significant role in determining the location of the SERS centroid.

Although the previous two examples support the hypothesis that the location of the SERS centroid must depend on the molecule as well as the plasmon modes of the nanoparticle

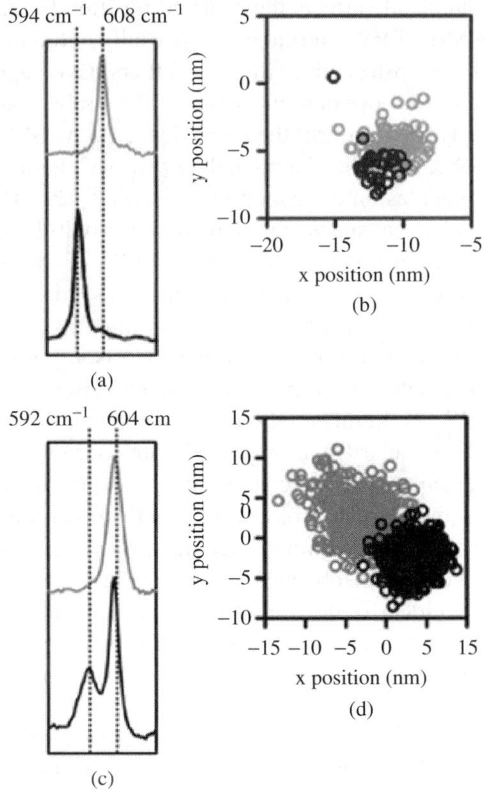

Figure 9.12 (a) Representative spectra of marker mode region from before and after analyte switch from R6G-d0 (gray) to R6G-d4 (black). (b) Scatter plot showing SERS emission centroids from each isotopolog ((0,0) indicates luminescence centroid). (c) Representative spectra showing a switch from emission from both analytes (black) to R6G-d0 (gray) emission only. (d) Scatter plot showing the change in SERS centroid after departure of R6G-d4. Adapted with permission from Ref. [29]. Copyright © 2012, American Chemical Society

substrate, the data presented cannot rule out the influence of molecular orientation on the localization of the SM-SERS centroid. To address this, we present the data in Figure 9.13, in which we observe evidence of molecular orientation but no corresponding shift in the SERS centroid. An intensity–time trace for this SERS-active nanoparticle is shown in Figure 9.13a, and we observe a stepwise drop in intensity at ~64 s. To understand the nature of this intensity drop, four representative spectra are shown in Figure 9.13b at the times indicated by the arrows in Figure 9.13a. As can be seen from this data, the ~598 cm^{-1} marker mode is present across all spectra, allowing us to assign the emitter as R6G-d4, but we also observe peak splitting during the later spectra. Taken together, the peak splitting and the intensity drop provide evidence of molecular reorientation within the hot spot [33, 48–51]. Figure 9.13a also shows the calculated x- and y-centroids as a function of time. The centroid position remains stable in both x and y over the time period when the intensity drop is

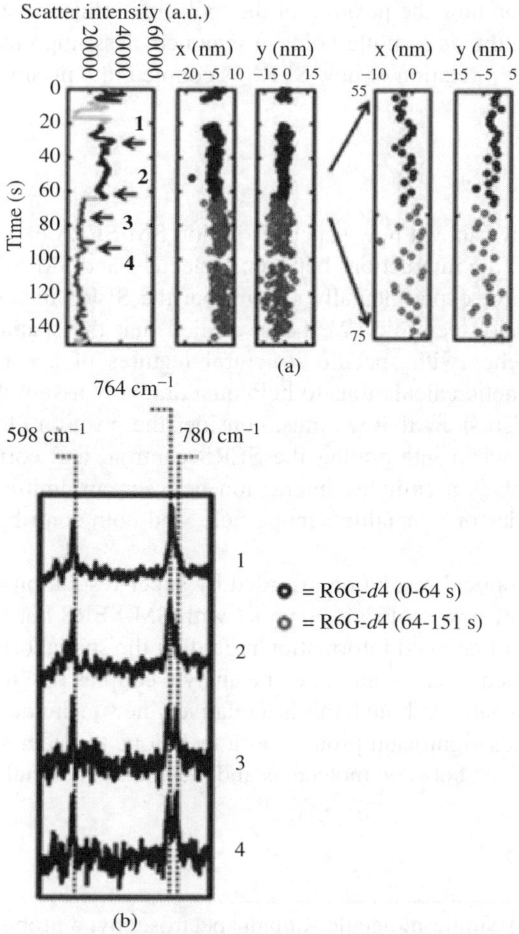

Figure 9.13 *(a) Intensity–time trace, centroid trajectories, and zoomed-in trajectories of proposed molecular reorientation within a hot spot. (b) Spectra (correlated to time points in (a)) showing peak splitting correlated with SERS intensity drop. Adapted with permission from Ref. [29]. Copyright © 2012, American Chemical Society*

observed, indicating that molecular reorientation is not the dominant mechanism for shifts in the SERS centroids.

The bi-analyte data verify that the molecule plays a role in dictating the location of the SERS centroid. Specifically, this study supports the hypothesis that emission from molecules at spatially distinct positions will exhibit different centroid locations, and that reorientation of the molecule within a hot spot does not cause a shift in the centroid. However, we emphasize again that the calculated centroid does not reflect the absolute position of the molecule. In SERS, the scattered light from the molecule still couples out through the plasmon modes of the underlying nanoparticle substrate, which means that the centroid is actually reporting on how the position of the molecule influences the plasmon-coupled emission [26]. While this is a subtle point, it is nonetheless important to our analysis of these data and our interpretation of how SERS is coupled into plasmonic hot spots.

9.5 Conclusions

In this chapter we have shown how super-resolution SM-SERS experiments can provide important insight into the interactions between molecules and SM-SERS hot spots. Using this technique, we have experimentally shown that the SERS intensity is related to the position of the emission centroid. We have verified that the spatial distribution of the SERS intensity matches with specific structural features of a nanoparticle aggregate, and used electromagnetic calculations to help quantitatively assign the centroid location. However, our data also show that we must consider the position of the molecule on the nanoparticle surface when interpreting the SERS centroid and corresponding emission intensity. SERS involves a complex interaction between an emitting molecule and the excited plasmon modes of a metallic nanoparticle, and both contribute to the position of the SERS centroid.

In conclusion, the optical resolution afforded by super-resolution imaging has allowed fundamental studies of how molecules interact with SM-SERS hot spots. This technique can provide a wealth of detailed information regarding the spatial origin of SM-SERS hot spots, their enhancement factors, and how the analyte couples into the plasmon modes of the nanoparticle aggregate. Although this is a relatively new technique as applied to SERS, we believe that it holds significant promise as a technique to understand the fundamental nature of the interactions between molecules and plasmonic materials.

References

1. Haran, G. (2010) Single-molecule Raman spectroscopy: a probe of surface dynamics and plasmonic fields. *Accounts of Chemical Research*, **43**, 1135–1143.
2. Pieczonka, N.P.W. and Aroca, R.F. (2008) Single molecule analysis by surfaced-enhanced Raman scattering. *Chemical Society Reviews*, **37**, 946–954.
3. Qian, X.M. and Nie, S.M. (2008) Single-molecule and single-nanoparticle SERS: from fundamental mechanisms to biomedical applications. *Chemical Society Reviews*, **37**, 912–920.
4. Kneipp, J., Kneipp, H. and Kneipp, K. (2008) SERS – a single-molecule and nanoscale tool for bioanalytics. *Chemical Society Reviews*, **37**, 1052–1060.
5. Etchegoin, P.G. and Le Ru, E.C. (2008) A perspective on single molecule SERS: current status and future challenges. *Physical Chemistry Chemical Physics*, **10**, 6079–6089.
6. Stranahan, S.M. and Willets, K.A. (2010) Super-resolution optical imaging of single-molecule SERS hot spots. *Nano Letters*, **10**, 3777–3784.
7. Dieringer, J.A., Lettan, R.B. II,, Scheidt, K.A. and Van Duyne, R.P. (2007) A frequency domain existence proof of single-molecule surface-enhanced Raman spectroscopy. *Journal of the American Chemical Society*, **129**, 16249–16256.

8. Le Ru, E.C., Blackie, E., Meyer, M. and Etchegoin, P.G. (2007) Surface enhanced Raman scattering enhancement factors: a comprehensive study. *Journal of Physical Chemistry C*, **111**, 13794–13803.
9. Michaels, A.M., Nirmal, M. and Brus, L.E. (1999) Surface enhanced Raman spectroscopy of individual rhodamine 6G molecules on large Ag nanocrystals. *Journal of the American Chemical Society*, **121**, 9932–9939.
10. Kneipp, K., Wang, Y., Kneipp, H. *et al.* (1997) Single molecule detection using surface-enhanced Raman scattering (SERS). *Physical Review Letters*, **78**, 1667–1670.
11. Nie, S. and Emory, S.R. (1997) Probing single molecules and single nanoparticles by surface-enhanced Raman scattering. *Science*, **275**, 1102–1106.
12. Willets, K.A., Stranahan, S.M. and Weber, M.L. (2012) Shedding light on surface-enhanced Raman scattering hot spots through single-molecule super-resolution imaging. *Journal of Physical Chemistry Letters*, **3**, 1286–1294.
13. Chen, C., Hutchison, J.A., Clemente, F. *et al.* (2009) Direct evidence of high spatial localization of hot spots in surface-enhanced Raman scattering. *Angewandte Chemie International Edition*, **48**, 9932–9935, S9932/9931-S9932/9938.
14. Andersen, P.C., Jacobson, M.L. and Rowlen, K.L. (2004) Flashy silver nanoparticles. *The Journal of Physical Chemistry B*, **108**, 2148–2153.
15. Betzig, E., Patterson, G.H., Sougrat, R. *et al.* (2006) Imaging intracellular fluorescent proteins at nanometer resolution. *Science*, **313**, 1642–1645.
16. Rust, M.J., Bates, M. and Zhuang, X. (2006) Sub-diffraction-limit imaging by stochastic optical reconstruction microscopy (STORM). *Nature Methods*, **3**, 793–796.
17. Fernandez-Suarez, M. and Ting, A.Y. (2008) Fluorescent probes for super-resolution imaging in living cells. *Nature Reviews Molecular Cell Biology*, **9**, 929–943.
18. Huang, B., Bates, M. and Zhuang, X. (2009) Super-resolution fluorescence microscopy. *Annual Review of Biochemistry*, **78**, 993–1016.
19. Bates, M., Huang, B., Dempsey, G.T. and Zhuang, X. (2007) Multicolor super-resolution imaging with photo-switchable fluorescent probes. *Science*, **317**, 1749–1753.
20. Yildiz, A., Forkey, J.N., McKinney, S.A. *et al.* (2003) Myosin V walks hand-over-hand: single fluorophore imaging with 1.5-nm localization. *Science (Washington)*, **300**, 2061–2065.
21. Biteen, J.S., Thompson, M.A., Tselentis, N.K. *et al.* (2008) Super-resolution imaging in live Caulobacter crescentus cells using photoswitchable EYFP. *Nature Methods*, **5**, 947–949.
22. Weber, M.L. and Willets, K.A. (2011) Correlated super-resolution optical and structural studies of surface-enhanced Raman scattering hot spots in silver colloid aggregates. *Journal of Physical Chemistry Letters*, **2**, 1766–1770.
23. Yildiz, A. and Selvin, P.R. (2005) Fluorescence imaging with one nanometer accuracy: application to molecular motors. *Accounts of Chemical Research*, **38**, 574–582.
24. Weber, M.L., Litz, J.P., Masiello, D.J. and Willets, K.A. (2012) Super-resolution imaging reveals a difference between SERS and luminescence centroids. *ACS Nano*, **6**, 1839–1848.
25. Kerker, M., Wang, D.-S. and Chew, H. (1980) Surface enhanced Raman scattering (SERS) by molecules adsorbed at spherical particles: errata. *Applied Optics*, **19**, 4159–4174.

26. Ausman, L.K. and Schatz, G.C. (2009) On the importance of incorporating dipole reradiation in the modeling of surface enhanced Raman scattering from spheres. *Journal of Chemical Physics*, **131**, 084708/084701–084708/084710.
27. Lee, P.C. and Meisel, D. (1982) Adsorption and surface-enhanced Raman of dyes on silver and gold sols. *Journal of Physical Chemistry*, **86**, 3391–3395.
28. Bell, S.E.J. and Sirimuthu, N.M.S. (2005) Surface-enhanced Raman spectroscopy as a probe of competitive binding by anions to citrate-reduced silver colloids. *The Journal of Physical Chemistry, A*, **109**, 7405–7410.
29. Titus, E.J., Weber, M.L., Stranahan, S.M. and Willets, K.A. (2012) Super-resolution SERS imaging beyond the single-molecule limit: an isotope-edited approach. *Nano Letters*, **12**, 5103–5110.
30. Michaels, A.M., Jiang, J. and Brus, L. (2000) Ag nanocrystal junctions as the site for surface-enhanced Raman scattering of single rhodamine 6G molecules. *Journal of Physical Chemistry B*, **104**, 11965–11971.
31. Blackie, E.J., Le Ru, E.C. and Etchegoin, P.G. (2009) Single-molecule surface-enhanced Raman spectroscopy of nonresonant molecules. *Journal of the American Chemical Society*, **131**, 14466–14472.
32. Moerner, W.E. and Fromm, D.P. (2003) Methods of single-molecule fluorescence spectroscopy and microscopy. *Review of Scientific Instruments*, **74**, 3597–3619.
33. Li, Z., Shegai, T., Haran, G. and Xu, H. (2009) Multiple-particle nanoantennas for enormous enhancement and polarization control of light emission. *ACS Nano*, **3**, 637–642.
34. Shegai, T., Brian, B., Miljkovic, V.D. and Kall, M. (2011) Angular distribution of surface-enhanced Raman scattering from individual Au nanoparticle aggregates. *ACS Nano*, **5**, 2036–2041.
35. Thompson, R.E., Larson, D.R. and Webb, W.W. (2002) Precise nanometer localization analysis for individual fluorescent probes. *Biophysical Journal*, **82**, 2775–2783.
36. Stranahan, S.M., Titus, E.J. and Willets, K.A. (2011) SERS orientational imaging of silver nanoparticle dimers. *Journal of Physical Chemistry Letters*, **2**, 2711–2715.
37. Stranahan, S.M., Titus, E.J. and Willets, K.A. (2012) Discriminating nanoparticle dimers from higher order aggregates through wavelength-dependent SERS orientational imaging. *ACS Nano*, **6**, 1806–1813.
38. Kaplan-Ashiri, I., Titus, E.J. and Willets, K.A. (2011) Subdiffraction-limited far-field Raman spectroscopy of single carbon nanotubes: an unenhanced approach. *ACS Nano*, **5**, 1033–1041.
39. Bolinger, J.C., Traub, M.C., Adachi, T. and Barbara, P.F. (2011) Ultralong-range polaron-induced quenching of excitons in isolated conjugated polymers. *Science (Washington)*, **331**, 565–567.
40. Hao, E. and Schatz, G.C. (2004) Electromagnetic fields around silver nanoparticles and dimers. *Journal of Chemical Physics*, **120**, 357–366.
41. Xu, H. and Käll, M. (2003) Polarization-dependent surface-enhanced Raman spectroscopy of isolated silver nanoaggregates. *ChemPhysChem*, **4**, 1001–1005.
42. Willets, K.A. and Stranahan, S.M. (2012) Super-resolution imaging of diffusing analyte in surface-enhanced Raman scattering hot-spots. *Proceedings of SPIE*, **8228**, 82280P/82281–82280P/82288.

43. Camden, J.P., Dieringer, J.A., Wang, T. *et al.* (2008) Probing the structure of single-molecule surface-enhanced Raman scattering hot spots. *Journal of the American Chemical Society*, **130**, 12616–12617.
44. Litz, J.P., Camden, J.P. and Masiello, D.J. (2011) Spatial, spectral, and coherence mapping of single-molecule SERS active hot spots via the discrete-dipole approximation. *The Journal of Physical Chemistry Letters*, **2**, 1695–1700.
45. Le Ru, E.C., Grand, J., Felidj, N. *et al.* (2008) Experimental verification of the SERS electromagnetic model beyond the |E|4 approximation: polarization effects. *Journal of Physical Chemistry C*, **112**, 8117–8121.
46. Geddes, C.D., Aslan, K., Gryczynski, I. *et al.* (2004) Noble-metal surfaces for metal enhanced fluorescence. *Reviews in Fluorescence*, **1**, 365–401.
47. Weiss, A. and Haran, G. (2001) Time-dependent single-molecule Raman scattering as a probe of surface dynamics. *Journal of Physical Chemistry B*, **105**, 12348–12354.
48. Moskovits, M. and Suh, J.S. (1984) Surface selection rules for surface-enhanced Raman spectroscopy: calculations and application to the surface-enhanced Raman spectrum of phthalazine on silver. *Journal of Physical Chemistry*, **88**, 5526–5530.
49. Fromm, D.P., Sundaramurthy, A., Kinkhabwala, A. *et al.* (2006) Exploring the chemical enhancement for surface-enhanced Raman scattering with Au bowtie nanoantennas. *Journal of Chemical Physics*, **124**, 061101/061101–061101/061104.
50. Papadopoulou, E. and Bell, S.E.J. (2011) Surface-enhanced Raman evidence of protonation, reorientation, and Ag + complexation of deoxyadenosine and deoxyadenosine-5′-monophosphate (dAMP) on Ag and Au surfaces. *Journal of Physical Chemistry C*, **115**, 14228–14235.
51. Wang, Z. and Rothberg, L.J. (2005) Origins of blinking in single-molecule Raman spectroscopy. *Journal of Physical Chemistry B*, **109**, 3387–3391.

10

Lithographically-Fabricated SERS Substrates: Double Resonances, Nanogaps, and Beamed Emission

Kenneth B. Crozier, Wenqi Zhu, Yizhuo Chu, Dongxing Wang, and Mohamad Banaee
School of Engineering and Applied Sciences, Harvard University, USA

10.1 Introduction

The continual drive to ever-decreasing minimum feature sizes has been a hallmark of the microelectronics industry [1]. This has been enabled in large part by continuous improvements in lithographic resolution. Similarly, high resolution lithography presents an interesting opportunity for SERS. As we will discuss in this chapter, enhancement factors (EFs) and angular emission profiles of SERS from molecules on nanoparticles can be modified and optimized by engineering their surrounding environments, namely other nanoparticles, and dielectric and metal films. Lithography permits SERS substrates to be fabricated for which metal nanoparticles can be situated in these optimal surrounding environments. It should be noted that this is not readily achievable using metal nanoparticles produced by colloidal synthesis [2], because unless special steps are taken, there is little or no control over their placement on the surfaces of SERS substrates.

This chapter is in three main parts. In Section 10.2, we discuss structures that we term "double resonance" SERS substrates. It is known that when SERS is performed with nanoparticles having a single plasmon resonance, maximum signals will be obtained when the resonance is situated between the laser and Stokes frequencies [3]. We present results from our studies [4–6] that demonstrate SERS substrates with double resonances, to enable strong intensity enhancements to be achieved at both the laser and Stokes

Frontiers of Surface-Enhanced Raman Scattering: Single Nanoparticles and Single Cells, First Edition.
Edited by Yukihiro Ozaki, Katrin Kneipp and Ricardo Aroca.

frequencies. In Section 10.3 we discuss a technique for forming nanogaps. It is well known that very small gaps between metal nanoparticles are a means of achieving very high EFs [7, 8]. Using electron-beam lithography, however, it is difficult to fabricate pairs of metal nanoparticles with gaps below 10 nm, unless special substrates or advanced lithography tools are employed [9, 10].We describe the results of our study [11], in which we demonstrate pairs of nanoparticles separated by gaps well below 10 nm by combining electron beam lithography with the use of sacrificial layers. In Section 10.4 we describe the results of our study [12] on the direct observation of the angular emission patterns of SERS from molecules on optical antennas. Appropriately designed surface plasmon nanostructures enable the emission patterns of surface-enhanced Raman scattering to be modified to facilitate efficient collection. We term this effect "beamed Raman scattering". We demonstrate the direct and unambiguous observation of this phenomenon by separating the Raman emission pattern from the background using energy momentum spectroscopy (EMS).

10.2 Double Resonance SERS Substrates

Broadly speaking, there are two major classes of surface plasmons: localized surface plasmons (LSPs) on metal nanoparticles and surface plasmon polaritons (SPPs) at the interfaces between metals and dielectrics. The device we consider (Figure 10.1) combines both classes. In this figure, a gold device is shown, although we have also demonstrated a silver device that functions along similar lines [4]. The device is shown schematically as Figure 10.1a, and comprises a gold film (100 nm) on which an SiO_2 spacer layer and a square array of gold disks is formed. The device achieves "double resonances" by combining the LSPs excited on the disks with the SPPs excited on the gold film. In Figure 10.1b, we show a scanning electron micrograph (SEM) of a portion of the device, whose overall extent is 100 μm × 100 μm. The device is fabricated by gold evaporation, deposition of SiO_2 by plasma-enhanced chemical vapor deposition (PECVD, thickness 23 nm), and formation of the gold disk array by electron-beam lithography, gold evaporation, and the lift-off process [13]. The gold disks are 40 nm thick, and have diameters ranging from 120 to 140 nm. The array periods vary from 350 to 780 nm.

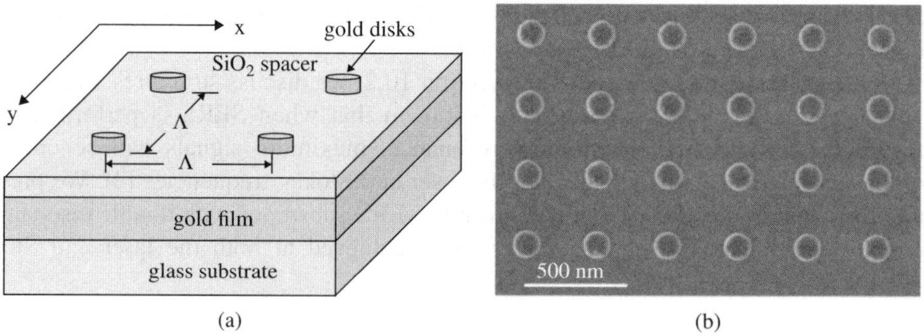

Figure 10.1 *(a) Schematic of gold double resonance structure. (b) Scanning electron micrograph (SEM) of gold disk array on SiO_2 spacer and gold film*

The frequency of the LSP resonance is determined by the size and shape of the gold disk and the refractive index of the surrounding medium [14]. The frequency at which the SPP is excited is determined by the component of the wavevector parallel to the gold interface and the refractive index of the surrounding medium [15]. The wavevector is determined by the array period, and by the angle of illumination. One should, therefore, anticipate that the illumination and collection conditions should influence the SERS performance. This motivates us to measure far-field reflection spectra under collimated and focused illumination conditions, using the set-ups shown as Figure 10.2a,b, respectively. In the collimated set-up, a collimated white-light beam illuminates the device through a beam splitter. The reflected light is collected by a microscope objective (10× magnification). In the focused set-up, light focused by a microscope objective impinges upon the device, with the reflected light also collected by it into a spectrometer. Microscope objectives with numerical apertures (NAs) of 0.1 (5× magnification) and 0.4 (20× magnification) are used. The light source used in these experiments is a Xenon lamp. To ensure that only the light reflected by the array is detected, and not that from the surrounding regions, an iris is used at the image plane of the objective. The spectrum of the light source and the spectral responses of the detector and other optical components are taken into account by making reference measurements at a location on the sample away from the gold disk array, that is, consisting of the SiO_2 spacer on the gold film. The normalized reflection spectra presented here are found by dividing the gold disk array spectra by the reference spectra. Figure 10.2c presents reflection spectra found in this way for a device of the type illustrated in Figure 10.1 with collimated and focused illumination. The disks have diameters of 130 nm, and the array period is 780 nm. These parameters are such that the LSPs and SPPs would occur at approximately the same frequency were they uncoupled. In the device, however, they are of course coupled, leading to hybridization of the resonances. The resultant splitting between the resonances is clearly visible in the collimated-illumination spectrum of Figure 10.2c. It can also be seen that this spectrum is similar to the reflection spectra measured under focused illumination. Additional features can be seen, however, between the two

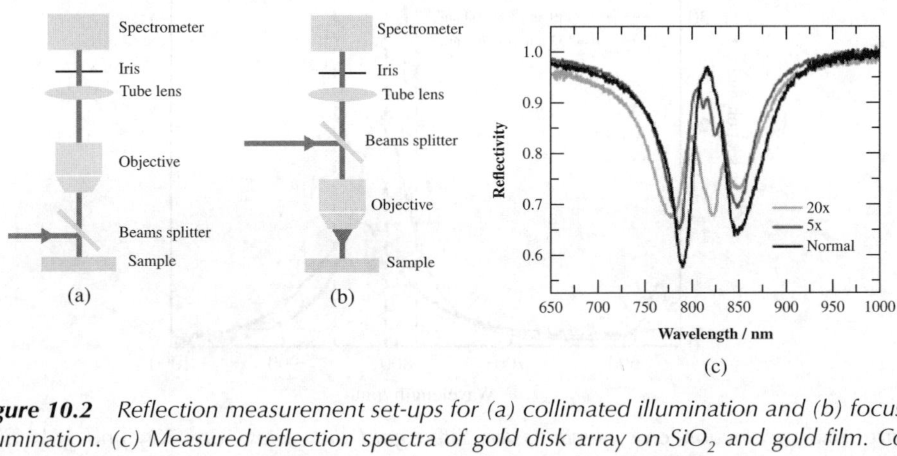

Figure 10.2 *Reflection measurement set-ups for (a) collimated illumination and (b) focused illumination. (c) Measured reflection spectra of gold disk array on SiO_2 and gold film. Collimated and focused illumination cases are denoted "normal" and "5×" or "20×," respectively. Gold disks have diameters of 130 nm, are 40 nm thick, and are in a square array (period: 780 nm)*

primary resonances. These originate from the angular spread of the illumination and are consequently more pronounced for the spectrum measured with the 20× objective lens as it has a greater numerical aperture.

We next consider the splitting of the resonances in more detail. We fabricate devices for which the disk diameters are similar, but with very different array periods. We measure reflection from these devices using the set-up of Figure 10.2b equipped with a microscope objective with a magnification of 5× and an NA of 0.1. These objective lens parameters are very similar to those used for the SERS measurements made on these devices that are presented later in this section. We convert the reflection spectra to the extinction cross sections associated with each gold disk by the relation $C_{ref} = (1 - R) \times \Lambda^2$, where R is the normalized reflectance and Λ is the period of the array. Extinction cross sections for three gold disk arrays on SiO_2 and gold film are shown in Figure 10.3. We first consider the device for which the disk diameters and array period are 135 and 350 nm, respectively. It can be seen that a resonance occurs at a wavelength of 833 nm. This originates from LSPs excited on the disks. We next consider the device for which the disk diameters and array period are 130 and 500 nm, respectively. Resonances at 568 and 813 nm can be seen. These correspond to the SPP and the LSP, respectively. We finally consider a device for which the disk diameters and array period are 133 and 780 nm, respectively. The larger array period red shifts the SPP so that it would be at approximately the same frequency as the LSP were they uncoupled. As discussed above, however, the LSP and SPP are of course coupled, leading to their hybridization and a splitting between them. Dashed vertical lines indicate the wavelengths of the laser ($\lambda = 783$ nm) used later in this section for SERS measurements and the wavelengths of the 420 and 1074 cm^{-1} Raman lines. It can be seen that the resonances of the device with period 780 nm match the wavelengths of laser and 1074 cm^{-1} Raman line reasonably well.

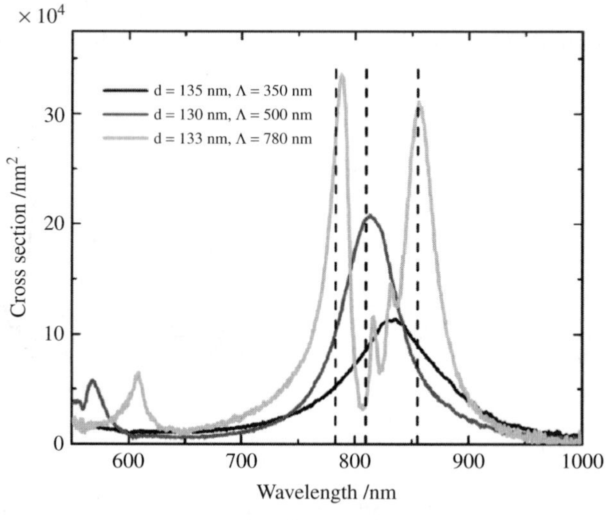

Figure 10.3 *Extinction cross sections for three gold disk arrays on SiO_2 and gold film. Black: disk diameter d = 135 nm, array period Λ = 350 nm; red: d = 130 nm, Λ = 500 nm; and green: d = 133 nm, Λ = 780 nm. Dashed lines: wavelengths of the excitation (used later in this section), and 420 and 1074 cm^{-1} Raman lines*

We next investigate the near-field properties of the device, including the nature of the hybridization, using electromagnetic simulations. The finite-difference time-domain (FDTD) method is used to calculate the near-field intensity distributions of structures with array periods of 500 and 780 nm. For both devices, the gold disks have diameters of 133 nm and are 40 nm thick. The SiO_2 spacer is 25 nm thick, and its refractive index is taken as 1.44. In Figure 10.4a, we plot the field intensity at the edge of the gold disk at the interface between the gold disk and SiO_2 spacer. In the spectrum of the 500 nm period structure, a resonance at $\lambda = 827$ nm (corresponding to the LSP) and a feature at $\lambda = 586$ nm (corresponding to the $(p, q) = (1,0)$ SPP mode) can be seen. Here, (p, q) refers to the reciprocal lattice vector of $G = {}^{2\pi}/_{a}\sqrt{p^2 + q^2}$ provided by the lattice, where a is the period. For the 780 nm period device, resonances at $\lambda = 794$ and 860 nm can be seen, and result from the LSP-SPP hybridization. An additional feature at $\lambda = 627$ nm can be seen, and corresponds to the (1,1) SPP mode. We next plot the field intensity ($|E|^2$) distributions on cross sections perpendicular to the substrate, whose positions are indicated in Figure 10.4b. These distributions are plotted for the illumination at $\lambda = 586$ and 827 nm for the 500 nm structure, and at $\lambda = 794$ and 860 nm for the 780 nm period structure. A fringe pattern above the gold film results from interference between the incident and reflected fields. This complicates physical interpretation of the field patterns arising from

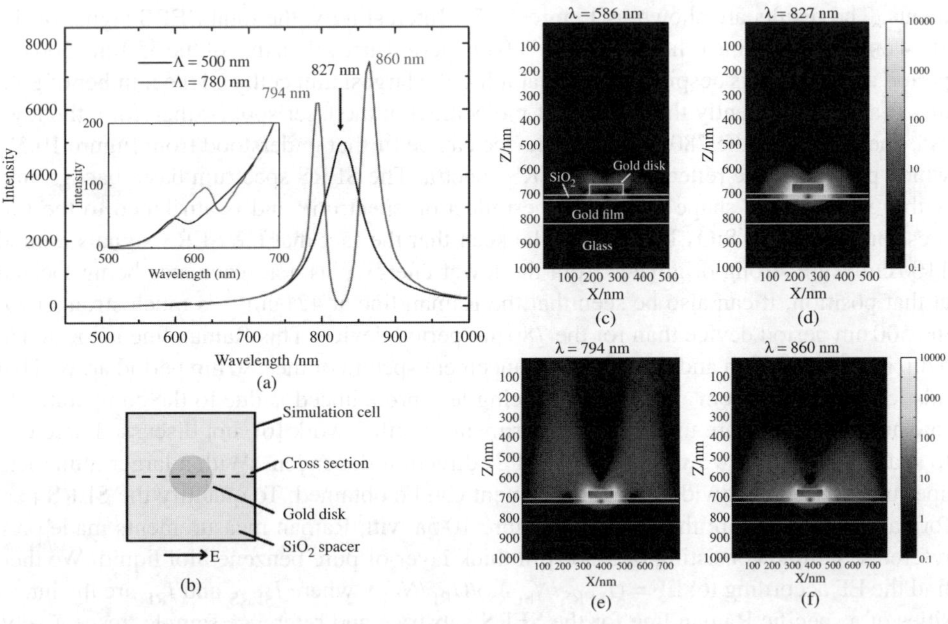

Figure 10.4 *(a) Simulated near-field intensity spectra. Gold disk diameter: 133 nm. SiO_2 spacer thickness: 25 nm. Black: device with array period of 500 nm. Red: device with array period of 780 nm. Inset: expanded view of near-field spectra over the range $\lambda = 500$–700 nm. (b) Top view of the structure in the simulation. (c–f) Intensity ($|E_s|^2$) distributions of scattered field on the cross section perpendicular to substrate for illumination at (c) $\lambda = 586$ nm and at (d) $\lambda = 827$ nm for the 500 nm period array, at (e) $\lambda = 794$ nm and at (f) $\lambda = 860$ nm for the 780 nm period array.* (See insert for color representation of the figure.)

the LSP and SPP resonances. To mitigate this, rather than the total field intensity we plot the scattered field intensity. To find this, we first perform reference calculations of structures consisting of an SiO_2 spacer on an unpatterned gold film. The scattered field E_s is then obtained by subtracting the field of the reference calculation from the field calculated for the SERS substrate, that is, with the addition of the gold disks. In this way, the fields arising from the LSP and SPP resonances can be visualized. The scattered fields found in this way are plotted as Figure 10.4c–f. For the 500 nm period device at $\lambda = 827$ nm, the field is mainly confined around the gold nanoparticle, that is, it follows the character of LSPs. On the other hand, at $\lambda = 586$ nm, there is an evanescent field above the gold film that indicates SPP excitation. For the 780 nm period device, enhanced near-fields about the gold nanoparticle are present at both $\lambda = 794$ and 860 nm. Extended evanescent fields above the film can also be seen, however. These features are characteristic of SPPs. The two resonances of the 780 nm period device, therefore, exhibit both SPP and LSP characters.

We next present results on the characterization of the SERS performance of the devices. Self-assembled benzenethiol monolayers are formed on the gold disks by immersing the devices in a 3 mM solution of benzenethiol in ethanol for 1 h, rinsing with neat ethanol, and then blow drying with nitrogen. A confocal Raman microscope (Renishaw) with near-infrared excitation ($\lambda = 783$ nm), a microscope objective lens with a magnification of 5× (0.12 NA) and a thermoelectrically-cooled CCD detector are used for the SERS measurements. The results are shown as Figure 10.5a. Interestingly, the total SERS signal of the 1074 cm^{-1} Raman line is higher for the 780 nm period array than that of the 350 and 500 nm period arrays. This is despite the fact that it has the largest unit cell, and the number of gold disks (and consequently the number of molecules) in the laser spot is, therefore, the lowest. The results for the 780 nm period device can be further understood from Figure 10.5b, which plots both the reflection and SERS spectra. The SERS spectrum has a background with a very similar shape to that of the reflection spectrum, and is attributed to the fluorescence from the SiO_2 layer. It can be seen that the fact that the SERS signals around 1100 cm^{-1} are strong originates from the lower energy plasmon resonance being located at that position. It can also be seen that the Raman line at 421 cm^{-1} is much stronger for the 500 nm period device than for the 780 nm period device. This Raman line is located at a dip in the extinction and near-field enhancement spectra of the 780 nm period array. That this leads to the 421 cm^{-1} Raman line being less pronounced is due to the comparatively low numerical aperture used in the experiment. In other work [6], not discussed here due to space limitations, we show that the EF is direction-dependent. With a larger numerical aperture, a wide bandwidth for enhancement can be obtained. To quantify the SERS performance, we compare the results of Figure 10.5a with Raman measurements made on a reference sample consisting of a 500 μm thick layer of pure benzenethiol liquid. We then find the EF according to: $EF = (I_{SERS}/N_{SERS})/(I_{RF}/N_{RF})$, where I_{SERS} and I_{RF} are the intensities of a specific Raman line for the SERS substrate and reference sample, respectively, and N_{SERS} and N_{RF} are the numbers of probed molecules in the laser spot for the SERS substrate and reference sample, respectively. The Raman and SERS intensities are found from the integral of the Raman (or SERS) line of the measured spectrum, from which the background has been subtracted, and normalized by the laser power and CCD integration time. For N_{SERS}, we assume the benzenethiol surface density is 6.8×10^{14} cm^{-2} [3]. The surface area taken for each gold disk accounts for the top surface as well as the side walls.

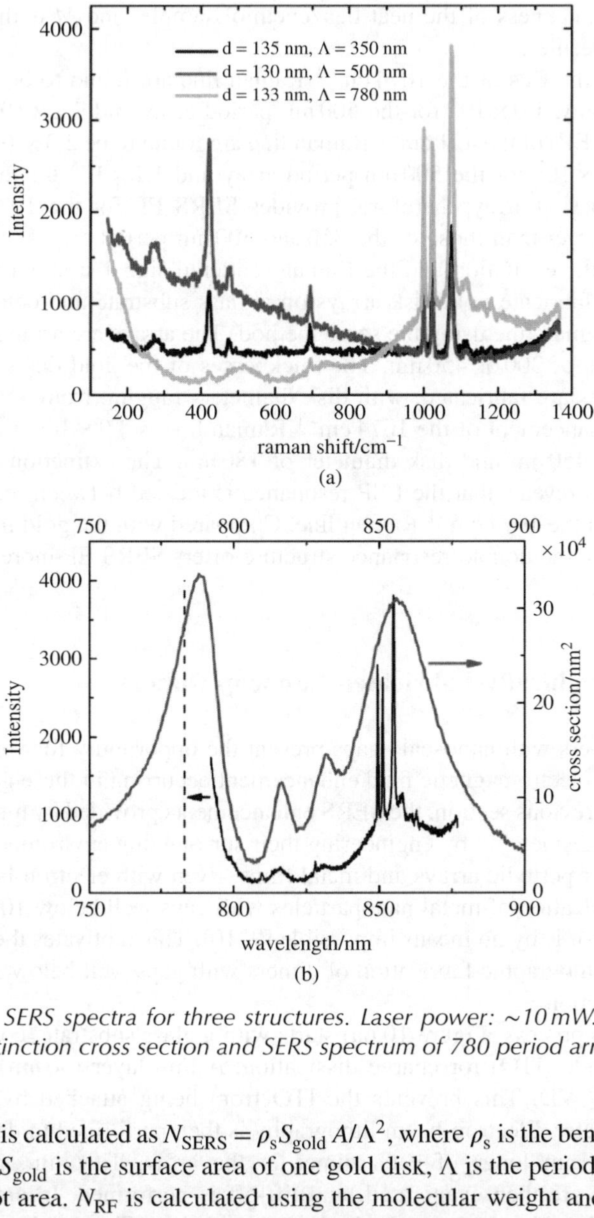

Figure 10.5 *(a) SERS spectra for three structures. Laser power: ~10 mW. CCD integration time: 30 s. (b) Extinction cross section and SERS spectrum of 780 period array*

Therefore N_{SERS} is calculated as $N_{SERS} = \rho_s S_{gold} A/\Lambda^2$, where ρ_s is the benzenethiol surface packing density, S_{gold} is the surface area of one gold disk, Λ is the period of the array, and A is the laser spot area. N_{RF} is calculated using the molecular weight and volume density of benzenethiol, together with the detection volume. Due to the variation in intensity and collection efficiency with position, the contribution of each molecule in the benzenethiol liquid to the measured Raman signal depends on its position. We find the effective confocal length by translating a reference Si sample through the focal volume, and recording the Raman signal at each position. The confocal length is found to be ~2.9 cm, that is, much larger than the thickness of the neat benzenethiol sample. Therefore, the thickness of the neat benzenethiol, 500 μm, is used to calculate the detection volume. Therefore, N_{RF} can be calculated from $N_{RF} = N_A \rho h A/M$, where N_A is Avogadro's number, ρ is the benzenethiol density, h is the thickness of the neat benzenethiol sample, and M is the molar mass of benzenethiol molecules.

From the data, the EFs of the $1074\,cm^{-1}$ Raman line are found to be 6.5×10^6 for the 350 nm period array, 1.0×10^7 for the 500 nm period array and 7.2×10^7 for the 780 nm period array. The EFs of the $421\,cm^{-1}$ Raman line are found to be 2.3×10^6 for the 350 nm period array, 1.1×10^7 for the 500 nm period array and 1.2×10^6 for the 780 nm period array. The 780 period array, therefore, provides SERS EF for the $1074\,cm^{-1}$ line that are 7–11 times larger than those of the 350 and 500 nm period arrays, as it offers strong enhancement of the excitation and the Raman scattered signal simultaneously. For comparison, we also fabricate gold disk arrays on a glass substrate and determine the EF of the $1074\,cm^{-1}$ Raman line using the same method. The arrays are arranged as square lattices with periods of 300 or 450 nm. The thicknesses of the gold disks are 40 nm and a variety of samples are fabricated, with disk diameters ranging from 170 to 200 nm. The largest SERS enhancement of the $1074\,cm^{-1}$ Raman line is 3.9×10^5, found on the array with a period of 450 nm and disk diameter of 180 nm. The extinction spectrum of this array (not shown) reveals that the LSP resonance is located between the wavelengths of the excitation and the $1074\,cm^{-1}$ Raman line. Compared with the gold nanoparticle array on glass substrate, the double-resonance structure offers SERS EF more than 2 orders of magnitude higher.

10.3 Lithographically-Fabricated Nanogap Dimers

Nanoparticle dimers with nanoscale gaps present the opportunity for very high EFs, due to the substantial electromagnetic field enhancement occurring in these gaps [7, 8]. As we have seen in the previous section, the SERS enhancements provided by metal nanoparticles can be boosted considerably by engineering their surrounding environments, in particular through the use of periodic arrays and metal films. Even with electron-beam lithography, however, the fabrication of metal nanoparticles with gaps well below 10 nm is very challenging, although it is by no means impossible [9, 10]. This motivates the development of a technique for lithographic fabrication of dimers with gaps well below 10 nm, and is the subject of this section.

The fabrication process (Figure 10.6a) starts with a glass substrate, coated with a layer of indium tin oxide (ITO) for charge dissipation. A thin layer (50 nm) of SiO_2 is then deposited by PECVD. This prevents the ITO from being attacked by the wet-etching step that occurs later. Electron-beam lithography is then performed to define the left-side nanoparticle (100 nm long) of each dimer in the array. Titanium, gold, silver, and chromium layers are then evaporated, and lift-off is performed. In ambient laboratory conditions, the chromium then expands due to oxidation. Electron-beam lithography is then performed to define the right-side nanoparticle of each dimer. In the exposed pattern, however, each nanoparticle is 144 nm long, and overlapped with the left-side dimer by 44 nm. As a consequence of the lateral expansion of the chromium that occurs upon oxidation, part of the titanium and gold that are next evaporated is formed on the left-side nanoparticle, and part on the substrate. The silver serves as a spacer layer: it forms an undercut structure beneath the chromium layer, in a similar manner to a bilayer resist in typical lift-off processes. The substrate is then immersed in chromium etchant, then rinsed

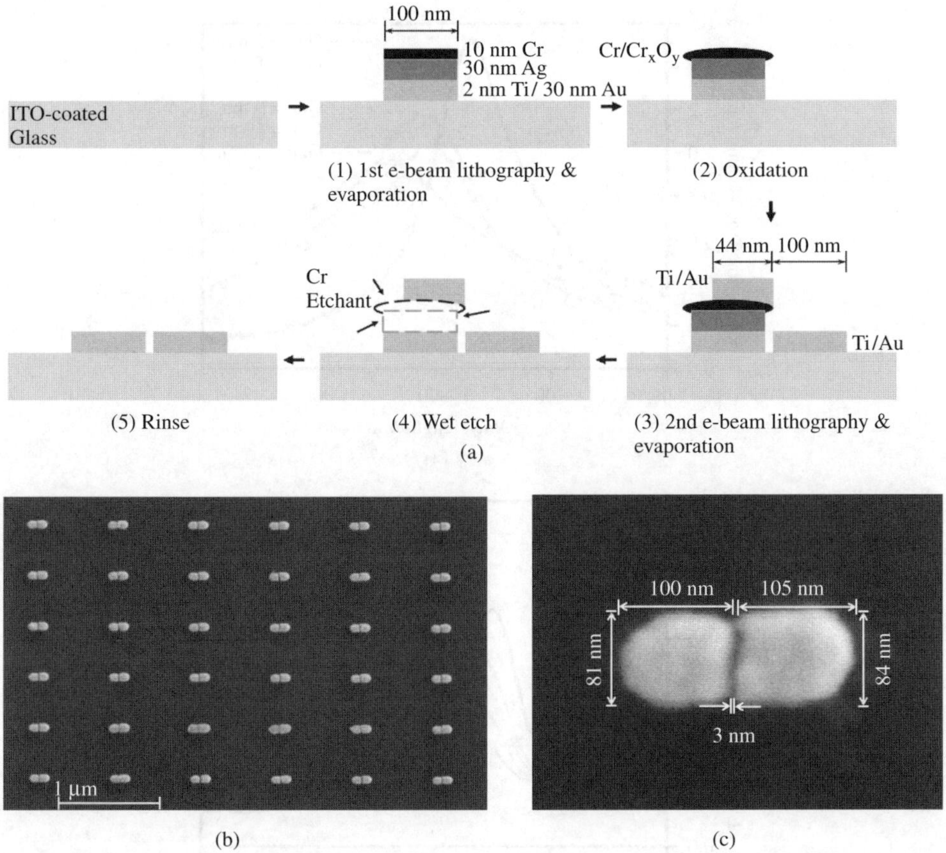

Figure 10.6 *(a) Schematic diagram of fabrication process, (b) SEM of fabricated dimer array, and (c) SEM of a dimer from array shown in panel (b)*

and blown dry with nitrogen. The results are shown as Figure 10.6b,c. The overall extent of the array is $75 \times 75\ \mu m^2$; only a portion is shown. It can be seen that the dimers consist of nanoparticles, each ~100 nm long, separated by gaps as narrow as ~3 nm.

The results clearly demonstrate an ability to fabricate dimer arrays with gaps well below 10 nm. We further demonstrate that the gap size can be controlled on the nanometer length scale. This is done by carrying out the oxidation of Step (2) in Figure 10.6a in an oxygen plasma at low power (20 W) and high flow rate (50 sccm). We find that each additional minute in the plasma chamber increases the gap width by 5.0 nm, which is sufficiently slow to permit sub-10 nm gaps to be made in a controllable manner. In Figure 10.7, we optically characterize arrays of dimers with gaps ranging from 3 to 18 nm by measuring the extinction cross section spectra. The measured trend of the resonances red-shifting for smaller gap size is consistent with the predictions of simulations (Figure 10.7b). However, the resonance peaks from simulation differ slightly from those from measurements. This could be due to the fact that the simulations take the dielectric constant of gold to be that of bulk gold, but this may not be the case, since it varies with grain structure [16].

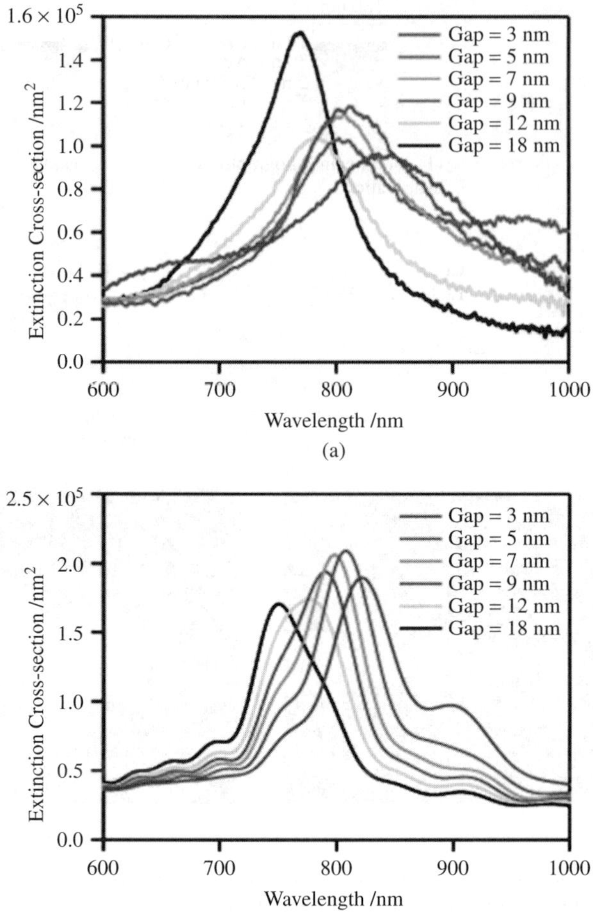

Figure 10.7 *Extinction cross-sections for dimer arrays with different gap sizes. (a) Measurements and (b) simulations. (*See insert for color representation of the figure.*)*

We also find that the simulated resonance linewidths (full-width half-maximum of extinction spectrum) are 50–70% smaller than those of the measurements, which could be due to the non-uniformity of gap sizes within each array. We also note that the measured peak extinction cross-sections vary between different samples; this may be due to differences in fabrication yield between samples. For example, the yield of the 3 nm-gap sample is estimated to be ~81%; while that of the 18 nm-gap structures is almost unity.

We next perform SERS measurements on arrays of dimers with 3 and 12 nm gaps. Monolayers of benzenethiol are formed on the dimers by soaking the substrates in benzenethiol solution (3 mM concentration) for 2 h, rinsing them in ethanol and then blowing them dry with nitrogen. SERS measurements are performed with a confocal Raman microscope (Renishaw inVia) equipped with a 20× objective lens (NA = 0.4), a spectrometer with a thermoelectrically-cooled CCD and a tunable titanium: sapphire laser.

In the SERS measurements, the excitation wavelength for each sample is selected such that the mean value of the laser wavelength and Stokes wavelength matches the peak resonance of the dimer array, as determined from the extinction spectroscopy measurements, in an attempt to maximize the EF [3].A SERS spectrum is measured for an array of 3 nm gap dimers (Figure 10.8a) with the following parameters: laser wavelength $\lambda = 784$ nm, power = 0.82 mW, and CCD integration time = 5 s. Similarly, a SERS spectrum is measured for an array of 12 nm gap dimers (Figure 10.8b) with the following parameters: laser wavelength $\lambda = 750$ nm, power = 1.14 mW, and CCD integration time = 5 s. Despite the fact that the laser power is ~40% higher for the 12 nm gap dimers, it can be seen that the SERS signals are smaller. This is due to the stronger field enhancement that is produced by the dimers with 3 nm gaps. Indeed, we find that the measured SERS EF for the 1072 cm^{-1} Raman line is 1.1×10^8 for the 3 nm gap sample; and 4.4×10^6 for the 12 nm gap sample, representing a ~25-fold improvement with the reduced gap size. These EFs are measured using the procedure discussed in the previous section. To study the uniformity of the SERS signal, we map the measured EF as a function of position on the arrays, which each have extents of 75 μm × 75 μm. The results are plotted as Figure 10.8c,d. To understand the observed SERS enhancement results, we next consider the field intensity distributions, as calculated by FDTD (Figure 10.8e,f). The highest field intensity (square of electric field, $|E|^2$) enhancement in the 3 nm gap is ~6 times larger than that in the 12 nm gap. A rough estimation of SERS EF is often based on $|E|^4$ approximation [7, 17] which indicates that the enhancement of the 3 nm gap antenna is ~36-fold higher than the 12 nm gap antenna. This is generally consistent with the improvement from the SERS measurement, while a more rigorous estimation of electromagnetic EF is presented next.

Dimer arrays with gaps ranging from 18 to 3 nm are fabricated, and SERS measurements are performed. This allows the SERS EF to be found as a function of gap size (Figure 10.9). The laser excitation wavelengths are again selected for each dimer array to ensure maximum EF, and 5 s integration times are again used. It can be seen that the gap size has a dramatic effect upon SERS enhancement. For gaps larger than 10 nm, the EFs are ~10^6. For gaps smaller than 10 nm, EFs of ~10^7 are obtained. When the gap size is further decreased to 5 nm or below, or even 3 nm as demonstrated here, EF can exceed 10^8, approaching the requirement for single molecule detection [18]. It can be seen from Figure 10.9 that samples with the same gap size exhibit different EFs. This is likely to be a consequence of differences in structural morphology between samples with the same gap size. FDTD simulations are also performed for structures with various gap sizes to predict the electromagnetic component of the EF. It can be seen from Figure 10.9 that the simulated EM EFs are ~75 times smaller than the measured SERS EFs. This difference could be in part to the chemical enhancement effect. A detailed explanation would require close examination of the individual dimer structure and is beyond the scope of this investigation. We note that the difference is on the same order as estimates of the chemical EF noted by Talley *et al.* [19] although Lombardi and Birke find that in other cases it can be much higher [20].

10.4 Beamed Raman Scattering

Appropriately-designed surface plasmon nanostructures enable the emission patterns of surface-enhanced Raman scattering to be modified to facilitate efficient collection, an effect

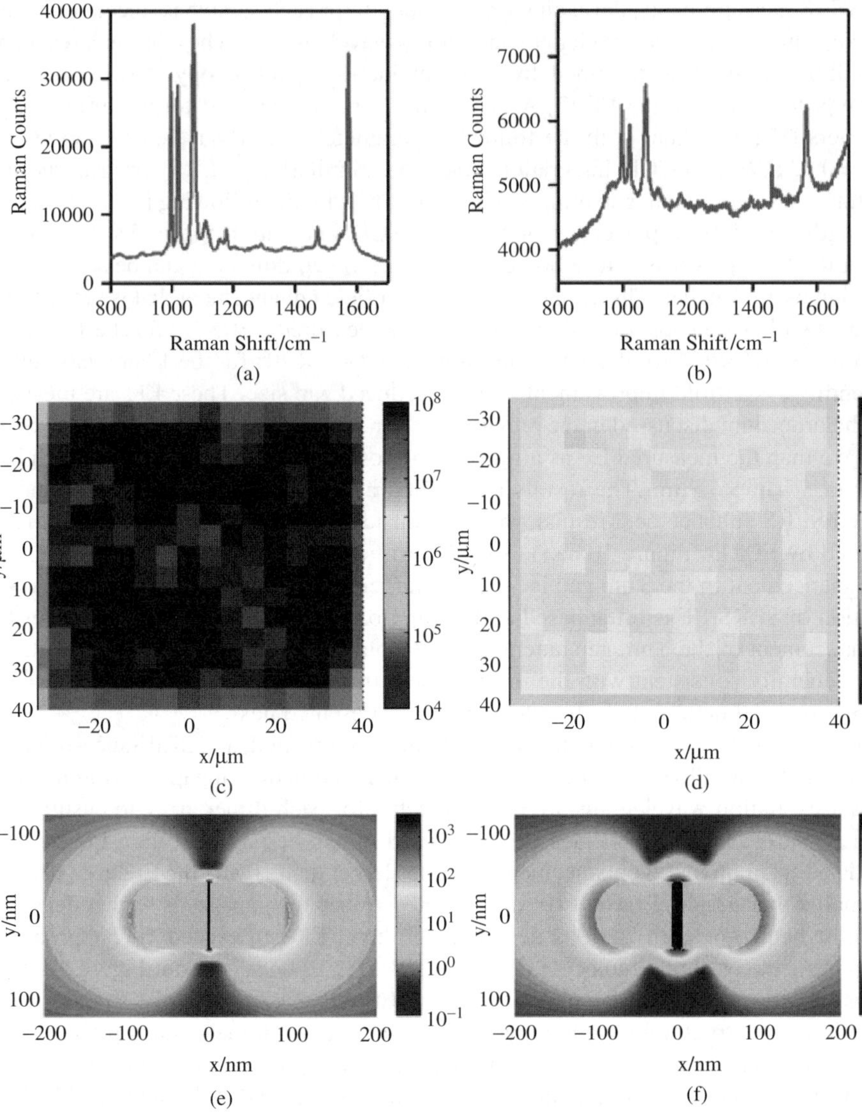

Figure 10.8 *(a). Benzenethiol SERS spectrum measured from 3 nm gap dimers. (b) Benzenethiol SERS spectrum measured from 12 nm gap dimers. (c) EF versus position on array with 3 nm gap dimers. (d) EF versus position on array with 12 nm gap dimers. (e) Simulated field intensity distribution ($|E|^2$) for 3 nm gap dimers, for plane wave illumination ($\lambda = 821$ nm) polarized along antenna axes. Field intensity distribution is shown over center plane of dimer, 15 nm above substrate. (f) Simulated field intensity distribution ($|E|^2$) for 3 nm gap dimers, for plane wave illumination ($\lambda = 776$ nm) polarized along antenna axes. Field intensity distribution is shown over center plane of dimer, 15 nm above substrate. (*See insert for color representation of the figure.*)*

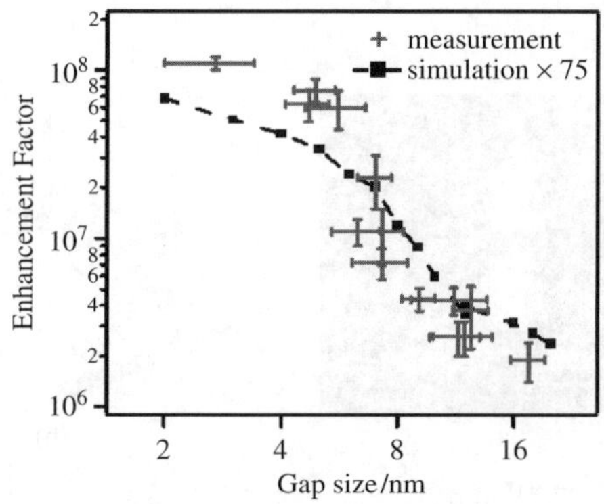

Figure 10.9 *SERS EF versus gap size: measurements and simulations. Each experimental data point is measured on a different sample, and has a vertical error bar representing EF standard deviation determined from measurements made on nine points on the sample. The horizontal error bar represents gap size standard deviation, determined from measurements made at nine points on the sample. Simulations model the electromagnetic component of EF, and are multiplied by 75 times to enable comparison to experimental data*

sometimes termed "beamed Raman scattering" [6, 21] or directional SERS (DERS) [22]. In this section, we describe recent work [12] where we demonstrate the direct and unambiguous observation of this phenomenon using energy-momentum spectroscopy. We here describe the observation of beamed Raman scattering using Yagi-Uda (YU) optical antennas. The emission patterns from different Raman lines are measured simultaneously.

Directional optical antennas have been recently employed in SERS substrates, achieving very high EFs (average value $\sim 10^{10}$) [23], and for single molecule sensitivity [24]. This motivates the development of a technique by which this phenomenon can be directly observed. Shegai *et al.* addressed the problem of measuring the angular distribution of SERS from Rhodamine 6G molecules adsorbed on individual gold nanoparticle aggregates [25]. A microscope objective was used both to focus laser light onto each aggregate and to collect the emission, with Rayleigh-scattered laser light removed with a long pass filter (LPF). The angular distribution of emitted light could then be monitored in the microscope's Fourier plane, but there was no provision to distinguish between the emission patterns of the Raman lines and that of the broad background continuum that can come from the molecules or from the substrate [26, 27]. To add this capability, we employ energy-momentum spectroscopy (EMS) [28]. In this method, a spectrometer measures the spectrum of each pixel in the back focal plane of the objective lens, enabling the emission pattern for each wavelength of interest to be determined. This enables us to determine the emission patterns of different Raman lines, separated from the background.

We start by using the EMS technique to measure the Raman emission patterns from optical dimer antennas of the type shown as Figure 10.10a. Each antenna comprises two gold nanorods (each 117 nm long and 65 nm wide) that are separated by a 13 nm gap. The antennas are fabricated by electron-beam lithography, evaporation and the lift-off process on a glass coverslip. The glass coverslip is coated with a layer of ITO for charge dissipation. We make measurements on arrays of optical dimer antennas, rather than single dimer antennas, to increase the signal-to-noise ratio. To minimize near-field coupling between the antennas, we choose the array's unit cell to be large (0.5 μm × 1.5 μm). SERS measurements are made

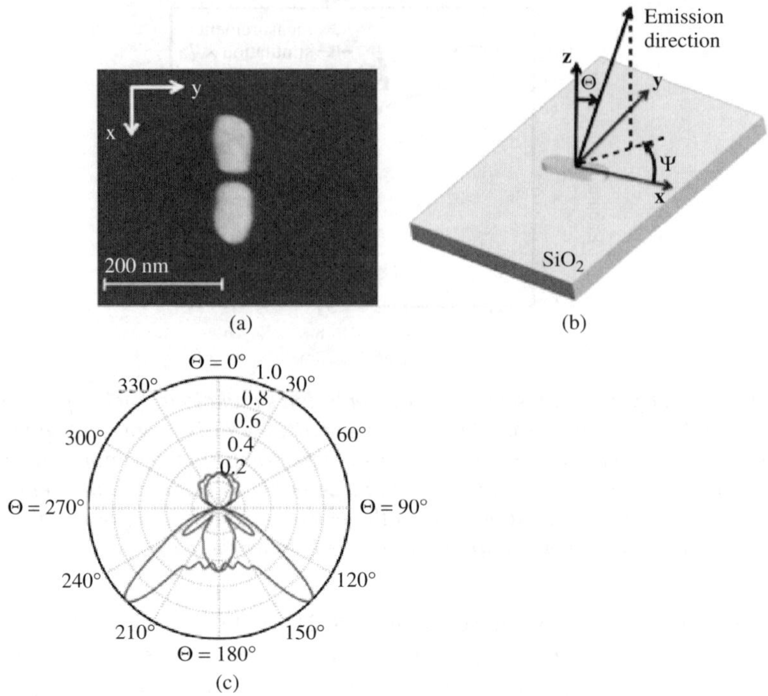

Figure 10.10 *Raman emission pattern from dimer antennas on dielectric substrate. (a) SEM image of dimer antenna. (b) Cartesian* xyz*-coordinate system that defines geometry of dimer structure and air/dielectric interface (*xy*-plane); and spherical Θ, Ψ-coordinate that defines emission direction. (c) FEM simulation of emission pattern, in which arbitrarily oriented Raman dipole is placed at gap center of dimer antenna. We specify plane of emission and allow Θ to range from 0° to 360°. Region below line from Θ = 90° to Θ = 270° represents dielectric substrate. Red: emission pattern in* yz*-plane (Ψ = 90°); blue: emission pattern in* xz*-plane (Ψ = 0°). (See insert for color representation of the figure.)*

on self-assembled monolayers of thiophenol formed on the antennas. The excitation wavelength used in both experiments and simulations is $\lambda = 785$ nm. The $1074\,\text{cm}^{-1}$ Raman line therefore occurs at $\lambda = 857$ nm. We also perform experiments and simulations for the Raman lines $415\,\text{cm}^{-1}$ ($\lambda = 811$ nm) and $1586\,\text{cm}^{-1}$ ($\lambda = 896$ nm), as a way of determining the bandwidths of our directional optical antennas.

In describing the antenna's emission pattern, we use the orientation and emission angle (Θ, Ψ) definitions that are schematically illustrated as Figure 10.10b. Our investigations begin with a simulation of the emission pattern at $\lambda = 857$ nm from an optical dimer antenna on glass. The finite-element method (FEM) is used. A point dipole (modeling a molecule undergoing Raman scattering) with random orientation is placed in the gap, and the far-fields are monitored. In the experiments, the thiophenol molecules coat the gold surfaces of the antenna uniformly. The fields are greatly enhanced in the gap, however, and the largest Raman dipole is induced there. It is, therefore, reasonable to model the emission pattern with excitation from a point dipole in the gap. In the emission process, the Raman dipole excites the dipolar LSP mode of the dimer antenna [11]. A match between the LSP frequency and the Raman dipole frequency therefore means that the Raman scattering emission pattern is dominated by the dimer antenna's dipole mode [29]. Therefore, despite the fact that the point dipole has an arbitrary orientation, the emission pattern (Figure 10.10c) follows the typical emission pattern of a horizontal dipole radiating above a dielectric substrate. Most of the radiated power goes into the glass substrate, mainly within the *yz*-plane that is perpendicular to the dipole orientation. This radiated power mainly occurs into two lobes that are just beyond the critical angle for total internal reflection at the air/dielectric interface.

We employ a home-built microscope set-up (Figure 10.11) to directly observe the beamed Raman phenomenon by imaging the Raman scattering radiation pattern onto the entrance slit of a spectrometer. An oil immersion microscope objective lens (Nikon 100×, NA = 1.40, infinity corrected, immersion oil $n = 1.5$) focuses light from a laser through the coverslip onto the antennas. The laser is polarized, continuous-wave (CW), has a wavelength of $\lambda = 785$ nm and the power is 9.3 mW. The Raman scattering from the thiophenol molecules is collected by the same objective lens. An LPF ($\lambda = 785$ nm) is placed in the optical path to block the Rayleigh scattered laser light. A convex lens (focal length: 100 mm), acting as a Bertrand lens, creates a first image plane of the objective lens back focal plane. A slit (120 μm wide) is placed at this back-focal-plane image plane. It is mounted on a translation stage, enabling it to be moved transverse to the optical path. A second convex lens (focal length 50 mm) creates a second back-focal-plane image plane at the spectrometer entrance slit. This second image is de-magnified by 3.2 times from the first image plane, enabling it to be commensurate with the size of the CCD in the spectrometer. The scanning slit enables the selection of a column from the back-focal-plane image, as described in further detail below. Each pixel of this column is dispersed horizontally by the spectrometer grating. This means that each row in the spectrometer CCD represents the spectrum of a pixel of the column of the back-focal-plane image. Scanning the slit enables different columns from the back-focal-plane image to be selected, and the SERS angular emission pattern to be found. This is described in further detail next.

The energy-momentum technique allows separation of the Raman signal from the background, and is described in Figure 10.12. A typical emission pattern measured using an LPF

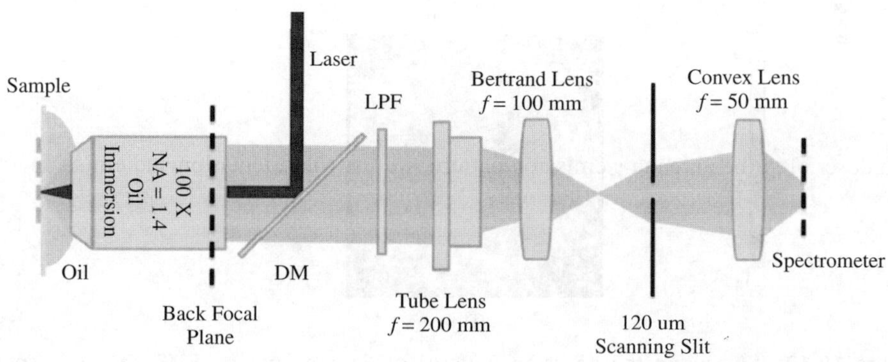

Figure 10.11 *Experimental set-up for energy momentum spectroscopy (EMS). Far-field emission pattern from sample is found by imaging back focal plane of objective lens. LPF: long-pass filter and DM: dichroic mirror*

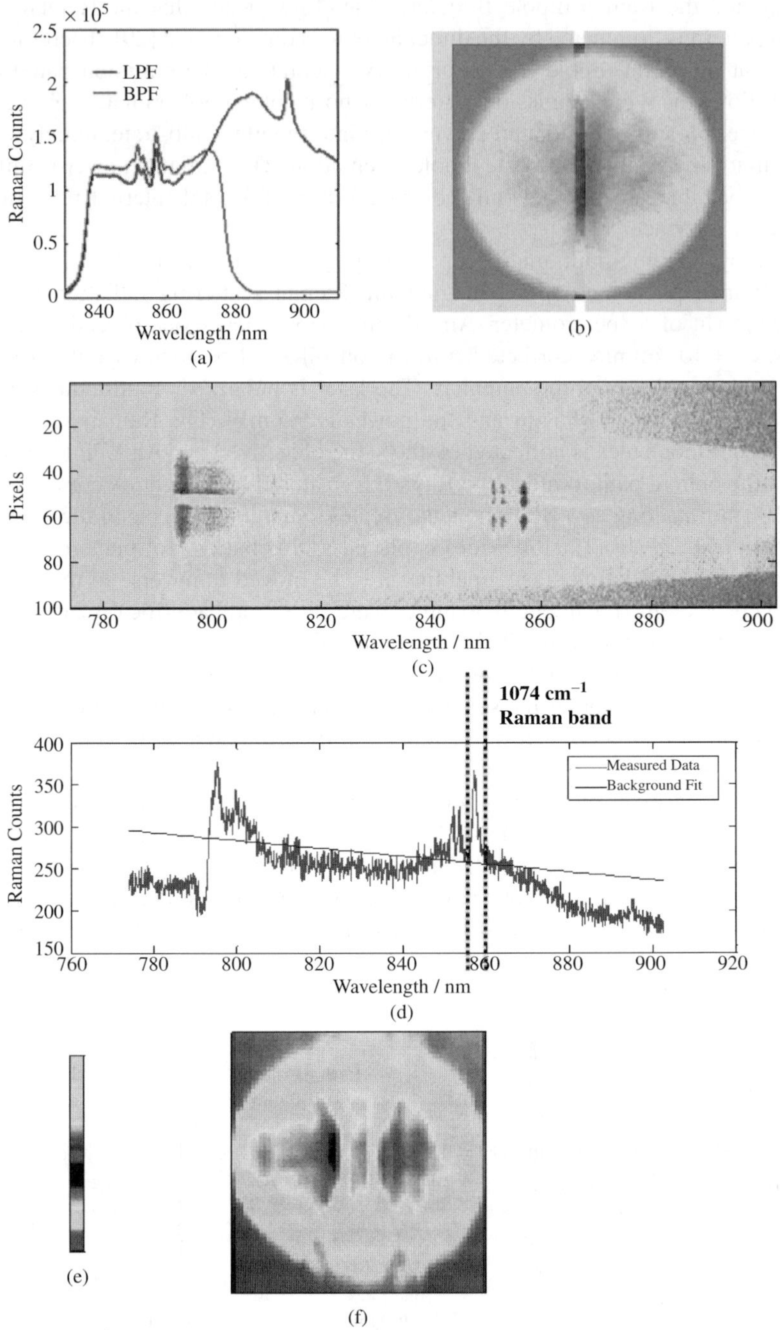

Figure 10.12 *Measuring the emission patterns of Raman lines by EMS. (a) SERS spectra obtained with LPF and BPF. (b) Second back-focal-plane image plane formed on the entrance slit of the spectrometer, using LPF and Bertrand lens. (c) Spectral map, as recorded by spectrometer CCD. (d) Raman spectrum of one row of pixels of (c). (e) Intensity of 1074 cm^{-1} Raman line at slit. (f) Reconstructed emission pattern of 1074 cm^{-1} Raman line. (*See insert for color representation of the figure.*)*

and a Bertrand lens is shown as Figure 10.12a. The sample used for these measurements consists of dimer antennas placed on an SiO_2/gold substrate [23]. For this sample, emission occurs into the region above the substrate, rather than into the substrate. An air objective (NA = 0.8) rather than an oil immersion objective is therefore used. We place a narrow slit on the first back-focal-plane image plane and create a second back-focal-plane image plane at the position of the entrance slit of the spectrometer (Figure 10.11). This process is shown as Figure 10.12b, with the slit denoted by gray shading. The entrance slit of the spectrometer is wide open, rather than stopped down, as there is no provision for moving it horizontally. This would be needed in order to sample the back-focal-plane image.

A typical image recorded on the spectrometer CCD camera is shown as Figure 10.12c. The image represents a two-dimensional (2D) spectral map, with the horizontal axis being the wavelength and the vertical axis being the vertical position on the CCD. The Raman peaks appear as sharp and bright lines. The lateral dispersion shown in the figure is mainly due to the toroidal mirrors of the spectrometer (Princeton Instruments®, SP-2300i). It can be reduced with a lower resolution grating or by using a region of the CCD with a smaller lateral extent. The vertical asymmetry is due to the slight rotational misalignment of the CCD camera equipped with the spectrometer and can be easily fixed in data analysis. One row of pixels of the image of Figure 10.12c is plotted as Figure 10.12d. It shows a typical Raman spectrum with Raman peaks and a background. The background is fitted with a line (black line in Figure 10.12d) via a linear fit of the detector counts over the $\lambda = 840–850$ nm and $\lambda = 860–870$ nm ranges. These are just away from the three Raman peaks around 857 nm. The signal for the Raman line of interest (e.g., 1074 cm^{-1}) is then found by subtracting the counts of the background fit from the spectrometer counts. This procedure is repeated for each pixel within the vertical column (Figure 10.12e). It is also repeated for each column of the back-focal-plane image by scanning the slit horizontally. This yields the emission pattern for the Raman line of interest (e.g., 1074 cm^{-1}), with a typical result shown as Figure 10.12f.

The emission patterns of several Raman lines from the dimer antennas of Figure 10.10a are measured with the EMS technique, with the results shown as Figure 10.13. As specified in Figure 10.10b, the dimer antennas are oriented such that their longer axes are parallel to the laser polarization ($\Psi = 0°$, x-axis). The emission pattern of the 1074 cm^{-1} Raman line (at $\lambda = 857$ nm) is shown as Figure 10.13a. This figure employs a polar coordinate system, with the origin at the center of the circular pattern [30]. The radial coordinate r corresponds to the polar angle Θ of Raman emission according to $r = |n \sin \Theta|$, where $n = 1.5$ for the glass cover-slip. The outline of the circle represents the numerical aperture of the objective lens (NA = 1.4). Two main emission lobes can be seen, and peak in the following directions: ($\Theta = 139.5°$, $\Psi = 100°$) and ($\Theta = 139.5°$, $\Psi = 260°$). It can be seen that the azimuthal angles are close to the expected values of $\Psi = 90$ and 270°. Both measured polar angles Θ are close to the critical angle at the air/glass interface (138.2°) within the measurement resolution. The full-width-at-half-maximum (FWHM) contours of the two beams have widths of ($\Delta\Theta = 22°$, $\Delta\Psi = 48°$) and ($\Delta\Theta = 15°$, $\Delta\Psi = 48°$). We define the front-to-back (F/B) ratio as the ratio of integrated emission power within these two FWHM contours. The ratio between the power radiated into $+y$ (defined as the forward direction) and $-y$ (defined as the backward direction) is 1.10, that is, very close to unity, as expected from the symmetry of the dimer antenna. We also measure the emission patterns of the 415 and 1586 cm^{-1} Raman lines, with the results shown as Figure 10.13b,c, respectively. For the Raman line

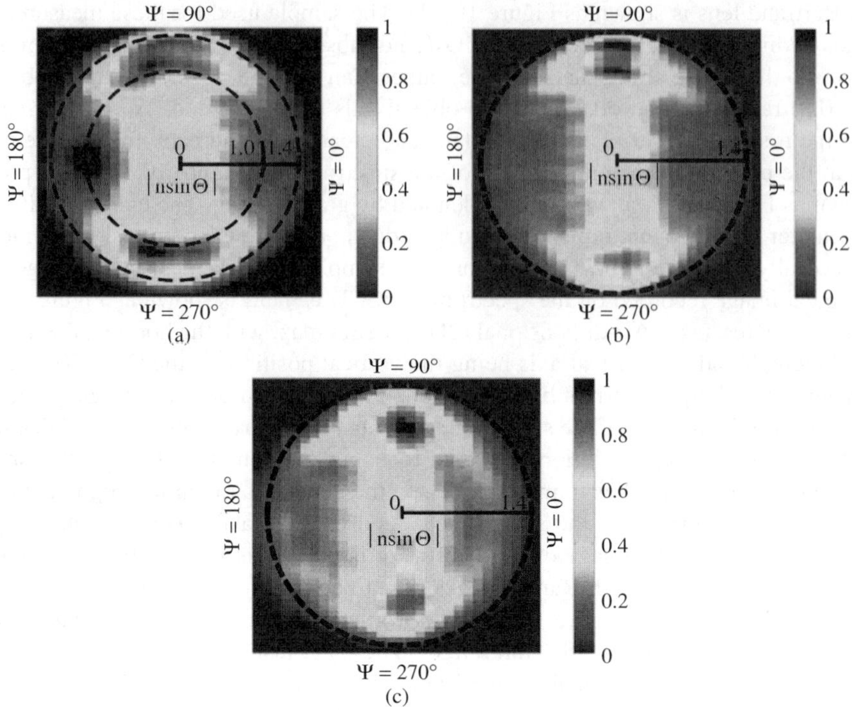

Figure 10.13 *(a) Measured emission pattern of thiophenol 1074 cm⁻¹ Raman line retrieved from EMS measurements. Map represents the normalized scattering intensity. Outer black dashed circle denotes NA of objective lens. Measured emission patterns of (b) 415 cm⁻¹ Raman line and (c) 1586 cm⁻¹ Raman line.* (See insert for color representation of the figure.)

at 415 cm^{-1}, the radiation peaks at ($\Theta = 139.1°$, $\Psi = 90.0°$) and ($\Theta = 139.1°$, $\Psi = 273°$), with a measured F/B ratio of 1.12. For the Raman line at 1586 cm^{-1}, the radiation peaks at $\Theta = 140.6°$, different Raman lines have similar radiation patterns. These Raman lines are enhanced by the same dipolar mode of the dimer antenna, whose radiation pattern varies little over the wavelengths of the different Raman lines [29].

YU antennas are often used at radio frequencies because of their directional properties. Nanoscale YU antennas [31, 32] have also been demonstrated at visible wavelengths, including for the realization of directional emission from quantum dots [33]. We demonstrate beamed Raman scattering using YU antennas. A nanoscale YU antenna, like its radio frequency counterpart, has three components: the feed, the reflector, and the directors. The feed enhances the excitation of, and emission from, nanoscale emitters. The reflector and directors provide antenna directivity. In previous nanoscale YU antenna designs, single nanorods were used for the feed element. Here, we instead use a dimer antenna, in order to use the gap mode [34] to further enhance Raman scattering [11]. We design the YU antenna to operate optimally at $\lambda = 857$ nm (i.e., the 1074 cm^{-1} Raman line).

An array of gold YU antennas is fabricated on an ITO-coated glass coverslip using the same techniques as used for the dimer antennas. The unit cell is 0.5 μm × 1.5 μm as before. Our YU antenna consists of a reflector, a dimer antenna feed, and three directors (Figure 10.14a).

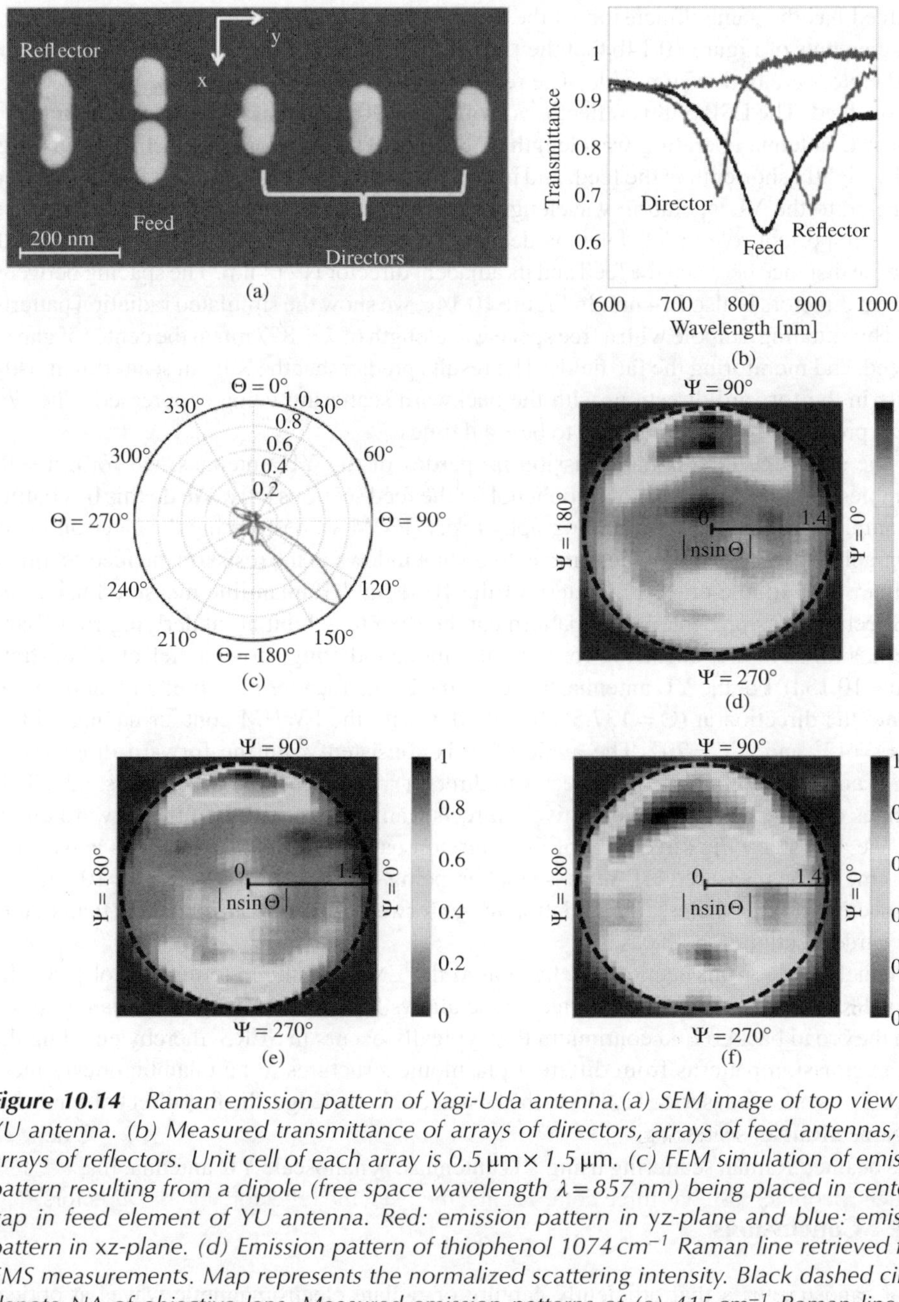

Figure 10.14 *Raman emission pattern of Yagi-Uda antenna.(a) SEM image of top view of a YU antenna. (b) Measured transmittance of arrays of directors, arrays of feed antennas, and arrays of reflectors. Unit cell of each array is 0.5 μm × 1.5 μm. (c) FEM simulation of emission pattern resulting from a dipole (free space wavelength $\lambda = 857$ nm) being placed in center of gap in feed element of YU antenna. Red: emission pattern in yz-plane and blue: emission pattern in xz-plane. (d) Emission pattern of thiophenol 1074 cm^{-1} Raman line retrieved from EMS measurements. Map represents the normalized scattering intensity. Black dashed circles denote NA of objective lens. Measured emission patterns of (e) 415 cm^{-1} Raman line and (f) 1586 cm^{-1} Raman line.* (See insert for color representation of the figure.)

The feed has the same dimension as the dimer of Figure 10.10a. It can be seen from the measurements of Figure 10.14b that the LSP of the feed peaks at $\lambda = 857$ nm. The reflector and directors are both 65 nm wide. The reflector has length 187 nm, that is, slightly longer than the feed. The LSP of this reflector is around $\lambda = 902$ nm, that is, red-shifted compared to the YU antenna operating wavelength of 857 nm. The directors are each 139 nm long, that is, slightly shorter than the feed, and have LSPs around $\lambda = 765$ nm, that is, blue-shifted compared to the YU operating wavelength. The distances between the nanoparticles also play an important role in YU antenna design [31]. The reflector is 187 nm from the feed, while the distance between the feed and its adjacent director is 214 nm. The spacing between adjacent directors is also 214 nm. In Figure 10.14c, we show the simulated radiation pattern, found by situating a dipole with a free space wavelength of $\lambda = 857$ nm in the center of gap of the feed, and monitoring the far-fields. The results predict that the Raman scattering mostly occurs in the forward direction, with the backward scattering being suppressed. The F/B ratio is predicted by the simulation to be ~4.4 times.

To the study the directional emission properties of the YU antennas, we form a self-assembled monolayer (SAM) of thiophenol on the feed sections only. We do this by coating the sample with electron-beam lithography resist (PMMA), performing the electron-beam lithography, then developing the sample, to open windows in the resist on the feed sections. In Figure 10.14d, the emission pattern of the 1074 cm^{-1} Raman line measured using the EMS technique is presented. The pattern can be seen to exhibit an underlying ring shape pattern, similar to that occurring for a dipolar mode radiating above a dielectric substrate (Figure 10.13a). For the YU antenna, however, the Raman scattering is predominantly emitted into the direction at ($\Theta = 137.5°$, $\Psi = 90.0°$), with the FWHM contour having widths of $\Delta\Theta = 42°$ and $\Delta\Psi = 70°$. The angle of Ψ is consistent with the forward direction of the YU antenna (pointing from reflector to director). The measured F/B ratio is 2.21. This indicates that the YU antenna effectively directs Raman scattering into the forward direction. We also study the directivity of the emission patterns of the Raman lines at 415 and 1586 cm^{-1}. The measured F/B ratio is smaller, being 1.34 and 1.15 for Figures 10.14e and f, respectively. Nonetheless, the radiation in the forward direction is still larger than that in backward direction in both cases.

In conclusion for this section, we have used the EMS technique to directly observe the beamed Raman phenomenon. This technique allows Raman scattering to be distinguished from the broad background continuum that typically occurs in SERS, thereby enabling the Raman emission patterns from different plasmonic structures to be unambiguously measured. The work we describe [12] is, to the best of our knowledge, the first direct observation of beamed Raman scattering.

10.5 Conclusions

Metal nanostructures that efficiently capture or radiate electromagnetic waves at optical frequencies offer a means to concentrate electromagnetic energy into deep subwavelength regions. Wessel noted that these structures can therefore be considered antennas [35]. More recently, investigations have focused on more efficient designs, termed "optical antennas," which employ small gaps or very sharp tips [36–38]. Optical antennas have been shown to present exciting opportunities for near-field scanning optical microscopy and compact subwavelength light sources [39]. In this chapter, we show two examples of optical antennas for SERS. In the first [4–6], we demonstrate SERS substrates with resonances at laser and Stokes wavelengths. We show that the SERS enhancement for the double-resonance structure is more than 2 orders of magnitude higher than that of a gold disk array on a glass substrate. In the second [12], we demonstrate a method by which SERS angular emission patterns can be directly observed. We apply this to measuring the SERS emission patterns from dimer antennas on a glass substrate. We then show that directional Raman scattering can be achieved by placing additional plasmonic structures around the feed dimer antennas to form YU antennas. In these and other optical antennas for SERS, nanoscale gaps are needed in order to achieve high EFs. With standard electron-beam lithography, however, the realization of gaps below 10 nm is difficult to achieve in a reproducible manner. As discussed, this motivates us to develop a procedure based on a sacrificial layer for the lithographic fabrication of optical antennas with gaps well below 10 nm [11]. We show that the measured SERS EFs increase by almost 2 orders of magnitude when the gap size is reduced from ≈ 20 to ≈ 3 nm.

References

1. International Technology Roadmap for Semiconductors (2011) www.itrs.net/reports.html (accessed 30 December 2012).
2. Lee, P.C. and Meisel, D. (1982) Adsorption and surface-enhanced Raman of dyes on silver and gold sols. *Journal of Physical Chemistry*, **86**, 3391–3395.
3. McFarland, A.D., Young, M.A., Dieringer, J.A. and Van Duyne, R.P. (2005) Wavelength-scanned surface-enhanced Raman excitation spectroscopy. *Journal of Physical Chemistry B*, **109**, 11279–11285.
4. Chu, Y., Banaee, M.G. and Crozier, K.B. (2010) Double-resonance plasmon substrates for surface-enhanced Raman scattering with enhancement at excitation and stokes frequencies. *ACS Nano*, **4**, 2804–2810.
5. Chu, Y., Wang, D., Zhu, W. and Crozier, K.B. (2011) Double resonance surface enhanced Raman scattering substrates: an intuitive coupled oscillator model. *Optics Express*, **19**, 14919–14928.
6. Chu, Y., Zhu, W., Wang, D. and Crozier, K.B. (2011) Beamed Raman: directional excitation and emission enhancement in a plasmonic crystal double resonance SERS substrate. *Optics Express*, **19**, 20054–20068.
7. Xu, H., Aizpurua, J., Käll, M. and Apell, P. (2000) Electromagnetic contributions to single-molecule sensitivity in surface-enhanced Raman scattering. *Physical Review E*, **62**, 4318–4324.
8. Aizpurua, J., Bryant, G.W., Richter, L.J. *et al.* (2005) Optical properties of coupled metallic nanorods for field-enhanced spectroscopy. *Physical Review B*, **71**, 235420.
9. Clark, A.W. and Cooper, J.M. (2010) Optical properties of multiple-split nanophotonic ring antennae. *Advanced Materials*, **22**, 4025–4029.
10. Fischbein, M.D. and Drndić, M. (2006) Nanogaps by direct lithography for high-resolution imaging and electronic characterization of nanostructures. *Applied Physics Letters*, **88**, 063116.
11. Zhu, W., Banaee, M.G., Wang, D. *et al.* (2011) Lithographically fabricated optical antennas with gaps well below 10 nm. *Small*, **7**, 1761–1766.

12. Zhu, W., Wang, D. and Crozier, K.B. (2012) Direct observation of beamed Raman scattering. *Nano Letters*, **12**, 6235–6243.
13. Chu, Y. and Crozier, K.B. (2009) Experimental study of the interaction between localized and propagating surface plasmons. *Optics Letters*, **34**, 244–246.
14. Kelly, K.L., Coronado, E., Zhao, L.L. and Schatz, G.C. (2003) The optical properties of metal nanoparticles: the influence of size, shape, and dielectric environment. *Journal of Physical Chemistry B*, **107**, 668–677.
15. Maier, S.A. (2007) *Plasmonics: Fundamentals and Applications*, Springer, Heidelberg.
16. Chen, K., Drachev, V.P., Borneman, J.D. *et al.* (2010) Drude relaxation rate in grained gold nanoantennas. *Nano Letters*, **10**, 916–922.
17. Le Ru, E.C. and Etchegoin, P.G. (2006) Rigorous justification of the $|E|^4$ enhancement factor in surface enhanced Raman Spectroscopy. *Chemical Physics Letters*, **423**, 63–66.
18. Le Ru, E.C., Blackie, E., Meyer, M. and Etchegoin, P.G. (2007) SERS enhancement factors: a comprehensive study. *Journal of Physical Chemistry C*, **111**, 13794–13803.
19. Talley, C.E., Jackson, J.B., Oubre, C. *et al.* (2005) Surface-enhanced Raman scattering from individual Au nanoparticles and nanoparticle dimer substrates. *Nano Letters*, **5**, 1569–1574.
20. Lombardi, J.R. and Birke, R.L. (2008) A unified approach to surface-enhanced Raman spectroscopy. *Journal of Physical Chemistry C*, **112**, 5605–5617.
21. Wang, D., Yang, T. and Crozier, K.B. (2011) Optical antennas integrated with concentric ring gratings: electric field enhancement and directional radiation. *Optics Express*, **19**, 2148–2157.
22. Ahmed, A. and Gordon, R. (2011) Directivity enhanced Raman spectroscopy using nanoantennas. *Nano Letters*, **11**, 1800–1803.
23. Wang, D., Zhu, W., Chu, Y. and Crozier, K.B. (2012) High directivity optical antenna substrates for surface enhanced Raman scattering. *Advanced Materials*, **24**, 4376–4380.
24. Ahmed, A. and Gordon, R. (2012) Single molecule directivity enhanced Raman scattering using nanoantennas. *Nano Letters*, **12**, 2625–2630.
25. Shegai, T., Brian, B., Miljković, V.D. and Käll, M. (2011) Angular distribution of surface-enhanced Raman scattering from individual Au nanoparticle aggregates. *ACS Nano*, **5**, 2036–2041.
26. Fromm, D.P., Sundaramurthy, A., Kinkhabwala, A. *et al.* (2006) Exploring the chemical enhancement for surface-enhanced Raman scattering with Au bowtie nanoantennas. *Journal of Chemical Physics*, **124**, 61101.
27. Beversluis, M., Bouhelier, A. and Novotny, L. (2003) Continuum generation from single gold nanostructures through near-field mediated intraband transitions. *Physical Review B*, **68**, 115433.
28. Taminiau, T.H., Karaveli, S., van Hulst, N.F. and Zia, R. (2012) Quantifying the magnetic nature of light emission. *Nature Communications*, **3**, 979.
29. Taminiau, T.H., Stefani, F.D., Segerink, F.B. and van Hulst, N.F. (2008) Optical antennas direct single-molecule emission. *Nature Photonics*, **2**, 234–237.

30. Lieb, M.A., Zavislan, J.M. and Novotny, L. (2004) Single-molecule orientations determined by direct emission pattern imaging. *Journal of the Optical Society of America B: Optical Physics*, **21**, 1210–1215.
31. Kosako, T., Kadoya, Y. and Hofmann, H.F. (2010) Directional control of light by nano-optical Yagi-Uda antenna. *Nature Photonics*, **4**, 312–315.
32. Dregely, D., Taubert, R., Dorfmüller, J. *et al.* (2011) 3D optical Yagi-Uda nanoantenna array. *Nature Communications*, **2**, 267.
33. Curto, A.G., Volpe, G., Taminiau, T.H. *et al.* (2010) Unidirectional emission of a quantum dot coupled to a nanoantenna. *Science*, **329**, 930–933.
34. Schnell, M., Garcia-Etxarri, A., Huber, A.J. *et al.* (2009) Controlling the near-field oscillations of loaded plasmonic nanoantennas. *Nature Photonics*, **3**, 287–291.
35. Wessel, J. (1985) Surface-enhanced optical microscopy. *Journal of the Optical Society of America B: Optical Physics*, **2**, 1538–1541.
36. Grober, R.D., Schoelkopf, R.J. and Prober, D.E. (1997) Optical antenna: towards a unity efficiency near-field optical probe. *Applied Physics Letters*, **70**, 1354–1356.
37. Crozier, K.B., Sundaramurthy, A., Kino, G.S. and Quate, C.F. (2003) Optical antennas: resonators for local field enhancement. *Journal of Applied Physics*, **94**, 4632–4642.
38. Novotny, L. (2007) Effective wavelength scaling for optical antennas. *Physical Review Letters*, **98**, 266802.
39. Mühlschlegel, P., Eisler, H.-J., Martin, O.J.F. *et al.* (2005) Resonant optical antennas. *Science*, **308**, 1607–1609.

11

Plasmon-Enhanced Scattering and Fluorescence Used for Ultrasensitive Detection in Langmuir–Blodgett Monolayers

Diogo Volpati[1], *Aisha Alsaleh*[2], *Carlos J. L. Constantino*[1], *and Ricardo Aroca*[2]
[1]*UNESP Universidade Estadual Paulista, Faculdade de Ciências e Tecnologia, Brazil*
[2]*Materials and Surface Science Group, University of Windsor, Canada*

11.1 Introduction

Phospholipids are a central architectural part of biological membranes. The artificial bilayer structures of lipids and phospholipids can be fabricated using the Langmuir–Blodgett (LB) technique [1], and the technique has been used extensively for analytical and structural studies of single monolayers, bilayers, and multilayers [2]. In particular, the analytical detection of lipids and phospholipids as bilayers is a subject of interest for fundamental studies in membrane biophysics, since phospholipid LB films are a good model system for biomembranes used in a variety of immobilization studies [3, 4]. Molecular vibrational fingerprints needed for the spectroscopic characterization are obtained by means of the complementary infrared and Raman techniques [5, 6]. However, high sensitivity and ultrasensitive detection can only be achieved using plasmon-enhanced spectroscopy, such as surface-enhanced Raman scattering (SERS) [7, 8]. Here, it is important to draw attention to two characteristics of the Raman spectra of lipid and phospholipids: they carry a very modest Raman scattering (RS) cross section and the fundamental vibrational frequencies

Frontiers of Surface-Enhanced Raman Scattering: Single Nanoparticles and Single Cells, First Edition.
Edited by Yukihiro Ozaki, Katrin Kneipp and Ricardo Aroca.

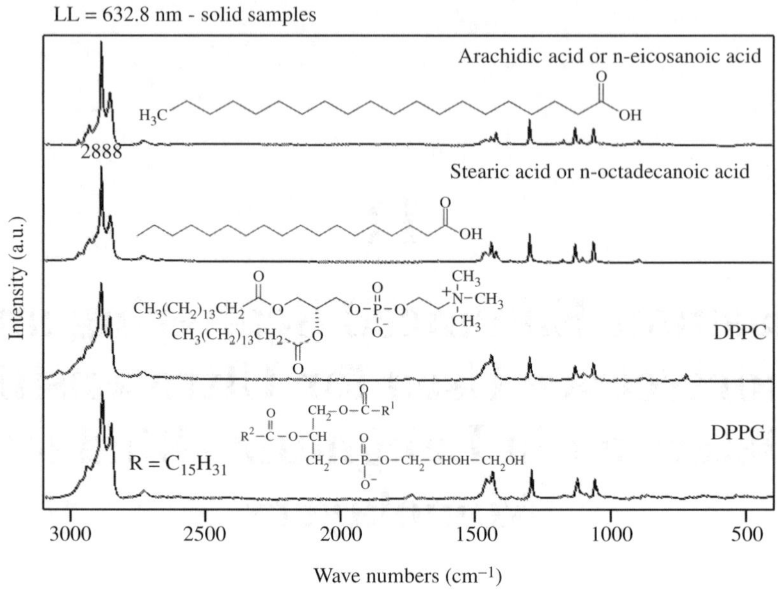

Figure 11.1 *Vibrational Raman spectra of lipids and phospholipids in solid samples*

are dominated by those of the lipid chain. The latter is illustrated in Figure 11.1, where the Raman spectra of solid samples of fatty acids and phospholipids are compared in the most common spectral window used in Raman. It can be seen that the spectra are dominated by the vibrational fundamentals of the poly(methylene): CH_2 symmetric-stretch (~2850 cm^{-1}) and antisymmetric stretch (~2888 cm^{-1}), and CH_2 scissoring, wagging, and twisting (in the 1460–1000 cm^{-1} region) [9]. The scattering cross section of these characteristic vibrational modes is on the order of 10^{-29} cm^2, and that presents a real challenge to the experimentalist to develop ultrasensitive detection for lipid and phospholipids.

Let us assume the necessary condition to detect the spectrum of a single molecule (SM) is a scattering cross section of 10^{-17} cm^2, then an enhancement factor (EF) [10] of 10^{12} would be needed, which is probably only available at the best of hot spots [11]. In practice, single molecule detection (SMD) of neat lipids or phospholipids has not been reported, and instead tagged phospholipids are used to take advantage of the large resonance Raman cross section (~10^{-25} cm^2) [12] of common chromophores allowing surface-enhanced resonance Raman scattering (SERRS), the most sensitive technique in plasmon-enhanced scattering. Correspondingly, the SMD work follows the lipid or phospholipid through the enhanced spectrum of other moieties with larger Raman cross sections. Notably, lipids are commonly used as the silent matrix during the detection of SERRS [13, 14]. However, many groups have undertaken the task of recording the SERS spectra of neat lipid and phospholipids [15, 16]. Recently, the SERS spectra for arachidic acid (AA) and dipalmitoylphosphatidylcholine (DPPC) layers on sapphire/Ag substrates were reported [17]. This work reports the (very weak) detection of average SERS of the target molecules. It should be pointed out that the SERS signal reported (SERS spectrum) is the average of millions of target molecules under the micro-Raman objective. By contrast, SMD completely removes the ensemble averaging, which brings about fluctuations and information other than that provided by average values. In Section 11.2 we describe the SMD detection of tagged-phospholipids using the LB technique and plasmon-enhanced scattering. In Section 11.3, we explore the plasmon-enhanced fluorescence using the LB technique and silver and gold shell-isolated nanoparticles (SHINs) [18, 19]. The main drawback of the ultrasensitive detection presented here is that SMD in SER(R)S or SHINEF (shell-isolated nanoparticle enhanced fluorescence) (plasmon-enhanced fluorescence with SHINs) necessitates attaching a chromophore to the phospholipid. Notwithstanding, interface properties such as diffusion, membrane interactions, protein folding, DNA sequencing, and so on are often investigated by SM spectroscopy tracking the behavior of individual molecules through an attached chromophore.

11.2 Surface-Enhanced Resonance Raman Scattering of Tagged Phospholipids

11.2.1 Experimental Details

N-(fluorescein-5-thiocarbamoyl)-1,2-dihexadecanoyl-*sn*-glycero-3-phosphoethanolamine, triethylammonium salt (fluorescein DHPE), and Oregon Green 488 1,2-dihexadecanoyl-sn-glycero-3-phosphoethanolamine (OG488-DHPE), were obtained from Invitrogen and used as received. All solutions were prepared using spectroscopic-grade chloroform as the solvent. AA (n-eicosanoic acid, the saturated fatty acid with a 20 carbon chain, $CH_3(CH_2)_{18}COOH$), chloroform, and ultrapure (99.9%) Ag shot (1–3 mm) were obtained from Aldrich, and used without further purification. A mixed monolayer of labeled phospholipid and AA was transferred onto a silver island film (SIF) and Corning glass microscope slides using an LB trough (Nima technology). After adding 200 μl of the mixture onto a water subphase at room temperature, the solvent is allowed to evaporate and the surface pressure is monitored by Wilhelmy sensor during film compression and deposition. Film transfer was carried out using Z deposition with a Nima Langmuir trough model 302 M electronically controlled dipping device at a constant surface pressure of 30 $mN\,m^{-1}$, corresponding to the condensed phase of Langmuir monolayers, and consistently resulting in transfer ratios near unity. The compressing was at 6 $mm\,min^{-1}$ until the pressure reached 30 $mN\,m^{-1}$ to transfer the monolayers onto the slides at a rate of 2 $mm\,min^{-1}$. A representative surface-pressure–area isotherm is shown in Figure 11.2. Langmuir monolayers containing OG488®-DHPE at various concentrations were prepared at the air/water interface of a Nima Langmuir trough and deposited onto the Ag nanostructured films. Spreading solutions were prepared as described in previous reports [20] from stock solutions of AA and tagged-phospholipids. Pure water subphase (18.2 MΩ cm) was maintained at a room temperature of 23 °C (surface pressure about 73.5 $mN\,m^{-1}$) during film fabrication. The ratio of AA/OG488®-DHPE molecules in the solutions that are spread onto the subphase is varied to achieve from 4×10^5 molecules down to 1 OG488®-DHPE molecule within 1 μm^2, an area that enclosed the scattering area being probed by the laser beam in our experimental micro-Raman set-up.

SIFs of 6 nm mass thickness were vacuum evaporated onto Corning glass microscope slides at a pressure of $\sim 10^{-6}$ Torr and a substrate temperature of 180 °C (maintained for

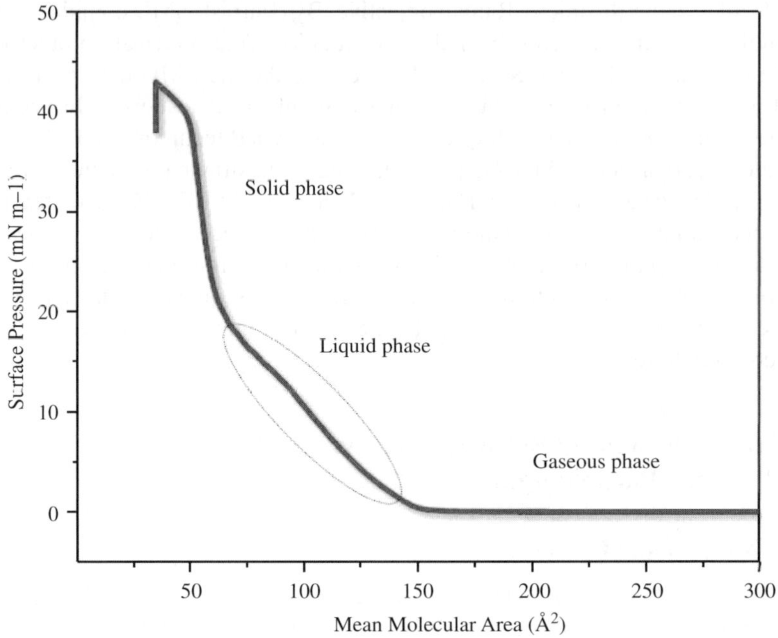

Figure 11.2 *Surface pressure versus mean molecular area (π–A) isotherms for a neat OG488®-DHPE dye-phospholipid monolayer at the air/water interface*

1 h after evaporation). UV–visible absorption spectra were recorded for all solutions and 6 nm Ag nanostructured films by employing a Cary 50 scan UV–visible spectrophotometer. Atomic force microscopy (AFM) images were recorded using a Digital Instruments Nano-Scope IV operating in tapping mode with an n^+ silicon tip. The plasmon and molecular absorption spectra and the AFM image of the SIF are shown in Figure 11.3. An OG488®-DHPE cast film on Ge was prepared from stock solution for transmission FTIR spectroscopy. Transmission infrared spectra were recorded at the mid-infrared region using a Bruker Equinox 55 spectrometer with 4 cm^{-1} of spectral resolution and 128 scans. All micro-RS experiments were conducted using a Renishaw InVia system with laser excitation at 514.5, 632.8, and 785 nm. All measurements were made in 180° backscattering geometry using a 50× microscope objective with a numerical aperture value of 0.75, providing a scattering area of about 1 μm^2. Single-point spectra were recorded at 4 cm^{-1} resolution and 10 s accumulation time, whereas 2D SERRS mapping results were collected through the rastering of a computer-controlled two-axis-encoded (XY) motorized stage, with a step of 3 μm, and low energy density to avoid the possibility of photodegradation.

11.2.2 Langmuir and LB films

Surface pressure–area per molecule (π–A) isotherm measurements of OG488®-DHPE surface monolayers were performed using the Langmuir technique [1, 21]. Figure 11.2 displays the π–A isotherm of an OG488®-DHPE Langmuir film. At molecular areas larger than 150 $Å^2$, fluid OG488®-DHPE monolayer is in the gaseous phase. A sudden increase in

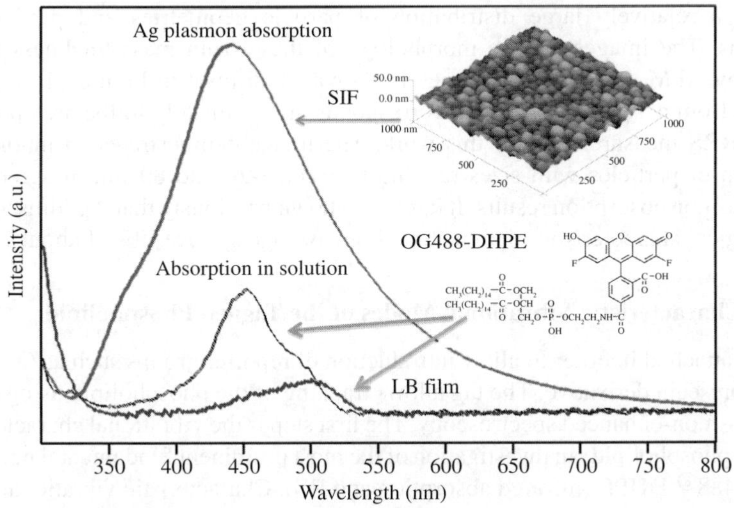

Figure 11.3 *Characterization of the Ag island film and OG488®-DHPE. Extinction spectrum and AFM of Ag island film, as well as the absorbance (1 cm optical path length) profiles for OG488®-DHPE in chloroform solution and monolayer LB. (*See insert for color representation of the figure*)*

surface pressure at a molecular area of about 150 Å^2 is observed, a feature for the beginning of a transition to a liquid phase. Further compression leads to a liquid-to-condensed region up to ~20 mN m^{-1}. A condensed phase of the neat monolayer giving about 75 Å^2/molecule, with collapse pressure around ~40 mN m^{-1} is observed with further monolayer compression. It is important to note the presence of a curvature of the isotherm at the liquid phase (belly), probably related to some reorientation of the phospholipid DHPE chains with the pressure increase. For all transference of films from aqueous (Langmuir films) to solid substrates (LB films), the pressure of 30 mN m^{-1} was used.

11.2.3 Electronic Absorption

The absorption spectra of 10^{-5} M solutions in chloroform and an LB monolayer of OG488®-DHPE, the surface plasmon absorption of the SIF, and a vacuum-evaporated silver film of 6 nm mass thickness, are shown in Figure 11.3. The spectrum of OG488®-DHPE solution reveals an absorption maximum around 451 nm followed by a shoulder around 492 nm, resulting from the $\pi-\pi^*$ electronic transitions of the OG488® chromophore (also shown in Figure 11.3). The LB film showed one broad absorption around 490 nm, also related to $\pi-\pi^*$ electronic transitions. It can be seen that the 514.5 nm laser excitation is in resonance with the molecular electronic absorption of LB OG488®-DHPE film and the localized surface plasmon resonance (LSPR) absorption of the SIF, effectively fulfilling the double-resonance condition of SERRS. The 632.8 nm excitation is still in resonance with the surface plasmon absorption of the Ag nanoparticles and out of resonance with the LB OG488®-DHPE film, resulting in an SERS condition. The SIF has a broad extinction spectrum, covering the entire visible range with a maximum at 445 nm, indicating a relatively large distribution of particle geometries and particle–particle interactions. The images for the morphology of these 6 nm mass thickness films were recorded by AFM, and a typical image is shown as an inset in Figure 11.3. This image was taken from a 1 μm^2 area, which corresponds approximately to the area probed in all of the SERRS measurements of this work. The image demonstrates an inhomogeneous distribution of particles with sizes ranging between ~20 and 80 nm, in agreement with surface plasmon absorption results. It has been shown previously that Ag films with similar morphological characteristics demonstrate high average surface EFs of about 10^4 [22].

11.2.4 Characteristic Vibrational Modes of the Tagged Phospholipid

The tag is attached in order to allow introduction of reporter groups such as Oregon Green 488, a fluorescein derivative. The tag allows tracking of the phospholipid using techniques such as plasmon-enhanced spectroscopy. The first step is the vibrational characterization of the tagged phospholipid, an investigation of the most prominent fundamental normal modes of the OG488®-DHPE (infrared absorption and RS). Characteristic vibrational modes are vibrational bands observed in the Raman (and the infrared) which carry the largest Raman cross section (or absorption cross section in the infrared). The infrared and resonance-Raman spectra (using 514.5 nm) for OG488®-DHPE powder are shown in Figure 11.4 for comparison. The FTIR spectrum in transmission mode shown in the figure was obtained for a cast film of OG488®-DHPE deposited onto a Ge plate. We have also carried out theoretical calculations using the DFT/6-31G basis in Gaussian to help with the vibrational assignment. The infrared and Raman spectra presented a relatively good agreement between the calculated and the experimental spectra (not shown). The 514.5 nm laser line is in resonance with the electronic absorption of the material (see Figure 11.3), providing

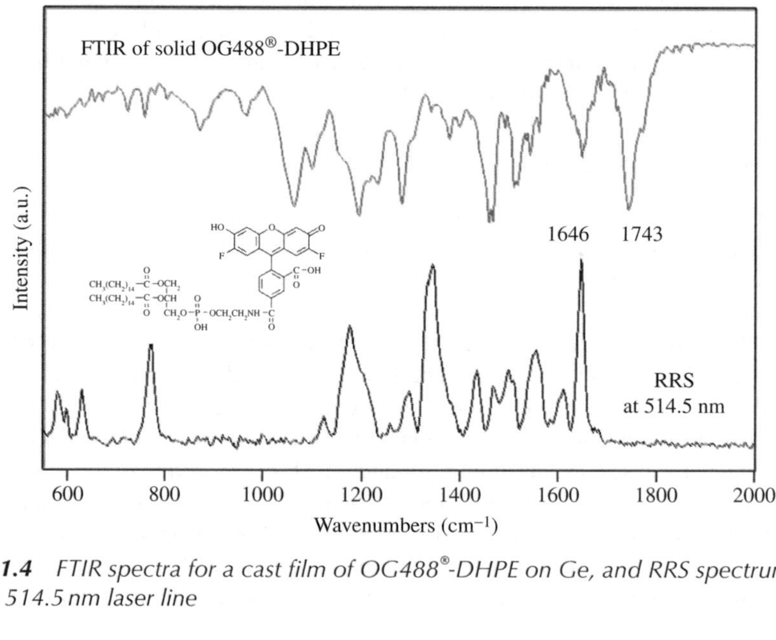

Figure 11.4 *FTIR spectra for a cast film of OG488®-DHPE on Ge, and RRS spectrum excited with the 514.5 nm laser line*

a resonance Raman scattering (RRS) spectrum on a strong fluorescence background. The fluorescence background has been subtracted from the spectrum presented in Figure 11.4. The difference between the FTIR and Raman spectra is striking. The infrared spectrum contains vibrational bands from the phospholipid and the chromophore moiety. For instance, the strongest bands in the IR are the C-H stretching modes that are not seen in the RRS spectrum. The C=O stretching band at 1743 cm^{-1} (ester carbonyl group phospholipid [23]) is not present in the RRS spectrum, instead we see the ring stretching of fluorescein at 1646 and 1335 cm^{-1} as the most intense bands in the RRS spectrum. None of the strong Raman vibrational bands seen in Figure 11.1 are prominent in the RRS spectrum of OG488®-DHPE. The 633 and 785 nm laser lines are out of resonance with the electronic absorption of the material, providing a conventional RS with peaks well defined for powder; but with different relative intensities compared to RRS. The RRS spectrum of the tagged phospholipid is completely dominated by the vibrations of the chromophore. In fact the LB-SERRS spectrum of OG488®-DHPE deposited onto Ag islands is almost the same as the reference RRS spectrum, as can be seen in Figure 11.5. We carried out parallel studies of another member of the same family of tagged phospholipid with a fluorescein chromophore (fluorescein DHPE). The LB-SERRS for the latter phospholipid for a single monolayer on SIF has also been included in Figure 11.5. The LB-SERRS is the average enhanced spectrum provided by the SIF substrate, and its reproducibility is reassuring for pursuing dilution studies for SMD. Average LB-SERS can also be seen with the 632.8 nm laser line, which is out of resonance with the LB monolayer electronic absorption, but is still in resonance with the SIF plasmon and SERS spectra presenting a well defined RS enhanced spectrum. It is important to note that LB-SERRS is an ultrasensitive detection technique. According

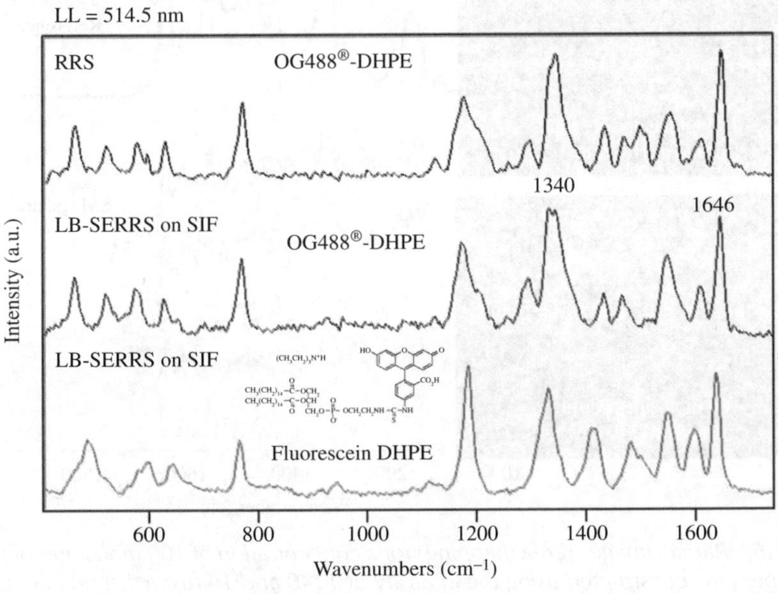

Figure 11.5 Resonance Raman scattering spectra obtained for powder OG488®-DHPE and an LB-SERRS of a monolayer of OG488®-DHPE and fluorescein-DHPE deposited on SIF using the 514.5 nm laser line

to the isotherm (Figure 11.2) the molecular surface area of the OG488®-DHPE (and also fluorescein-DHPE) is 75 Å^2, the total number of molecules in 1 μm^2 of surface area probed by the microscope objective (50×) is 1.3×10^6. Assuming that every single molecule contributes to the backscattering signal detected in LB-SERRS, the sensitivity is 2 attomole.

11.2.5 Single Molecule Detection

The LSPR spectrum of the 6 nm evaporated Ag nanoparticle films used in this work, and the 10^{-5} M solution absorption spectrum of OG488®-DHPE, are shown in Figure 11.3. The former shows a broad extinction covering the entire visible range, with a maximum centered at 445 nm, indicating a broad distribution of nanoparticle sizes. This maximum is centered quite near the laser excitation wavelength of 514.5 nm used in this work, explaining quite well the plasmon enhancement. However, below the zepto-mole and approaching the yocto-mole detection (SMD), it is likely that only highly localized local fields or *hot spots* [11, 24, 25] are responsible for the observation of single molecule SERS or SERRS spectra. 2D SERRS spectra were recorded from mixed LB monolayers deposited on SIF, having various concentrations of OG488®-DHPE, fabricated following the technique developed for LB-SERRS of dyes on a silent lipid matrix [22]. The mapping images shown in Figure 11.6 for a concentration of 100 molecules per μm^2 are constructed using the intensity of the strong 1340 and 1646 cm^{-1} fundamental modes of OG488®-DHPE mapped as a function

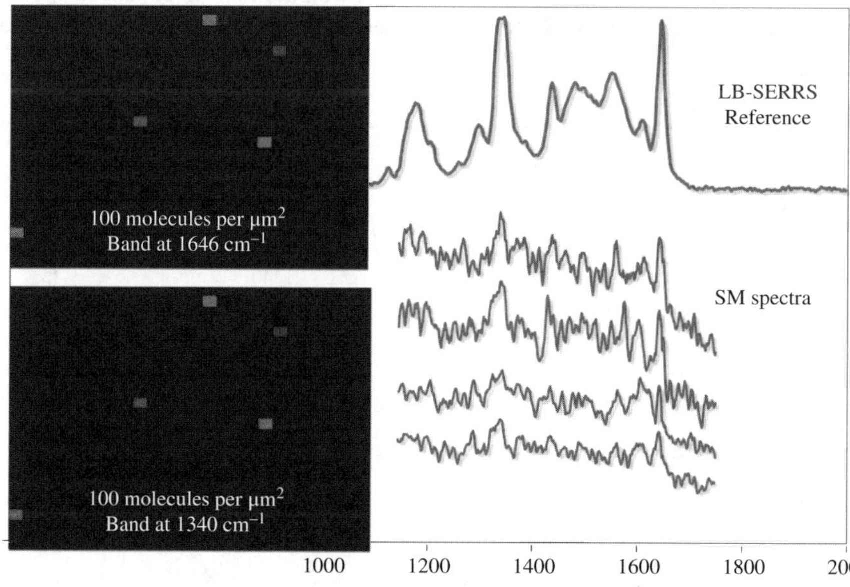

Figure 11.6 Raman images (area mapping) for a concentration of 100 molecules of tag-phospholipid per μm^2 constructed using the intensity of 1340 and 1646 cm^{-1} fundamental modes of OG488®-DHPE as a function of the x–y area. Each pixel in the figure corresponds to a single recorded spectrum from a total of 676 spectra. (See insert for color representation of the figure)

of $x-y$ area components, and each pixel corresponds to a single recorded spectrum from a total of 676 spectra. For attomole concentrations, strong SERRS signals are detected at all spots on the sample. As the concentration of OG488®-DHPE in these monolayers is decreased, signals in the brighter regions are reduced compared to signals of the bright spots on images of more highly concentrated films, and darker areas are observed. The maps showing single events, as seen in Figure 11.6, are obtained for concentrations from 1000, 100, 10 to 1 molecule per μm^2, indicating the occurrence of a rare event where a molecule is located at a hot spot in the SIF. We can interpret this result as an indication that below zepto-mole concentration levels, average surface-enhancement is insufficient for molecular detection, and all signals are detected from local electromagnetic hot spots.

11.3 Shell-Isolated Nanoparticle Enhanced Fluorescence (SHINEF)

In SERRS, the process benefits from the close proximity to the enhancing nanostructure: maximum enhancement for the first layer. The case in fluorescence is quite different. In the immediate vicinity of the nanoparticle, the molecule, due to the intensified electric field, will see an enhancement in its absorption and a reduction in its radiative life time and, most importantly, re-emission of the fluorescence to the far field through nanoparticle scattering; however, at the same time the metal surface provides new pathways of energy de-excitation, resulting in fluorescence quenching. Therefore, the average enhancement in surface-enhanced fluorescence (SEF – which is sometimes called metal-enhanced fluorescence (MEF) [26].) follows a unique molecule–nanostructure distance dependence with a maximum enhancement at a few nanometers from the surface. After Wokaun demonstrated this distance dependence experimentally in 1983 [27], many other groups have confirmed the same trend. In addition to quenching effects, in SEF only the incident field is enhanced ($|E|^2$ dependence) [28], whereas SERS enhancement gains amplification from both the incident and scattered fields ($|E|^4$ dependence) [29]. Correspondingly, average fluorescence enhancements are significantly lower than those of SERS. The enhancement at a hot spot may require an additional discussion; but here we are dealing with the average SEF. Average SEF (or MEF) enhancement is commonly of two digits. However, fluorescence cross sections could be in the 10^{-19} cm^2 range, so a small enhancement is significant for analytical applications. In addition, plasmon enhancement can be applied selectively in different regions of the electromagnetic spectrum, broadening the range of potential applications. Here we used shell-isolated nanoparticles [19], silver and gold enhancers coated with SiO_2 to achieve SHINEF [18]. In previous reports SHINEF experiments were carried out in solution or by direct casting of SHINs on the surface containing the analyte. Here, we examine the silver and gold SHINEF results obtained in solution, by casting, and the formation of a two-dimensional SHINs surface to be used, in SHINEF experiments.

11.3.1 Tuning the Enhancement Factor in SHINEF

Synthesis of silver SHINs: A citrate silver colloid dispersion was prepared via reduction of silver in aqueous $AgNO_3$ solution with trisodium citrate [30]. All glassware used was thoroughly cleaned with aqua regia (three parts HCl and one part HNO_3). The water used in the experiments was doubly-distilled, purified using a Milli-Q system from Millipore. 36 mg of $AgNO_3$ were dissolved in 200 ml milli-Q water in a round bottom flask equipped with a condenser and heated in a sand bath to 220 °C, while vigorously stirred. When the solution started to boil, 4 ml of 1% (w/v) trisodium citrate ($Na_3C_6H_5O_7$) was added to the solution, boiling was continued for another hour. The solution was lifted from the heat and an absorption measurement was taken. The plasmon peak was very broad with a maximum at about 413 nm. At this point, the pH of the colloids was adjusted to 10.5 using NaOH. While the solution was still warm, 12 ml of 1 mM APTMS (3-amino propyl trimethoxysilane) was added and stirring was continued for another 15 min. The solution was heated again to 95 °C in a sand bath, 36 ml of active 0.54% (w/v) sodium silicate solution was added and the coating process was carried out for 3 h. All reagents, and also octadecyl rhodamine B chloride R18 ($C_{46}H_{67}ClN_2O_3$), AA ($C_{20}H_{40}O_2$), and dichloromethane (CH_2C_{l2}) were obtained from Aldrich and used without any further purification or processing. Gold SHINs were prepared according to the protocol given in Ref. [18]. In order to illustrate the role of SHINs aggregation on the EF, the dispersed silver SHINs in solution were transferred to silanized glass for SEF measurements. For surface silanization, Corning 2048 glass microscope slides 75 cm × 25 cm were cleaned with an aqueous solution of 50% H_2SO_4 for 3–4 h, then washed with Milli-Q water several times by sonication, dried at ambient conditions while protecting from dust, and finally dipped into a 2% APTMS solution in acetone for about 10 min. After silanization, the glass slides were washed with Milli-Q water to remove excess silane on the surface and plunged into a citrate silver colloidal dispersion for deposition of SHINs nanoparticles by self-assembly. The adhesion of SHINs on the silanized surface is time dependent. The self-assembly of SHINs on silanized glass was tested for different time periods of deposition from 3 to 16 h, and the surface coverage was monitored by AFM imaging. The SHINEF spectra of the R18 monolayer on SHIN substrate after 6 and 16 h are shown in Figure 11.7. The intrinsic quantum yield of R18

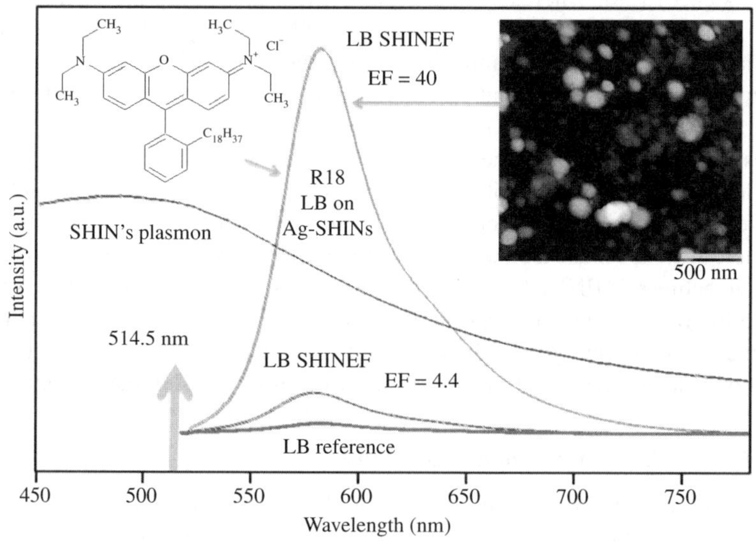

Figure 11.7 SHINEF spectra of the R18 monolayer on Ag-SHIN substrate after 6 and 16 h. (See insert for color representation of the figure)

is about 0.7 [31]. The maximum average EF attained was 40 and the AFM image for the 16 h deposition shown in the inset indicates the formation of aggregates. Good quality EFs are also obtained by casting, the advantage of the SHIN surface is the homogeneity of the enhancement in different spatial locations. Similar results are achieved with gold-SHINs [18]. However, we found that the coating of Au-SHINs is more resilient than that of the Ag-SHINs, thereby extending the lifetime for practical applications.

11.3.2 SHINEF of Fluorescein-DHPE

The SHINEF results for tagged phospholipid: fluorescein-DHPE with gold-SHIN are presented. The fabrication of Langmuir and LB monolayers is no different from that of OG488®-DHPE, and we limit the discussion here to the SHINEF results. Fluorescein is a fluorophore with a high quantum yield; QY ~0.95 (at 496 nm) [32]. The extinction spectra of fluorescein and fluorescein-DHPE, together with the reference emission spectra of fluorescein-DHPE in solution and from an LB film are shown in Figure 11.8. The fluorescence spectra were collected by exciting with the 514.5 nm laser line. SHINEF of fluorescein-DHPE is obtained with both Ag and Au-SHINs. The Au-SHINs, with a plasmon maximum at 524 nm (inset in Figure 11.9), are a good match for the fluorescein-DHPE emission, as can be seen in Figure 11.8.

SEF has rarely been measured in solution (with the exception of work in our group [29]), and measurements are carried out by depositing both fluorophore molecules and the enhancing nanostructure on solid substrates. However, with SHINs the work in solution is viable, and to a solution of fluorescein-DHPE in chloroform were added microliters of Au-SHIN solution while monitoring the fluorescence of the mixed solution. A maximum enhancement of 2 is achieved, a very modest EF compared with solid substrate experiments.

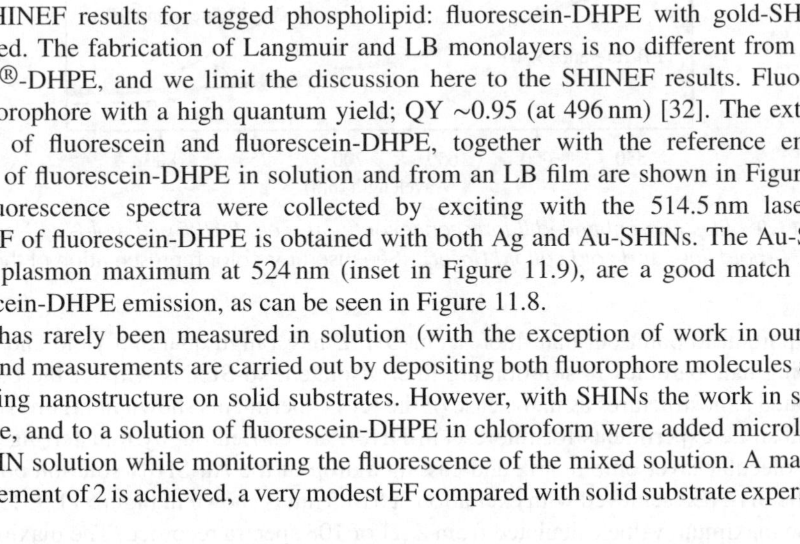

Figure 11.8 *Extinction and emission spectra of fluorescein-DHPE. The extinction spectrum of fluorescein is included for comparison.* (See insert for color representation of the figure)

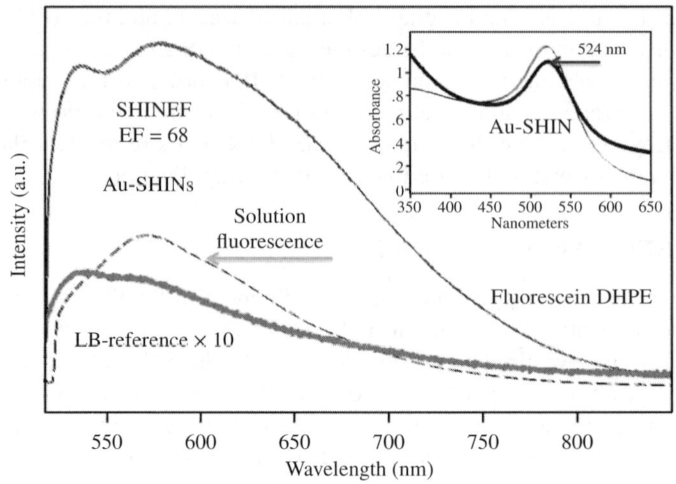

Figure 11.9 *Plasmon-enhanced fluorescence for fluorescein-DHPE with gold-SHINs. LSPR of Au-SHIN (bold line) and gold colloid (inset).* (See insert for color representation of the figure)

It is important to point out that fluorescein-DHPE has a high quantum yield and that the enhancing nanoparticles in solution are likely isolated Au-SHINs without the benefit of aggregated nanostructures as in the case of the R18 experiments shown in Figure 11.7. The solid substrate experiments for fluorescein-DHPE are carried out by transferring a single monomolecular layer onto quartz and casting a drop of the Au-SHIN solution on the LB film. The SHINEF achieved with excitation at 514.5 nm is shown in Figure 11.9. The EF of 68 is the maximum value calculated from a set of 108 spectra recorded. The maximum EF is found in the spatial region where there are large aggregates of SHIN particles. The LSPR of Au-SHINs has a maximum at 524 nm and full width at half maximum of 61 nm which overlaps with the main emission of the fluorophore. The EF allows ultrasensitive detection, even when we have 1–10 molecules in the field of view of the objective. However, the fluorescence spectra are broad and cannot provide the specific narrow vibrational fingerprints of SERS. Therefore, the technique will be challenged by mixed dyes with close emission. However, when the molecular emissions are well separated, the technique allows tuning of the SHINEF by matching the LSPR to a particular molecular emission.

11.4 Conclusions

Using tagged phospholipids as the target molecules, the plasmon-enhanced scattering and fluorescence are seen in LB films. Plasmon-enhanced fluorescence (SHINEF) of a high quantum yield chromophore is reported in solution and LB films. The molecular fingerprints with narrow FWHM, observed in SERRS down to SMD, provide a way to distinguish the target molecular moiety even in the presence of other absorbing chromophores in the same spectral region. The highly sensitive but broad enhanced fluorescence can be tuned for selective enhancement of fluorescent signals that are well separated in the electromagnetic spectrum.

Acknowledgments

Financial assistance from the Natural Science and Engineering Research Council of Canada (NSERC) is gratefully acknowledged.

DV acknowledges financial support from Brazilian agency CAPES (grant 5739-11-0).

References

1. Roberts, G. (1990) *Langmuir-Blodgett Films*, Plenum Press, New York.
2. Mowald, H. (1993) Surfactant layers at water interfaces. *Reports on Progress in Physics*, **56**, 653–685.
3. Caseli, L., Crespilho, F.N., Nobre, T.M. *et al.* (2008) Using phospholipid Langmuir and Langmuir-Blodgett films as matrix for urease immobilization. *Journal of Colloid and Interface Science*, **319**, 100–108.
4. Nobre, T.M., De Sousa, E., Silva, H. *et al.* (2009) Molecular view of the interaction between i-Carrageenan and a phospholipid film and its role in enzyme immobilization. *Journal of Physical Chemistry B*, **113**, 7491–7497.
5. Herzberg, G. (1950) *Molecular Spectra and Molecular Structure. I Spectra of Diatomic Molecules*, D.Van Nostrand Co., Inc, Princeton, NJ.
6. Wilson, E.B. Jr.,, Decius, J.C. and Cross, P.C. (1955) *Molecular Vibrations; The Theory of Infrared and Raman Vibrational Spectra*, McGraw-Hill, New York.
7. Aroca, R. (2006) *Surface-Enhanced Vibrational Spectroscopy*, John Wiley & Sons, Ltd, Chichester.
8. Le Ru, E.C. and Etchegoin, P.G. (2009) *Principles of Surface Enhanced Raman Spectroscopy (and Related Plasmonic Effects)*, Elsevier, Amsterdam.
9. Shimanouchi, T. (1972) *Tables of Molecular Vibrational Frequencies. Consolidated*, National Standard Reference Data Series, Vol. I, National Bureau of Standards (US).
10. Le Ru, E.C., Blackie, E., Meyer, M. and Etchegoin, P.G. (2007) Surface enhanced Raman scattering enhancement factors: a comprehensive study. *Journal of Physical Chemistry C*, **111**, 13794–13803.
11. Cang, H., Labno, A., Lu, C.G. *et al.* (2011) Probing the electromagnetic field of a 15-nanometre hotspot by single molecule imaging. *Nature*, **469**, 385–389.
12. Le Ru, E.C., Schroeter, L.C. and Etchegoin, P.G. (2012) Direct measurement of resonance Raman spectra and cross sections by a polarization difference technique. *Analytical Chemistry*, **84**, 5074–5079.
13. Moula, G. and Aroca, R.F. (2011) Plasmon-enhanced resonance Raman scattering and fluorescence in Langmuir-Blodgett monolayers. *Analytical Chemistry*, **83**, 284–288.
14. Pieczonka, N.P.W., Moula, G. and Aroca, R.F. (2009) SERRS for single-molecule detection of dye-labeled phospholipids in Langmuir-Blodgett monolayers. *Langmuir*, **25**, 11261–11264.
15. Dai, S., Zhang, X., Du, Z. *et al.* (2005) Structural properties and Raman spectroscopy of lipid Langmuir monolayers at the air-water interface. *Colloids and Surfaces, B: Biointerfaces*, **42**, 21–28.
16. Saint-Pierre Chazalet, M., Masson, M., Bousquet, C. *et al.* (1994) Surface-enhanced Raman scattering studies of lipid planar bilayers in water. *Thin Solid Films*, **244**, 852–856.

17. Sweetenham, C.S., Larraona-Puy, M. and Notingher, I. (2011) Simultaneous Surface-Enhanced Raman Spectroscopy (SERS) and Atomic Force Microscopy (AFM) for Label-free physicochemical analysis of lipid bilayers. *Applied Spectroscopy*, **65**, 1387–1392.
18. Guerrero, A.R. and Aroca, R.F. (2011) Surface-enhanced fluorescence with shell-isolated nanoparticles (SHINEF). *Angewandte Chemie International Edition*, **50**, 665–668.
19. Li, J.F., Huang, Y.F., Ding, Y. *et al.* (2010) Shell-isolated nanoparticle-enhanced Raman spectroscopy. *Nature*, **464**, 392–395.
20. Constantino, C.J.L., Lemma, T., Antunes, P.A. and Aroca, R. (2001) Single-molecule detection using surface-enhanced resonance Raman scattering and Langmuir-Blodgett monolayers. *Analytical Chemistry*, **73**, 3674–3678.
21. Petty, M.C. (1996) *Langmuir-Blodgett Films: An Introduction*, Cambridge University Press, Cambridge.
22. Goulet, P.J.G., Pieczonka, N.P.W. and Aroca, R.F. (2005) Mapping single-molecule SERRS from Langmuir-Blodgett monolayers on nanostructured silver island films. *Journal of Raman Spectroscopy*, **36**, 574–580.
23. Cieslik-Boczula, K., Szwed, J., Jaszczyszyn, A. *et al.* (2009) Interactions of dihydrochloride fluphenazine with DPPC liposomes: ATR-IR and P-31 NMR studies. *Journal of Physical Chemistry B*, **113**, 15495–15502.
24. Xu, H., Bjerneld, E.J., Kall, M. and Borjesson, L. (1999) Spectroscopy of single hemoglobin molecules by surface enhanced Raman scattering. *Physical Review Letters*, **83**, 4357–4360.
25. Michaels, A.M., Jiang, J. and Brus, L. (2000) Ag nanocrystal junctions as the site for surface-enhanced Raman scattering of single Rhodamine 6G molecules. *Journal of Physical Chemistry B*, **104**, 11965–11971.
26. Geddes, C.D. and Lakowicz, J.R. (2002) Metal-enhanced fluorescence. *Journal of Fluorescence*, **12**, 121–129.
27. Wokaun, A., Lutz, H.P., King, A.P. *et al.* (1983) Energy transfer in surface enhanced luminescence. *Journal of Chemical Physics*, **79**, 509–14.
28. Moskovits, M. (1985) Surface-enhanced spectroscopy. *Reviews of Modern Physics*, **57**, 783–826.
29. Guerrero, A.R., Zhang, Y. and Aroca, R.F. (2012) Experimental confirmation of local field enhancement determining far-field measurements with shell-isolated silver nanoparticles. *Small*, **8**, 2964–2967.
30. Lee, M. and Meisel, D. (1982) Adsorption and surface-enhanced Raman of dyes on silver and gold sols. *Journal of Physical Chemistry*, **86**, 3391–3395.
31. Demas, J.N. and Crosby, G.A. (1971) Measurement of photoluminescence quantum yields –review. *Journal of Physical Chemistry*, **75**, 991–1024.
32. Lakowicz, J.R. (2006) *Principles of Fluorescence Spectroscopy*, Springer Science, Singapore.

12

SERS Analysis of Bacteria, Human Blood, and Cancer Cells: a Metabolomic and Diagnostic Tool

W. Ranjith Premasiri, Paul Lemler, Ying Chen, Yoseph Gebregziabher, and Lawrence D. Ziegler
Department of Chemistry and the Photonics Center, Boston University, USA

12.1 Introduction

The ability to observe surface enhanced Raman scattering (SERS) at near single cell levels offers great opportunities for biomedical applications, most broadly defined, and for learning about biochemical processes of living organisms at the molecular level. The work summarized here illustrates the ongoing efforts in our laboratory to exploit SERS for these two purposes. In the context of a book focused on the frontiers of SERS we forgo any summary of this well-established phenomenon [1–5] but highlight the attributes of this effect that contribute to its value as a successful probe of cellular activity.

SERS-based methodologies can exploit the advantages of speed, sensitivity, multiplexing capabilities, specificity, ease of use, cost, portability, and lack of need for extrinsic labeling that this optical approach offers. The multiplexing, ease-of-use, relatively low cost, no labeling requirement and, to some considerable degree, portability are inherent to the use of spontaneous Raman scattering as an analytical probe for chemical or biological applications. The vibrational linewidths as compared to electronic spectral breadths, commercial availability of low cost laser diodes and development of compact optical designs contribute to these factors for detection and identification of analytes in relatively high concentrations. However, as illustrated here, the enhanced specificity and sensitivity, with

Frontiers of Surface-Enhanced Raman Scattering: Single Nanoparticles and Single Cells, First Edition.
Edited by Yukihiro Ozaki, Katrin Kneipp and Ricardo Aroca.

resulting improvement in speed and portability capabilities, result from the effective Raman cross section enhancement that is the hallmark characteristic of SERS. However, for all the biomedical results summarized here, an added advantage of SERS is that it gives access to different molecular species or vibrational transitions than seen in normal Raman spectra. In part this accounts for the enhanced specificity found in SERS as compared to normal Raman and thus provides a more robust diagnostic assay, as demonstrated by the SERS studies of vegetative bacterial cells discussed below. Furthermore, new biological processes are uncovered by SERS measurements of cells and biological fluids containing cells than are evident via normal Raman scattering or infrared observations and thus SERS uniquely allows vibrational probing of selected cellular activity in complex biological mixtures. The aging of blood, and the metabolic profiling of bacterial and cancer cells illustrate this attribute. As shown here, not only are these qualities of SERS found to be valuable for developing diagnostic protocols but they also provide a new tool for learning about the biological activity taking place at the outer layers and in the extracellular metabolome of single cells in real time.

More specifically, the state of our current progress in developing an SERS-based platform for bacterial diagnostics at near single cell levels is described below. These efforts have been directed toward the identification and detection of whole bacterial cells enriched from human body fluids, for example, blood and urine, for the development of a rapid SERS-based diagnostic of bacteremia and urinary tract infections (UTIs), respectively, in point-of-care settings. When coupled with procedures for the enrichment of bacterial cells from human body fluids, powerful multivariate analysis methods and readily expandable reference libraries, a SERS-based diagnostic offers several advantages as compared to traditional cell growth or genotypic-based approaches such as polymerase chain reaction (PCR) methods, as discussed further below. In addition, the current status of efforts to employ SERS for studies of biochemical processes related to whole blood components is given here. As recently shown, SERS of whole blood, and red blood cells (RBCs) in particular, yield robust vibrational signatures, and the potential for SERS-based blood disease diagnoses, such as early malaria detection, will be described. Finally, some preliminary results on the exploitation of SERS metabolic profiling for cancer cell identification is summarized, broadening the emerging picture that SERS of single cells is a new tool for the study of metabolomics [6–8].

Assays based on single cell levels SERS capabilities can also have significant impact outside the area of biomedical diagnostics, the focus of this review. For example, the need for rapid bacterial identification methods for food and water safety applications has been illustrated by the several high profile incidents reported nationwide over the past few years, and applications in forensics, in addition to biowarfare agent detection, can benefit from the successful development of this optical approach for single cell level detection and identification.

12.2 SERS of Bacterial Cells: Methodology and Diagnostics

Traditional phenotypic techniques for bacterial cell diagnostics based on cell morphology, staining character, motility, spore-forming ability, growth condition dependence, or chemical reactivity are cell culture growth dependent. Consequently, this methodology is slow, requiring 24–72 h typically for species and strain identification. Furthermore, distinguishing closely related strains may be difficult and it is not a point-of-care procedure. The current "best" methods of bacterial identification are genotypic-based approaches, most notably "real-time" PCR [9, 10]. If no culturing is required, 2–4 h is the best typical timeframe for identification. However, there are some potential limitations for these approaches. Contamination or amplification inhibitory factors can be a problem for PCR because it relies on sample RNA or DNA amplification [11]. SERS, on the other hand, is a signal amplification technique. Other potential limitations of PCR-based methods are infectious mixture resolution, the need for the required primer set for the pathogen of interest, cost, questions of trained personnel use, and point-of-care capability. The increasing proliferation of drug resistant strains places an increasing premium on being able to identify these microbes with strain specificity in a timeframe useful for narrow spectrum antibiotic drug prescription [12].

Given the potential advantages of speed, sensitivity, specificity, multiplexing, and ease of use that SERS has to offer, it is not surprising that there has been considerable interest by many researchers in exploiting these attributes for the detection and identification of whole vegetative bacterial cells [13–36]. Our work in this area was initially inspired by the development of new technological approaches for identification of potential biowarfare agents. However, it soon became clear that this technology could have a significant impact in the area of clinical diagnostics where the rapid identification of infectious agents from human body fluids could have a significant positive impact on disease outcomes.

The envisioned SERS-based platform for detection and identification of bacteria for diagnostic purposes at the single cell level requires four key components:

1. *Sample preparation device/procedure*. Although often overlooked, sample preparation is a key ingredient for the effective use of novel pathogen detection methods. The ability to obtain SERS spectra capable of providing cellular level identification cannot be fully exploited without developing procedures for collecting or enriching these cells from biological samples. For example, we are developing a two stage approach for bacterial enrichment from human blood. Detecting bacteria in blood is extremely challenging because RBC concentrations are $\sim 10^9$/ml in human blood and other particles, such as white blood cells and platelets are present at $\sim 10^6$/ml levels. In contrast medically relevant concentrations corresponding to bacteremic presentations, that is, blood infections, are in the $10^2 - 10^3$/ml range, and the surface area of a RBC is about an order of magnitude larger than that of a typical bacterial cell. However, via selective lysis, centrifugation, and micro-evaporation [37] sufficient enrichments of $\sim 10^5$ can be achieved, allowing acquisition of SERS spectra in 100 nl volumes starting with 10 ml of infected blood. Sensitivities down to 10^3/ml have already been achieved in a bench top procedure and an automated device is currently being developed in our lab. The sample preparation procedures required for UTI detection are simpler than those needed for bacteremia diagnostics because the concentration of other cells in urine, such as white blood cells or epithelial cells, is normally lower than in human blood and the clinical levels of bacterial infection concentrations are much higher in urine ($\geq 10^5$/ml) than blood, as described further below. Similarly, early detection of diseases such as cancer or malaria can only be accomplished in conjunction with the development of techniques that can bring the few suspected cells to the focus of a Raman microscope for identification.

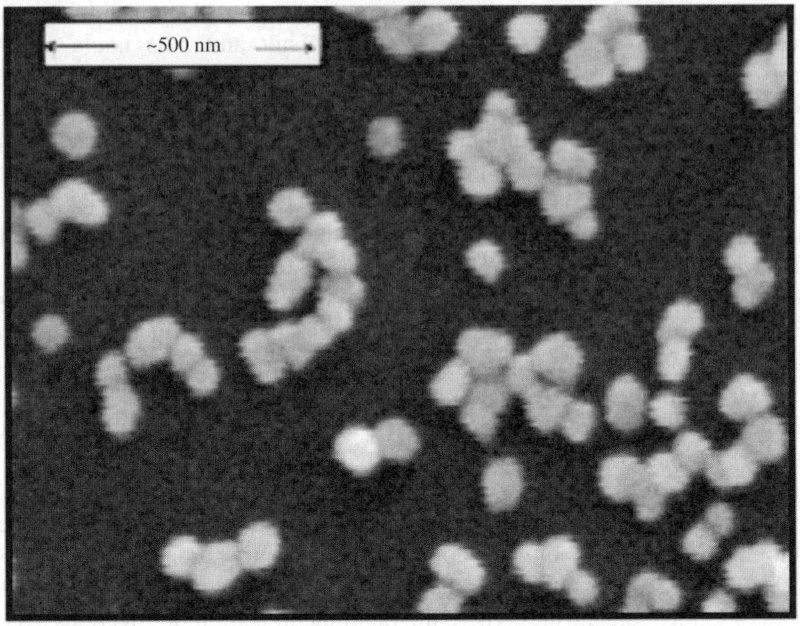

Figure 12.1 *A scanning electron micrograph image of the Au nanoparticle-covered SERS substrate produced by an Au ion-doped sol-gel process. Clusters of 2–15 ~80 nm Au particles are evident on the surface of the SiO_2 SERS chip. Reprinted with permission from [18]. Copyright © 2005, American Chemical Society*

2. *SERS substrate.* Most of our SERS data has been acquired from bacterial cells placed on a substrate produced by an Au ion-doped sol-gel procedure. The details for the production of these substrates have been given previously [18]. They are found to yield a strongly enhanced, reproducible SERS signal for vegetative bacterial cells when excited by 785 nm excitation. As shown in Figure 12.1, these substrates have clusters of 2–15 80-nm Au nanoparticles physi-adsorbed to the outer surface of an SiO_2 matrix. An ensemble-averaged per molecule SERS enhancement factor of $\sim 5 \times 10^7$ has been estimated for this substrate, as measured by the Raman scattering of glycine [18]. The production cost for these Au nanoparticle-covered SERS substrates is very low and, by insulating the chips in an inert Ar environment, storage times of up to five months have been achieved. These attributes are central to point-of-care use and the portability of this methodology. Ag can be substituted in the recipe for the production of these *in situ* grown substrates, however, the Au nanoparticle-covered chips have a longer lifetime and exhibited very comparable SERS intensities with 785 nm excitation [24]. Further improvements in SERS chip performance will only enhance the value of this technique. We have collaborated on the design and production of several engineered substrates [38–40], however, the sol-gel Au nanoparticle "chips" have proven to provide the best signal/noise for bacterial identification purposes of those we have tested thus far.
3. *A portable SERS signal acquisition instrument.* Although several hand-held Raman devices with nearly point and click capabilities are commercially available, true

microscopic capabilities are necessary for this SERS application, as illustrated further below. Although such less expensive options for portable Raman signal acquisition can currently be purchased, the ability to reliably and routinely bring the focus of a 50× objective to a SERS-active surface, and provide high S/N for reproducible and robust specificity via SERS spectral signatures requires more demanding instrumentation specifications than available via low resolution hand-held devices. Thus, we designed and built several prototype high-performance (cooled CCD array, ~5 cm^{-1} spectral resolution, piezo-driven sample positioning stages, submicron imaging resolution) portable Raman microscopes for this purpose. This unique instrument allows positioning of the 785 nm Raman excitation on a specific cell or sample region on the SERS substrate for Raman analysis and excellent signal quality (S/N and resolution). Imaging also provides additional morphology information that can aid identification procedures. Our most recent version of the Raman microscope is built in collaboration with BioTools, Inc. (Jupiter FL).

4. *Bacterial identification software*. As described further below we have developed a novel multivariate data analysis technique that provides accurate distinction between the SERS spectra of closely related bacterial species and strains when combined with a previously developed reference library [41].

12.3 Characteristics of SERS Spectra of Bacteria

SERS spectra of eight different bacterial species corresponding to six different genera acquired with ~2 mW of incident 785 nm diode laser power in ~10 s (50× objective RM2000 Renishaw Raman microscope) are shown in Figure 12.2. These bacteria have

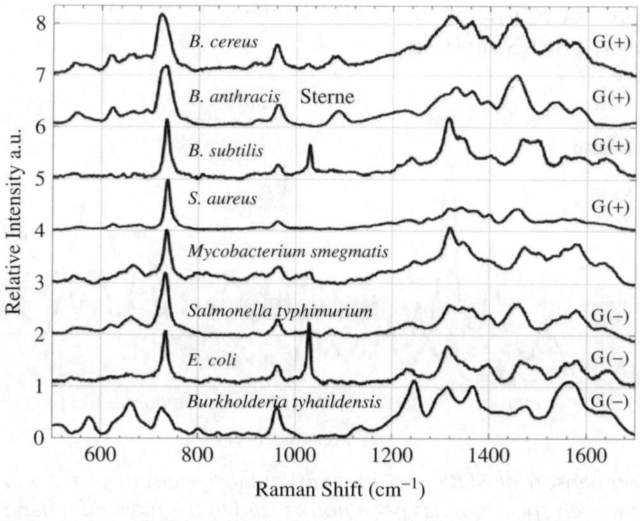

Figure 12.2 *SERS spectra of eight bacterial species obtained on gold aggregate coated SiO_2 chips (Figure 12.1). An incident laser power of ~2 mW and a data accumulation time of 10 s were used to obtain these spectra. Spectra are offset vertically for display purposes and top-to-bottom ordered according to their phylogenetic relationship*

been cultured in growth media, and washed four times with de-ionized water to remove all traces of the growth media before being placed on the SERS substrate [42]. The displayed spectra result from ~4 single scans and have been arranged in accordance with their phylogenic proximity. Gram positive/negative staining designation, which distinguishes two broad categories of outer cell layer structure, is indicated. The spectra are close to the single cell limit and result from illumination of ~10 cells in the Raman data collection region. Single cell SERS capabilities have been previously reported for a two cell *Bacillus* organism [18]. As indicated by these typical bacterial results, each species (and strain) exhibits a unique SERS spectrum with excellent signal-to noise. Gram positive versus Gram negative differences are not obvious from the SERS spectra. In fact, the SERS spectral features are only moderately correlated with phylogenic proximity although a number of common vibrational features appear in many of these SERS spectra of vegetative bacterial cells. The molecular origins of these vibrational features will be discussed below.

The SERS spectra of bacteria have virtually no correspondence with the normal, that is, non-SERS, spectra of the identical bacterial cells. For example, the SERS and normal Raman spectra of a strain of *E. coli* excited at 785 nm are compared in Figure 12.3 and no common vibrational bands are found in these two spectra. This result is not surprising because the normal Raman spectrum is given by the number density weighted Raman polarizabilities of scatters in the Raman sample region and is thus dominated by molecular species in the cell's cytoplasm. On the other hand, owing to the distance dependence of the plasmonic enhancement mechanism, only those molecules at the outer layers of the bacterial cell (<~10 nm) will contribute to the SERS spectrum as the bacterium sits on the SERS-active substrate. Judging by the most intense feature in each spectrum, the SERS spectrum is enhanced by $10^4 - 10^5$ on a per bacterium basis relative to the normal Raman signal [18]. Since different molecular species contribute to these two types of spectra this factor is, however, not equivalent to the SERS enhancement factor characterizing the Au nanoparticle-covered SERS substrate.

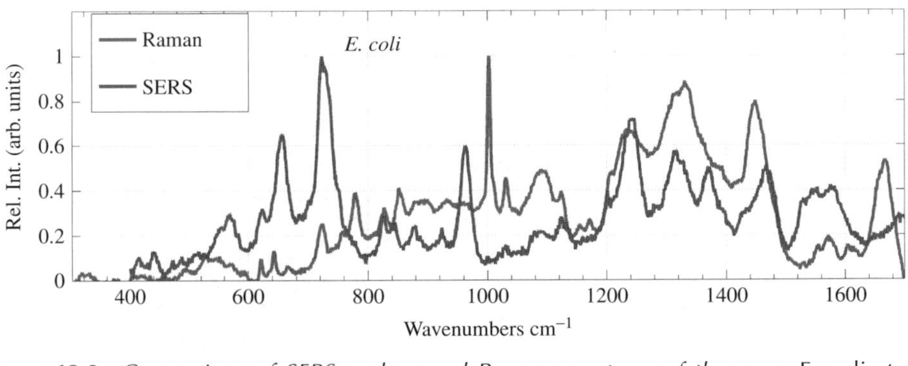

Figure 12.3 *Comparison of SERS and normal Raman spectrum of the same* E. coli *strain excited at 785 nm. Each spectrum has been normalized to the maximum band intensity. There is no correspondence between the bands that appear in the "normal" (non-SERS) and SERS spectrum. (See insert for color representation of the figure.)*

12.4 PCA Barcode Analysis

In order to quantify the reproducibility and specificity of these SERS spectra, and to use these vibrational signatures for a diagnostic we have developed a unique multivariate data analysis approach based on principal component analysis (PCA) [41]. PCA is a widely used and well known linear projection technique for maximizing the variance in a multivariate data set. What makes our procedure unique is that, after removing the high frequency noise by Fourier filtering, we convert the normalized SERS spectrum into a barcode, that is, a series of "ones" or "zeros" based on the sign of the second derivative of the spectrum as a function of Raman frequency. An example of a SERS-based "barcode" derived from an observed SERS spectrum of *B. Cereus* is shown in Figure 12.4. These barcodes, or arrays of ones and zeros, color-coded as black (curvature down) or white (curvature up) per wavenumber interval in this figure, are the input vectors for standard PCA analyses and related clustering methods, such as discriminant function analysis (DFA) and hierarchical cluster analysis (HCA). The barcode methodology results in significantly enhanced diagnostic specificity [41] and is the analytical tool used for quantitative measures of reproducibility.

An example of the effectiveness of this barcode methodology, and its use for bacterial diagnostics and measures of specificity and reproducibility are shown in Figure 12.5. The PCA plot (PC2 vs. PC3) in Figure 12.5a results when six SERS spectra of four closely related *Bacillus* bacteria (*B. thuringiensis, B. cereus, B. anthracis* Sterne, *B. anthracis* Ames) are the input vectors to a standard PCA procedure. Each spectrum corresponds to a "dot" in this plot, which has been color coded to indicate the organism it represents and the rings correspond to 2D standard deviations. The tightness of the clusters for a given species/strain characterizes the reproducibility of the spectra and the distance between the clusters represents the specificity offered by these measurements. *B. anthracis* Ames and

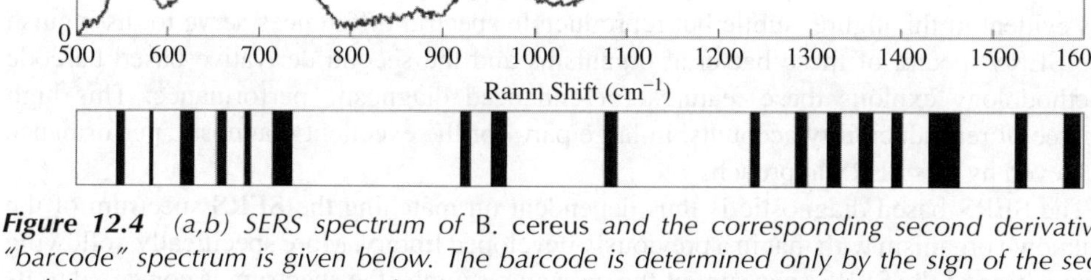

Figure 12.4 *(a,b) SERS spectrum of* B. cereus *and the corresponding second derivative "barcode" spectrum is given below. The barcode is determined only by the sign of the second derivative (curvature up = white, curvature down = black). These barcodes are the inputs for the principal component analysis (PCA) used to identify unknown bacteria via SERS.* © *L. Ziegler*

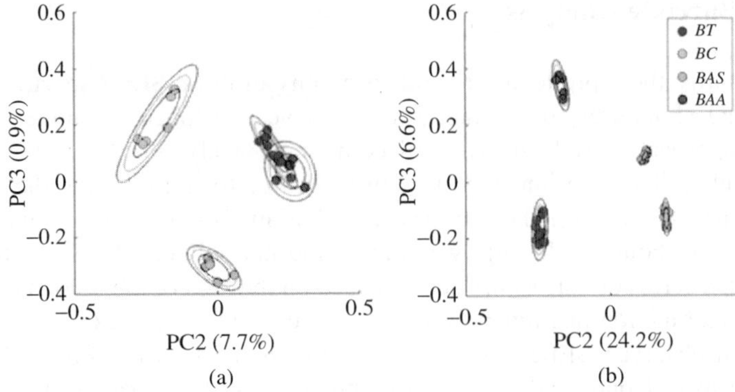

Figure 12.5 *Comparison of PCA plots (PC3 vs. PC2) for SERS spectra of four closely related* Bacillus *organisms (*B. thuringiensis, B. cereus, *B.* anthracis *Sterne,* B. anthracis *Ames). When the observed SERS spectra are the inputs to the PCA, the contour plot shown in (a) results. By contrast, when the input to the PCA are the barcode representations of the observed SERS spectra, the contour plot shown in (b) results. The enhancement in specificity resulting from the barcode spectra is clearly evident by this comparison. © L. Ziegler. (*See insert for color representation of the figure.*)*

B. thuringiensis, for example, cannot be separated in this treatment based on input vectors given by the SERS spectra . However, when the input to the PCA treatment are the barcode spectra determined by the second derivatives, as described above, all species/strains are well-separated, and much tighter clusters are obtained, as seen in the PCA plot in Figure 12.5b. The contrast between these two panels illustrates the enhanced specificity that results from this barcoding procedure and is key to the most effective use of this methodology for bacterial diagnostics.

The success of this second derivative barcode methodology for bacterial cell diagnostics results, in part, from its obvious insensitivity to broad baseline variations inherent to these backscattering experimental configurations, but also to the highly reproducible, complex spectral bandshapes that are observed on these Au nanoparticle substrates coupled with high quality signal acquisition (S/N and spectral resolution). For example, a strong spectral feature in the 720–740 cm^{-1} range is observed in all SERS spectra of vegetative cells (Figure 12.2). An expanded view of this region for 10 spectra of three different species, *E. faecalis, K. pneumoniae*, and *S. aureus*, is shown in Figure 12.6. As evident in this figure, subtle but reproducible spectral differences serve to distinguish the SERS spectra of these bacterial organisms and the second derivative based barcode methodology exploits these features for enhanced diagnostic performance. This high degree of reproducibility accounts, in large part, for the excellent diagnostic performance achieved by this SERS approach.

The SERS-based diagnostic is thus dependent on matching the SERS spectrum of the unknown organism with that in a previously developed library. More specifically, following acquisition of the SERS spectrum of the unknown sample, the spectrum is converted to its corresponding barcode and PCA analysis is performed with selected groups of barcode SERS spectra of known bacterial identity from a previously developed library in order to identify the bacterial unknown with strain and species specificity. The species/strain group that the unknown barcoded spectrum clusters with determines the identity of the unknown.

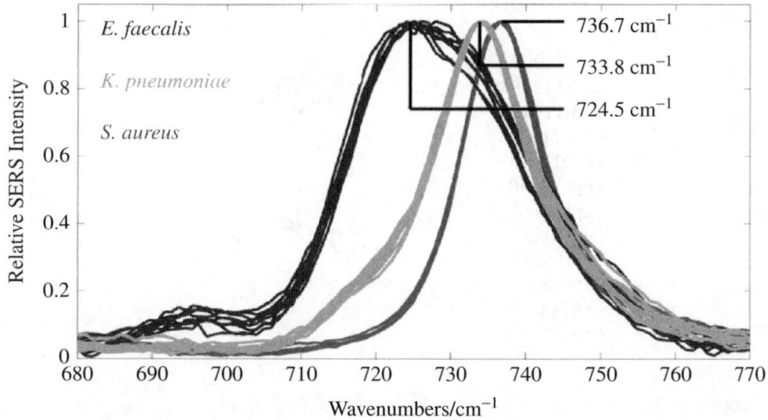

Figure 12.6 *Ten SERS spectra of* E. faecalis, K. pneumoniae, *and* S. aureus *bacterial strains in the 680–770 cm*$^{-1}$ *region demonstrating the reproducibility of spectral shapes in this important part of the SERS spectrum. The Raman shift of the spectral peak in this region is indicated for each species. (*See insert for color representation of the figure.*)*

The identification method is only limited by the library size, which can be continuously and readily expanded, and is not dependent upon specific primer sets or antibody specificity as for PCR or immunological- based methods. This barcode approach is quite general [41] and, in principle, could be expanded to other pathogens, such as fungi or viral particles or specific chemical signatures, for improved identification purposes.

12.5 Biological Origins of Bacterial SERS Signatures

Evidence that the observed SERS spectra can be correlated with genomic factors, and hence have biological significance, is illustrated in Figure 12.7. The result of a barcode analysis of 12 SERS spectra for each of 12 different organisms is displayed in this figure. Three clinical strains of four different species, *Staphylococcus aureus, Staphylococcus epidermidis, Escherichia coli*, and *Klebsiella pneumoniae* were used in this study. As seen in the barcode-based DFA plot, all strains of the *staphylococcus* genus cluster overlap at the same place, whereas each of the *E. coli* and *K. pneumonia* strains exhibit distinct clusters. Thus, these bacterial SERS spectra must be indicative of some inherent biological significance that accounts for the observed vibrational features. Furthermore, when the overlapping *staphylococcus* spectra were clustered separately all strains could be separately resolved.

The molecular origins of the vibrational bands observed in the SERS spectra of vegetative bacterial cells are usually attributed to the diversity of lipoproteins, proteins, lipids, and other cell wall components at the outer membrane layer of these bacterial cells [16, 18, 22, 25, 30, 32, 43]. The distance dependence of the SERS enhancement mechanism supports this hypothesis. However, we have discovered that nearly all of the observed SERS vibrational features are due to small metabolites which are concentrated at the outer region of vegetative bacterial cells [44]. In particular, the largest bands in the SERS spectra are generally products in the purine degradation pathway, such as hypoxanthine, adenine, xanthine, and so on. As compared to all the molecular species in the extracellular metabolome,

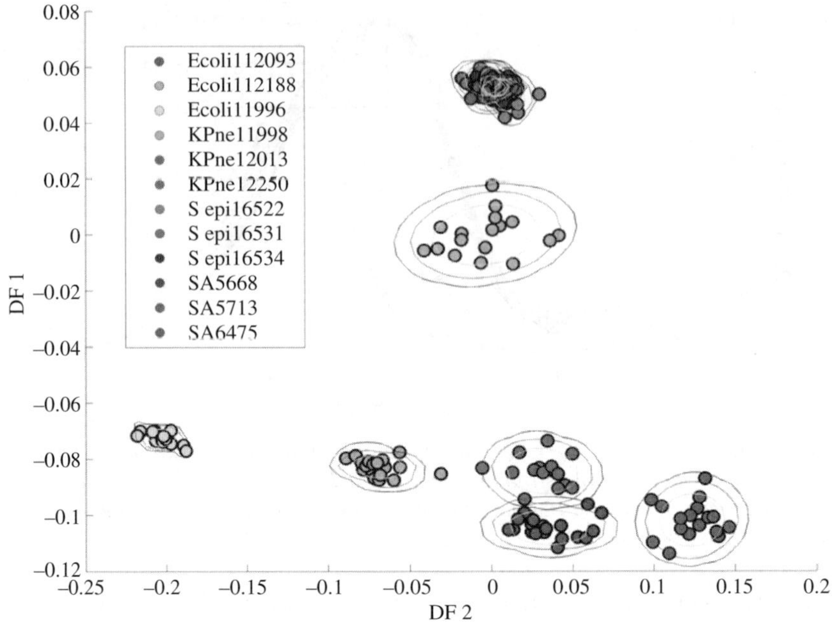

Figure 12.7 Barcode-based discriminant function analysis (DFA) plot derived from twelve SERS spectra for three strains for each of four different species (144 SERS spectra); E. coli, K. pneumoniae, S. epidermidis, *and* S. aureus. *All six staphylococcus strains are found to cluster together whereas the other six strains cluster separately indicating the biological significance of these SERS spectra. The six staphylococcus spectra separate in PCA treatments when analyzed separately. (*See insert for color representation of the figure.*)*

these compounds appear to have the largest SERS cross sections, at least on the SERS substrates used in these studies, and mass spectrometry has been used to confirm their presence in these biological samples [44]. The large range and unique reproducible signatures found for the SERS bacterial spectra result from the large number of enzymatic variations in these metabolic pathways. The absence or presence of these various enzymes results in different relative concentration of these small molecules found in the near-cell extracellular region. Thus, SERS offers not only reproducible molecular signatures for analytical detection and identification applications, as shown here, but also a new tool for studying biochemical activity in real time at the outer layers of these organisms. A more complete description of the chemical origins of the SERS spectral components of vegetative whole bacterial cells will be given shortly [44].

12.6 SERS Bacterial Identification in Human Body Fluids: Bacteremia and UTI Diagnostics

The ability to rapidly detect and identify vegetative bacterial cells in human body fluids at relatively low cost and in point-of-care settings is a continuing need for health care providers worldwide. Efforts in our lab have been directed toward developing a

SERS-based platform for strain specific bacteremia and UTI diagnostics. Bacteremia, the presence of bacteria in blood, results from severe infections at sites in the body, surgical wounds, or contaminated implanted devices and may lead to the potentially fatal condition of sepsis. Owing to a ~40% mortality rate, sepsis ranks 13th among the causes of death overall in the United States [45, 46]. As described above, enriching bacteria from blood is a challenging task owing to the relative concentrations of blood cells (10^9/ml) and bacterial cells (10^3/ml) in the infected blood of bacteremic patients. Efforts in our lab are ongoing to develop an automated selective lysis, centrifugation and micro-evaporation procedure to achieve the required enrichment to observe SERS spectra of whole bacterial cells. Preliminary evidence is encouraging that once these levels of enrichment are attained, detection and identification at clinically relevant concentration will be successful. Following a bench-top selective lysis and centrifugation procedure, SERS spectra of *B. anthracis* Sterne were obtained from blood spiked with vegetative cells at concentrations as low as 10^5/ml [24]. Furthermore, as shown in Figure 12.8, when closely related *Bacillus* bacteria were recovered by this procedure from spiked blood samples, barcode-based PCA clusters were obtained. Thus the effects of blood exposure and the blood cell lysing procedure had minimal impact on the resulting SERS spectra and their ability to be used as a clinical diagnostic.

UTIs, evidenced by the appearance of bacteria in urine, are among the most common types of infections in humans. Approximately 50% of all women will have at least one UTI in their lifetime and a sizable percentage of those will suffer from chronic UTIs [47, 48]. In the US alone, UTIs are responsible for more than 7 million doctors office visits and over 1 million hospital admissions, at a cost of approximately $1 billion per year. We have recently begun efforts in our laboratory to develop this SERS-based approach for UTI diagnostics. With relatively routine centrifugation- and filtration-based techniques bacterial concentration enrichments of $10^3 - 10^4$ can be achieved in urine samples that have been spiked at the clinically relevant concentration of 10^5 cfu/ml which is the minimal level for a UTI diagnosis and the typical threshold for antibiotic treatment [49]. Figure 12.8 shows the PCA plot resulting from the acquisition of SERS spectra of three different *E. coli* strains that were spiked into urine at this minimal clinical diagnostic level, enriched for signal acquisition and then placed on the SERS substrate for data acquisition. *E. coli* is the most common causative agent of UTI. Distinct clusters are readily obtained for the bacterial cells after recovery from urine, indicating that strain-specific SERS spectral signatures can be identified by this SERS barcode procedure. Several weeks later this procedure was repeated for one of these strains, designated as *E. coli* 11996, and when the SERS spectra of this "unknown" were acquired following enrichment from urine, they were correctly identified by our SERS-based diagnostic from this albeit limited library. Such tests were repeatedly successful for the "unknown" samples we have tested thus far. This prototype, bench-top UTI test took about 40 min to perform. Thus, at the very least, these preliminary results illustrate the potential value of this promising SERS-based methodology for rapid, low cost UTI diagnostics.

12.7 Red Blood Cells and Hemoglobin: Blood Aging and Disease Detection

As a result of our interest in developing a SERS based platform for detection and identification of vegetative bacterial cells enriched from human blood as described briefly above,

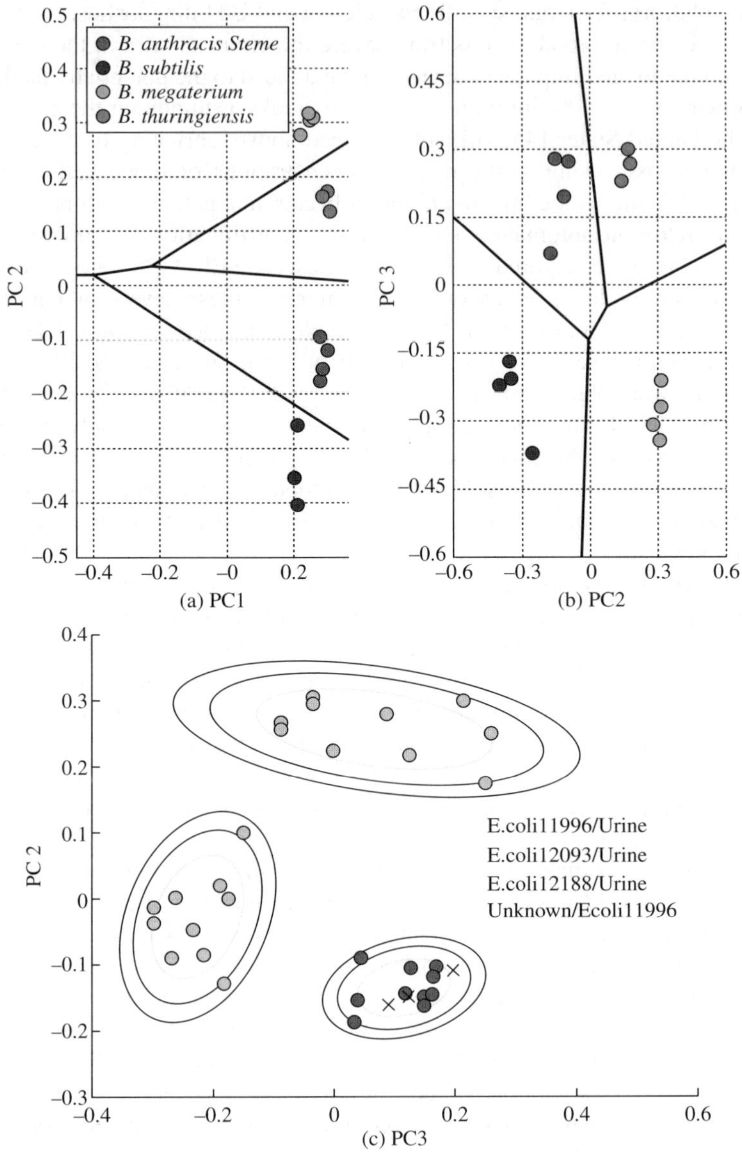

Figure 12.8 *(a) Barcode PCA plots for four Bacillus bacteria that have been spiked into human blood and then recovered, demonstrating that specificity is maintained via SERS after blood exposure. Black lines are equidistant to the mean of each cluster (Voroni diagram). (b) Barcode PCA plot for three* E. coli *strains spiked into urine at 10^5/ml and recovered by centrifugation. "X" corresponds to use of this limited library to identify the "unknown" in a test of the use of this limited library following enrichment of this bacterium from a spiked urine sample. (*See insert for color representation of the figure.*)*

it was necessary to develop an understanding and characterization of the SERS signals due to whole blood components that may remain after the enrichment process. This was our initial motivation for our SERS studies of blood and, in particular, its cellular components. However, in the course of learning about potential human blood contributions to SERS signatures for bacteremia diagnostics, the experimental results suggested some novel biomedical as well as forensic applications for SERS spectroscopy of whole blood and its components. Those related to whole blood cell biochemical processes and diagnostics in human blood are discussed below.

12.8 SERS of Whole Blood

We recently reported the SERS of whole human blood, blood plasma and RBCs on our Au nanoparticle-covered SiO_2 substrates excited at 785 nm [50]. Despite the extensive interest in exploiting SERS for a wide variety of bioanalytical applications and the central biological role of blood, there had been no prior 785 nm excited SERS studies of whole human blood or RBCs. Only a single SERS study of whole blood [51], which resulted from Ag colloid enhancement excited at 514.5 nm, had been given previously. The SERS spectrum of fresh whole blood excited at 785 nm that we observe is just that of blood plasma for the simple drop coat sample preparation procedure employed. When a thin layer of whole blood is placed on the *in situ* grown Au nanoparticle-covered SiO_2 substrates, an SERS spectrum due to the proteins, lipoproteins, carbohydrates, and other small molecules, including metabolites, that are found in blood plasma is observed [50]. For this sample preparation procedure, no cellular components of blood (RBCs, white blood cells, or platelets) contribute to the observed 785 nm SERS spectrum of whole blood. This is in contrast to the normal Raman spectrum of whole blood excited at 785 nm and open to ambient air, which is exclusively due to the vibrational scattering of oxyhemoglobin [52–54].

In the course of these whole blood SERS measurements we observed an unanticipated important effect that is central to the use of whole blood SERS signatures for bioanalytical and diagnostic applications [50]. The SERS spectra of the same blood sample stored at 8 °C are shown as a function of storage time over a period of 24 h in Figure 12.9a. These spectra have been normalized by the intensity of the largest feature in the 24 h spectrum at 724 cm^{-1}. As seen in this figure, the 785 nm excited SERS spectrum of whole blood dramatically changes over this time period. For example, new vibrational bands appear at 724, 965, 1070, 1332, 1365, and 1453 cm^{-1} and increase in relative importance as a function of whole blood storage time. The 24 – 1 h difference SERS spectrum of whole blood is displayed in this figure as well and demonstrates that the new features appearing in the SERS spectrum of whole blood with time are due to hypoxanthine, the well-known purine degradation product mentioned above in the context of the origins of the SERS bacterial spectra. In Figure 12.9b, the intensity of the 724 cm^{-1} band of hypoxanthine relative to the 490 cm^{-1} band of the fresh whole blood spectrum is plotted as a function of time. The hypoxanthine signal is not linear with time but increases dramatically after ~12 h of storage before saturating at ~20 h when it completely dominates the 785 nm SERS spectrum of whole blood. The hypoxanthine concentration in this whole blood sample increased by ~50 – 100 times its value in fresh blood and reached a maximum value of ~200 μM by

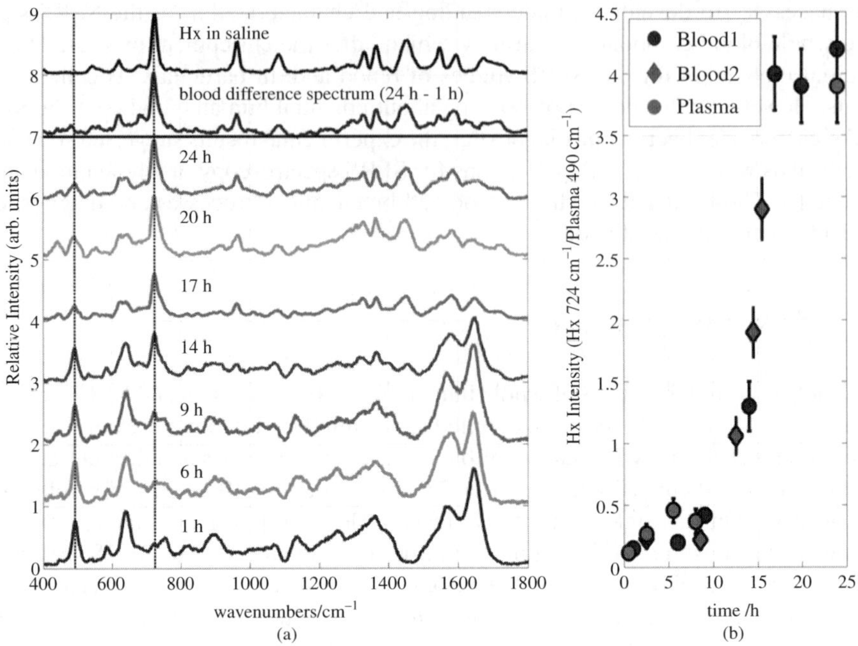

Figure 12.9 *SERS spectrum of whole blood as a function of storage time (4 °C), 24 h – 1 h SERS blood difference spectrum and the SERS spectrum of hypoxanthine (Hx) in normal saline solution are shown in (a). The relative intensities of the time dependent whole blood SERS spectra have been normalized by the observed signal size intensity of the largest feature in the 24 h spectrum at 724 cm^{-1}. The SERS intensity of the 724 cm^{-1} band of Hx relative to the 490 cm^{-1} band of whole blood, and the 724 cm^{-1} band of Hx relative to the 490 cm^{-1} band of a blood plasma sample as a function of storage time are shown in (b). Reprinted with permission from [50]. Copyright © 2012, American Chemical Society. (*See insert for color representation of the figure.*)*

20 h. It is also important to note that this blood component (hypoxanthine), and hence this biological activity, is not evident in the normal, that is, non-SERS Raman spectra, of whole human blood excited at 785 nm during this 24 h period.

We subsequently found that hypoxanthine appears in the SERS spectrum of plasma separated from whole blood at each time interval over a period of ~24 h with a rate that is nearly identical to that seen for the whole blood samples [50]. Thus, the hypoxanthine signal results from the leakage or secretion of this species into the extracellular fluid from one or more cellular components during this time-frame. Preliminary evidence suggests that this time-dependent purine metabolite is being released by either polymorphonuclear or mononuclear leukocytes or platelets under these non-ideal biological conditions. Studies are ongoing in order to further understand the origin and time-dependent characteristics of the SERS spectra of human whole blood. Aside from revealing that SERS can be a sensitive probe of biological activity in this complex body fluid, and offering a potential means for understanding the fundamental biochemical processes that account for this time-dependent hypoxanthine signal, this observation highlights the potential difficulties that can be encountered when trying to develop SERS-based whole blood assays for disease detection and identification due to the "age" of the blood sample alone. The inherent SERS signature of whole blood is *in vitro* time dependent.

The ability to accurately measure levels of hypoxanthine in blood or plasma is important for purine metabolism disorders, such as xanthine oxidase deficiency [55, 56], hyperuricemia (gout) [57], and fetal hypoxia [58]. Elevated hypoxanthine has also been used in the early detection of acute cardiac ischemia [59, 60] and may be a marker of prelabor fetal development problems [61]. Traditional techniques for determining hypoxanthine concentrations in whole blood or blood plasma are usually based on high-performance liquid chromatography (HPLC) [55, 62, 63]. The substitution of an SERS-based methodology for detection of hypoxanthine and related small molecules offers advantages relative to HPLC approaches. SERS detection is fast, thus allowing essentially "real time" concentration measures to be determined for both rapid diagnostic capabilities and aiding fundamental studies of the biochemical mechanisms involving these species. Secondly, the SERS optical approach requires minimal sample handling as compared to the multistep preparation procedure required for HPLC. Finally SERS sensitivity may be greater than HPLC-based approaches. We have found that hypoxanthine in plasma can be detected at the ~1 nM level, which appears to be more than an order of magnitude lower than that reported for HPLC methods [56, 59, 64]. Thus, SERS should provide a potentially valuable substitute for HPLC-based methods for the detection, identification, and quantification of purine compounds in blood or blood plasma for a variety of medical applications. Furthermore, the continuing characterization of the SERS activity of human blood and its components may have potential importance additionally in the developing area of forensic science where rapid, on-site blood age determinations may be a relevant factor [65].

12.9 SERS of RBCs

Erythrocytes or RBCs are the most abundant cells in the human body and are primarily responsible for delivering O_2 throughout the body. They are present in blood at concentrations of $(4–6) \times 10^9$/ml. Furthermore, about 33% by volume of RBCs are hemoglobin . RBCs are a convenient target for Raman studies owing to the well-known visible electronic transitions of the heme group [66], allowing resonance Raman access to the structurally sensitive vibrational spectrum of this protein [67]. Only a single SERS study of RBCs excited at 514.5 nm and hence resulting from both electronic as well as plasmonic enhancement had been previously reported [68]. The ability to acquire SERS spectra of RBCs at the single cell level could have biomedical applications, for instance as a diagnostic for early malaria or for detection of other oxygen carrying deficiency diseases, as a probe of blood aging, which is crucial to blood transfusion and storage capabilities, and for forensic applications requiring high sensitivity.

The 785 nm SERS spectra of RBCs isolated from whole human blood and resuspended in normal saline solution exhibit a very strong, robust, and reproducible spectrum on the Au nanoparticle-covered chips [50], as shown in Figure 12.10. The observed spectra result from essentially a single RBC given the relative sizes of the Raman illumination region (Renishaw RM2000 Raman microscope, 50× objective) and the RBC cell (6–8 μm diameter). For contrast, the normal Raman of RBCs excited at 785 nm are also shown in this figure. The Raman spectrum of RBCs excited at 785 nm is dominated by oxygenated hemoglobin vibrational modes [50, 54, 69]. Comparison of these two RBC spectra show

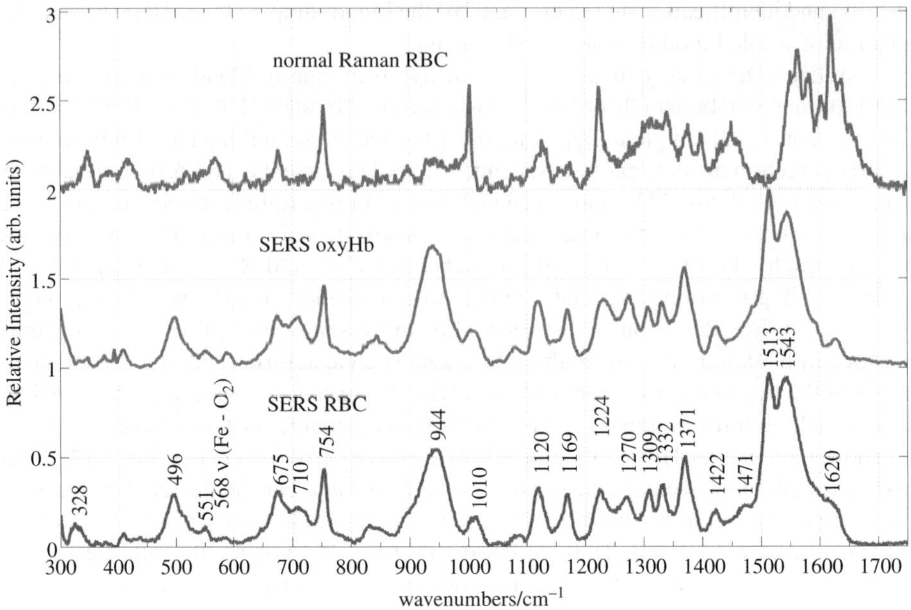

Figure 12.10 *Comparison of the SERS and normal Raman spectra of RBCs with the SERS spectrum of oxy-hemoglobin excited at 785 nm. The observed frequencies and tentative assignment of the RBC SERS vibrational bands are indicated*

that the vibrational frequencies for some transitions are unchanged for several bands (e.g., 754 cm^{-1} ν_{15}, 675 cm^{-1} ν_7, 1224 cm^{-1} ν_{42}/ν_{13}), whereas several appear to be slightly red shifted in the SERS spectrum (e.g., 1375 to 1371 cm^{-1} ν_4, 1127 to 1120 cm^{-1} ν_5, and 1551 to 1543 cm^{-1} ν_{11}). Furthermore, other bands in the 785 nm RBC Raman spectrum are not evident in the corresponding SERS spectrum: 1654 cm^{-1} (amide I), 1450 cm^{-1} (CH_2/CH_3 amino acid deformation modes), 1399 cm^{-1} (ν_{20}), and 568 cm^{-1} (ν_{Fe-O2}), and a few new bands at 1513, 551, 1471, and 496 cm^{-1}, whose parentage in the normal Raman spectrum is unclear, appear prominently in the SERS spectrum. The extent to which these new vibrational bands could be attributed to RBC membrane components, such as glycoproteins, glycolipids, spectrin, cholesterol, phospholipids, or transmembrane proteins which are closer to the SERS-active substrate than the hemoglobins, can be tested by comparison of RBC and hemoglobin SERS spectra. The SERS spectrum of oxyhemoglobin is shown in Figure 12.10 and is found to be identical with that of the single cell RBC SERS spectrum. Thus, the SERS spectrum of RBCs is entirely due to hemoglobin despite the closer proximity of cell membrane components. Conversion of oxyHb to the high spin ferric heme form, met-hemoglobin, does not appear to be the origin of these new features, based on detailed comparisons with SERS spectra of metRBCs [50]. The vibrational assignment of the "new" features in the RBC SERS spectra and the origin of the slightly red-shifted vibrational bands will be the subject of future studies. However, it is interesting to note that, just as found for the bacterial work summarized above, the cell wall components play a minor role at best in contributing to the SERS vibrational signatures for these two intact whole cell samples.

12.10 Malaria Detection

The early detection of malaria and the study of anti-malarial drug interactions are examples of potential applications based on the RBC and hemoglobin SERS capabilities described above. Malaria results from a parasite, carried by mosquitoes, that ultimately invades RBCs [70]. Digestion of hemoglobin in the food vacuole of the malaria parasite produces high quantities of the redox-active free heme known as hematin which is the iron(III) OH adduct [71–73]. This compound, however, is toxic to the parasite and thus it has developed a unique process to detoxify free heme by converting it first to the dimer, hemozoin, or β-hematin, and then to the polymer, often referred to as the malaria pigment. Hemozoin formation is a target for most of the well-known existing antimalarial drugs and continues to be a target for the development of new antimalarials [74].

The 785 nm SERS spectra of hemoglobin, hematin (monomer), and β-hematin (dimer) obtained on the Au nanoparticle-covered substrates are shown in Figure 12.11. A seen in this figure the SERS spectra of each of these species are qualitatively very similar, including some of the so-called "new" SERS features seen in the RBC SERS vibrational signature at 785 nm. However, some robust differences are also evident and the "barcode" PCA analysis can be used to clearly distinguish these closely related species [75]. Thus, the potential for SERS to detect and distinguish the malaria related heme by-products of parasite invasion in human RBCs is established. Furthermore, vibrational analysis of hematin-drug complexes that have demonstrated antimalarial activity may be an added tool for learning about how these drugs interact with the hematin to prevent dimer and subsequent polymer formation. Ongoing studies in our laboratory are currently addressing these questions.

12.11 Cancer Cell Detection: Metabolic Profiling by SERS

Cancer is the second-leading cause of mortality in the United States and is responsible for one in every four deaths [76]. Hence, there is a continuing need to develop rapid, accurate diagnostic capabilities for the detection and identification of cancer cells. Given the routine

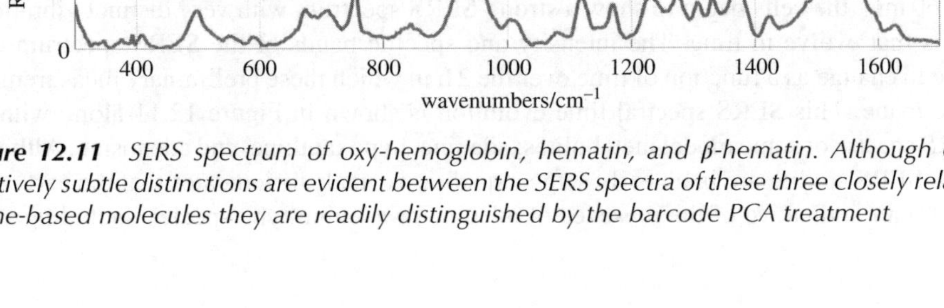

Figure 12.11 SERS spectrum of oxy-hemoglobin, hematin, and β-hematin. Although only relatively subtle distinctions are evident between the SERS spectra of these three closely related heme-based molecules they are readily distinguished by the barcode PCA treatment

availability of micro-Raman instrumental capabilities and the previously discussed and demonstrated attributes that SERS has to offer for cell level analysis, SERS is a potentially powerful technique for clinical single-cell, *in vivo* or *in vitro* clinical analysis of cancer.

The use of SERS-based methodologies for the detection and identification of cancer has been the subject of many investigators, particularly over the past decade [77–89]. The attributes of sensitivity, speed, and ease of use, coupled with strategies to achieve specificity combine to make SERS an analytical approach that could be developed for successful for real-time, cancer diagnostics. Whether this SERS-based approach is ultimately developed into a rapid, real-time *in vivo* cancer diagnostic is uncertain at best, however, the results described here demonstrate that at the very least the use of SERS has the potential to improve our molecular level understanding of cellular processes in these diseased tissues. Most of the SERS-based approaches for the identification of cancer cells have centered on the use of antibody-labeled metal nanoparticles that exhibit selective binding for cancer cell surface biomarkers and subsequently allow specific cellular imaging due to bright SERS reporter molecules [77–84]. The advantages of this optical approach are the sensitivity and multiplexing capabilities inherent to SERS. High specificity is determined by that associated with known antibodies. Hence the SERS signal of a reporter molecule plays the role of a fluorescent tag. However, this imaging technique is dependent on the identification of overexpressed proteins at the cell's surface and the development of procedures to exploit conjugate specific antibody interactions.

In an alternative approach, SERS spectra of cultured cancer cells or blood plasma of cancer patients have been examined and shown to phenomenologically offer vibrational signatures that, when coupled with multivariate data analysis techniques, serve to distinguish cancer and health cells [85–89]. The SERS vibrational features that have been associated with diagnostic capabilities have been assigned to anomalous cell surface features or due to altered levels of nucleic acids, proteins, saccharides, or lipids in circulating body fluids. Interestingly, most of these studies cite anomalous intensities of a vibration spectral band at ~725 cm^{-1}, attributed to adenine or to the C-N bond of quaternary ammonium groups in membrane lipid bilayer components [85–88]. Other anomalous band intensities have also been noted at 655 or 638 cm^{-1}, and have been assigned to the C-S stretch in cysteine or a tyrosine mode, respectively [85–88].

In an effort to extend our SERS studies of intact cells to additional potential biomedical applications, we have recently obtained SERS spectra of human renal cancer cells (ATCC# CRL-1933). These cells were grown in basal medium (ATCC# RPI-1640) supplemented with 10% FBS (fetal bovine serum) and 1% penicillin/streptomycin in a humid incubator at 37 °C with 5% carbon dioxide for three days. The harvested cells were washed with saline solution prior to SERS measurements. The SERS spectra of these cancer cells as a function of time in saline solution before signal acquisition are shown in Figure 12.12. Initially, these cells exhibit only a weak relatively featureless spectrum. However, within 30–60 min, the cells begin to show a strong SERS spectrum with very distinct vibrational bands that evolve in time. The intensity and specific bands of the SERS spectrum continue to change as a function of time over the 2 h in which these preliminary measurements were made. This SERS spectral time evolution is shown in Figure 12.11 along with the SERS spectra of some model metabolites; adenine, hypoxanthine, and adenosine. Although these SERS studies are in an early stage a qualitative analysis indicates that the early time spectra are indicative of the presence of adenine or adenine-type species, including some

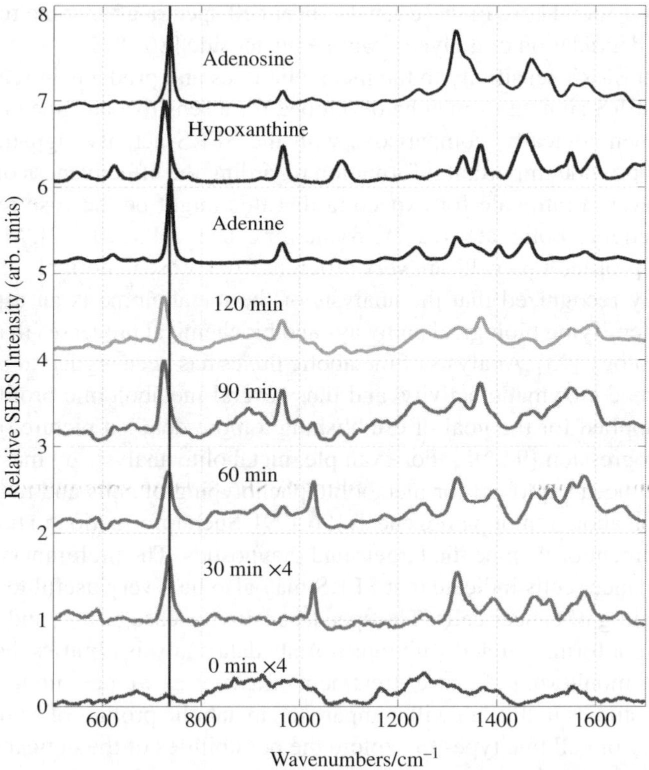

Figure 12.12 *SERS spectra of renal cancer cells as a function of time in saline solution. SERS spectra of adenine, hypoxanthine, and AMP are shown for comparison. The vertical black line is at 736 cm^{-1} and is a guide that highlights the shift of the band in this region from an adenine-like frequency to the hypoxanthine frequency*

adenosine or adenosine monophosphate (AMP), which then changes under these conditions to a hypoxanthine-like spectrum. For example, the vertical black line in Figure 12.12 is at 736 cm^{-1} and is essentially coincident with the strongest band in the adenine and adenosine SERS spectrum (and AMP, not shown) and the peak position of a prominent band in the 30 and 60 min spectra of this renal cancer cell line. In the 90 and 120 min spectra, however, this band is red shifted to 725 cm^{-1} which is the frequency of the ring bending mode of hypoxanthine. Other spectral features in the 30 and 60 min spectra at 1320, 1397, 1470, 1505, 1580, and 1654 cm^{-1}, suggest the presence of a nucleoside such as adenosine. Local concentrations of these species are in the 10–100 μM regime. Consistent with results discussed previously, the SERS spectra of these whole renal cells appear dominated by the small molecules that are secreted from these cells as a consequence of metabolic processes and responses to the current environmental conditions. In contrast to bacterial cells these SERS signatures are strongly time-dependent but again suggest a pattern of organism survival strategy enhanced by degradation of the purine energy currency in low nutrient environments. The ~725 cm^{-1} band repeatedly reported in earlier SERS studies of cancer detection and identification is most likely due to the metabolite hypoxanthine or a close analog found here in these renal cancer cell spectra after some time, apparently resulting from degradation of a glycosylamine nucleoside [86–88].

This apparent SERS sensitivity to the metabolic rates and products in renal cancer cells suggests that SERS signatures may be developed for a sensitive and specific cancer diagnostic. Phenomenologically, comparisons with the SERS activity signatures of healthy renal cells will provide important information regarding the effectiveness of this proposed strategy. However, a rationale for expecting that this might be the case results from the known enhanced metabolic rates that are found for cancer cells [90–92]. Metabolic activities in afflicted proliferating cells are very different from those in normal cells. It is becoming increasingly recognized that the analysis of the metabolome is an especially useful approach for identifying biological pathways and biochemical processes that are perturbed in a given pathology [93]. Analysis of metabolic fluxes has been argued to be a better indicator of perturbed enzymatic activity, and thus several metabolomic profiling tools have already been applied for the goal of establishing a more detailed picture of tumor development and progression [94, 95]. For example, metabolite analysis by mass spectrometry and NMR have been exploited for metabolite phenotyping of cells and tissues to provide new information about tumor pathogenesis [96–99]. Such information in turn may be used for the development of therapeutic targets and diagnostics. The preliminary results shown here for renal cancer cells indicate that SERS may also be a very useful technique for the metabolic profiling of cancer cells. The speed, sensitivity, ease of use, and cost argue that a SERS-based platform, coupled with multivariate data analysis libraries, has the potential for noninvasive monitoring of cancer treatment efficacies as well as single cell level diagnostics. Future studies in this lab will compare the metabolic profiles of healthy and cancer cells in a variety of cell line types to explore the possibilities of the optical approach.

12.12 Conclusions

The results discussed here demonstrate that SERS of living cells has the potential to become very useful in disease diagnostics and as a probe of biochemical activity taking place in the extracellular metabolome. When coupled with appropriate hardware and software, preliminary results suggest that a SERS-based platform may provide a rapid, sensitive, specific, point-of-care methodology for the diagnosis of bacterial infections in human body fluids without the requirement for the slow cell growth step. Initial studies of blood and urine bacterial infections have been our initial targets but the SERS-based platform can be extended to bacterial diagnostics in cerebrospinal fluid, broncoalveolar lavage fluid, or sputum. As will be discussed in depth in a subsequent publication, the basis for this bacterial identification is the use of SERS for rapid metabolic profiling. RBCs offer a strong and reproducible SERS spectrum on the Au nanoparticle-covered chips. That spectrum is shown to be entirely due to the heme protein. Aside from the use of SERS for understanding and monitoring aging processes in blood, such as the release of the metabolite hypoxanthine from blood particles, it may be a useful tool for detecting early signs of malarial infection and other blood diseases and for learning about mechanisms of anti-malarial drug action. Finally, it was shown that using SERS for metabolic profiling may be a sensitive and relatively rapid methodology for cancer cell identification. Interestingly for each of these three cell types, bacteria, RBC, and renal cancer cells, probed by SERS in our studies the cell wall makes virtually no contribution to the observed signals. Thus, despite the presumably close contact to the nanostructured metal surface, the cellular SERS signals discussed here are due to small molecules excreted or collecting in the cell outer layers, or result from highly Raman polarizable molecules located at the inside of the cell wall (RBC) and are not due to the cell wall structures. Subsequent work from this laboratory will continue to develop these SERS-based applications for diagnostics and metabolomics as well as extending these studies for forensics.

Acknowledgment

The support of the National Institute of Health (Grant # 1R01AI090815-01) and the National Science Foundation I/UCRC program are gratefully acknowledged. We thank BD Technologies, Research Triangle Park, NC for supplying some of the bacterial strains used here.

References

1. Jeanmaire, D.L. and Van Duyne, R.P. (1977) Surface Raman spectroelectrochemistry: Part I. Heterocyclic, aromatic, and aliphatic amines adsorbed on the anodized silver electrode. *Journal of Electroanalytical Chemistry*, **84**, 1–20.
2. Tian, Z.Q. (2005) Surface-enhanced Raman spectroscopy: advancements and applications. *Journal of Raman Spectroscopy*, **36**, 466–470.
3. Haynes, C.L., Yonzon, C.R., Zhang, X. and Van Duyne, R.P. (2005) Surface-enhanced Raman sensors: early history and the development of sensors for quantitative biowarfare agent and glucose detection. *Journal of Raman Spectroscopy*, **36**, 471–484.
4. Camden, J.P., Dieringer, J.A., Zhao, J. and Van Duyne, R.P. (2008) Controlled plasmonic nanostructures for surface-enhanced spectroscopy and sensing. *Accounts of Chemical Research*, **41**, 1653–1661.
5. Moskovits, M. (1985) Surface enhanced spectroscopy. *Reviews of Modern Physics*, **57**, 783–826.
6. Kaderbhai, N.N., Broadhurst, D.I., Ellis, D.I. *et al.* (2003) Functional genomics via metabolic footprinting: monitoring metabolite secretion by Escherichia coli tryptophan metabolism mutants using FT-IR and direct injection electrospray mass spectrometry. *Comparative and Functional Genomics*, **4**, 376–91.
7. Maharjan, R.P. and Ferenci, T. (2005) Metabolomic diversity in the species Escherichia coli and its relationship to genetic population structure. *Metabolomics*, **1**, 235–242.
8. Resmer, K.L. and White, R.L. (2011) Metabolic footprinting of the anaerobic bacterium Fusobacterium varium using 1H NMR spectroscopy. *Molecular BioSystems*, **7**, 2220–2227.
9. Nadkarni, M.A., Martin, F.E., Jacques, N.A. and Hunter, N. (2002) Determination of bacterial load by real-time PCR using a broad-range (universal) probe and primers set. *Microbiology*, **148**, 257–266.
10. Cheng, J.-C., Huang, C.-L., Lin, C.-C. *et al.* (2006) Rapid detection and identification of clinically important bacteria by high-resolution melting analysis after broad-range ribosomal RNA real-time PCR. *Clinical Chemistry*, **52**, 1997–2004.

11. Aslanzadeh, J. (2004) Preventing PCR amplification carryover contamination in a clinical laboratory. *Annals of Clinical and Laboratory Science*, **34**, 389–396.
12. Yu, X., Susa, M., Weile, J. *et al.* (2007) Rapid and sensitive detection of fluoroquinolone-resistant Escherichia coli from urine samples using a genotyping DNA microarray. *International Journal of Medical Microbiology*, **297**, 417–29.
13. Efrima, S., Bronk, B.V. and Czege, J. (1999) Surface enhanced Raman spectroscopy of bacteria coated by silver. *Proceedings of the SPIE*, **3602**, 164–171.
14. Fell, N.F. Jr.,, Smith, A.G.B., Vellone, M. and Fountain, A.W. III, (2002) Optimization of substrates for surface enhanced Raman spectroscopy of bacteria. *Proceedings of the SPIE*, **4577**, 174–181.
15. Guzelian, A.A., Sylvia, J.M., Janni, J. *et al.* (2002) SERS of whole cell bacteria and trace levels of biological molecules. *Proceedings of the SPIE*, **4577**, 182–192.
16. Jarvis, R.M. and Goodacre, R. (2004) Discrimination of bacteria using surface-enhanced Raman spectroscopy. *Analytical Chemistry*, **76**, 40–7.
17. Jarvis, R.M., Brooker, A. and Goodacre, R. (2004) Surface-enhanced Raman spectroscopy for bacterial discrimination utilizing a scanning electron microscope with a Raman spectroscopy interface. *Analytical Chemistry*, **76**, 5198–5202.
18. Premasiri, W.R., Moir, D.T., Klempner, M.S. *et al.* (2005) Characterization of the Surface Enhanced Raman Scattering (SERS) of bacteria. *Journal of Physical Chemistry B*, **109**, 312–320.
19. Sengupta, A., Laucks, M.L. and Davis, E.J. (2005) Surface-enhanced Raman spectroscopy of bacteria and pollen. *Applied Spectroscopy*, **59**, 1016–1023.
20. Zeiri, L., Bronk, B.V., Shabtai, Y. *et al.* (2002) Silver metal induced surface enhanced Raman of bacteria. *Colloids and Surfaces A: Physicochemical and Engineering Aspects*, **208**, 357–362.
21. Sengupta, A., Mujacic, M. and Davis, E.J. (2006) Detection of bacteria by surface-enhanced Raman spectroscopy. *Analytical and Bioanalytical Chemistry*, **386**, 1379–1386.
22. Jarvis, R.M., Brooker, A. and Goodacre, R. (2006) Surface-enhanced Raman scattering for the rapid discrimination of bacteria. *Faraday Discussions*, **132**, 281–292; discussion 309–319..
23. Kahraman, M., Yazici, M.M., Sahin, F. *et al.* (2007) Reproducible surface-enhanced Raman scattering spectra of bacteria on aggregated silver nanoparticles. *Applied Spectroscopy*, **61**, 479–85.
24. Premasiri, W.R., Moir, D.T., Klempner, M.S. and Ziegler, L.D. (2007) in *New Approaches in Biomedical Spectroscopy* (eds K. Kneipp, R. Aroca, H. Kneipp and E. Wentrup-Byrne), Oxford University Press, New York, p. 164.
25. Kahraman, M., Yazici, M.M., Sahin, F. and Culha, M. (2008) Convective assembly of bacteria for surface-enhanced Raman scattering. *Langmuir*, **3**, 894–901.
26. Kahraman, M., Zamaleeva, A.I., Fakhrullin, R.F. and Culha, M. (2009) Layer-by-layer coating of bacteria with noble metal nanoparticles for surface-enhanced Raman scattering. *Analytical and Bioanalytical Chemistry*, 2559–2567.
27. Guicheteau, J. and Christesen, S.D. (2006) Principal component analysis of bacteria using surface-enhanced Raman spectroscopy. *Proceedings of the SPIE*, **6218**, 62180G-1–62180G-8.

28. Ivleva, N.P., Wagner, M., Horn, H. *et al.* (2008) In situ surface-enhanced Raman scattering analysis of biofilm. *Analytical Chemistry*, **80**, 8538–8544.
29. Liu, Y., Chen, Y.-R., Nou, X. and Chao, K. (2007) Potential of surface-enhanced Raman spectroscopy for the rapid identification of Escherichia coli and Listeria monocytogenes cultures on silver colloidal nanoparticles. *Applied Spectroscopy*, **61**, 824–831.
30. Liu, T.-T., Lin, Y.-H., Hung, C.-H. *et al.* (2009) A high speed detection platform based on surface-enhanced Raman scattering for monitoring antibiotic-induced chemical changes in bacteria cell wall. *PLoS ONE*, **4**, e5470.
31. Walter, A., Marz, A., Schumacher, W. *et al.* (2011) Towards a fast, high specific and reliable discrimination of bacteria on strain level by means of SERS in a microfluidic device. *Lab on a Chip*, **11**, 1013–1021.
32. Kahraman, M., Keseroglu, K. and Culha, M. (2011) On sample Preparation for Surface-Enhanced Raman Scattering (SERS) of bacteria and the source of spectral features of the spectra. *Applied Spectroscopy*, **65**, 500–506.
33. Culha, M., Kahraman, M., Cam, D. *et al.* (2010) Rapid identification of bacteria and yeast using surface-enhanced Raman scattering. *Surface and Interface Analysis*, **42**, 462–465.
34. Culha, M., Yazici, M.M., Kahraman, M. *et al.* (2012) Surface-enhanced Raman scattering of bacteria in microwells constructed from silver nanoparticles. *Journal of Nanotechnology*, 7, Article ID 297560.
35. Felix-Rivera, H., Gonzalez, R., Rodriguez, G. *et al.* (2011) Improving SERS Detection of Bacillus thuringiensis using silver nanoparticles reduced with hydroxylamine and with citrate capped borohydride. *International Journal of Spectroscopy*, 1–9, Article ID 989504.
36. Chu, H., Huang, Y. and Zhao, Y. (2008) Silver nanorod arrays as a surface-enhanced Raman scattering substrate for foodborne pathogenic bacteria detection. *Applied Spectroscopy*, **62**, 922–931.
37. Zhang, J.Y., Do, J., Premasiri, W.R. *et al.* (2010) Rapid point-of-care concentration of bacteria in a disposable microfluidic device using meniscus dragging effect. *Lab on a Chip*, **10**, 3265–3270.
38. B. Yan, A. Thubagere, W. R. Premasiri, L. D. Ziegler, L. Dal Negro and B. M. Reinhard: Engineered SERS substrates for bacterial pathogen detection: nanoparticle cluster arrays with multiscale signal enhancement *Nano Letters* 3 (2009) 1190–1202.
39. Yang, L., Yan, B., Premasiri, W.R. *et al.* (2010) Engineering nanoparticle cluster arrays for bacterial biosensing: the role of the building block in multiscale SERS substrates. *Advanced Functional Materials*, **20**, 2619–2628.
40. Gopinath, A., Boriskina, S.V., Premasiri, W.R. *et al.* (2009) Plasmonic nanogalaxies: multiscale aperiodic arrays for surface-enhanced Raman sensing. *Nano Letters*, **9**, 3922–3929.
41. Patel, I.S., Premasiri, W.R., Moir, D.T. and Ziegler, L.D. (2008) Barcoding bacterial cells: a SERS based methodology for pathogen identification. *Journal of Raman Spectroscopy*, **39**, 1660–1672.
42. Premasiri, W.R., Gebregziabher, Y. and Ziegler, L.D. (2011) On the difference between surface-enhanced Raman scattering (SERS) spectra of cell growth media and whole bacterial cells. *Applied Spectroscopy*, **65**, 493–9.

43. Zeiri, L., Bronk, B.V., Shabtai, Y. *et al.* (2004) Surface-enhanced Raman spectroscopy as a tool for probing specific biochemical components in bacteria. *Applied Spectroscopy*, **58**, 33–40.
44. Premasiri, W.R., Lee, J.C., Theberge, R. *et al.* On the Molecular Origin Of Bacterial SERS Spectra, in preparation.
45. Bearman, G.M. and Wenzel, R.P. (2005) Bacteremias: a leading cause of death. *Archives of Medical Research*, **36**, 646–659.
46. Angus, D.C., Linde-Zwirble, W.T., Lidicker, J. *et al.* (2001) Epidemiology of severe sepsis in the United States: analysis of incidence, outcome, and associated costs of care. *Critical Care Medicine*, **29**, 1303–1310.
47. Schappert, S.M. and Rechtsteiner, E.A. (2008) Ambulatory Medical Care Utilization Estimates for 2006. National Health Statistics Reports No. 8, National Center for Health, Statistics, Hyattsville, MD.
48. Griebling, T.L. (2007) in *Urologic Diseases in America* (eds M.S. Litwin and C.S. Saigal), Department of Health and Human Services, Public Health Service, National Institutes of Health, National Institute of Diabetes and Digestive and Kidney Diseases, Washington, DC, pp. 587–619.
49. Premasiri, W.R., Gebregziabher, Y., Lemler, P., and Ziegler, L.D. SERS for antibiotic specific urinary tract infection diagnostics (in preparation).
50. Premasiri, W.R., Lee, J.C. and Ziegler, L.D. (2012) Surface-enhanced Raman scattering of whole human blood, blood plasma, and red blood cells: cellular processes and bioanalytical sensing. *The Journal of Physical Chemistry B*, **116**, 9376–9386.
51. Casella, M., Lucotti, A., Tommasini, M. *et al.* (2011) Raman and SERS recognition of beta-carotene and haemoglobin fingerprints in human whole blood. *Spectrochimica Acta, Part A: Molecular and Biomolecular Spectroscopy*, **79**, 915–919.
52. Sato, H., Chiba, H., Tashiro, H. and Ozaki, Y. (2001) Excitation wavelength-dependent changes in Raman spectra of whole blood and hemoglobin: comparison of the spectra with 514.5-, 720-, and 1064-nm excitation. *Journal of Biomedical Optics*, **6**, 366–370.
53. Buschman, H.P., Marple, E.T., Wach, M.L. *et al.* (2000) In vivo determination of the molecular composition of artery wall by intravascular Raman spectroscopy. *Analytical Chemistry*, **72**, 3771–5.
54. Wood, B.R., Caspers, P., Puppels, G.J. *et al.* (2007) Resonance Raman spectroscopy of red blood cells using near-infrared laser excitation. *Analytical and Bioanalytical Chemistry*, **387**, 1691–703.
55. Boulieu, R., Bory, C., Baltassat, P. and Gonnet, C. (1983) Hypoxanthine and xanthine levels determined by high-performance liquid chromatography in plasma, erythrocyte, and urine samples from healthy subjects: the problem of hypoxanthine level evolution as a function of time. *Analytical Biochemistry*, **129**, 398–404.
56. Boulieu, R., Bory, C., Baltassat, P. and Gonnet, C. (1982) High-performance liquid chromatographic determination of hypoxanthine and xanthine in biological fluids. *Journal of Chromatography*, **233**, 131–140.
57. Yamamoto, T., Moriwaki, Y., Takahashi, S. *et al.* (1995) Determination of plasma purine nucleoside phosphorylase activity by high-performance liquid chromatography. *Analytical Biochemistry*, **227**, 135–139.
58. O'Connor, M.C., Harkness, R.A., Simmonds, R.J. and Hytten, F.E. (1981) The measurement of hypoxanthine, xanthine, inosine and uridine in umbilical cord blood and

fetal scalp blood samples as a measure of fetal hypoxia. *British Journal of Obstetrics and Gynaecology*, **88**, 381–390.

59. Himmel, H.M., Sadony, V. and Ravens, U. (1991) Quantitation of hypoxanthine in plasma from patients with ischemic heart disease: adaptation of a high-performance liquid chromatographic method. *Journal of Chromatography*, **568**, 105–115.
60. Farthing, D., Sica, D., Gehr, T. *et al.* (2007) An HPLC method for determination of inosine and hypoxanthine in human plasma from healthy volunteers and patients presenting with potential acute cardiac ischemia. *Journal of Chromatography B: Analytical Technologies in the Biomedical and Life Sciences*, **854**, 158–164.
61. Sawa, R., Yoneyama, Y., Otsubo, Y. *et al.* (1997) Fetal plasma hypoxanthine level in growth-retarded fetuses before labor. *Journal of Clinical Endocrinology and Metabolism*, **82**, 4028–4031.
62. Harmsen, E., de Jong, J.W. and Serruys, P.W. (1981) Hypoxanthine production by ischemic heart demonstrated by high pressure liquid chromatography of blood purine nucleosides and oxypurines. *Clinica Chimica Acta*, **115**, 73–84.
63. Kojima, T., Nishina, T., Kitamura, M. *et al.* (1986) Reverse phase liquid chromatographic determination of purine compounds in serum applied to studies of hypouremia. *Clinical Chemistry*, **32**, 287–290.
64. Liu, Z., Li, T. and Wang, E. (1995) Simultaneous determination of guanine, uric acid, hypoxanthine and xanthine in human plasma by reversed-phase high-performance liquid chromatography with amperometric detection. *Analyst*, **120**, 2181–2184.
65. Virkler, K. and Lednev, I.K. (2010) Raman spectroscopic signature of blood and its potential application to forensic body fluid identification. *Analytical and Bioanalytical Chemistry*, **396**, 525–534.
66. Antonini, E. and Brunori, M. (1971) *Hemoglobin and Myoglobin in their Reactions with Ligands*, North Holland, Amsterdam.
67. Spiro, T.G. (1988) *Biological Applications of Raman Spectroscopy*, Resonance Raman Spectra of Heme and Metalloproteins, Vol. **3**, John Wiley & Sons, Inc., New York.
68. Brazhe, N.A., Abdali, S., Brazhe, A.R. *et al.* (2009) New insight into erythrocyte through in vivo surface-enhanced Raman spectroscopy. *Biophysical Journal*, **97**, 3206–14.
69. Venkatesh, B., Ramasamy, S., Mylrajan, M. *et al.* (1999) Fourier transform Raman approach to structural correlation in hemoglobin derivatives. *Spectrochimica Acta A*, **55**, 1691–1697.
70. Francis, S.E., Sullivan, D.J. Jr., and Goldberg, D.E. (1997) Hemoglobin metabolism in the malaria parasite plasmodium falciparum. *Annual Review of Microbiology*, **51**, 97–123.
71. Tripathi, A.K., Khan, S.I., Walker, L.A. and Tekwania, B.L. (2003) Spectrophotometric determination of de novo hemozoin/b-hematin formation in an in vitro assay. *Analytical Biochemistry*, **325**, 85–89.
72. Hempelmann, E. (2007) Hemozoin biocrystallization in Plasmodium falciparum and the antimalarial activity of crystallization inhibitors. *Parasitology Research*, **100**, 671–676.
73. Kumar, S., Guha, M., Choubey, V. *et al.* (2007) Antimalarial drugs inhibiting hemozoin (beta-hematin) formation: a mechanistic update. *Life Sciences*, **80**, 813–828.

74. Pagola, S., Stephens, P.W., Bohle, D.S. *et al.* (2000) The structure of malaria pigment beta-haematin. *Nature*, **404**, 307–310.
75. Lemler, P., Premasiri, W.R., and Ziegler, L.D. SERS of hemoglobin and related derivatives (in preparation).
76. American Cancer Society (2009) Cancer Facts and Figures.
77. Lee, S., Kim, S., Choo, J. *et al.* (2007) Biological imaging of HEK293 cells expressing PLCgamma1 using surface-enhanced Raman microscopy. *Analytical Chemistry*, **79**, 916–922.
78. Chon, H., Lee, S., Son, S.W. *et al.* (2009) Highly sensitive immunoassay of lung cancer marker carcinoembryonic antigen using surface-enhanced Raman scattering of hollow gold nanospheres. *Analytical Chemistry*, **81**, 3029–3034.
79. Huang, X., El-Sayed, I.H., Qian, W. and El-Sayed, M.A. (2007) Cancer cells assemble and align gold nanorods conjugated to antibodies to produce highly enhanced, sharp, and polarized surface Raman spectra: a potential cancer diagnostic marker. *Nano Letters*, **7**, 1591–1597.
80. Jun, B.-H., Noh, M.S., Kim, J. *et al.* (2010) Multifunctional silver-embedded magnetic nanoparticles as SERS nanoprobes and their applications. *Small*, **6**, 119–125.
81. Qian, X., Peng, X.-H., Ansari, D.O. *et al.* (2008) In vivo tumor targeting and spectroscopic detection with surface-enhanced Raman nanoparticle tags. *Nature Biotechnology*, **26**, 83–90.
82. Sha, M.Y., Xu, H., Natan, M.J. and Cromer, R. (2008) Surface-enhanced Raman scattering tags for rapid and homogeneous detection of circulating tumor cells in the presence of human whole blood. *Journal of the American Chemical Society*, **130**, 17214–17215.
83. Tang, H.-W., Yang, X.B., Kirkham, J. and Smith, D.A. (2007) Probing intrinsic and extrinsic components in single osteosarcoma cells by near-infrared surface-enhanced Raman scattering. *Analytical Chemistry*, **79**, 3646–3653.
84. Wang, X., Qian, X., Beitler, J.J. *et al.* (2011) Detection of circulating tumor cells in human peripheral blood using surface-enhanced Raman scattering nanoparticles. *Cancer Research*, **71**, 1526–1532.
85. Feng, S., Chen, R., Lin, J. *et al.* (2010) Nasopharyngeal cancer detection based on blood plasma surface-enhanced Raman spectroscopy and multivariate analysis. *Biosensors and Bioelectronics*, **25**, 2414–2419.
86. Yan, B. and Reinhard, B.r.M. (2010) Identification of tumor cells through spectroscopic profiling of the cellular surface chemistry. *The Journal of Physical Chemistry Letters*, **1**, 1595–1598.
87. Feng, S.Y., Pan, J., Wu, Y. *et al.* (2011) Study on gastric cancer blood plasma based on surface-enhanced Raman spectroscopy combined with multivariate analysis. *Science in China Life Sciences*, **54**, 828–834.
88. Lin, D., Feng, S., Pan, J. *et al.* (2011) Colorectal cancer detection by gold nanoparticle based surface-enhanced Raman spectroscopy of blood serum and statistical analysis. *Optics Express*, **19**, 13565–77.
89. Lin, J., Chen, R., Feng, S. *et al.* (2012) Surface-enhanced Raman scattering spectroscopy for potential noninvasive nasopharyngeal cancer detection. *Journal of Raman Spectroscopy*, **43**, 497–502.

90. Bustamante, E., Morris, H.P. and Pedersen, P.L. (1981) Energy metabolism of tumor cells. Requirement for a form of hexokinase with a propensity for mitochondrial binding. *Journal of Biological Chemistry*, **256**, 8699–8704.
91. DeBerardinis, R.J., Lum, J.J., Hatzivassiliou, G. and Thompson, C.B. (2008) The biology of cancer: metabolic reprogramming fuels cell growth and proliferation. *Cell Metabolism*, **7**, 11–20.
92. Weber, G. (1977) Enzymology of cancer cells. *New England Journal of Medicine*, **296**, 542–551.
93. Ramanathan, A., Wang, C. and Schreiber, S.L. (2005) Perturbational profiling of a cell-line model of tumorigenesis by using metabolic measurements. *Proceedings of the National Academy of Sciences*, **102**, 5992–5997.
94. Griffin, J.L. and Shocker, J.P. (2004) Metabolic profiles of Cancer cells. *Nature Reviews*, **4**, 551–561.
95. Boros, L.s.G., Cascante, M. and Paul Lee, W.-N. (2002) Metabolic profiling of cell growth and death in cancer: applications in drug discovery. *Drug Discovery Today*, **7**, 364–372.
96. Griffiths, J.R., McSheehy, P.M., Robinson, S.P. *et al.* (2002) Metabolic changes detected by in vivo magnetic resonance studies of HEPA-1 wild-type tumors and tumors deficient in hypoxia-inducible factor-1beta (HIF-1beta): evidence of an anabolic role for the HIF-1 pathway. *Cancer Research*, **62**, 688–695.
97. Howe, F.A., Barton, S.J., Cudlip, S.A. *et al.* (2003) Metabolic profiles of human brain tumors using quantitative in vivo 1H magnetic resonance spectroscopy. *Magnetic Resonance in Medicine*, **49**, 223–232.
98. Chan, E.C.Y., Koh, P.K., Mal, M. *et al.* (2008) metabolic profiling of human colorectal cancer using high-resolution magic angle spinning nuclear magnetic resonance (HR-MAS NMR) spectroscopy and gas chromatography mass spectrometry (GC/MS). *Journal of Proteome Research*, **8**, 352–361.
99. Denkert, C., Budczies, J., Kind, T. *et al.* (2006) Mass spectrometry-based metabolic profiling reveals different metabolite patterns in invasive ovarian carcinomas and ovarian borderline tumors. *Cancer Research*, **66**, 10795–10804.

13

SERS in Cells: from Concepts to Practical Applications

Janina Kneipp and Daniela Drescher
Humboldt-Universität zu Berlin, Department of Chemistry, Brook-Taylor-Str. 2, 12489 Berlin, Germany

13.1 Introduction

Raman scattering microscopy, in non-resonant, resonant, and surface-enhanced experiments is extremely versatile regarding the many alterations in the spectroscopic fingerprints of biological materials, specifically cells, that can be observed also over time [1–5]. Moreover, several questions in tumor diagnostics and cell biology can be answered by *in vivo* spectroscopic measurements. The molecular information in the Raman spectroscopic fingerprints is connected with a multitude of cellular parameters. In addition to providing general "phenotypic" information on cellular composition, the Raman data of cells reveal details about molecular structure and interactions.

In recent examples, surface-enhanced Raman scattering (SERS) in the local optical fields of gold or silver nanoparticles has been used in cells to probe the local cellular environment, and to elucidate aspects of metal nanoparticle–biomaterial interactions [6, 7]. Due to the confinement of the local fields to the immediate proximity of the nanoparticles, the excitation and scattering volumes are drastically reduced compared to the actual resolution of the optical microscope. Only molecules that reside in the nanometer-scaled vicinity of the nanoparticles contribute to the observed spectrum, even if the focal volume defined by the excitation laser and geometry of the set-up is much larger. Molecules that serve as reporters or influence the uptake and cellular recognition processes can be introduced with the nanoparticles into cells and organisms, and thereby the nanoparticles can be traced as

Frontiers of Surface-Enhanced Raman Scattering: Single Nanoparticles and Single Cells, First Edition.
Edited by Yukihiro Ozaki, Katrin Kneipp and Ricardo Aroca.

labels also *in vivo*. Furthermore, cellular parameters that are of importance to the process of the nanoparticle–cell interaction and to its consequences for the cellular biochemistry may be identified.

Here, we will discuss some aspects of SERS experiments inside cells of animal origin. We focus on the properties of the nanoparticles that are used for probing, and on their interaction with the cells. First, some basic concepts of typical SERS nanoparticle–cell experiments will be introduced. Then, the great potential of SERS for studies of cells and organisms will be illustrated by discussion of some of their applications.

13.2 SERS Labels and SERS Nanoprobes: Different Approaches to Obtain Different Information

13.2.1 Highlighting Cellular Substructures with SERS Labels

When gold and silver nanoparticles are introduced into cells, the high local fields in their close vicinity upon excitation of localized surface plasmons lead to SERS from molecules residing in the nanoparticles' very surroundings. These can be biological molecules contained in the cell, depending on the location of the nanoparticles, or molecules that are deliberately attached to the nanoparticles. Apart from specific functionalization of the nanoparticles' surface it is possible to attach so-called reporter molecules to them. If the SERS spectrum of such nanoparticles contains contributions from a reporter molecule, the structures can serve as SERS labels (sometimes also referred to as SERS tags).

SERS labels are, typically, combinations of gold or silver nanoaggregates and reporter molecules, that can be used to visualize their cellular distribution and to give further information on intracellular parameters by their specific Raman signature. Dye reporters include Indocyanine Green [8], Crystal Violet [9, 10], Methylene Blue [9, 11], Rhodamine 6G [12], Eosin [11], Giemsa [11], and others. "Colorless" molecules have also been used as Raman reporters. Often, they are attached to gold or silver nanoparticles by a thiol group. Popular candidates are mercaptopyridine, aminothiophenol, *para*-mercaptobenzoic acid (*p*MBA), or thiocresol [10, 13–22].

In general, there are two types of SERS labels. In the first type, the SERS label (metal nanoparticle plus reporter) is covered by a protective layer of glass or polyethylene glycol [117,118]. The advantages of these coated SERS labels include physical robustness, stable signals, and immunity to their biological and chemical environment. Overall, despite these essential technical advantages, this type of SERS label exhibits the same function as a fluorescence tag, that is, highlighting and imaging a biological structure based on the optical signature of the reporter.

The second type of SERS label does not use such a protective cover, here we call these labels SERS hybrid probes, relating to the hybrid nature of the spectroscopic information they transmit [8, 9]. The main advantage of SERS hybrid probes lies in their multifunctionality: SERS hybrid probes can be identified by the signature of a reporter molecule. At the same time, they deliver sensitive molecular structural information on their cellular nanoenvironment. Therefore, hybrid probes are different from typical SERS tags, as the latter do not provide information from other molecules except the reporter [23–25].

In many cases, reporter molecules are fluorescent dyes that can have effective SERS cross sections up to 10^{-16} cm^2, that is, of the level of fluorescence cross sections of good fluorophores [26]. However, compared to fluorescence, SERS labels provide several advantages for live cell use [27]. Instead of a broad and relatively nonspecific fluorescence signal, the optical signature of a SERS label is comprised of several sharp Raman lines within a spectral range comparable to the bandwidth of a fluorescent dye. Because of the high molecular specificity of a Raman spectrum, spectral overlap between different labels is very unlikely, even when their chemical structures are very similar [15, 23, 28]. Spectral correlation techniques using different SERS lines can also improve the contrast of a SERS label relative to the cellular background [8, 15].

By excitation out-of-resonance with electronic transitions of a reporter molecule, photodegradation can be avoided, resulting in high stability of the reporter molecules (and the signals). Consequently, in the non-resonant process, the excitation wavelength is a free parameter that can be set in the near-infrared, where relatively low-energy photons do not cause photodecomposition of the probed object, and where possible interfering biomaterial autofluorescence is weak.

When SERS labels are used to highlight cellular structures, analogously to dyes, or quantum dots in fluorescence labeling, they can be followed and imaged inside a cell based on the SERS spectrum of their reporter molecule. In all these experiments, the distribution of reporter SERS signals can provide an indication of the interaction of the nanostructure with the cell. For example, as illustrated in Figure 13.1, a mixture of different SERS nanoprobe populations carrying different reporters, upon subsequent incubation with cells, can be imaged using multivariate analysis [15].

The possibility of obtaining intrinsic spectral signatures from live cells using hybrid SERS labels made from gold nanoparticles and well-studied dyes, such as Rose Bengal, Crystal Violet [9], and Indocyanine Green [8], as reporters has also been investigated. Figure 13.2 displays examples of labels bearing Crystal Violet as reporter.Trace A shows a spectrum of Crystal Violet. As can be seen in traces C and D it becomes clear that, in

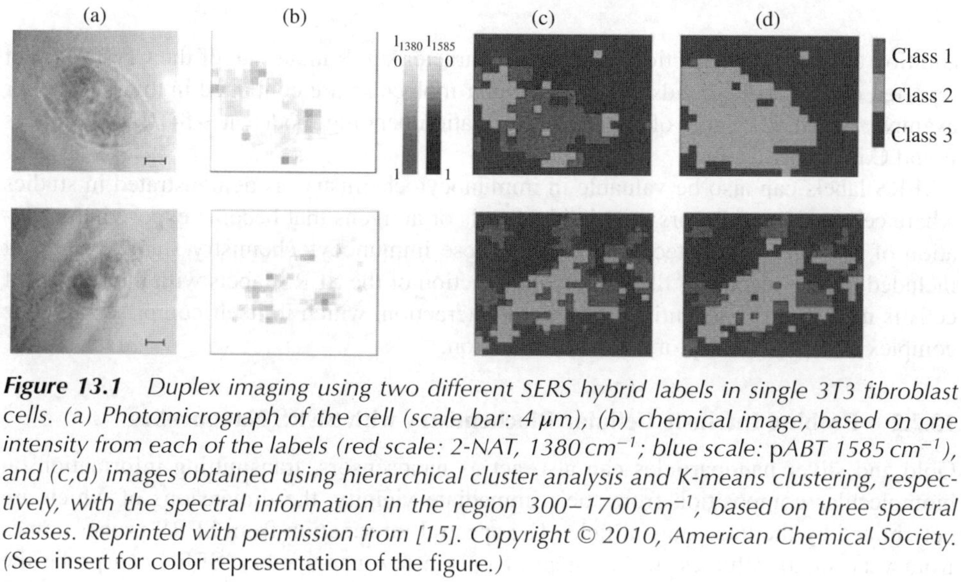

Figure 13.1 Duplex imaging using two different SERS hybrid labels in single 3T3 fibroblast cells. (a) Photomicrograph of the cell (scale bar: 4 μm), (b) chemical image, based on one intensity from each of the labels (red scale: 2-NAT, 1380 cm^{-1}; blue scale: pABT 1585 cm^{-1}), and (c,d) images obtained using hierarchical cluster analysis and K-means clustering, respectively, with the spectral information in the region 300–1700 cm^{-1}, based on three spectral classes. Reprinted with permission from [15]. Copyright © 2010, American Chemical Society. (See insert for color representation of the figure.)

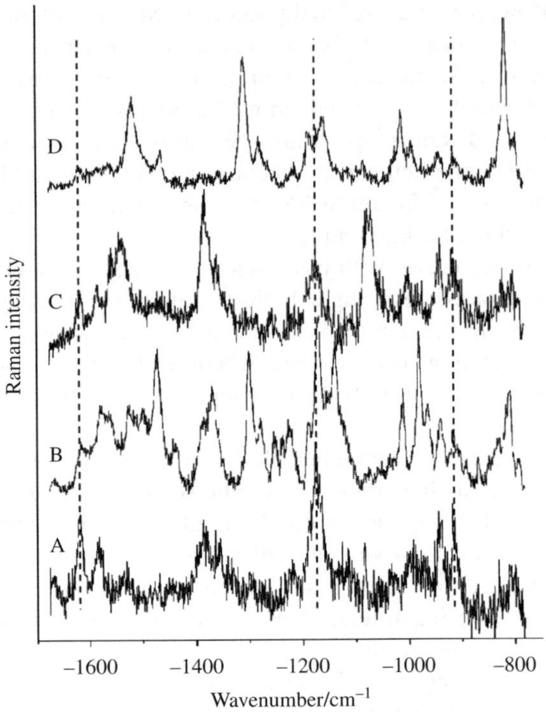

Figure 13.2 *SERS spectra measured in 3T3 cells after 2 h incubation with Crystal Violet SERS labels. SERS spectra collected in a live cell are composed of Raman lines from the reporter along with Raman signatures originating from surrounding native cell molecules. (Three Raman lines of CV are marked by the dashed lines.) Reprinted with permission from Ref. [9]. Copyright © 2008 John Wiley & Sons, Ltd*

addition to their own detection based on a characteristic Raman line of the Crystal Violet reporter, characteristic bands from the cellular molecules are contained in the spectra, for example, the SERS signal of the C-H deformation/bending modes at $\sim$1470 cm^{-1} (traces B and C in Figure 13.2).

SERS labels can also be valuable in immunocytochemistry, as demonstrated in studies where cell surface receptors in individual cells or antigens that become exposed after fixation of a tissue were detected [29–36]. These immunocytochemistry studies were not included in this chapter as this targeted interaction of the SERS labels with the surface of cells is mediated by the antibody–antigen interaction, which in itself comprises a whole complex of topics worthy of detailed discussion.

13.2.2 Probing Intrinsic Cellular Biochemistry with SERS Nanoprobes

Gold and silver nanoparticles can also act as nanosensors, transmitting information on biomolecular composition from their immediate vicinity, the dimensions of which are determined by the local optical field. The spectral information from SERS probes comes from very small volumes in the range of only a few nanometers. SERS can probe the intracellular environment very locally. Thus, the hybrid labels discussed above should not only be viewed from the perspective of the reporter SERS spectrum but also with regard to their ability to provide information on cellular biochemistry. They also serve as nanoprobes.

To understand the spectra that are obtained from nanoparticles contained in cells of a cell culture, the investigation of the uptake mechanism and the nanoparticle processing by the cell is the first crucial step. In fact, the uptake of a material will determine its fate in the cell, as it influences the position in the cellular architecture, the routing within the cellular ultrastructure, and the potential processing pathways. In some Raman experiments, nanoparticle injection was chosen to administer metal nanoparticles into cells for SERS experiments, to avoid interaction of the nanoparticles with components of the cell culture medium and to achieve direct targeting of the cytosol [37], or to reach one specific cell [38]. Others have used electroporation for the direct uptake of nanoparticles [39–41].

As we will discuss in detail in the next sections, in most facile experiments with cells, nanoparticles enter cells by uptake via the fluid phase, that is, from the cell culture medium. Therefore, the SERS spectra of these experiments provide information on the composition of endosomal structures inside the cells [2, 8, 9, 12, 42–46]. Some of these studies also give electron microscopic evidence of the intra-endosomal localization of the nanoparticles in samples that have to be prepared parallel to those used for SERS.

SERS spectra from cellular molecules have been reported in several different cell types, using either only gold nanoparticles [2, 42, 45–47], or reporter-carrying gold or silver nanoparticles in a reporter concentration that permits the detection of intrinsic SERS spectra from the cellular biochemistry in addition to the reporter signature [8, 9, 12, 15, 43].

High SERS signals are mainly caused by high local optical fields around the nanostructures [48–53]. The dependence on the local fields implies that the SERS enhancement depends strongly on the morphology (e.g., the size, shape, and aggregation) of the nanoparticles. Therefore, the spectral information that can be obtained for the biological component is coupled very strongly to the physicochemical properties of the nanoparticles. In a SERS experiment, this strong connection expresses in the following characteristics: (i) The observed SERS intensities vary with the enhancement factor, which depends critically on the ability of the nanoparticles to form aggregates [54, 55]; (ii) the enhancement depends on the type of aggregates, particularly on the size of the gap between nanoparticles and their geometries [56–59]; and (iii) the average intensity will depend on the number of nanoparticles or nanoaggregates in the probed volume. As we will discuss in the next section, the plasmonic properties of noble metal nanoparticles can be greatly altered in the course of the interaction of the nanoparticles with the cell.

13.3 Consequences of Endocytotic Uptake and Processing for Intrinsic SERS Probing in Cells

In the natural situation, nanomaterials can enter eukaryotic cells of animal origin in a variety of ways [60–63]. The physicochemical properties of the nanomaterial, such as size, shape, surface charge, and modification, determine their uptake efficiency and localization [64–67]. The surface charge of metal nanoparticles is changed, for example, when reporter species are attached to them. The altered surface charge can, in principle, result in different uptake behavior of the cells, as was suggested based on the results of recent SERS experiments [13, 68].

Small polyethylene-glycol-functionalized gold nanoparticles were observed to roam free in the cytosol, suggesting their direct uptake through the cell membrane [60]. An important uptake mechanism for nanomaterial is endocytosis. Endocytosis is sensitively regulated, and serves both the uptake of nutrients as well as signaling in the cell. By endocytosis, the cell can take up macromolecules, liquids and pathogens, and also most types of artificial nanoparticles [69, 70]. When the material approaches the surface of the cell, the cellular membrane invaginates and forms a membrane vesicle inside the cell, the endosome. The schematic in Figure 13.3a illustrates the invagination of the cellular membrane around a nanoparticle. Inside the cell, the endosomes including their cargo are transported along a network of actin filaments and microtubules. Endosome formation varies, depending on the type of the cargo and on its size (Figure 13.3a): after interacting with surface receptors or specific regions in the membrane of phagocytic cells, large, micron-sized particles are taken up by phagocytosis, liquids by macropinocytosis. The endocytic process of nanoparticles <300 nm is usually mediated by proteins in the cell membrane, the most important mediators being clathrin and caveolin [71, 72]. The nanoparticles interact with receptor molecules in the membrane, or with membrane rafts displaying a specific lipid composition [69]. Inside the cell, most of the nanomaterial resides in endolysosomal structures, as shown for gold and silver nanoparticles in the transmission electron micrographs in Figure 13.3b,c, respectively.

The major challenge encountered in the design of intracellular SERS probes is concerned with the enhanced local optical fields that the high SERS enhancement levels were shown to be associated with [52, 53]. So far, the highest SERS enhancement factors have been

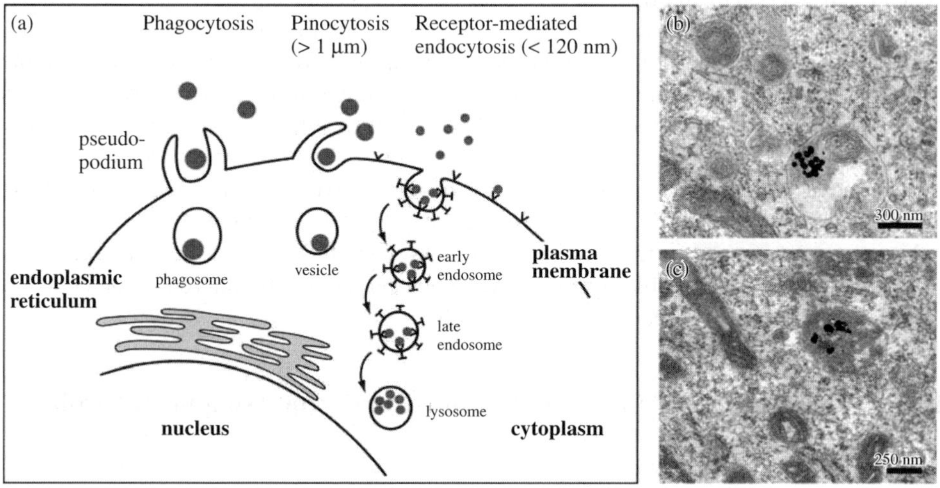

Figure 13.3 Uptake of metal nanoparticles into cells by endocytosis. (a) Schematic: Depending on cell type and material properties, the particle uptake occurs via phagocytosis, pinocytosis, or receptor-mediated endocytosis. (b,c) Transmission electron micrographs of ultramicrotome slices of fibroblast cells show gold (b) and silver (c) nanoparticles enclosed in endosomes

obtained exploiting extremely high field enhancement on aggregates or clusters formed by individual silver or gold nanoparticles [26]. As experiments and theory show, the clusters can vary in size, ranging from dimers [59, 73] to fractal structures [58, 74, 75]. Because of the important role of nanoaggregates and clusters for electromagnetic enhancement, the formation of gold nanoparticle aggregates has to be controlled in the intracellular SERS experiments. Imaging of the nanoparticles in the biological context, using a method that can reveal the cellular substructures (such as electron microscopy), in particular membranes as boundaries of compartments, is crucial for understanding intracellular behavior.

The existence of endosomes inside the cell has been evidenced by several different kinds of results in SERS experiments:

1. Variation in SERS intensities that indicate changes in the plasmonic properties of gold and silver nanoparticles related to processing and aggregation in aging endosomal vesicles [1, 2].
2. The ability to monitor endosomal transport of the metal nanoparticles over time [42, 76].
3. Evidence of a biochemical composition and molecular processing that are characteristic of endosomes and lysosomes, for example, [2, 77].
4. pH values that are typical of different endosomal stages obtained in SERS-based measurements of pH [14, 78]

Points (1–3) result from an interrogation of the biological component using SERS nanoprobes without reporter molecules. In contrast, the measurement of the local pH (4) relies on the spectral signature of a reporter molecule (see Section 13.6 for more details).

As an example of plasmonic conditions changing due to variation of nanoaggregate morphology inside the cell after endocytosis, the results of an experiment where gold nanoprobes were applied in an epithelial cell line as a pulse are shown in Figure 13.4. The nanoparticles were added to the cell culture medium and withdrawn again after a defined time of 30 min. Then, in a series of samples taken at different time points, the SERS signal from the vicinity of the nanoprobes was observed over time (Figure 13.4a–c). The signal strength of the SERS was found to vary significantly for different time points after endocytosis of the nanoparticles had taken place. These changes in SERS enhancement indicate morphological differences of the gold nanostructures inside the endosomal vesicle, that were explained by the formation of nanoparticle aggregates [2]. Recently, in accordance with theoretical predictions, we have shown experimentally that the enhancement of individual gold nanospheres is very low, around 10^2, but that the formation of aggregates can change the enhancement factor by many orders of magnitude [54].

The signal strength as a function of nanoaggregate morphology and aggregation was investigated by a transmission electron microscopy (TEM) study carried out in parallel samples (compare example in Figure 13.4d) [2]. In our experiments, the spectra obtained after 120 min (Figure 13.4b) exhibited the strongest signals. This improvement in signal strength correlates with the formation of small gold aggregates, mostly dimers and trimers [2]. TEM images (example in Figure 13.4d) indicate that later in the time course, after 180 min, interparticle distances in the nanoparticle accumulations were greater than in the dimers and trimers observed after 120 min (see arrow in Figure 13.4d). The higher interparticle distances, which are probably due to multivesicular inclusions, may lead to a decrease in the overall SERS intensity again (Figure 13.4c).

Considering the function of endolysosomes as the "cell's waste disposal" and the abundance of proteases and other hydrolyzing enzymes taking care of the degradation of biomacromolecules, it seems obvious that nanomaterials will also be modified in these cellular compartments. Raman experiments employing SERS provide evidence of molecular surface modification of gold and silver nanoparticles. For example, Oyelere *et al.* introduced peptide-modified nanorods into cells, where they monitored the intensity of the C-N stretch band of a thioalkyl-triazole linker for a targeting peptide at $760\,cm^{-1}$ [77]. From a decrease in intensity they conclude conformational changes of the linker when the nanoparticles with the attached peptide enter the cells [77].

Similarly, the molecules that are used as reporters in SERS labels may be modified or displaced from the nanoparticle surface. Sirimuthu and coworkers reported the degradation of signal from reporter molecules on the surface of silver nanoparticles, depending on whether the reporter molecules were attached by chemisorption to the silver by a mercapto function, as with *p*MBA, or by physisorption, as in the case of the dye Rhodamine 6G [18]. They observed that the signal of Rhodamine 6G physisorbed to silver nanoaggregates disappeared after 8 h in the cell culture medium, and after ~12 h in Chinese hamster ovary and MG-63 cells. It is remarkable that the SERS signature from the dye was not replaced by SERS signals from the cellular molecules in these experiments but instead disappeared completely [18], contradictory to work with labeled and unlabeled silver nanoparticles in cells, where signals from cellular molecules have been reported [1, 13].

13.4 Quantification of Metal Nanoparticles in Cells

Due to the impact of the plasmonic conditions created by nanoparticle aggregation inside cells, and the variance regarding nanoparticle uptake and processing between individual cells as well as cell lines, the results from SERS experiments cannot be considered quantitatively reliable regarding the number of metal nanoparticles in a cell.

Therefore it is necessary to achieve quantification of the nanoparticles by other means. Inductively coupled plasma mass spectrometry (ICP-MS) is a well-established method to determine the metal content per cell [79, 80]. Inorganic nanoparticles can be quantified by elemental analysis, in particular by ICP-MS or ICP-AES (inductively coupled plasma atomic emission spectroscopy) upon extraction or after digestion of the cell suspension [64, 81, 82]. Gschwind *et al.* and Franze *et al.* coupled a droplet on demand sample introduction system to an ICP-MS for the direct analysis of nanoparticles with particles sizes below 100 nm [83, 84]. All these techniques have the disadvantage that the spatial information on the distribution of nanoparticles in an individual cell is lost.

A technique that provides several advantages over most ICP-MS studies, is metal nanoparticle quantification by laser ablation inductively coupled plasma mass spectrometry (LA-ICP-MS). We have recently shown that spatially resolved LA-ICP-MS can be employed to visualize the distribution of silver and gold nanoparticles in single cells (Figure 13.5) [85]. Beside the ease of calibration in element quantification, the multi-element detection with high sensitivity and high spatial resolution makes LA-ICP-MS well-suited for single-cell analysis [79, 86–88].

Based on a matrix-matched calibration, proof-of-principle could be provided that quantification of nanoparticles in individual cells can be achieved using an LA-ICP-MS

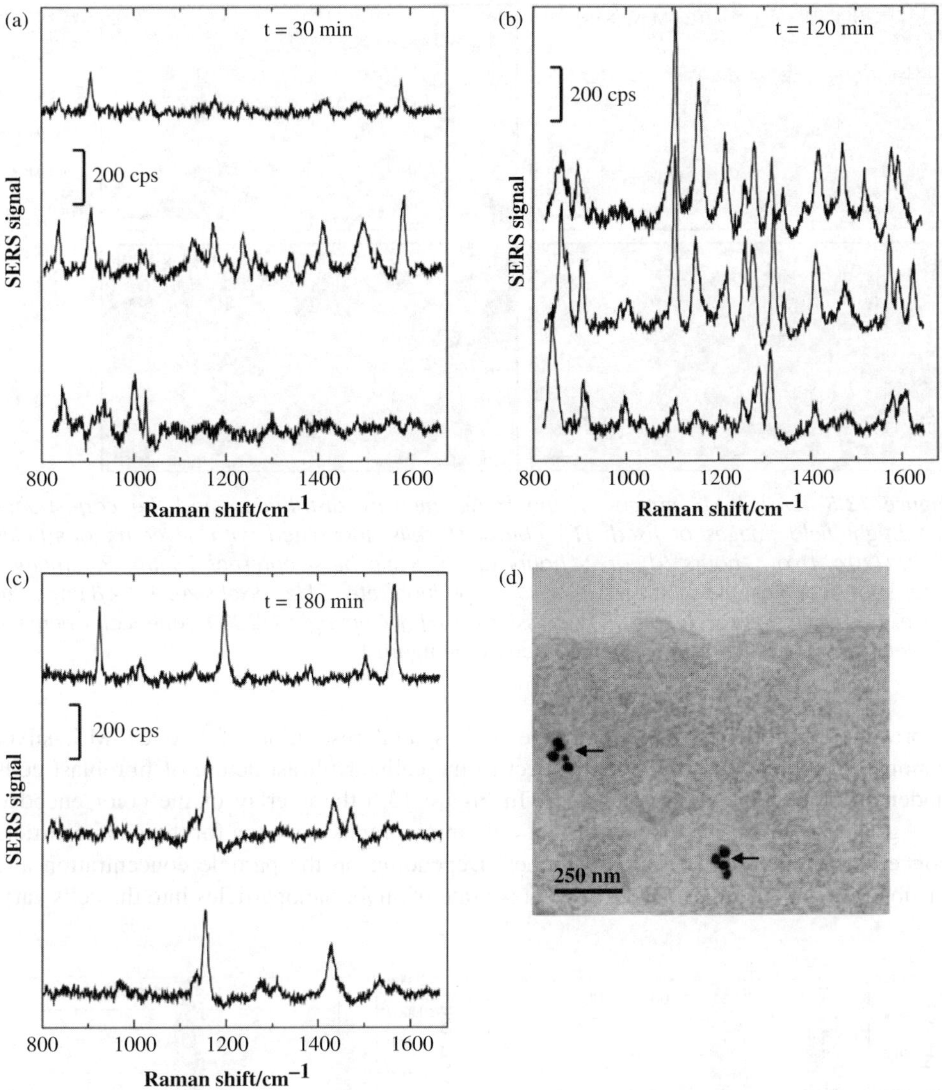

Figure 13.4 *(a–c) Typical SERS spectra from cells of the epithelial cell line IRPT after incubation with gold nanoparticles, excited with $<3 \times 10^5\ W\,cm^{-2}$ at 785 nm, collection time 1 s. The incubation time, including a 30 min nanoparticle pulse, is indicated in each panel. (d) Transmission electron micrograph of IRPT cell showing gold nanoparticles in endosomal structures. Arrows point to large distances between the nanoparticles, probably brought about by multivesicular inclusions. Reproduced with permission from Ref. [2]. Copyright © 2009, American Chemical Society*

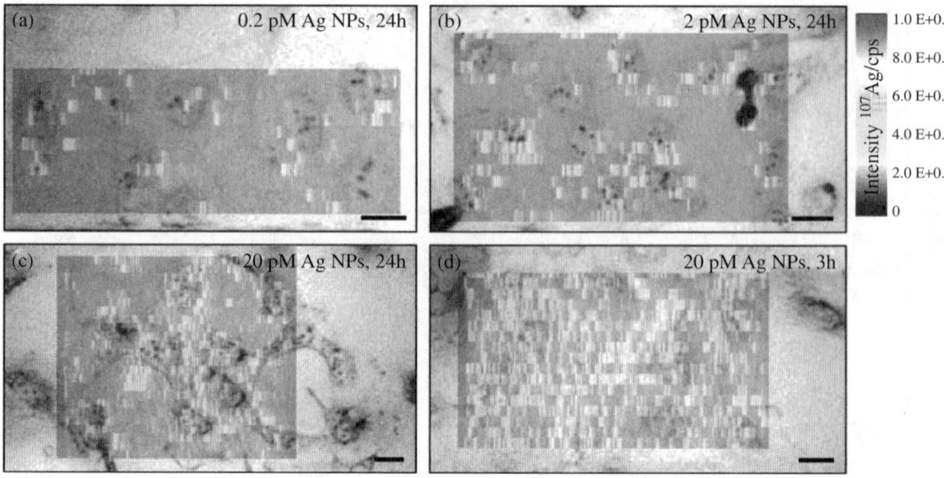

Figure 13.5 LA-ICP-MS images of the ^{107}Ag intensity distribution and the corresponding bright field images of fixed 3T3 fibroblast cells, incubated with silver nanoparticles (0.2/2/20 pM) or 3 hours (d) or 24 hours (a–c). Scale bars represent 25 μm. Parameters: laser spot size, 8 μm; scan speed, 8 μm s^{-1}; repetition rate, 5 Hz; pixel size, 1.2 × 8 μm^2; fluence, 1.5 J cm^{-2}. Reproduced with permission from [2]. Copyright © 2009, American Chemical Society. (See insert for color representation of the figure.)

approach (Figure 13.6). Owing to the high spatial resolution of LA-ICP-MS, silver nanoparticles are visualized with respect to the cellular ultrastructure of fibroblast cells under different incubation conditions. In Figure 13.6 the overlay of the color encoded $^{107}Ag^+$ local intensity maps with bright field micrographs is shown for three nanoparticle concentrations and two incubation times. Depending on the particle concentration and incubation time, the distribution and uptake rate of silver nanoparticles into the cells vary.

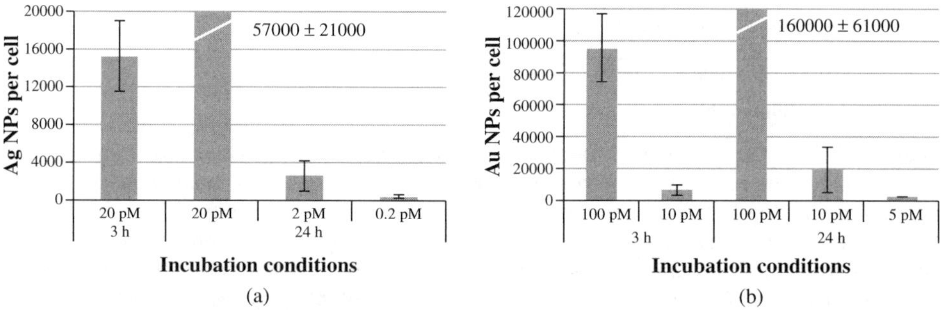

Figure 13.6 Number of silver (a) and gold (b) nanoparticles per fibroblast cells as determined by LA-ICP-MS based on matrix-matched calibration data (mean value of 6–10 cells). Cells were incubated with silver and gold nanoparticles in different concentrations for 3 and 24 h. The standard deviation bars represent the differences in nanoparticle uptake between cells to the other. Parameters: laser spot size, 8 μm; scan speed, 8 μm s^{-1}; repetition rate, 5 Hz; fluence, 1.5 J cm^{-2}

The 2D maps in Figures 13.5a–c show a high quantity of ^{107}Ag in the perinuclear region of the cells, which is in contrast to the 3 h-exposure where the particles are homogenously distributed in the cytoplasm. Only low signal intensity was found in the nucleus region since nanoparticles above 20 nm could not enter through the nuclear membrane. These observations are in good agreement with results of other studies from microscopic experiments of cells incubated with silver nanoparticles [89, 90].

For the quantification, single fibroblast cells as well as a nitrocellulose membrane doped with the gold and silver nanoparticle suspensions were ablated and analyzed by ICP-MS. Due to the lack of appropriate internal standards, we must rely here on an absolute approach, which is valid for fully ablated samples. A limit of detection of 20 particles for silver and 190 particles for gold nanoparticles allows the determination of small aggregates in the cellular interior. The number of nanoparticles per cell in Figure 13.6 is given as mean value of 6–10 cells for each incubation condition. The standard deviation bars illustrate the differences in nanoparticle uptake from one cell to another due to the variability in cell size, morphology, and phase. Different uptake rates were found for gold and silver nanoparticles depending on the cell experimental conditions. With increasing particle concentration and incubation time, the number of gold and silver nanoparticles per cell is increased. Fibroblast cells incubated with 0.2–2 pM silver nanoparticles for 24 h contain 380 and 2500 particles per cell, respectively (Figure 13.6a). A silver nanoparticle concentration of 20 pM, that was proven to be cytotoxic, yields 57 000 particles per cell. Compared to the silver nanoparticles, the number of gold nanoparticles per cell is much higher for the different incubation conditions (Figure 13.6b).

The high sensitivity and spatial resolution of LA-ICP-MS allows quantification of nanoparticles at the single-cell level. These quantitative results can be used to correlate the factors of influence, for example, number of nanoparticles per laser spot and intracellular particle localization, with SERS data. The combination of LA-ICP-MS and SERS can give valuable insight into the interaction of gold and silver nanoparticles with individual cells. Associated with the quantification and localization of nanoparticles, which constitute information on the interaction of nanomaterials with cells, are the problems related to cytotoxicity.

13.5 Toxicity Considerations

In the context of nanosafety and nanomedicine, investigations of potential cytotoxicity and design of new intracellular nanosensors based on plasmonic properties of noble metal nanostructures are required. The cytotoxic behavior of nanoparticles (Figure 13.7) depends strongly on the material's properties, namely size, morphology, crystallinity, and surface modification, as well as the composition of the biomolecular corona. The toxicity of gold [91–97] and silver nanoparticles [76, 98–100] has been extensively investigated.

For gold nanoparticles various cytotoxic effects have been described, depending on size, particle concentration, and surface modification. Pernodet *et al.* reported a decrease in cell proliferation, adhesion, and motility of dermal fibroblast cells after exposure to 14 nm gold nanoparticles [95]. In contrast, others found no toxic impact of gold nanoparticles on K562 leukemia cells [101] and RAW264.7 macrophage cells [102], respectively, despite the nanoparticles being taken up efficiently. Contrary results such as these could arise from different surface charge or surface modifications used for stabilization of the nanoparticles. Several studies suggest adverse effects on cell viability for smaller gold nanoparticles that have been attributed to mitochondrial stress and DNA damage [100, 103, 104].

Especially in the case of silver nanoparticles, cytotoxicity should be considered when designing SERS nanoprobes. In the last few years silver nanotoxicity has been heavily studied, mainly due to the ubiquity of silver nanoparticles in our everyday life, for example, in disinfectants and as additives in textiles. The cytotoxicity of silver nanoparticles is attributed to membrane disruption, the formation of reactive oxygen species and DNA damage (Figure 13.7) [98, 100, 105, 106]. These effects are more pronounced for smaller particles since the surface area of the particles and the uptake efficiency are strongly increased. In red blood cells, addition of silver nanoparticles, different from gold or platinum nanoparticles, results in hemolysis, that is, destruction of the erythrocyte membrane and leakage of the hemoglobin from the cells [107].

An important issue that has, however, only been rarely addressed so far is the toxicity of SERS labels and hybrid labels containing reporter molecules, and of the reporter molecules themselves. So far, in many cases, biocompatible dyes, such as Indocyanine

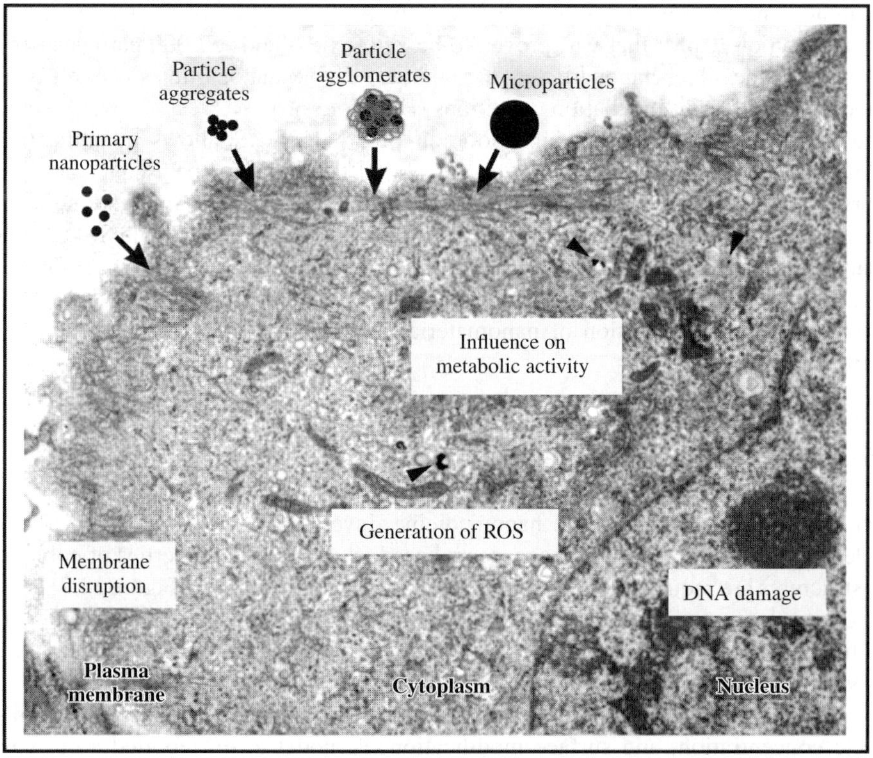

Figure 13.7 *Schematic of potential nanoparticle/cell interactions. Aspects that are of interest in the interaction of silver and gold nanoparticles and SERS labels with cells, with emphasis on potential nanotoxicity. Toxicity of the nanomaterial may depend strongly on the particle properties, and involves many biochemical parameters*

Green or Rose Bengal have been applied and assumed to be non-toxic [8, 9]. However, creating great arsenals of different reporters for live cell multiplex approaches will require an assessment of their potential cytotoxicity. The toxicity of aromatic thiols (benzenethiol and *para*-aminobenzenethiol, *p*ABT), that are often used as SERS reporters was shown in early work in blood cells [108]. There, oxidative stress was induced in the cells using *p*ABT at a concentration of 0.25 mM in cell culture medium [108].

We investigated the viability of cells after administration of reporter molecules and SERS labels in reporter concentrations of 10^{-7} to 10^{-6} M that are typical of many SERS experiments. Cell viability can be studied using an XTT (2,3-bis-(2-methoxy-4-nitro-5-sulfophenyl)-2H-tetrazolium-5-carboxanilide) assay, a standard assay where the metabolic activity of the cells is determined based on the activity of intra-mitochondrial dehydrogenase [109] (see the schematic in Figure 13.8a). The tetrazolium salt XTT is reduced via NAD(P)H-dependent cellular oxidoreductase enzymes to yield a water-soluble formazan dye (Figure 13.8a). The concentration of the formazan salt, quantified by measuring the absorbance at a wavelength of 460 nm using a plate reader, correlates with the number of viable cells in the culture. Cell viability is expressed as the ratio of the result obtained with cells exposed to the reporter molecules to that for negative controls (untreated cells). Positive controls can be generated by exposing them to other known noxes, such as surfactants that destroy the cells by solubilizing cell membranes, for example, Triton X.

Figure 13.8b displays the viability of fibroblast cells after 24 h incubation with two reporter molecules, *p*ABT and 2-naphthalene thiol (2-NAT), at five different concentrations, and with SERS labels carrying the reporters (10^{-10} M particle concentration in the culture medium). The cells were washed with buffer once and incubated with the XTT reagent in culture medium. After an incubation time of 24 h with both *p*ABT and 2-NAT

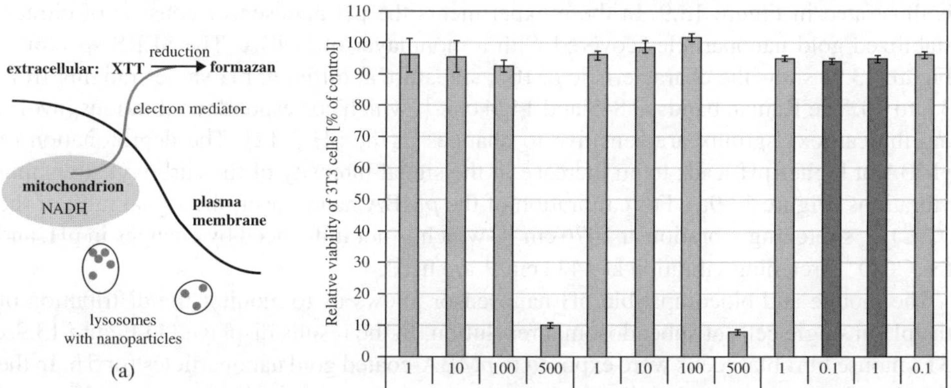

Figure 13.8 *(a) Principle of XTT assay (schematic). (b) Viability of 3T3 cells after 24 h-exposure to different concentrations of* pABT *and 2-NAT in culture medium as determined by XTT assay (mean value ± margin of deviation of four replicates). The cytotoxic effect of the gold nanoprobes with the corresponding reporter on 3T3 cells, as used in SERS experiments, was also assigned (bars are highlighted). Reproduced with permission from [15]. Copyright © 2010, American Chemical Society*

in concentrations of $\leq 10^{-4}$ M, the cells display >90% viability. Reporter concentrations $>10^{-4}$ M have a significant influence on the metabolic activity of the cells, and they reduce the cell viability dramatically (Figure 13.8b). After incubation for 24 h with the SERS labels, the viability of fibroblast cells was observed to be 95% of that of control cells. This is in good agreement with results for plain gold nanoparticles [91, 94, 101].

13.6 Applications

13.6.1 pH Nanosensors for Studies in Live Cells

A parameter that varies drastically over time in the endosomes containing nanoparticles is the pH [110]. In many cell types, significant acidification compared to earlier endosomal stages takes place in the lysosome. There, to ensure for example, proper enzymatic function, pH can be well below 5, and even lower (<4) [111].

The SERS signatures of particular reporter molecules as part of SERS probes were found to be pH-dependent. SERS spectra of *p*MBA on silver and gold electrodes and the dependence of the spectra on the pH value have been reported and discussed in terms of the adsorption geometry on the metal and on the state of dissociation of the carboxyl group as a consequence of the surrounding pH [112]. The pH-dependent SERS spectra of *p*MBA adsorbed on gold nanoshells bound to a silicon substrate were used to create a pH meter working over the range pH 5.8–7.6 pH units [113]. Using 4-mercaptopyridine as a reporter, the pH range detected on the nanoshells can be extended from 3 to 7 [114]. Similar experiments demonstrated that hollow gold nanospheres with *p*MBA were responsive over a pH range of 3.5–9 [115].

The application of a *p*MBA-based SERS nanosensor for pH imaging of single live cells is illustrated in Figure 13.9. In these experiments the pH nanosensor consists of citrate-stabilized gold nanoparticles covered with a monolayer of *p*MBA. The SERS spectra in Figure 13.9a show the characteristic *p*MBA signature at different pH values ranging from 3.8 to 10. The Raman bands at 849 and 1424 cm^{-1}, which correspond to vibrations involving the carboxyl group, are sensitive to changes in the pH [112]. The deprotonation of *p*MBA at higher pH leads to an increase in the signal intensity of the carboxyl stretching vibrations (Figure 13.9a). For calibration of the *p*MBA nanosensor, the signal ratio of the $(C\text{-}C)_{ring}$ stretching vibration at 1076 cm^{-1}, which is not influenced by changes in pH, and the COO^- stretching vibration at 1424 cm^{-1} are used.

The mobile and biocompatible pH nanosensor allows us to monitor the distribution of the pH in single cells at subendosomal resolution. In the results displayed in Figure 13.9c, 3T3 mouse fibroblast cells were exposed to *p*MBA-coated gold nanoparticles for 5 h. In the Raman microspectroscopic cell experiments, spectra were acquired by raster scanning over a single live cell in steps of 2 μm. The SERS spectra measured in the cellular environment of the pH nanosensor (Figure 13.9a) exhibit different *p*MBA signatures, which indicate different pH values inside the cell.

The Raman chemical map of an individual fibroblast cell in Figure 13.9c visualizes the intracellular distribution of the pH. Obviously, after a few hours incubation, the incorporated gold nanoparticles have accumulated inside the cell. Depending on the localization of the pH nanosensor, the intraorganelle pH differs within the cells. Close to the cellular membrane the pH is higher than in the perinuclear region. This is attributed to the maturation

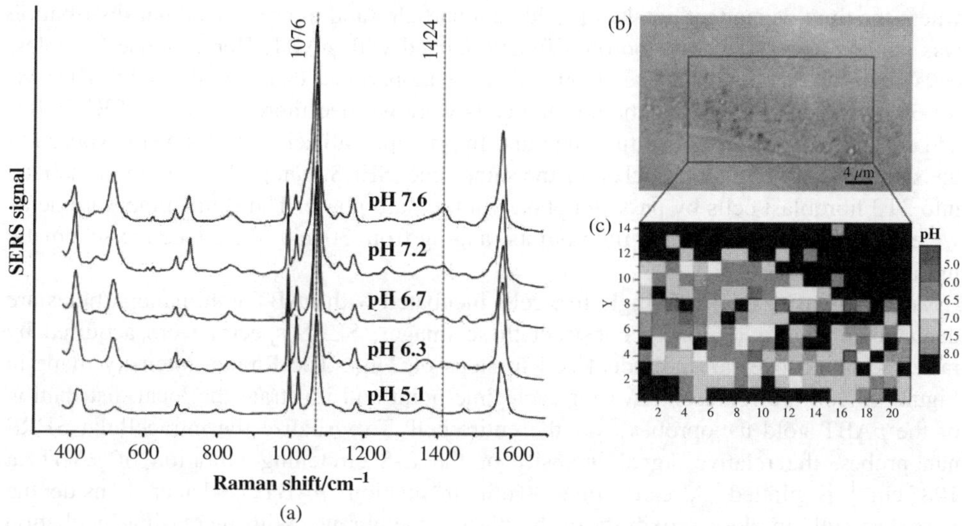

Figure 13.9 *SERS investigation of the intraendosomal pH value. (a) Selected SERS spectra of the pH nanosensor, composed of* para-mercaptobenzoic acid *(*p*MBA) coated gold nanoparticles, in live cells. Mouse fibroblast cells were exposed to* p*MBA gold nanoparticles for 5 h (pMBA concentration 1 mM). (b) Bright field image of an individual fibroblast cell. (c) Raman chemical map indicating the local distribution of the intracellular pH value* in vitro *(intensity ratio of the* p*MBA Raman bands at 1076–1424 cm^{-1}). An excitation wavelength of 785 nm, a laser intensity of 8×10^4 $W cm^{-2}$ and an acquisition time of 1 s were applied [89]. (*See insert for color representation of the figure.*)*

of the vesicles containing the pH nanosensor. In the course of cellular processing the pH becomes more acidic, decreasing from pH 7.4 in the endocytic vesicle formed during particle uptake to pH 4.5–5 in the lysosomes. The latter often accumulate near the cell nucleus. Since particles are continuously taken up during the incubation time of 5 h, early endosomes and lysosomes coexist in the same cell. Ultrastructural studies of this cell line (see Figure 13.3) prove, that at this stage, many particles are enclosed in late endosomes and lysosomes near the cell nucleus.

The distribution of sampled spots of different pH in the cells, as displayed in Figure 13.9c, can also be analyzed statistically, and can be used to follow the dynamics of the pH change in a population of endosomes [78]. We also demonstrated the application of a *p*MBA SERS sensor using two-photon excited Raman scattering (surface-enhanced hyper Raman scattering, SEHRS) [1] [14]. SEHRS spectra of *p*MBA show the same bands as SERS spectra, but the relative signal strengths are changed. These differences in the SEHRS signature of *p*MBA are useful for creating a pH sensor that allows measurements over a more extended range, from pH 2 to 8 [14].

13.6.2 Following Cell Division with SERS

The uptake, accumulation, and re-distribution of SERS labels in cells can be monitored over the course of many hours and days. In Figure 13.10, the results of an experiment are shown where the time-dependent uptake of gold nanoparticles and their intracellular distribution was studied using SERS nanoprobes functionalized with *p*ABT. Here, mouse fibroblast cells were incubated with citrate-stabilized gold nanoparticles coated with *p*ABT (100 nM) for 30 min. After particle incubation, the cells were washed thoroughly with PBS buffer and covered with fresh cell culture medium. In principle, all cell samples were exposed to the same amount of nanoparticles for the same time. SERS nanoprobes were internalized into 3T3 fibroblast cells by passive uptake via the fluid phase. The Raman measurements were conducted after different time-points ranging from 30 min to 5 d (see the schematic in Figure 13.10a).

Raman intensity maps of single live cells incubated with *p*ABT gold nanoparticles are depicted in Figure 13.10. To construct these images, SERS spectra were acquired by raster scanning over an individual cell in steps of 2 μm. The Raman intensity maps in Figure 13.10b are representative for each time-point and illustrate the local distribution of the *p*ABT gold nanoprobes over the entire cell. To visualize the intracellular SERS nanoprobes, the relative signal intensity of the C-S stretching vibration of *p*ABT at 1085 cm^{-1} is plotted. Already after 30 min incubation, *p*ABT is detected inside the fibroblast cells in close proximity to the plasma membrane. With increasing incubation time (3 h to 2 d), particles are evenly distributed in the cytosol and accumulate in the perinuclear region in the course of endo-lysosomal processing. The Raman intensity map acquired after incubation for 2 d does not show signals in the region of the cell nucleus. After incubation for 5 d and several cell divisions: the intracellular signal intensity of

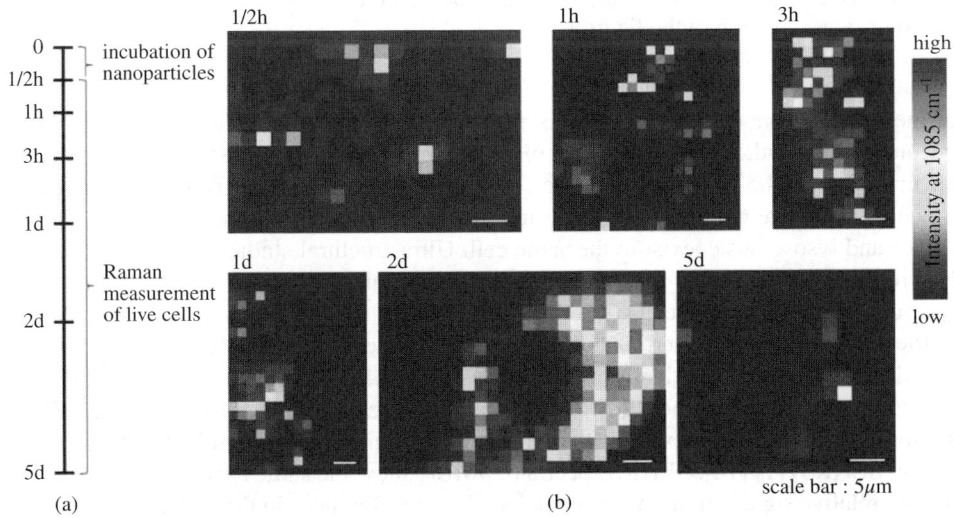

Figure 13.10 *SERS nanoprobes for time-dependent analysis of nanoparticle uptake and their intracellular distribution in live cells. (a) Experimental design: Fibroblast cells were incubated with* para-aminobenzenethiol *coated gold nanoparticles (*pABT*, 100 nM) for 30 min. SERS measurements were conducted after different time points. (b) Raman chemical maps illustrate the nanoparticle distribution inside fibroblast cells for different stages of incubation. The intensity of the C-S stretching vibration of* pABT *at 1085* cm^{-1} *was used for imaging. (*See insert for color representation of the figure.*)*

*p*ABT and the amount of detected Raman spectra per cell decrease significantly since the primary vesicles are distributed over more cells. During cell division, the cytoplasm is divided between the two daughter cells. This dilution of SERS nanoprobe-containing endosomes is a trivial effect in cell lines, and has also been observed in other studies [68]. However, observing particle distribution between differentiating cells in developing tissues may lead to better understanding of the distribution of cytoplasmic and nanomaterial in developing organisms [38, 116].

SERS spectra of different stages of the endo-lysosomal pathway show the characteristic Raman bands of *p*ABT. After incubation for 1, 2 and 5 d, cell-specific Raman bands are detected in addition to the *p*ABT Raman signature. This is attributed to the replacement of particle-bound *p*ABT by cellular molecules with increasing incubation times. Furthermore, the signal intensity increases with increasing incubation time from 30 min to 48 h, which is correlated with a higher SERS enhancement. This can be assigned to changes in the particle aggregation with increasing incubation times due to multivesicular fusion (see also TEM images of gold nanoparticles in Figure 13.4). The SERS nanoprobes used in these experiments induce no cytotoxic effects, as evidenced by XTT assay (Figure 13.8) [15].

Acknowledgment

Both authors acknowledge funding by ERC Starting Grant MULTIBIOPHOT.

References

1. Kneipp, J., Kneipp, H. and Kneipp, K. (2006) Two-photon vibrational spectroscopy for biosciences based on surface-enhanced hyper-Raman scattering. *Proceedings of the National Academy of Sciences of the United States of America*, **103**, 17149–17153.
2. Janina Kneipp, H., Kneipp, M., McLaughlin, D. *et al.* (2006) In vivo molecular probing of cellular compartments with gold nanoparticles and nanoaggregates. *Nano Letters*, **6**, 2225–2231.
3. Okada, M., Smith, N.I., Palonpon, A.F. *et al.* (2012) Label-free Raman observation of cytochrome C dynamics during apoptosis. *Proceedings of the National Academy of Sciences of the United States of America*, **109**, 28–32.
4. Peticolas, W.L., Patapoff, T.W., Thomas, G.A. *et al.* (1996) Laser Raman microscopy of chromosomes in living eukaryotic cells: DNA polymorphism in vivo. *Journal of Raman Spectroscopy*, **27**, 571–578.
5. van Manen, H.-J., Kraan, Y.M., Rosos, D. and Otto, C. (2005) Single-cell Raman and fluorescence microscopy reveal the association of lipid bodies with phagosomes in leukocytes. *Proceedings of the National Academy of Sciences of the United States of America*, **102**, 10159–10164.
6. Chourpa, I., Lei, F.H., Dubois, P. *et al.* (2008) Intracellular applications of analytical SERS spectroscopy and multispectral imaging. *Chemical Society Reviews*, **37**, 993–1000.
7. Kneipp, J., Kneipp, H. and Kneipp, K. (2008) Sers – a single-molecule and nanoscale tool for bioanalytics. *Chemical Society Reviews*, **37**, 1052–1060.

8. Kneipp, J., Kneipp, H., Rice, W.L. and Kneipp, K. (2005) Optical probes for biological applications based on surface enhanced Raman scattering from indocyanine green on gold nanoparticles. *Analytical Chemistry*, **77**, 2381–85.
9. Kneipp, J., Kneipp, H., Rajaduraj, A. *et al.* (2009) Optical probing and imaging of live cells using SERS labels. *Journal of Raman Spectroscopy*, **40**, 1–5.
10. Tan, X., Wang, Z., Yang, J. *et al.* (2009) Polyvinylpyrrolidone- (PVP-) coated silver aggregates for high performance surface-enhanced Raman scattering in living cells. *Nanotechnology*, **20**, 445102.
11. Stokes, R.J., McKenzie, F., McFarlane, E. *et al.* (2009) Rapid cell mapping using nanoparticles and SERRS. *Analyst*, **134**, 170–175.
12. Tang, H.-W., Yang, X.B., Kirkham, J. and Smith, D.A. (2007) Probing intrinsic and extrinsic components in single osteosarcoma cells by near-infrared surface-enhanced Raman scattering. *Analytical Chemistry*, **79**, 3646–3653.
13. Gregas, M.K., Yan, F., Scaffidi, J. *et al.* (2011) Characterization of nanoprobe uptake in single cells: spatial and temporal tracking via SERS labeling and modulation of surface charge. *Nanomedicine-Nanotechnology Biology and Medicine*, **7**, 115–122.
14. Kneipp, J., Kneipp, H., Wittig, B. and Kneipp, K. (2007) One- and two-photon excited optical pH probing for cells using Surface-enhanced Raman and Hyper-Raman nanosensors. *Nano Letters*, **7**, 2819–2823.
15. Matschulat, A., Drescher, D. and Kneipp, J. (2010) Surface-enhanced Raman scattering hybrid nanoprobe multiplexing and imaging in biological systems. *ACS Nano*, **4**, 3259–3269.
16. Park, J.H., Park, J., Dembereldorj, U. *et al.* (2011) Raman detection of localized transferrin-coated gold nanoparticles inside a single cell. *Analytical and Bioanalytical Chemistry*, **401**, 1631–1639.
17. Shamsaie, A., Heim, J., Yanik, A.A. and Irudayaraj, J. (2008) Intracellular quantification by surface enhanced Raman spectroscopy. *Chemical Physics Letters*, **461**, 131–135.
18. Narayana, M., Sirimuthu, S., Syme, C.D. and Cooper, J.M. (2011) Investigation of the stability of labelled nanoparticles for Se(R)Rs measurements in cells. *Chemical Communications*, **47**, 4099–4101.
19. Syme, C.D., Sirimuthu, N.M.S., Faley, S.L. and Cooper, J.M. (2010) SERS mapping of nanoparticle labels in single cells using a microfluidic chip. *Chemical Communications*, **46**, 7921–23.
20. Talley, C.E., Jusinski, L., Hollars, C.W. *et al.* (2004) Intracellular pH sensors based on surface-enhanced Raman scattering. *Analytical Chemistry*, **76**, 7064–68.
21. Wang, Z.Y., Wu, H., Wang, C.L. *et al.* (2011) Gold aggregates- and quantum dots-embedded nanospheres: switchable dual-mode image probes for living cells. *Journal of Materials Chemistry*, **21**, 4307–4313.
22. Xie, J., Zhang, Q., Lee, J.Y. and Wang, D.I.C. (2008) The synthesis of SERS-active gold nanoflower tags for in vivo applications. *ACS Nano*, **2**, 2473–2480.
23. Qian, X.M., Peng, X.H., Ansari, D.O. *et al.* (2008) In vivo tumor targeting and spectroscopic detection with surface-enhanced Raman nanoparticle tags. *Nature Biotechnology*, **26**, 83–90.
24. Cao, Y.W.C., Jin, R.C. and Mirkin, C.A. (2002) Nanoparticles with Raman spectroscopic fingerprints for DNA and RNA detection. *Science*, **297**, 1536–1540.

25. Gellner, M., Koempe, K. and Schlücker, S. (2009) Multiplexing with SERS labels using mixed SAMs of Raman reporter molecules. *Analytical and Bioanalytical Chemistry*, **394**, 1839–1844.
26. Kneipp, K., Wang, Y., Kneipp, H. *et al.* (1997) Single molecule detection using surface-enhanced Raman scattering (SERS). *Physical Review Letters*, **78**, 1667.
27. J. Kneipp, Nanosensors based on SERS for applications in living cells, *Surface-Enhanced Raman Scattering: Physics and Applications*, Topics in Applied Physics, Vol. **103**, Springer, Berlin 2006, pp. 335–349.
28. Kneipp, K., Kneipp, H., Itzkan, I. *et al.* (1999) Ultrasensitive chemical analysis by Raman spectroscopy. *Chemical Reviews*, **99**, 2957–2967.
29. Hering, K., Cialla, D., Ackermann, K. *et al.* (2008) SERS: a versatile tool in chemical and biochemical diagnostics. *Analytical and Bioanalytical Chemistry*, **390**, 113–124.
30. Hu, Q., Tay, L.-L., Noestheden, M. and Pezacki, J.P. (2007) Mammalian cell surface imaging with nitrile-functionalized nanoprobes: biophysical characterization of aggregation and polarization anisotropy in SERS imaging. *Journal of the American Chemical Society*, **129**, 14–15.
31. Kennedy, D.C., Hoop, K.A., Tay, L.-L. and Pezacki, J.P. (2010) Development of nanoparticle probes for multiplex SERS imaging of cell surface proteins. *Nanoscale*, **2**, 1413–1416.
32. Kennedy, D.C., McKay, C.S., Tay, L.-L. *et al.* (2011) Carbon-bonded silver nanoparticles: alkyne-functionalized ligands for SERS imaging of mammalian cells. *Chemical Communications*, **47**, 3156–3158.
33. Leen Koh, A., Shachaf, C.M., Elchuri, S. *et al.* (2008) Electron microscopy localization and characterization of functionalized composite organic–inorganic SERS nanoparticles on leukemia cells. *Ultramicroscopy*, **109**, 111–121.
34. Lutz, B.R., Dentinger, C.E., Nguyen, L.N. *et al.* (2008) Spectral analysis of multiplex Raman probe signatures. *ACS Nano*, **2**, 2306–2314.
35. Schlücker, S., Küstner, B., Punge, A. *et al.* (2006) Immuno-Raman microspectroscopy: in situ detection of antigens in tissue specimens by surface-enhanced Raman scattering. *Journal of Raman Spectroscopy*, **37**, 719–721.
36. Sun, L., Sung, K.-B., Dentinger, C. *et al.* (2007) Composite organic–inorganic nanoparticles as Raman labels for tissue analysis. *Nano Letters*, **7**, 351–356.
37. Candeloro, P., Tirinato, L., Malara, N. *et al.* (2011) Nanoparticle microinjection and Raman spectroscopy as tools for nanotoxicology studies. *Analyst*, **136**, 4402–4408.
38. Wang, Y.L., Seebald, J.L., Szeto, D.P. and Irudayaraj, J. (2010) Biocompatibility and biodistribution of surface-enhanced Raman scattering nanoprobes in zebrafish embryos: in vivo and multiplex imaging. *ACS Nano*, **4**, 4039–4053.
39. Lin, J., Chen, R., Feng, S. *et al.* (2009) Rapid delivery of silver nanoparticles into living cells by electroporation for surface-enhanced Raman spectroscopy. *Biosensors and Bioelectronics*, **25**, 388–394.
40. Yun, Y., Lin, J.-Q., Huang, H. *et al.* (2011) Effect of silver colloid concentration on electroporation for intracellular surface enhanced Raman scattering. *Progress in Biochemistry and Biophysics*, **38**, 961–966.
41. Yun, Y., Lin, J., Wu, Y. *et al.* (2011) Optimizing electroporation assisted silver nanoparticle delivery into living C666 cells for surface-enhanced Raman spectroscopy. *Spectroscopy – An International Journal*, **25**, 13–21.

42. Ando, J., Fujita, K., Smith, N.I. and Kawata, S. (2011) Dynamic SERS imaging of cellular transport pathways with endocytosed gold nanoparticles. *Nano Letters*, **11**, 5344–5348.
43. Gregas, M.K., Scaffidi, J.P., Lauly, B. and Vo-Dinh, T. (2010) Surface-enhanced Raman scattering detection and tracking of nanoprobes: enhanced uptake and nuclear targeting in single cells. *Applied Spectroscopy*, **64**, 858–866.
44. Kneipp, J., Kneipp, H., Wittig, B. and Kneipp, K. (2010) Novel optical nanosensors for probing and imaging live cells, nanomedicine: nanotechnology. *Biology and Medicine*, **6**, 214–226.
45. Kneipp, K., Haka, A.S., Kneipp, H. *et al.* (2002) Surface-enhanced Raman spectroscopy in single living cells using gold nanoparticles. *Applied Spectroscopy*, **56**, 150–154.
46. Rao Sathuluri, R., Yoshikawa, H., Shimizu, E. *et al.* (2011) Gold nanoparticle-based surface-enhanced Raman scattering for noninvasive molecular probing of embryonic stem cell differentiation. *PLoS ONE*, **6**, e22802.
47. Moody, B., Haslauer, C.M., Kirk, E. *et al.* (2010) In situ monitoring of adipogenesis with human-adipose-derived stem cells using surface-enhanced Raman spectroscopy. *Applied Spectroscopy*, **64**, 1227–1233.
48. Blatchford, C.G., Campbell, J.R. and Creighton, J.A. (1982) Plasma resonance - enhanced Raman scattering by absorbates on gold colloids: the effects of aggregation. *Surface Science*, **120**, 435–55.
49. Hao, E. and Schatz, G.C. (2004) Electromagnetic fields around silver nanoparticles and dimers. *Journal of Chemical Physics*, **120**, 357–366.
50. Hildebrandt, P., Keller, S., Hoffmann, A. *et al.* (1993) Enhancement factor of surface-enhanced Raman scattering on silver and gold surfaces upon near-infrared excitation. Indication of an unusual strong contribution of the chemical effect. *Journal of Raman Spectroscopy*, **24**, 791–796.
51. Kerker, M., Siiman, O. and Wang, D.S. (1984) Effect of aggregates on extinction and surface-enhanced Raman scattering spectra of colloidal silver. *Journal of Physical Chemistry*, **88**, 3168–70.
52. Kneipp, K., Kneipp, H. and Kneipp, J. (2006) Surface-enhanced Raman scattering in local optical fields of silver and gold nanoaggregates – from single-molecule Raman spectroscopy to ultrasensitive probing in live cells. *Accounts of Chemical Research*, **39**, 443–450.
53. Moskovits, M. (2005) Surface-enhanced Raman spectroscopy: a brief retrospective. *Journal of Raman Spectroscopy*, **36**, 485–496.
54. Joseph, V., Matschulat, A., Polte, J. *et al.* (2011) SERS enhancement of gold nanospheres of defined size. *Journal of Raman Spectroscopy*, **42**, 1736–1742.
55. Otto, A., Mrozek, I., Grabhorn, H. and Akemann, W. (1992) Surface-enhanced Raman scattering. *Journal of Physics: Condensed Matter*, **4**, 1143–1212.
56. Goulet, P.J.G., dos Santos, D.S., Alvarez-Puebla, R.A. *et al.* (2005) Surface-enhanced Raman scattering on dendrimer/metallic nanoparticle layer-by-layer film substrates. *Langmuir*, **21**, 5576–5581.
57. Li, K.R., Stockman, M.I. and Bergman, D.J. (2003) Self-similar chain of metal nanospheres as an efficient nanolens. *Physical Review Letters*, **91**, 227402.

58. Stockman, M.I., Shalaev, V.M., Moskovits, M. *et al.* (1992) Enhanced Raman scattering by fractal clusters: scale-invariant theory. *Physical Review B*, **46**, 2821.
59. Xu, H.X., Aizpurua, J., Kall, M. and Apell, P. (2000) Electromagnetic contributions to single-molecule sensitivity in surface-enhanced Raman scattering. *Physical Review E*, **62**, 4318–4324.
60. Brandenberger, C., Muhlfeld, C., Ali, Z. *et al.* (2010) Quantitative evaluation of cellular uptake and trafficking of plain and polyethylene glycol-coated gold nanoparticles. *Small*, **6**, 1669–1678.
61. Chithrani, D.B. (2010) Intracellular uptake, transport, and processing of gold nanostructures. *Molecular Membrane Biology*, **27**, 299–311.
62. Nativo, P., Prior, I.A. and Brust, M. (2008) Uptake and intracellular fate of surface-modified gold nanoparticles. *ACS Nano*, **2**, 1639–1644.
63. Oh, E., Delehanty, J.B., Sapsford, K.E. *et al.* (2011) Cellular uptake and fate of pegylated gold nanoparticles is dependent on both cell-penetration peptides and particle size. *ACS Nano*, **5**, 6434–6448.
64. Chithrani, B.D., Ghazani, A.A. and Chan, W.C.W. (2006) Determining the size and shape dependence of gold nanoparticle uptake into mammalian cells. *Nano Letters*, **6**, 662–668.
65. Jin, H., Heller, D.A., Sharma, R. and Strano, M.S. (2009) Size-dependent cellular uptake and expulsion of single-walled carbon nanotubes: single particle tracking and a generic uptake model for nanoparticles. *ACS Nano*, **3**, 149–158.
66. Lunov, O., Syrovets, T., Loos, C. *et al.* (2011) Differential uptake of functionalized polystyrene nanoparticles by human macrophages and a monocytic cell line. *ACS Nano*, **5**, 1657–1669.
67. Nel, A.E., Madler, L., Velegol, D. *et al.* (2009) Understanding biophysicochemical interactions at the nano-bio interface. *Nature Materials*, **8**, 543–557.
68. Sirimuthu, N.M.S., Syme, C.D. and Cooper, J.M. (2010) Monitoring the uptake and redistribution of metal nanoparticles during cell culture using surface-enhanced Raman scattering spectroscopy. *Analytical Chemistry*, **82**, 7369–7373.
69. Canton, I. and Battaglia, G. (2012) Endocytosis at the nanoscale. *Chemical Society Reviews*, **41**, 2718–2739.
70. Conner, S.D. and Schmid, S.L. (2003) Regulated portals of entry into the cell. *Nature*, **422**, 37–44.
71. Li, W., Chen, C.Y., Ye, C. *et al.* (2008) The translocation of fullerenic nanoparticles into lysosome via the pathway of clathrin-mediated endocytosis. *Nanotechnology*, **19**.
72. Rejman, J., Oberle, V., Zuhorn, I.S. and Hoekstra, D. (2004) Size-dependent internalization of particles via the pathways of clathrin-and caveolae-mediated endocytosis. *Biochemical Journal*, **377**, 159–169.
73. Inoue, M. and Ohtaka, K. (1983) Surface enhanced Raman scattering by metal spheres. I. Cluster effect. *Journal of the Physical Society of Japan*, **52**, 3853–3864.
74. Shalaev, V.M. (1996) Electromagnetic properties of small-particle composites. *Physics Reports*, **272**, 61–137.
75. Yamaguchi, Y., Weldon, M.K. and Morris, M.D. (1999) Fractal characterization of SERS-active electrodes using extended focus reflectance microscopy. *Applied Spectroscopy*, **53**, 127–132.

76. Fujita, K., Ishitobi, S., Hamada, K. *et al.* (2009) Time-resolved observation of surface-enhanced Raman scattering from gold nanoparticles during transport through a living cell. *Journal of Biomedical Optics*, **14**, 024038.
77. Oyelere, A.K., Chen, P.C., Huang, X. *et al.* (2007) Peptide-conjugated gold nanorods for nuclear targeting. *Bioconjugate Chemistry*, **18**, 1490–1497.
78. Kneipp, J., Kneipp, H., Wittig, B. and Kneipp, K. (2010) Following the dynamics of pH in endosomes of live cells with SERS nanosensors. *The Journal of Physical Chemistry C*, **114**, 7421–7426.
79. Becker, J.S., Matusch, A., Palm, C. *et al.* (2010) Bioimaging of metals in brain tissue by laser ablation inductively coupled plasma mass spectrometry (La-Icp-Ms) and metallomics. *Metallomics*, **2**, 104–11.
80. Krystek, P., Ulrich, A., Garcia, C.C. *et al.* (2011) Application of plasma spectrometry for the analysis of engineered nanoparticles in suspensions and products. *Journal of Analytical Atomic Spectrometry*, **26**, 1701–1721.
81. Ojea-Jimenez, I., Garcia-Fernandez, L., Lorenzo, J. and Puntes, V.F. (2012) Facile preparation of cationic gold nanoparticle-bioconjugates for cell penetration and nuclear targeting. *ACS Nano*, **6**, 7692–702.
82. Xia, T., Kovochich, M., Liong, M. *et al.* (2008) Comparison of the mechanism of toxicity of zinc oxide and cerium oxide nanoparticles based on dissolution and oxidative stress properties. *ACS Nano*, **2**, 2121–2134.
83. Franze, B., Strenge, I. and Engelhaardt, C. (2012) Single particle inductively coupled plasma mass spectrometry: evaluation of three different pneumatic and piezo-based sample introduction systems for the characterization of silver nanoparticles. *Journal of Analytical Atomic Spectrometry*, **27**, 1074–1083.
84. Gschwind, S., Flamigni, L., Koch, J. *et al.* (2011) Capabilities of inductively coupled plasma mass spectrometry for the detection of nanoparticles carried by monodisperse microdroplets. *Journal of Analytical Atomic Spectrometry*, **26**, 1166–1174.
85. Drescher, D., Giesen, C., Traub, H. *et al.* (2012) Quantitative imaging of gold and silver nanoparticles in single eukaryotic cells by laser ablation ICP-MS. *Analytical Chemistry*, **84**, 9684–9688.
86. Giesen, C., Wäntig, L., Mairinger, T. *et al.* (2011) Iodine as an elemental marker for imaging of single cells and tissue sections by laser ablation inductively coupled plasma mass spectrometry. *Journal of Analytical Atomic Spectrometry*, **26**, 2160–2165.
87. Qin, Z.Y., Caruso, J.A., Lai, B. *et al.* (2011) trace metal imaging with high spatial resolution: applications in biomedicine. *Metallomics*, **3**, 28–37.
88. Trouillon, R., Passarelli, M.K., Wang, J. *et al.* (2012) Chemical analysis of single cells. *Analytical Chemistry*, **79**, 3554–3560.
89. Drescher, D. and Kneipp, J. (2012) Nanomaterials in complex biological systems: insights from Raman spectroscopy. *Chemical Society Reviews*, **41**, 5780–5799.
90. Greulich, C., Diendorf, J., Simon, T. *et al.* (2011) Uptake and intracellular distribution of silver nanoparticles in human mesenchymal stem cells. *Acta Biomaterialia*, **7**, 347–354.
91. Khan, J.A., Pillai, B., Das, T.K. *et al.* (2007) Molecular effects of uptake of gold nanoparticles in HeLa cells. *ChemBioChem*, **8**, 1237–1240.

92. Khlebtsov, N. and Dykman, L. (2011) Biodistribution and toxicity of engineered gold nanoparticles: a review of in vitro and in vivo studies. *Chemical Society Reviews*, **40**, 1647–1671.
93. Pan, Y., Neuss, S., Leifert, A. *et al.* (2007) Size-dependent cytotoxicity of gold nanoparticles. *Small*, **3**, 1941–1949.
94. Patra, H.K., Banerjee, S., Chaudhuri, U. *et al.* (2007) Cell selective response to gold nanoparticles. *Nanomedicine*, **3**, 111–119.
95. Pernodet, N., Fang, X., Sun, Y. *et al.* (2006) Adverse effects of citrate/gold nanoparticles on human dermal fibroblasts. *Small*, **2**, 766–773.
96. Uboldi, C., Bonacchi, D., Lorenzi, G. *et al.* (2009) Gold nanoparticles induce cytotoxicity in the alveolar type-II cell lines A549 and Ncih441. *Particle and Fibre Toxicology*, **6**, 18.
97. Wang, Z.Y., Bonoiu, A., Samoc, M. *et al.* (2008) Biological pH sensing based on surface enhanced Raman scattering through a 2-aminothiophenol-silver probe. *Biosensors and Bioelectronics*, **23**, 886–891.
98. Asharani, P.V., Mun, G.L.K., Hande, M.P. and Valiyaveettil, S. (2009) Cytotoxicity and genotoxicity of silver nanoparticles in human cells. *ACS Nano*, **3**, 279–290.
99. Carlson, C., Hussain, S.M., Schrand, A.M. *et al.* (2008) Unique cellular interaction of silver nanoparticles: size-dependent generation of reactive oxygen species. *Journal of Physical Chemistry B*, **112**, 13608–13619.
100. Yen, H.J., Hsu, S.H. and Tsai, C.L. (2009) Cytotoxicity and immunological response of gold and silver nanoparticles of different sizes. *Small*, **5**, 1553–1561.
101. Connor, E.E., Mwamuka, J., Gole, A. *et al.* (2005) Gold nanoparticles are taken up by human cells but do not cause acute cytotoxicity. *Small*, **1**, 325–327.
102. Shukla, R., Bansal, V., Chaudhary, M. *et al.* (2005) Biocompatibility of gold nanoparticles and their endocytotic fate inside the cellular compartment: a microscopic overview. *Langmuir*, **21**, 10644–10654.
103. Schaeublin, N.M., Braydich-Stolle, L.K., Schrand, A.M. *et al.* (2011) Surface charge of gold nanoparticles mediates mechanism of toxicity. *Nanoscale*, **3**, 410–420.
104. Soenen, S.J., Manshian, B., Montenegro, J.M. *et al.* (2012) Cytotoxic effects of gold nanoparticles: a multiparametric study. *ACS Nano*, **6**, 5767–83.
105. Liu, W., Wu, Y.A., Wang, C. *et al.* (2010) Impact of silver nanoparticles on human cells: effect of particle size. *Nanotoxicology*, **4**, 319–330.
106. Maiorano, G., Sabella, S., Sorce, B. *et al.* (2010) Effects of cell culture media on the dynamic formation of protein-nanoparticle complexes and influence on the cellular response. *ACS Nano*, **4**, 7481–7491.
107. Asharani, P.V., Sethu, S., Vadukumpully, S. *et al.* (2010) Investigations on the structural damage in human erythrocytes exposed to silver, gold, and platinum nanoparticles. *Advanced Functional Materials*, **20**, 1233–42.
108. Amrolia, P., Sullivan, S.G., Stern, A. and Munday, R. (1989) Toxicity of aromatic thiols in the human red blood cell. *Journal of Applied Toxicology*, **9**, 113–8.
109. Scudiero, D.A., Shoemaker, R.H., Paull, K.D. *et al.* (1988) Evaluation of a soluble tetrazolium formazan assay for cell-growth and drug sensitivity in culture using human and other tumor-cell lines. *Cancer Research*, **48**, 4827–4833.
110. Murphy, R.F., Powers, S. and Cantor, C.R. (1984) Endosome pH measured in single cells by dual fluorescence flow-cytometry – rapid acidification of insulin to ph-6. *Journal of Cell Biology*, **98**, 1757–1762.

111. Montcourrier, P., Mangeat, P.H., Valembois, C. *et al.* (1994) Characterization of very acidic phagosomes in breast-cancer cells and their association with invasion. *Journal of Cell Science*, **107**, 2381–2391.
112. Michota, A. and Bukowska, J. (2003) Surface-Enhanced Raman Scattering (SERS) of 4-Mercaptobenzoic acid on silver and gold substrates. *Journal of Raman Spectroscopy*, **34**, 21–25.
113. Bishnoi, S.W., Levin, C.S., Rozell, C.J. *et al.* (2006) All-optical nanoscale pH meter: a plasmonic nanodevice with quantifiable output. IEEE LEOS Annual Meeting Conference Proceedings, Vols. 1 and 2, pp. 300–301.
114. Jensen, R.A., Sherin, J. and Emory, S.R. (2007) Single nanoparticle based optical pH probe. *Applied Spectroscopy*, **61**, 832–838.
115. Schwartzberg, A.M., Oshiro, T.Y., Zhang, J.Z. *et al.* (2006) Improving nanoprobes using surface-enhanced Raman scattering from 30-Nm hollow gold particles. *Analytical Chemistry*, **78**, 4732–4736.
116. Charan, S., Chien, F.-C., Singh, N. *et al.* (2011) Development of lipid targeting Raman probes for *in vivo* imaging of *caenorhabditis elegans*. *Chemistry-A European Journal*, **17**, 5165–5170.
117. B. Küstner, M. Gellner, M. Schütz, F. Schöppler, A. Marx, P. Ströbel, P. Adam, C. Schmuck, and S. Schlücker. *SERS labels for red laser excitation: silica-encapsulated SAMs on tunable gold/silver nanoshells Angewandte Chemie International Edition* 2009, **48**, 1.
118. Mulvaney, S. P.; Musick, M. D.; Keating, C. D.; Natan, M. J. Glass-coated, analyte-tagged nanoparticles: a new tagging system based on detection with surface-enhanced Raman scattering *Langmuir* 2003, **19**, 4784.

Index

Frontiers of Surface-Enhanced Raman Scattering: Single Nanoparticles and Single Cells, First Edition.
Edited by Yukihiro Ozaki, Katrin Kneipp and Ricardo Aroca.